Environmental Systems and Societies

for the IB Diploma

Second edition
Paul Guinness and Brenda Walpole

Cambridge University Press's mission is to advance learning, knowledge and research worldwide.

Our IB Diploma resources aim to:
- encourage learners to explore concepts, ideas and topics that have local and global significance
- help students develop a positive attitude to learning in preparation for higher education
- assist students in approaching complex questions, applying critical-thinking skills and forming reasoned answers.

CAMBRIDGE
UNIVERSITY PRESS

Shaftesbury Road, Cambridge CB2 8EA, United Kingdom

One Liberty Plaza, 20th Floor, New York, NY 10006, USA

477 Williamstown Road, Port Melbourne, VIC 3207, Australia

314–321, 3rd Floor, Plot 3, Splendor Forum, Jasola District Centre, New Delhi – 110025, India

103 Penang Road, #05–06/07, Visioncrest Commercial, Singapore 238467

Cambridge University Press & Assessment is part of the University of Cambridge.

It furthers the University's mission by disseminating knowledge in the pursuit of education, learning and research at the highest international levels of excellence.

Information on this title: education.cambridge.org

© Cambridge University Press & Assessment 2012, 2016

This publication is in copyright. Subject to statutory exception and to the provisions of relevant collective licensing agreements, no reproduction of any part may take place without the written permission of Cambridge University Press & Assessment.

First published 2012
Second edition 2016

20 19 18 17 16 15 14 13 12 11 10 9 8

Printed in Great Britain by Ashford Colour Press Ltd.

A catalogue record for this publication is available from the British Library

ISBN   978-1-107-55643-0   Paperback

Cambridge University Press & Assessment has no responsibility for the persistence or accuracy of URLs for external or third-party internet websites referred to in this publication, and does not guarantee that any content on such websites is, or will remain, accurate or appropriate. Information regarding prices, travel timetables, and other factual information given in this work is correct at the time of first printing but Cambridge University Press & Assessment does not guarantee the accuracy of such information thereafter.

The material has been developed independently by the publisher and the content is in no way connected with nor endorsed by the International Baccalaureate Organization.

....

NOTICE TO TEACHERS IN THE UK

It is illegal to reproduce any part of this work in material form (including photocopying and electronic storage) except under the following circumstances:
(i) where you are abiding by a licence granted to your school or institution by the Copyright Licensing Agency;
(ii) where no such licence exists, or where you wish to exceed the terms of a licence, and you have gained the written permission of Cambridge University Press;
(iii) where you are allowed to reproduce without permission under the provisions of Chapter 3 of the Copyright, Designs and Patents Act 1988, which covers, for example, the reproduction of short passages within certain types of educational anthology and reproduction for the purposes of setting examination questions.

The website accompanying this book contains further teacher resources.
Visit cambridge.org/go and register for access.

Separate website terms and conditions apply.

Contents

Introduction v

Acknowledgements vii

How to use this book ix

Topic 1 – Foundations of environmental systems and societies 1

    1.01 Environmental value systems 2
    1.02 Systems and models 16
    1.03 Energy and equilibria 24
    1.04 Sustainability 31
    1.05 Humans and pollution 39
        End-of-topic questions 48

Topic 2 – Ecosystems and ecology 50

    2.01 Species and populations 51
    2.02 Communities and ecosystems 62
    2.03 Flows of energy and matter 74
    2.04 Biomes, zonation and succession 86
    2.05 Investigating ecosystems 103
        End-of-topic questions 116

Topic 3 – Biodiversity and conservation 118

    3.01 An introduction to biodiversity 119
    3.02 Origins of biodiversity 123
    3.03 Threats to biodiversity 135
    3.04 Conservation of biodiversity 147
        End-of-topic questions 160

Topic 4 – Water and aquatic food production systems and societies 161

    4.01 Introduction to water systems 162
    4.02 Access to fresh water 171
    4.03 Aquatic food production systems 188
    4.04 Water pollution 198
        End-of-topic questions 211

Topic 5 – Soil systems and terrestrial food production systems and societies 212

    5.01 Introduction to soil systems 213
    5.02 Terrestrial food production systems and food choices 223
    5.03 Soil degradation and conservation 239
        End-of-topic questions 254

Contents

Topic 6 – Atmospheric systems and societies — 255

    6.01 Introduction to the atmosphere — 256
    6.02 Stratospheric ozone — 265
    6.03 Photochemical smog — 279
    6.04 Acid deposition — 290
    End-of-topic questions — 298

Topic 7 – Climate change and energy production — 299

    7.01 Energy choices and security — 300
    7.02 Climate change – causes and impacts — 317
    7.03 Climate change – mitigation and adaptation — 336
    End-of-topic questions — 348

Topic 8 – Human systems and resource use — 349

    8.01 Human population dynamics — 350
    8.02 Resource use in society — 372
    8.03 Solid domestic waste — 389
    8.04 Human population carrying capacity — 402
    End-of-topic questions — 414

Answers to self-assessment questions — 415

Answers to end-of-topic questions — 433

Answers to case study questions — 441

Glossary — 449

Index — 457

Introduction

This book covers the syllabus for the IB Diploma Programme Environmental Systems and Societies, which is offered at Standard Level only. Our understanding of the environment and its importance to our lives has grown rapidly over recent decades and the Environmental Systems and Societies course, which is a transdisciplinary subject combining the knowledge and techniques associated with a group 4 science subject and the social and cultural aspects of the more anthropocentric approach of a group 3 subject.

The book follows the sequence of the syllabus in terms of the eight topics and the sub-sections within these topics. The overall objective of this book is to provide comprehensive coverage of all the topics in the syllabus in an up-to-date and interesting format. Each topic is covered in a separate section and the significant ideas and key questions are listed at the start of each section. Case studies have been chosen to represent a wide range of geographical locations and biological examples, so that as you read, you can reflect on the essential international element of this course. Examples from across the environmental and economic spectrums highlight how our impact on the environment is not just an issue for one country or section of society but something important to us all. The book considers a range of environmental issues from small-scale local events to large-scale global issues. The use of ICT and technology in general has made all of us more aware than ever before of what is happening elsewhere in the world and of the implications that changes in other parts of the world can have on us.

Topic 1 *Foundations of environmental systems and societies* explains the environmental value systems that drive societies to protect and value the natural world. It outlines the essential systems approach to the study of this subject, identifying some of the underlying principles that can be applied to living systems. Your syllabus advocates a holistic approach to the analysis of environmental systems so that you can arrive at informed personal viewpoints while being aware of the values of others. The topic introduces important concepts of energy, sustainability and the impact of humans on the environment.

Topic 2 *Ecosystems and ecology* presents much of the basic scientific knowledge and understanding for the topics that follow. Techniques to measure and evaluate components of systems and how they can change are central to this topic which also covers the key concepts of species, populations, biomes and succession.

Topic 3 *Biodiversity and conservation* addresses issues of how biodiversity has arisen and how it is now under threat, mainly because of human interference in natural systems. Humans are attempting to redress the balance and some conservations options are covered here.

Topic 4 *Water and aquatic food production systems and societies* considers how access to fresh water is crucial to the survival of all living things. Humans use aquatic systems to harvest and produce food; as our population increases so do our demands on these resources. Fish farming and aquaculture may help feed future generations. Pollution of water is a problem discussed here.

Topic 5 *Soil systems and terrestrial food production systems and societies* provides detailed coverage of the planet's soils in terms of systems, structure, how they are used for human benefit, and how their misuse is storing up major problems for the future. Understanding all aspects of soil systems around the world is fundamental to ensure food security for present and future populations. The topic begins with an introduction to soil systems, followed by consideration of terrestrial food production systems and food choices, and ends with analysis of soil degradation and conservation.

Topic 6 *Atmospheric systems and societies* begins with an introduction to the atmosphere which provides a basis for this topic and also for Topic 7. The three following sub-topics in Topic 6 consider major atmospheric issues which impact severely on people and the environment and may cumulatively threaten the future liveability of the planet. This topic examines the extent of these atmospheric problems and considers progress made in their management.

Topic 7 *Climate change and energy production* examines what is generally considered to be the number one problem facing planet Earth. The opening sub-topic sets the scene by acknowledging that production and consumption of energy are by far the most important factors in climate change. The concept of 'security' again comes to the fore, as it also does in Topic 4 (water) and

Introduction

Topic 5 (food). If significant further progress is to be made in tackling climate change, we will need to be reliant on a higher level of international cooperation than has been the case in the past along with significant advances in science and technology.

Topic 8 *Human systems and resource use* begins with an analysis of human population dynamics which considers the models and indicators used to quantify human populations, and the range of factors which affect human population growth. The topic then examines resource use in society, solid domestic waste, and human population carrying capacity. The important concept of the ecological footprint (EF) is discussed. The key concept of sustainability is central to this topic as it is to the study of environmental systems and societies in general.

The phrase 'Think globally, act locally' was first used by Scottish town planner and social activist Patrick Geddes, who wrote *Cities in Evolution* in 1915. It has since become a concept widely used by environmentalists and taken into consideration by governments, educators and communities. It is an idea on which we can all reflect in our daily lives and at the end of this book.

Full details of the Assessment Objectives and examination requirements for the Diploma Programme Environmental Systems and Societies course can be found in the relevant IBO guide.

Paul Guinness
Brenda Walpole

Acknowledgements

The authors and publishers acknowledge the following sources of copyright material and are grateful for the permissions granted. While every effort has been made, it has not always been possible to identify the sources of all material used, or to trace all copyright holders. If any omissions are brought to our notice, we will be happy to include the appropriate acknowledgements on reprinting.

p 374 text from www.rncalliance.org used with permission of the RNC Alliance; p381 'The cultural value of forests' by Lara Barbier, April 2011 from the TEEB website bankofnaturalcapital.com; Figures 5.03 and 5.04 redrawn and reproduced with permission of Nelson Thornes Ltd from *Geography: An integrated Approach* (4th edn), David Waugh, 978-1-4085-0407, first published in 2009; Figure 5.20 adapted from figures by Floor Anthoni, seafriends.org.nz; p253–4 Table 5.05 from Soil Management and Agrodiversity: a case study from Arumeru, Arusha, Tanzania by FBS Kaihura, M Stocking and E Kahembe, 2008; Figures 4.01 and 4.02 redrawn from Digby et al: *A2 Geography for Edexcel Student Book* (OUP, 2009), reprinted by permission of Oxford University Press; Figure 4.16 redrawn from *GCSE Geography*, Garret Nagle, Hodder Education, Fig 22, p.148 reproduced with permission of Hodder Education; p144 extract from 'The terrible lesson of the bee orchid' by Richard Mabey, published by *The Guardian*, copyright © Richard Mabey, 2005 reproduced by permission of Sheil Land Associates Ltd; Figures 4.29 and 4.30 Citizen Monitoring Biotic Index data recording form and key reproduced by permission of the Board of Regents of the University of Wisconsin System; Figures 8.21 and 8.22 from 'Taking out the rubbish: municipal waste composition, trends and futures' by *Resource Futures*, 2009; Figure 6.03 redrawn after Dr TR Oke, University of British Columbia in *Atmosphere, Weather & Climate*, Routledge, 1998; Figures 6.07, 6.09 and 6.10 redrawn from *Atmospheric process and human influence*, P Warburton, Collins Educational; figure 6.18 after Professor Richard Foust, Northern Arizona University; Figure 6.27 'What are scientists dong to better understand acid deposition?' © Ecological Society of America; Table 7.09, reproduced by permission of Phillip Allan Updates; Figure 7.19 reprinted from *The Lancet*, Vol 367, Anthony J McMichael, Rosalie E Woodruff, Simon Heals, 'Climate change and human health; present and future risks', Figure 1, © 2006, with permission from Elsevier; Figure 7.20 from *Geography*, Vol 96, Spring 2011 by permission of the Geographical Association www.geography.org.uk; Figure 7.21 from 'The IPCC messed up over Amazongate' by George Monbiot, guardian.co.uk, 2nd July 2010, copyright Guardian News & Media Ltd 2010; Figure 1.01 adapted from Figure 10.1 '*The Environmentalist objectives and strategies in the seventies*', page 372, first published in O'Riordan, T *Environmentalism*, 1981, Pion Ltd, London www.pion.co.uk/www.

Thanks to the following for permission to reproduce photographs

p1, chapter opening page SPL; p4*l* Bettmann/Corbis; p4*r* Ria Novosti/SPL; Daniel Beltra/Greenpeace/naturepl.com; p9 Aslund/Greenpeace; p12*t* SPL; p12 SPL*b*; p14 George Holton/SPL; p36 David South/Alamy; p37*l* Ron Nickel/Design Pics/Still Pictures: p37*r* Nazrul Islam/Majority World/Still Pictures; p39*r* SPL; p41 Ria Novosti/TopFoto/TopFoto.co.uk; p47*t* John Stanmeyer/VII/Corbis; p47*b* SPL; p50 SPL; p53*t* SPL; p53*b* SPL; p54 Bob Gibbons/SPL; p56*l* SPL; p56*r* SPL; p60 Peter Bird, Dept of Primary Industries and Regions, South Australia/PIRSA; p72 Topham/AP/TopFoto.co.uk; p78 NASA Goodard Space Flight Center; p85 SPL; p88*l* SPL, p88*r* SPL; p90*l* SPL, 90*r* SPL; p91*t* SPL, 91*m* Gerry Ellis/Minden Pictures/FLPA, 91*b* Beth Davidow, Visuals Unlimited/SPL; p97*t* Biosphoto/Cyrill ruoso/BIOSphoto/Still Pictures, 97*b* Geoff Dore/naturepl.com; p99 blickwinker/Alamy; p101*l*/Nick turner/naturepl.com, p67*r* Robin Chittenden/FLPA, p67*br* Chien Lee/Minden Pictures/FLPA; p102*l* David R. Frazier Photolibrary Inc/SPL, p102*r* Niall Benvie/naturepl.com; p106 David Hosking/FLPA; p109 Chris Linder/Visuals Unlimited/SPL; p110 Art Directors & Trip/Alamy; p118 SPL; Francois Savigny/naturepl.com; p124*l* SPL, FLPA/Alamy; p127*t* SPL; p128*l* © PR Grant and BR Grant; p138*t* ARCO/naturepl.com, 138*b* Lynn M Stone/naturepl.com; p139 *tl* © Santiago Ron, p139*tr* Alex Cortes, p139*m* zoonar.com/NicoSmit/zoonar.de Specialist Stock, p139*b*© D Hal Cogger; p140 Spencer Sutton/SPL; p141 Corbis; p142 SPL; p146 NASA EARTH Observatory/Jesse Allen/

Acknowledgements

Robert Simmon/NASA EO-1 team; p149 Prisma/Superstock; Alan Sirulnikoff/SPL; p154 Aaron Ferster/SPL; p163*l* Paul Guinness; p165 Paul Guinness; p173*r* and *l* Paul Guinness; p175l Shutterstock, p175 www.wateraid.org; p177 Alamy; p179 Shut terstock; p189 SPL; p189 NASA/Robert Simmon and Jesse Allen; p208*l* Ulrike Welsch/SPL, p208*r* SPL; pp213, 214 *r* and *l*, 216, 219, 220, 224*l* and *r*, 226, 240, 246*r* and *l*, 248, 250, 255, 259*l* and *r*, 261, 270, 282, 286, 287, 288, 292*l* and *r*, 301, 306, 309, 312, 313, 316, 322, 326, 327, 330*l* and *r*, 338, 347, 349, 356, 357, 359, 363, 365, 374, 378, 379, 380, 391, 397, 398, 406, Paul Guinness; p230 Bruce Davidson/naturepl.com; p290 SPL; p350 Chris Guinness; p372 Mark Bowler/SPL; p375 Alamy; p375 James Steinberg/SPL; p377 TopFoto/TopFoto.co.uk; p381 SumioHarada/Minden Pictures/Corbis; p383 Topham Picturepooint/TopFoto/TopFoto.co.uk; p389 Photofusion Picture Library/Alamy; p397*t* Silvere Teutsch/Eurelios/SPL; p403 TopFoto/TopFoto.co.uk.

Artwork on page 14*r* by John MacNeill; artworks on pages 13, 17, 82, 83, 94, 95, 106, 167, 181 and 185 are by Kathy Baxendale.

Abbreviations key: SPL= Science Photo Library; *r* = right; *l* = left; *t* = top; *b* = bottom.

The publisher would like to thank Dharmendra Dan Dubay of Shanghai American School, Shanghai, Dr Andrea Peoples-Marwah of The Quarry Lane School, Dublin, California and Anthony Brewer of Victoria Shanghai Academy, Hong Kong for reviewing the content of this second edition.

How to use this book

Learning objectives and key questions – set the scene of each chapter, help with navigation through the book and give a reminder of what's important about each topic.

Self-assessment questions – check your own knowledge and see how well you're getting on by answering questions. Each set of self-assessment questions includes a discussion point or research idea offering the opportunity for more extensive investigation and group work.

Consider this – particularly interesting aspects of each topic are highlighted throughout each chapter, providing extra opportunities for discussion in class.

Key terms – clear and straightforward explanations of the most important words in each topic.

How to use this book

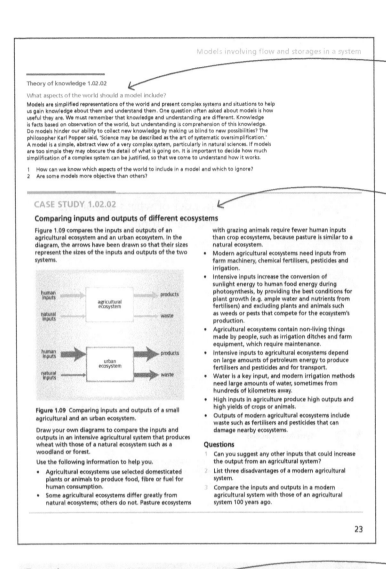

Theory of knowledge - allow you to reflect on the central role of Theory of knowledge on our knowledge and understanding of environmental philosophy. Each one asks you to consider a question which could form the basis of group discussion or a homework task.

Case studies - fascinating real-world settings are described and discussed to illustrate environmental phenomena that are relevant internationally. Questions allow you to check your knowledge and understanding.

Extension - further information can be found using these links.

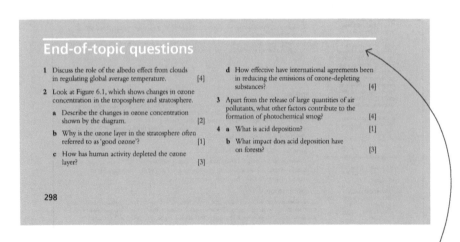

End-of-topic questions - use the questions at the end of each topic to check your knowledge and understanding of the whole topic and to practise answering questions similar to those you will encounter in your exams.

Topic 1
FOUNDATIONS OF ENVIRONMENTAL SYSTEMS AND SOCIETIES

1.01 Environmental value systems

LEARNING OBJECTIVES

After reading this chapter you should be able to:

- define what is meant by 'environmental value system (EVS)'
- outline the main factors that define different environmental philosophies
- describe the key historical influences on the development of the modern environmental movement
- describe how these philosophies influence decision-making on environmental issues
- justify your own view on environmental issues.

KEY QUESTIONS

How do historical events affect the development of environmental value systems (EVSs)?

What other influences are important to the development of environmental movements and EVSs?

What are the viewpoints of different EVSs?

1.01.01 Historical influences on the modern environmental movement

An **environmental value system (EVS)** is a particular worldview or set of paradigms that shapes the way individuals or societies perceive and evaluate environmental issues.

There are many different **environmental value systems (EVSs)**. Some of them consider humans to be more important than ecology and nature, while others prioritise the rights of living things and urge humans to behave more responsibly. Three examples of EVSs are described in this chapter. Any EVS is influenced by cultural (including religious), economic and sociopolitical contexts, for example whether individuals are from a less economically developed country (LEDC) or a more economically developed country (MEDC), or a democratic or an authoritarian society. Historical events also affect the development of EVSs and environmental movements in different societies.

Many of our current concerns for the environment have their origins in the reaction that some people had to the growth of cities and industries from the mid-1800s onwards. Industrialisation brought with it worsening air quality and pollution. During the 20th century, technology advanced and human influence on the environment became more widespread as the world population and industry grew. With improvements in communication, such as newspapers, television, the internet and social networking, people have become more aware of environmental issues more quickly. A number of significant milestones have punctuated the development of the modern environmental movement, and the idea of human 'stewardship' of the natural world has developed.

One of the first pollution disasters that became well publicised was in 1956 at Minamata Bay in Japan. People living in the area were poisoned by mercury that was discharged into the bay by a local factory. You can read more details about this in Topic 2, page 72. There have been other significant events over the last 60 years which have also influenced the development of people's awareness of the environment. Some of these are described here, but there are many others. Events like these have led not only to the formation of environmental movements and political parties but also to the establishment of EVSs that define what people believe are the best ways for us to live in harmony with our environment.

The environmental accidents and incidents that have occurred since the middle of the last century have motivated politicians and members of the general population to become more involved in environmental and conservation issues – that is, to develop their own EVSs.

1962 – *Silent Spring*

Rachel Carson (1907–1964) was a US writer and ecologist who became concerned about the overuse of synthetic chemical pesticides in agriculture. Her book *Silent Spring*, published in 1962, challenged modern agricultural practices and called for a change in the way the natural world is viewed. She was one of the earliest writers to highlight the effects of bioaccumulation of pesticides on populations of predatory birds (see Topic 2). Some dismissed her work as alarmist but she continued to speak out. *Silent Spring* brought environmental concerns to public attention, led to significant changes in policies towards pesticides, and later led to the ban on DDT (dichlorodiphenyltrichloroethane). The book was a major influence and is credited with inspiring American environmental movements which eventually resulted in the formation of the Environmental Protection Agency.

1975 – Save the Whales

Greenpeace, founded in 1971, is an organisation that campaigns for the environment and conservation of natural resources.

In 1975, Greenpeace launched the world's first anti-whaling campaign by taking direct action against Soviet whaling ships in the Pacific Ocean. The voyage of its ship *Rainbow Warrior* sparked an international outcry after pictures and videos taken by the Greenpeace crew shocked the world. The campaign 'Save the Whales' was taken up in many countries. After nearly a decade of intense lobbying the International Whaling Commission finally declared a moratorium on commercial whaling in 1986.

Greenpeace followed this success with several campaigns against nuclear testing in the 1980s, and the organisation is now recognised as a major player in the environmental movement.

1984 – Bhopal disaster

In 1984, a serious gas leak occurred at the Union Carbide pesticide factory in the Indian city of Bhopal. Even today, it is often cited as the world's worst industrial disaster. It is estimated that more than 500 000 people were exposed to methyl isocyanate and other chemicals from the plant. Of these, nearly 4000 died within weeks and a further 8000 died soon afterwards. Many thousands were partially or severely disabled, and since the tragedy as many as 25 000 deaths have been attributed to exposure to the gas (see Image 1.01).

Since the Bhopal incident, the chemical industry has been put under pressure to develop and implement strict safety and environmental standards to ensure that such an accident never happens again. In Bhopal, 3 December, the day of the incident, is still an official day of mourning.

1986 – Chernobyl disaster

The nuclear disaster at Chernobyl began during a routine test on a reactor at the plant. An emergency shutdown failed and a sequence of events followed which resulted in explosions that sent clouds of highly radioactive smoke into the atmosphere and destroyed one of the reactors (see Image 1.02). The smoke was carried over the nearby city of Prypiat and extensive areas of the western Soviet Union and Europe, which received high levels of radioactive fallout. More than 350 000 people had to be evacuated and resettled from the most severely contaminated parts of Belarus, Russia and Ukraine.

1.01 Environmental value systems

Russian data has suggested that over 900 000 premature cancer deaths have occurred as a result of radioactive contamination from Chernobyl. The accident raised concerns about the safety of nuclear power all over the world, and the expansion of nuclear power plants in many countries was halted or slowed.

Image 1.01 At the Bhopal factory, important safety systems were not working on the night that toxic gases leaked affecting the lungs and eyes of thousands of people who lived nearby.

Image 1.02 Explosions destroyed one of the reactors at the Chernobyl power station. Eventually the plant was enclosed in a concrete 'sarcophagus' that kept dangerous radioactive material inside.

2006 – *An Inconvenient Truth*

Promoted by former US Vice President Al Gore, the film *An Inconvenient Truth* put the issues of climate change (see Topic 7) at the front of the agenda for a wide audience. Publicity by a well-known politician and the accessible format of the film, produced by Davis Guggenheim, brought information about global warming to many and raised awareness of environmental issues worldwide. The film was supplied to schools and colleges, which also received additional material to help teachers explain the ideas to young people.

2010 – *Deepwater Horizon oil spill*

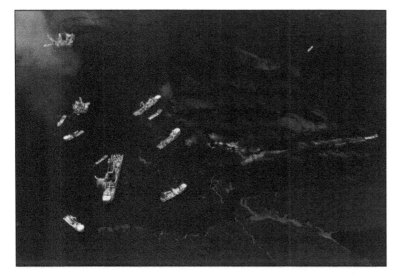

Image 1.03 Oil gushed from the *Deepwater Horizon* oil well into the sea from 20 April 2010 until the well was finally capped on 15 July 2010.

In 2010, oil poured out of a ruptured drilling rig deep below the sea in the Gulf of Mexico for almost three months. It was the largest marine oil spill in the history of the petrochemical industry (see Image 1.03). The oil

caused extensive damage to marine and wildlife ecosystems, as well as
to fishing and tourism in the area. Attempts to contain and disperse the oil had limited
success.

A report into the accident concluded that reforms were needed in both industry practices and government policy to prevent a similar event ever happening again. Images of the sea covered by an oily floating mass and of seabirds with oil-covered feathers spread rapidly via the internet and news media, and many environmentalists questioned the need to drill for oil in insecure or dangerous locations.

2011 – Fukushima nuclear accident

The damage to the Fukushima nuclear power station in Japan caused the largest nuclear disaster since Chernobyl. The power station was hit by waves of seawater almost 15 m high from a tsunami that followed a severe undersea earthquake near the island of Honshu. The plant was designed to withstand waves of 5.7 m but was overwhelmed by the ferocity of the tsunami. Japan was well prepared with evacuation procedures, and nuclear fallout was limited by prompt action. By the end of 2011, Japanese authorities were able to declare the plant to be stable, although it may take decades to decontaminate the surrounding areas and to decommission the power station.

As the disaster unfolded, it was watched on 24-hour news bulletins by millions of people all over the world, which is a further example of the influence of media coverage on public awareness of environmental issues.

1.01.02 An outline of the range of environmental philosophies

EVSs are divided into three general categories which form a continuous spectrum (see Figure 1.01):

- **ecocentric** (nature centred)
- **anthropocentric** (people centred).
- **technocentric** (based on technology).

Ecocentrists are likely to distrust modern technology and large-scale production and prefer to maintain natural environmental systems on a small scale. Ecocentrists view themselves as being under nature's control rather than controlling it. They foresee a limit to the Earth's resources. Deep ecologists are extreme ecocentrists and include people who believe that nature has more value than humanity.

Technocentrists believe that the brain power and resourcefulness of humans will enable us to control the environment. Such people have an optimistic worldview. They consider that natural processes must be understood to be controlled and that scientific research is important in policy making.

Anthropocentrists tend to include both viewpoints in their value system.

> An **ecocentric** system is a nature-centred value system that views people as being under nature's control rather than in control of it.
>
> An **anthropocentric** system is a human-centred value system that places humans as the central species and assesses the environment from an exclusively human perspective.
>
> A **technocentric** system is a technologically-based value system that believes the brain power of humans will enable us to control the environment.

1.01 Environmental value systems

ENVIRONMENTAL
⟵──────────────────────────────────────⟶

ECOCENTRISM (nature centred)	ANTHROPOCENTRISM (people centred)	TECHNOCENTRISM (technology centred)
Holistic world view. Minimum disturbance of natural processes. Integration of spiritual, social and environmental dimensions. Sustainability for the whole Earth. Self-reliant communities within a framework of global citizenship. Self-imposed restraint on resource use.	People as environmental managers of sustainable global systems. Population control given equal weight to resource use. Strong regulation by independent authorities required.	Technology can keep pace with and provide solutions to environmental problems. Resource replacement solves resource depletion. Need to understand natural processes in order to control them. Strong emphasis on scientific analysis and prediction prior to policy-making. Importance of market, and economic growth.

Deep ecologists

1. Intrinsic importance of nature for the humanity of man.
2. Ecological (and other natural) laws dictate human morality.
3. Biorights – the right of endangered species or unique landscapes to remain unmolested.

Self-reliance soft ecologists

1. Emphasis on smallness of scale and hence community identity in settlement, work and leisure.
2. Integration of concepts of work and leisure through a process of personal and communal improvement.
3. Importance of participation in community affairs, and of guarantees of the rights of minority interests. Participation seen as both a continuing education and a political function.

Environmental managers

1. Belief that economic growth and resource exploitation can continue assuming:
   a. suitable economic adjustments to taxes, fees, etc.
   b. improvements in the legal rights to a minimum level of environmental quality
   c. compensation arrangements satisfactory to those who experience adverse environmental and/or social effects.
2. Acceptance of new project appraisal techniques and decision review arrangements to allow for wider discussion or genuine search for consensus among representative groups of interested parties.

Cornucopians

1. Belief that man can always find a way out of any difficulties, whether political, scientific or technological.
2. Acceptance that pro-growth goals define the rationality of project appraisal and policy formulation.
3. Optimism about the ability of man to improve the lot of the world's people.
4. Faith that scientific and technological expertise provides the basic foundation for advice on matters pertaining to economic growth, public health and safety.
5. Suspicion of attempts to widen the basis for participation and lengthy discussion in project appraisal and policy review.
6. Belief that all impediments can be overcome given a will, ingenuity and sufficient resources arising out of growth.

4. Lack of faith In modern large-scale technology and its associated demands on elitist expertise, central state authority and inherently anti-democratic institutions.
5. Implication that materialism for its own sake is wrong and that economic growth can be geared to providing for the basic needs of those below subsistence levels.

Figure 1.01 Categories of Environmental Value Systems showing the range of beliefs held by different groups.

Ecocentrism

Ecocentrism takes a nature-centred, holistic view of the world. It proposes that we know very little about living things and their complex relationships, so we cannot have the ability to manage the environment in the way that technocentrists suggest. Biocentric or life-centred philosophers consider that all life is inherently valuable and is not simply for use by humans. They consider that people should not harm any species, whether it is useful or not, and that we should preserve ecosystems so that life will thrive. Humans are just one species which is no more important than any other. Some ecocentrists stress the holistic nature of our ethical obligation to the Earth, highlighting the need to limit our use of its resources.

One group of ecocentrists are the self-reliant or soft technologists, who believe small-scale, local and individual actions, such as recycling, can make a difference. At the other end of the spectrum are deep ecologists who value nature over humanity and believe that all species and ecosystems have values and rights that humans should not interfere with. They believe that the human population should decrease so that humans consume less of the Earth's resources (see Theory of knowledge 1.01.02).

Technocentrism

Technocentrism proposes that humans and technology will always be able to provide a solution to difficulties, whether they are scientific or political. In ecological systems, technocentrists believe that technology will always be able to solve environmental problems, even when humans push resources to the limit.

At one extreme of the range, some technocentrists, known as the cornucopians, view the world as a place with infinite resources to benefit humans. They believe that growth will provide wealth to improve the lives of everyone. They propose that a free-market economy can achieve this.

At the other extreme, another group – the environmental managers – see the world in terms analogous to a garden that needs care and attention, or 'stewardship'. They hold that legislation is needed to protect the environment and that, if an environment is damaged, those who suffer should receive compensation. They believe that, if humans take care of the Earth, it will take care of them.

Anthropocentrism

Anthropocentrism is the people-centred view of the world that includes viewpoints from both ecocentrism and techocentrism. People living in MEDCs are more likely to have this world view. Humans are viewed as a dominant species which manages the environment for its own requirements. Anthropocentrists' views include some of those of self-reliant soft ecologists and some of those of environmental managers.

It is important to remember that an EVS is individual and it is impossible to say that any EVS is wrong. Every individual and each **society** will have its own EVS.

Society is defined as an arbitrary group of individuals who share some common characteristics such as location, cultural background, religion or value system.

1.01.03 EVSs and the systems approach

The EVS of an individual or society can be considered as a system because, like all other systems, it has inputs such as education, experience, media influences and religious doctrines, and outputs such as courses of action and decisions, which are determined by processing of the inputs.

Information flows to individuals in societies are processed into changes in perceptions of the environment and changes in decisions about how to react to environmental issues. Some inputs will have no obvious immediate effect on an individual or group, while others will lead to direct actions (see Case study 1.01.01) in response to concerns about the environment. Information comes from ideas in films, books, newspapers and so on. Some people liken those who originate

1.01 Environmental value systems

The Ten Key Values of the Greens

- Ecological Wisdom
- Social Justice
- Grassroots Democracy
- Nonviolence
- Decentralization
- Community-Based Economics
- Feminism
- Respect for Diversity
- Personal and Global Responsibility
- Future Focus/Sustainability

Image 1.04 The key values of green politics are shared throughout the world.

this information to the 'producers' in an ecosystem. Other members of society then become the 'consumers' of the new ideas.

Ecological issues are rarely confined to local areas. Ecosystems often cross international boundaries, so that differences in EVSs can lead to conflict. Whaling is an example of such a conflict. Nations with different EVSs have very different perspectives on the exploitation and conservation of whales (see Case study 1.01.02). Similarly, the importance of fishing and conservation of fish stocks is viewed differently in societies in different parts of the world (see Chapter 4.04).

1.01.04 Politics and EVSs – the green revolution

One way in which people can express their own EVS is through political parties and discussion. The green movement is one EVS that encourages people to influence decisions made about the environment. Green movements have been set up in many parts of the world. Green politics is an ecocentric ideology, and its aim is for an ecologically sustainable society that protects the environment.

Green politics began in the 1970s. Among the first active green parties was one in Australia that contested elections in 1972. In Europe, the Popular Movement for the Environment was founded in 1972 in Switzerland, and the Green Party in the UK began to develop in 1973. There are now green politicians who share similar values (see Image 1.04) in many countries throughout the world.

One important focus of green politics is to reduce deforestation, particularly in the rainforest, and to support efforts to plant more trees. The Green Party in New Zealand called for a reduction in the destruction of rainforest and drew attention to the loss of species in biodiversity hotspots. It highlighted loss of homes and livelihoods of people who live in rainforests, and the impact of deforestation on greenhouse emissions and climate change. The party called on the government to stop the import of illegal and unsustainable timber products, and to ensure that all biofuel used in the country was from sustainable sources.

In the UK, green campaigners have called for international agreement to stop global deforestation in consultation with the local and indigenous communities so that traditional land rights are recognised. It has also demanded a global moratorium on logging and burning of old forests and ecological restoration of degraded ancient forests.

Many individuals and organisations support green issues, and publicity, petitions to governments and political movements have helped raise awareness and educate more and more people. Green parties encourage individual actions such as buying ethically produced goods. In the last 40 years, most have tended to grow from small-scale, local beginnings. With more support, they have gradually gained influence and now participate fully in the national politics of many countries. Green movements are now firmly established in Europe, Australia and New Zealand, as well as in the USA, Brazil and Colombia.

SELF-ASSESSMENT QUESTIONS 1.01.01

1. How important is green politics in rainforest conservation?
2. Have green politicians overemphasised the threat to rainforest ecosystems?
3. **Discussion point:** Can you think of any examples of issues where green politics influenced the EVS of people in your area?

CASE STUDY 1.01.01

Greenpeace – can direct action bring results?

Greenpeace is an ecocentric non-governmental organisation (NGO) founded by a small group of activists in 1971, which now has a presence in more than 40 countries. Greenpeace campaigns for positive change through action to protect the environment. This action takes many forms, from publicising environmental issues and lobbying governments, to promoting environmentally responsible solutions and taking non-violent direct action. Greenpeace is funded by supporters who donate money and sometimes take part in direct action themselves.

In 2011, Greenpeace began a campaign to prevent drilling for oil and gas in the Arctic, to protect the environment and its wildlife. As climate change melts the Arctic ice, oil companies are investigating the possibility of extracting fossil fuels from beneath the ice. Greenland is considering opening up an untouched area of its north-eastern waters to oil companies.

Greenpeace maintains that above the Arctic Circle, freezing temperatures, a narrow drilling window and a remote location mean that an oil spill would be almost impossible to deal with and would leave the habitat and its wildlife under threat. Greenpeace took direct action to subvert a meeting of oil industry leaders which was being held to discuss the issue. Activists greeted the oil industry leaders with a red carpet drenched in oil, and huge floating banners, one of which that read 'Protect the Arctic: No License to Drill' (see Image 1.05). Activists also gave delegates at the meeting their own alternative presentation about drilling in the Arctic, in the same building as the conference.

The campaign to prevent drilling in the Arctic continues. Another ecocentric organisation, World Wide Fund for Nature (WWF), has added its support to Greenpeace and, although it does not favour direct action, it has asked governments to handle Arctic development responsibly by:

- improving safety through the use of risk-lowering technology and higher standards for spill prevention and cleanup
- moving to renewable energy wherever possible
- protecting valuable species and areas of biological, economic and cultural importance.

Nevertheless, oil companies are pressing for drilling permits and started drilling at test sites in summer 2015.

Questions

1. This example of peaceful direct action allowed Greenpeace to lobby industry leaders directly. Do you think that the method used is an acceptable way to communicate an EVS?
2. Environmental campaigns by Greenpeace and other organisations have not always been trouble-free. In 1985, Greenpeace was involved in a campaign against nuclear testing in the Pacific Ocean. A Greenpeace vessel, *Rainbow Warrior*, was bombed while in harbour in New Zealand. Can protests that involve damage to property or that interrupt the lives of ordinary citizens ever be justified?
3. Research other environmental campaigns that have featured either direct action, such as the Greenpeace campaign, or lobbying, like the WWF approach.

Image 1.05 Greenpeace activists campaigning in 2011.

1.01 Environmental value systems

CASE STUDY 1.01.02

Whale hunting

Different nations and cultures have different views on the hunting of whales, even though many whale species are now endangered. Like all EVSs, these views depend on history and tradition. Nations that support whaling are likely to hold an anthropocentric viewpoint, whereas organisations that oppose it hold an ecocentric viewpoint. Greenpeace has run its campaign 'Save the Whales' since the 1970s to protect whales.

You can read more about the 'Save the Whales' campaign on the Greenpeace website: www.greenpeace.org/international/en/campaigns/oceans/fit-for-the-future/whaling

Historical background

By the 1930s more than 50 000 whales were being killed each year. The International Whaling Commission was set up to protect the whales, and in 1986 commercial whaling was banned in an effort to increase whale numbers. Today, there are only three nations remaining with whaling industries: Norway, Iceland, and Japan, whose industry is the largest, claiming up to 1000 whales annually (see Image 1.06). In the mid-1990s the International Whaling Commission considered easing its ban on commercial whaling to allow Japan to hunt whales off its coast, if Japan promised to kill fewer whales in the Antarctic. International reaction to that proposal was summed up by Captain Paul Watson of the Sea Shepherd Conservation Society, quoted in the *Los Angeles Times* on 27 January 2009: 'It's sort of like saying to bank robbers that you can't rob a bank in the city, but we'll let you do it in the country.'

The situation today

The following article was published by *The Guardian* on 18 November 2014.

Japan cuts Antarctic whale quota after UN court ruling

'Japan has reduced the quota of whales it plans to catch by two-thirds after UN court called the controversial "research whaling"', programme a commercial hunt masquerading as science.

The country now has a plan to kill 333 minke whales in the Southern Ocean next year as part of its push to resume whaling following a legal setback instigated by Australia. The whales will be hunted in a vast sweep of Antarctic waters, including ocean claimed by Australia.

This figure is a sharp reduction in the previous quota Japan awarded itself last year, when it aimed to take 855 minke whales, 50 humpback whales and 10 fin whales. Japan ended up taking fewer than this due to the disruptive tactics of anti-whaling activists Sea Shepherd.

Japan suspended its 2014 whale hunt after the ruling at the UN international court of justice. The case brought by Australia and supported by New Zealand successfully argued that Japan's program was not scientific and was simply a façade for commercial whaling. Japan has indicated that it is committed to starting a new whaling programme in the Southern Ocean at the end of 2015.'

Questions

1. What do the newspaper articles tell you about the EVS of whaling nations and Australia?
2. How are the EVSs of the different societies formed?
3. How important is international legislation in upholding EVSs of the majority of nations?
4. Discuss how important you think education and cultural influences are in forming a view on whaling.

Image 1.06 Most species of whale are listed on the WWF endangered list, but hunting still continues in some parts of the world.

Theory of knowledge 1.01.01

The importance of religion

Religion has been a significant influence on ethics and how they are applied to the environment. The concepts of 'dominion' and 'stewardship' are important in both Muslim and Christian belief systems. In both faiths, humans are called on to act as stewards of nature in a way that emphasises human moral superiority over non-human (biotic and abiotic) factors. The value of other species and objects is defined by the pleasure and profit they bring to humans. This anthropocentric view is based on the external and instrumental value of such factors for humans, and is known as the ethic of 'instrumentalism'.

An alternative interpretation of the anthropocentric worldview is one that emphasises environmental conservation for the benefit of humans and is based on our moral responsibility to ensure that future generations inherit all possible natural resources. It has been suggested that this 'conservation ethic' was the dominant viewpoint held by delegates to the United Nations (UN) Conference on Environment and Development (the World Summit) in Rio in 1992. The key aim of the conference was to preserve world biodiversity, and the meeting influenced all subsequent UN conferences, which have examined the relationship between human rights, population, social development, women and human settlements – and the need for environmentally sustainable development.

1 To what extent do you think that a personal EVS is dependent on a person's religious faith or other strongly held view?
2 How important are events like the World Summit in informing and influencing a person's EVS?

Theory of knowledge 1.01.02

Deep ecologists

The founder of deep ecology was the Norwegian philosopher Arne Næss (1912–2009). Næss believed that, if you do not know how the outcomes of your actions will affect other beings, you should not act. People who adhere to the precautionary principle hold a similar view.

Deep ecologists are at one extreme of the range of the EVSs continuum. They believe that the world does not exist as a resource to be freely exploited by humans. The ethics of deep ecology hold that a whole system is superior to any of its parts. Deep ecologists summarise their values with eight key points.

- The well-being of human and non-human life on Earth has value in itself which is independent of the usefulness of the non-human world to humans.
- The richness and diversity of life contribute to these values.
- Humans have no right to reduce this richness and diversity except to satisfy vital human needs.
- For human life to flourish a substantial reduction in the human population is needed, and for non-human life to thrive such a decrease is required.
- Human interference with the non-human world is excessive, and the situation is becoming worse.
- A substantial change in policy towards economics, technology and ideology is needed to change the direction of human progress.
- Humans should appreciate the quality of life rather than always seeking higher standards of living. People should understand the difference between 'big' and 'great'.
- People who hold ecocentric views should do all they can to help make the changes necessary to improve the well-being of the Earth.

1 Think about your personal viewpoint. How many of the values listed above do you share with deep ecologists?
2 Can the actions of individuals lead to substantial changes in the policies of governments?

1.01 Environmental value systems

1.01.05 Decision-making and the influence of environmental philosophy

All decisions that are made about environmental issues are influenced by the philosophical standpoint of those taking the decisions. It is important to evaluate the implications of different viewpoints. Important environmental issues that are addressed in this course include acid rain, use of water resources, fossil fuels, climate change and ecological footprints. You can read more about all these issues in later topics of this book. All of them can be considered from different viewpoints. Here we compare how people with an ecocentric view and people with a technocentric view apply their different EVSs to aspects of the biosphere.

The demand for water resources

An ecocentric view of water management involves conservation and recycling so that water can be used sustainably without harm to the environment (see Topic 4). An ecocentrist would encourage the use of meters and monitoring so that water use was kept to a minimum and used for essential purposes only. On the other hand, a technocentrist would seek to provide water for the future by using technology and seeking new or innovative methods. A technocentrist would use technology to limit water use in the home and in industry. Technological solutions such as seeding clouds to produce rain, desalination of seawater (see Image 1.07) and iceberg capture, or breeding and developing crops that can grow using less water, would all be options that technocentrists would favour. Technocentrists would also advocate greater use of purified wastewater and the extraction of water from sources deep beneath the Earth.

Image 1.07 Desalination plants like this one on Lake Mead are very efficient at producing drinking water from seawater.

Climate change

Technocentrists would favour plans such as the one devised by scientists at the National Center for Atmospheric Research in the USA. Scientists here propose using a fleet of unmanned, wind-powered ships to spray salt water up into low-lying clouds through 20 m high cylinders (see Image 1.08). Tiny particles of salt within each droplet act as centres of condensation, leading to a greater concentration of water droplets within each cloud, and thus a greater albedo. Clouds with increased albedo can cool the Earth by reflecting radiation back into space.

Ecocentrists would prefer to see a limit on the consumption of fossil fuels. They stress the need to modify farming methods and reduce human dependence on livestock such as cattle which increase the level of greenhouse gases in the atmosphere. They would favour replanting trees and increasing the level of phytoplankton in the sea as a means to increase carbon dioxide uptake by the environment.

Ecocentrists favour the use of renewable and carbon-neutral sources of energy, such as biomass, solar and wind power.

(For more about climate change, see Topic 7.)

Fossil fuels

Overuse of fossil fuels has caused environmental problems which include pollution and global warming.

A technocentric or more extreme cornucopian solution would call on science to refine and extend new technologies such as alternative energy sources using wind, waves and hydrogen fuel cells as a solution to the issue. Rather than changing lifestyles, technocentrists would change technology to reduce carbon dioxide emissions and use science to endeavour to remove carbon dioxide from the atmosphere rather than reducing industry.

Carbon sequestration could provide a useful technocentric solution (see Figure 1.02). Carbon sequestration involves the capture and storage of carbon dioxide that would otherwise be present in the atmosphere. Carbon dioxide can be removed from the atmosphere and retained by plants and the soil that supports the plants. Alternatively, carbon dioxide can be captured (either before or after fossil fuel is burnt) and then stored (sequestered) within the Earth.

An ecocentric view would focus on the reduction in carbon dioxide by limiting the emissions that industry is permitted to make, even if this did limit economic activity and growth. Schemes such as carbon trading, which would allow large companies, such as airlines, a limited number of carbon 'credits', would be favoured by ecocentrists. To exceed its allowance, a company would have to buy additional credits from other organisations that were low-emitters or used their credits more efficiently.

(Resource use is discussed in Topic 8.)

Image 1.08 Artist's impression of proposed salt-spraying vessel.

Ecological footprints

A society's ecological footprint is the hypothetical amount of land that it requires to satisfy all its need for resources and to assimilate all its wastes. Many societies now consume more than is sustainable, and we are beginning to realise that this cannot continue. An increase in reliance on fossil fuels and increasing use of technology all increase a society's ecological footprint, whereas actions such as recycling, limiting pollution and reducing the use of resources all reduce the footprint (see Chapter 1.04). It is these actions that help reduce consumption that are being encouraged today.

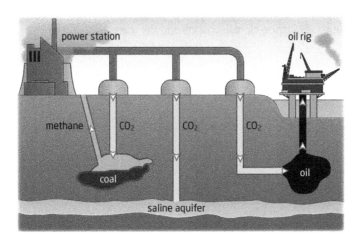

Figure 1.02 Carbon capture and storage involves capturing the carbon dioxide and preventing it entering the atmosphere by storing it deep underground.

1.01 Environmental value systems

CONSIDER THIS

In 2010, data showed that the United Arab Emirates (UAE) had the world's largest ecological footprint. Its population of 6.25 million people had an average footprint of 10.68 global hectares per capita (gha). The lowest footprint was found in Puerto Rica, where the population of 3.94 million had an average of 0.04 gha.

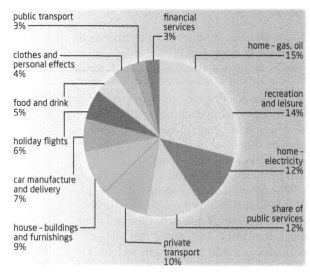

Figure 1.03 This pie chart shows the main elements of a typical individual's carbon footprint in an MEDC. A carbon footprint is one aspect of the ecological footprint. It is an estimate of the total greenhouse gas emission of this individual.

A **carbon footprint** is the total set of greenhouse gas emissions caused directly and indirectly by an individual, organisation, event or product (UK Carbon Trust, 2008).

The idea of 'ecological deficit' is linked to the availability of a biologically productive area in a country. When the ecological footprint of a society or country is greater than the biologically productive area, it has an ecological deficit. An ecocentrist would argue that, if ecological deficit happens, economic growth has caused the society to live unsustainably. Humans should therefore change their lifestyle and reduce their demands on the land. A technocentrist would counter that technology can solve the problems and, with economic growth, prosperity will help to redress the imbalances and ensure that deficits are removed.

A typical **carbon footprint** for an individual in an MEDC is shown in Figure 1.03. How do you think an ecocentrist would reduce this person's carbon footprint?

SELF-ASSESSMENT QUESTIONS 1.01.02

1. Define an 'environmental value system (EVS)'. How is this system similar to other systems?
2. How does a technocentric worldview differ from an ecocentric view?
3. **Discussion point:** How do environmental philosophies influence the decisions which are made about issues such as climate change and the use of fossil fuels?

1.01.06 Your personal viewpoint on environmental issues

Where do you stand on the continuum of environmental philosophies? Your personal value system, influenced by your background, education, culture and the society in which you live, will form your personal view of the world. You will have your own attitudes to the environment based on the influences you have had, the assumptions you have made and the conclusions you have come to. These factors will help form your own EVS. Personal value systems are principles that guide your behaviour and help you determine what is meaningful and important to you. A personal value system helps you express who you are and what you stand for. If you are unaware of your values, you may end up making choices out of impulse or for instant gratification rather than basing them on reason and responsible decision-making. This is why it is so important to know what you value and what is important to you.

Knowledge of the environment and the problems it faces will help you understand and become involved with environmental issues. You will probably also be influenced by those around you; so, for example, you may be more likely to join recycling schemes if your friends and neighbours do so and encourage you to do the same. The seriousness of an environmental problem and how close it is to you may also influence your behaviour. If the coastline near where you live is polluted by plastic waste (see Image 1.09), you are more likely to consider your own use of plastic shopping bags. Perhaps you are concerned about air pollution in your town; this may influence you to cycle or encourage your family to buy a hybrid car. Emotions also affect people's behaviour, so that anger or disgust at environmental damage such as burning forests or slaughtering whales can be a strong influence on your EVS.

As people become more knowledgeable about the environment, their awareness and sense of urgency of the need to deal with environmental problems also increases. More people realise that they can make a difference as individuals and that science can help to solve problems too.

Image 1.09 How does an image like this influence your view of the environment and human activity?

SELF-ASSESSMENT QUESTIONS 1.01.03

1   Which factors in your own community have an influence on your EVS?
2   How does an EVS affect the way people respond to environmental issues?
3   Which of these human activities is most likely to have a *negative* impact on the stability of global ecosystems?
    A   Decreasing water pollution levels
    B   Increasing recycling programmes
    C   Decreasing habitat destruction
    D   Increasing world population growth
4   **Discussion point:** The philosopher Socrates said, 'Not life, but good life, is to be chiefly valued.' Discuss what this statement might mean today.

1.02 Systems and models

LEARNING OBJECTIVES

After reading this chapter you should be able to:

- outline the concepts and characteristics of systems
- apply the concepts of a system to a range of different scales
- define open, closed and isolated systems
- understand how the systems approach can help in the study of a complex environmental issue
- understand how the systems approach enables us to take a more holistic view.

KEY QUESTIONS

How does the systems approach help us study complex environmental issues?

What are the key features of any system?

How are models used to help us understand how a system works?

1.02.01 The characteristics of systems

Our awareness of the impact of humans on the environment has grown over the last century as urbanisation and the growth of cities have transformed the natural world. In the 1960s, people became aware of the harm that pesticides could do. Since that time climate change, energy security, water supplies and conservation of biodiversity have all been high on the international agenda. The **systems approach** considers whole ecosystems and examines the best ways to protect our natural heritage.

The environmental systems that we are studying in this book are examples of the type of complex **systems** that can be studied using a systems approach. Others include biological systems, the systems that make up a society, transport systems and communication systems, or mechanical systems such as those in a bicycle. The systems approach looks at the environment or another complex system as a set of components that work together as integrated units. In ecology, we may study plants, animals or the atmosphere separately, but using the systems approach we consider them together as components of the complex environments in which they are found. Using this approach, we can obtain an integrated picture of the environment and the relationships and interactions within it. Integrated study is what makes the systems approach very different from the separate study of botany, zoology or geography. If we study a bicycle as a system, we can investigate how the chain, pedals, wheels and frame work together, rather than studying each component separately.

> The **systems approach** is a way of visualising a complex set of interactions in ecology, society or another system.
>
> A **system** is defined as an assemblage of parts and the relationships between them that enable them to work together to form a functioning whole.
>
> **Biomes** are groups of ecosystems with similar climates (see Topic 3).

Systems may be living or non-living, large or small. A single cell, a whole body, an ecosystem, or non-living examples, such as a banking system or a social system, are all examples of systems. In this course, you will be studying natural systems which include individual organisms, ecosystems (communities of organisms and their environment) and **biomes**. You will have the opportunity to examine the interactions within systems, which are often represented as diagrams with inputs, flows and outputs.

All systems have components, which are represented in diagrams showing their interconnections (see Figure 1.04). Components and their commonly used representations are shown in Table 1.01.

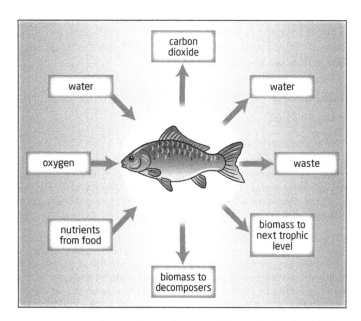

Figure 1.04 The systems approach can be used to consider an individual fish in a pond ecosystem. A systems diagram can be drawn to show storages, inputs and outputs.

Component of system	Shown as
storages (stores of matter or energy)	boxes
flows (into, within and out of the system)	arrows
inputs	arrows into the system
outputs	arrows out of the system
processes (which transfer or transform energy or matter from one storage to another)	labels such as respiration, consumption or photosynthesis

Table 1.01 System components and their representation.

The scale of a system

The scale of a system can range from a small part of a larger ecosystem, such as a tree in a forest, to, on a larger scale, the whole ecosystem or, on a significantly larger scale, the Earth itself could be regarded as a system. Whatever their size, all these systems have inputs, stores and outputs.

One example of a small-scale natural system is a pond within a woodland. The pond has inputs which include the light energy, water and nutrients needed to sustain the system, and also outputs such as the oxygen, decomposing material and nutrients which are generated in the pond. The pond in turn forms part of the whole woodland ecosystem that also has inputs, stores and outputs. On a much bigger scale, the woodland can also be thought of as part of an even larger system that includes all the woodland biomes in different parts of the world. These woodlands share the same climatic conditions and we can study them together as one large-scale system.

1.02 Systems and models

Emergent properties

Interactions within a system lead to the production of more complex **emergent properties**. To understand this, consider a single instrument such as a flute playing alone; it produces simple tune. But if we add more instruments, such as violins and drums, to make an orchestra, the same tune can be played in a far more complex way by the orchestra. We can think of the orchestra as a system, and the sound it produces as an emergent property. In a similar way, in natural science, emergent properties appear when a number of simple components operate together to form something more complex. For example, atoms combine to form molecules such as polypeptides that fold to become complex proteins. These molecules interact to build cells, tissues and organs of the body. On a larger scale, the behaviour of flocks of birds or shoals of fish shows emergent properties, and all the biological communities in the world form the **biosphere**, and this produces many complex interactions and emergent properties

> **Emergent properties** are features of a system that cannot be present in the individual component parts.

> The **biosphere** is the part of the Earth inhabited by organisms, and it extends from the upper atmosphere to the depths of the Earth's crust.

Theory of knowledge 1.02.01

Reductionist versus systems approach

A reductionist view of a natural system looks at a single object that can be clearly recognised and identified by its properties. The organisms found within a pond are individually described in reference books, which describe them in terms of the characteristics that they have, such as whether they are a plant or an animal. At the next level, animals may be described as invertebrates or vertebrates, and so on. This **reductionist approach** does not try to consider how the pond works as a dynamic system.

A systems approach gives a holistic view of the pond. The reductionist view does not allow interconnections and interrelationships that go on in the pond to be taken into account. The systems approach considers any system as a set of interrelated objects. In the pond, the most obvious interrelationship between the plants and animals is that some plants and animals are food for other animals; this relationship is called a food chain and, without it, animals would die of starvation. Imagining the pond system as flows of energy and matter (food) between objects (plants and animals) means that a picture of the pond system's structure (the objects and their relationships) and function (the purpose of the various interactions) can be built up.

1. What are the advantages and disadvantages of the systems approach compared with a reductionist approach to the study of an ecosystem?
2. In science, the reductionist and the systems approach may use similar methods of study. What is the most important difference between the philosophies of the two approaches?

> **Reductionist approach** to a system reduces the complex interactions within it to their constituent parts, in order to study them; whereas a systems approach considers the whole system and the interactions between the various components.

Open, closed and isolated systems

Systems are divided into three types: **open**, **closed** and **isolated systems** (see Figure 1.05).

Most living systems and all ecosystems are open systems which exchange energy, new matter and wastes with their environment. These open systems and the exchanges which take place can be seen in any living environment, even in remote locations such as Antarctica or tiny isolated islands.

In a woodland ecosystem, the main inputs include light and carbon dioxide, which plants use for photosynthesis. Further inputs come from woodland herbivores that return mineral nutrients to the soil in faeces, and bacteria in the soil that fix nitrogen from the air. Outputs may include water that is lost during respiration and transpiration, nutrients that flow away in waterways, and heat that is exchanged with the environment around the woodland.

In a closed system, energy but not matter is exchanged across the boundaries of the system. Closed systems are very rare in nature. Most examples are used for experiments and are artificial. A bottle garden or an aquarium can be set up so that light and heat are exchanged across its boundaries but matter cannot be exchanged or leave the system. In most cases, these systems do not survive, because they become unbalanced. Organisms may die as oxygen is depleted or as food runs out or waste matter builds up to toxic levels.

No isolated system is known to exist, although some people regard the entire Universe as an isolated system.

> An **open system** exchanges both matter and energy within its surroundings across the boundaries of the system.

> A **closed system** exchanges energy but not matter across the boundaries of the system.

> An **isolated system** exchanges neither energy nor matter with its environment.

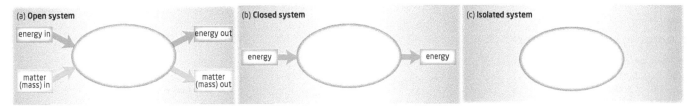

Figure 1.05 Open systems exchange energy and matter, closed systems exchange only energy, and isolated systems do not exchange energy or matter.

CASE STUDY 1.02.01

Biosphere 2 – a model of a closed system

Biosphere 2 was an experimental system set up in 1991. Eight people were sealed into a 1.27 hectare glass enclosure in Arizona (see Image 1.10). The aim of the project was to test whether a small group could live in a self-sustaining way inside the biosphere. It was hoped that they could produce their own food and recycle all their waste and water. But things began to go wrong when, after about six months, the inhabitants had lost a lot of body weight and the oxygen levels inside the sealed system began to fall. The inhabitants weren't able to produce enough food for their needs, partly because unusually cloudy external conditions prevented crops from growing and also because animals inside the biosphere were not doing well. Hens were not producing enough eggs and pigs were consuming too much food. Eventually the animals had to be killed and the people began to eat seeds that were supposed to be used for planting. After 18 months, the oxygen levels fell to 14%, well below the safe level of 195% for human breathing and the people suffered breathing problems and lethargy. For medical reasons, air was allowed into the biosphere, and thus the project to maintain a completely closed system had to be considered a failure.

Since the first experiments, Biosphere 2 has taken on a new role, not as a closed system but as a research institute and visitor centre. Its aims are to study and teach about the Earth and its living systems, for example how the water cycle is related to ecology, atmospheric science, soil geochemistry and climate change.

Image 1.10 Biosphere 2 was set up as a closed system enclosed by sealed glass.

Questions

1. What were the inputs into the closed system of Biosphere 2?
2. What were the two main reasons for terminating the project after two years?
3. Research idea: Find out more about the current research carried out in Biosphere 2.

A useful, interesting website for further reading: Jane Poynter's discussion on Biosphere 2

www.ted.com/talks/jane_poynter_life_in_biosphere_2

1.02 Systems and models

CONSIDER THIS

For many centuries, people have discussed ideas of a holistic view of the Earth as an integrated, living whole. In ancient Greek mythology, Gaia was the goddess who personified the Earth. James Lovelock, a scientist and environmentalist, is well known for proposing the 'Gaia hypothesis'. This hypothesis proposes that the biosphere is self-regulating and is able to keep the Earth healthy by controlling the interactions between the chemical and physical aspects of the environmental system. Lovelock gave the name Gaia to his hypothesis at the suggestion of novelist William Golding. Gea is an alternative spelling for the name of the goddess Gaia that is reflected in the prefix in geology, geophysics and geochemistry. The Gaia hypothesis became well known during the 1960s at the time of the space race between the Soviet Union and the USA, when people first saw images of the whole Earth taken from space.

SELF-ASSESSMENT QUESTIONS 1.02.01

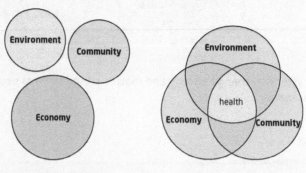

Figure 1.06 The difference between the systems approach and the conventional approach to the study of a system.

1. Figure 1.06 outlines the difference between the systems approach and the conventional approach to the study of a society. What are the benefits of using the systems approach in the study of: **a** an ecosystem and **b** other subjects such as economics or engineering?
2. Construct a table to compare the exchange of matter and energy in open, closed and isolated systems.
3. **Discussion point:** Do you think it is useful to have the concept of an isolated system which does not exchange energy or matter with its surroundings?

Transfers and transformations

Transfers

Both matter and energy pass through ecosystems, and if their movement involves only a change in location and does not involve any change of form (or state), the movement is called a **transfer**. A trophic level is a group of organisms that are all the same number of energy transfers from the producer (plant) in a food chain or food web (see Topic 2). Energy flows through an ecosystem: as biomass found in the bodies of organisms in one trophic level is eaten, so biomass and energy pass to the next trophic level.

Some examples of transfers include:

- transfer of matter through a food chain as one animal eats another
- transfer of energy as wind carries heat energy from one part of the world to another
- transfer of matter as water flows from a river to the sea.

Transformations

A **transformation** occurs when a flow in a system involves a change of form or state, or leads to an interaction within the system. The evaporation of water from a river is an example of a transformation, because water is changed in form from a liquid to a vapour in the atmosphere. In ecosystems, energy is transformed from sunlight into chemical energy in the bonds of molecules in plants during the process of photosynthesis. As organisms respire, chemical energy is transformed into heat and kinetic energy.

Some examples of transformations include:

- energy to energy – light energy to electrical energy in a solar panel (photovoltaic cell)
- matter to matter – decomposition of leaf litter into inorganic materials

---

Transfers involve flow through a system and involve a change in location.

Transformations lead to an interaction within a system and the formation of a new end product, or they may involve a change of state.

- matter to energy – burning coal to produce heat and light
- energy to matter – light energy converted by photosynthesis to produce glucose molecules.

Flows (inputs and outputs) and storages (stock) in a system

Energy and matter are the inputs and outputs that flow through ecosystems, but they are also stored within the system as storages (or stock).

In an ecosystem, the energy input is sunlight, which is transformed into chemical energy in the bonds of glucose formed during photosynthesis. Energy flows from one part of an ecosystem to another as one organism eats another. Some energy is used to drive the life processes of these organisms, and energy leaves the system in the form of heat which is released as a result of respiration.

Matter flows from one trophic level to the next as plants or animals are eaten. Eventually, matter is recycled through the decomposition and decay of dead organisms and of their waste products. In any ecosystem, there are storages linked by different flows. Carbon and nitrogen are two elements which are cycled around an ecosystem and pass between storages in living organisms, the atmosphere and the land (see Topic 2).

As you saw in Table 1.01, storages are represented by boxes, which can be drawn to be proportional to the size of the storage. Likewise, arrows that indicate flows between storages can be drawn so that their width is in proportion to the size of the flow.

1.02.02 Models involving flow and storages in a system

Environmental scientists use models to show the flows, storages and links within an ecosystem. A model, in this case, is a diagram that uses different symbols to represent each part of the system. Arrows are used to represent the flow of energy or materials, and different boxes are often used to represent storages in producers, consumers and so on. Figure 1.07 illustrates the general flows and storages in an ecosystem.

In Figure 1.08, the flow of energy is shown by the red arrows. The blue arrows show the cycling of nutrients, and the boxes represent storages.

Models like these enable environmental scientists to draw comparisons between different ecosystems by representing the different inputs, outputs and storages as boxes or arrows that are in proportion to their sizes.

Evaluation of models' strengths and weaknesses

Models are drawn to represent situations found in real systems, but in reality they can only be approximations and predictions in most cases. Computer modelling and simulations are used to predict outcomes such as the pattern of the weather and the likely course of climate change.

Models have many strengths and weaknesses, and it is important to bear these in mind when models are used. Computer modelling of climate change is a good example of how modelling can lead to controversy as well as consensus. Not everyone agrees on the scale of projected inputs and outputs, or on the interpretation of the models. Table 1.02 shows the advantages and disadvantages of modelling.

1.02 Systems and models

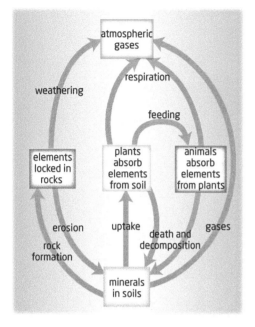

Figure 1.07 The biogeochemical cycle showing flows (arrows) and storages (boxes).

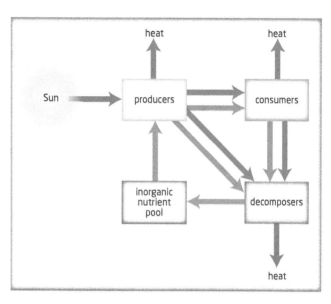

Figure 1.08 Energy flow, red arrows, and nutrient cycling, blue arrows, in a system.

Advantages	Disadvantages
Modelling allows complex systems to be simplified.	Models may be oversimplified so that accuracy is lost.
Modelling allows predictions to be made about future events.	Models and predictions depend on the skills and experience of the people making them.
Different scenarios can be considered by changing inputs and calculating likely outcomes.	Models may be interpreted differently by different scientists. Different models may predict different outcomes.
A model can form the basis for discussion and consultation with others who are interested in the system being modelled.	Data may not be accurate and models can be manipulated for financial or political gain.

Table 1.02 Modelling – the pros and cons.

SELF-ASSESSMENT QUESTIONS 1.02.02

1. What is the difference between a transfer and a transformation in an ecosystem?
2. Give one example of each of the following in an ecosystem: an input, an output, a storage.
3. Give three advantages to drawing a model of climate change, and suggest three weaknesses of the modelling process.
4. **Discussion point:** Why do you think that scientists are keen to use models to communicate their ideas to the general public and politicians? What are the merits of presenting information in this way?

Theory of knowledge 1.02.02

What aspects of the world should a model include?

Models are simplified representations of the world and present complex systems and situations to help us gain knowledge about them and understand them. One question often asked about models is how useful they are. We must remember that knowledge and understanding are different. Knowledge is facts based on observation of the world, but understanding is comprehension of this knowledge. Do models hinder our ability to collect new knowledge by making us blind to new possibilities? The philosopher Karl Popper said, 'Science may be described as the art of systematic oversimplification.' A model is a simple, abstract view of a very complex system, particularly in natural sciences. If models are too simple, they may obscure the detail of what is going on. It is important to decide how much simplification of a complex system can be justified, so that we come to understand how it works.

1   How can we know which aspects of the world to include in a model and which to ignore?
2   Are some models more objective than others?

CASE STUDY 1.02.02

Comparing inputs and outputs of different ecosystems

Figure 1.09 compares the inputs and outputs of an agricultural ecosystem and an urban ecosystem. In the diagram, the arrows have been drawn so that their sizes represent the sizes of the inputs and outputs of the two systems.

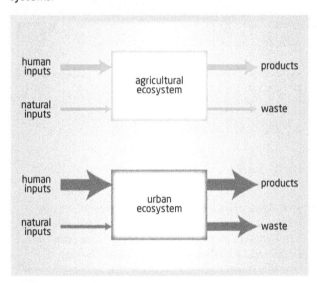

Figure 1.09 Comparing inputs and outputs of a small agricultural and an urban ecosystem.

Draw your own diagrams to compare the inputs and outputs in an intensive agricultural system that produces wheat with those of a natural ecosystem such as a woodland or forest.

Use the following information to help you.

- Agricultural ecosystems use selected domesticated plants or animals to produce food, fibre or fuel for human consumption.
- Some agricultural ecosystems differ greatly from natural ecosystems; others do not. Pasture ecosystems with grazing animals require fewer human inputs than crop ecosystems, because pasture is similar to a natural ecosystem.
- Modern agricultural ecosystems need inputs from farm machinery, chemical fertilisers, pesticides and irrigation.
- Intensive inputs increase the conversion of sunlight energy to human food energy during photosynthesis, by providing the best conditions for plant growth (e.g. ample water and nutrients from fertilisers) and excluding plants and animals such as weeds or pests that compete for the ecosystem's production.
- Agricultural ecosystems contain non-living things made by people, such as irrigation ditches and farm equipment, which require maintenance.
- Intensive inputs to agricultural ecosystems depend on large amounts of petroleum energy to produce fertilisers and pesticides and for transport.
- Water is a key input, and modern irrigation methods need large amounts of water, sometimes from hundreds of kilometres away.
- High inputs in agriculture produce high outputs and high yields of crops or animals.
- Outputs of modern agricultural ecosystems include waste such as fertilisers and pesticides that can damage nearby ecosystems.

Questions

1   Can you suggest any other inputs that could increase the output from an agricultural system?
2   List three disadvantages of a modern agricultural system.
3   Compare the inputs and outputs in a modern agricultural system with those of an agricultural system 100 years ago.

1.03 Energy and equilibria

LEARNING OBJECTIVES

After reading this chapter you should be able to:

- outline the laws of thermodynamics
- explain how laws of thermodynamics relate to environment systems and govern the flow of energy in a system
- describe how a system can exist in alternative states of equilibrium
- describe how positive destabilising feedback mechanisms can drive a system to a tipping point
- describe how negative feedback can stabilise a system.

KEY QUESTIONS

How do the laws of thermodynamics govern energy flow?

What is negative feedback?

How are stable systems driven to tipping points?

1.03.01 Laws of thermodynamics and why they are relevant to environmental systems

> **CONSIDER THIS**
>
> Claude Lévi-Strauss was a French anthropologist and ethnologist who defined a scientist as someone who asks the right questions, not someone who gives all the right answers.

The first law of thermodynamics states that energy cannot be created or destroyed but can be converted from one form to another. Energy exists in the form of light, heat, chemical energy, electrical energy, sound and kinetic energy. Different forms of energy are interconvertible but, in a *living* system, heat energy cannot be converted to other forms. (The most obvious *non-living* conversion of heat energy is seen in a steam engine when heat energy is converted to kinetic energy.)

In an ecosystem, useful energy enters the system as light energy, which is converted to chemical energy during photosynthesis and used to build the bonds found in plant biomass (biological organic matter). This energy in plant biomass is passed along a food chain in a series of transfers (see Figure 1.10) as organisms feed on plants and are themselves eaten. At each stage in the transfer processes, some energy is transformed (see Figure 1.10) to other forms, including heat, as organisms respire and use the energy from their food to fuel their life processes. Energy leaves the system as heat, because heat energy cannot be transformed in a living process. In living systems no new energy has been created, but the input energy has been converted from one form to another.

Although the total amount of energy in the system does not change, the amount that is available to living things gradually reduces as energy is used for growth, movement, reproduction and other processes. Energy transfer and transformation are not very efficient in living systems, and at each transfer there is less energy available after the transfer than at the start; less than 10 per cent of usable energy is passed from one organism to the next in a food chain such as when a rabbit eats grass and is then eaten by a fox (see Figure 1.10).

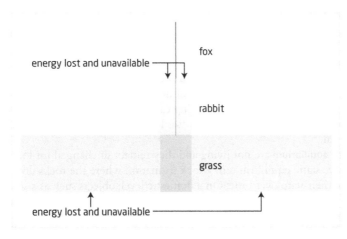

Figure 1.10 Less than 10 per cent of the energy is transferred at each link in a food chain.

The **second law of thermodynamics** states that, in isolated systems, **entropy** tends to increase with time because the system becomes disorganised. Entropy can be explained as the evenness of energy distribution in a system. Energy is used to create order and hold molecules together. This means that entropy (disorder) increases if less energy is available. As entropy increases, energy and matter change from a concentrated to a more dispersed form. The availability of energy to carry out processes becomes reduced and the system becomes less orderly.

The most concentrated form of energy is that from the Sun, and the most dispersed form is heat. If the Universe is considered as an isolated system, the level of entropy is gradually increasing as energy is distributed within it. Eventually, in billions of years' time, energy may run out. But unlike isolated systems, natural and environmental systems are never isolated. Living systems require a constant input of energy from the Sun to maintain their order and structure, and to replace energy that is lost.

Second law of thermodynamics In isolated systems entropy tends to increase.

Entropy is a measure of the evenness of energy distribution in a system. It is also defined as a measure of the disorder of a system: the greater the disorder, the higher the level of entropy.

CONSIDER THIS

It is generally accepted that the biosphere is getting more complex and organised (entropy is decreasing), and this seems to contradict the second law of thermodynamics. But if we consider examples such as the production of cars, clothes and chemicals, we see that the entropy of this part of a system is decreased as the components that make them up are organised. However, this decreased entropy (increased organisation) is due to the manufacturing system whose entropy is increased as products are made. Thus, one subsystem (the manufacturing system) is able to influence the entropy of another (the products). But eventually, over time, the entropy of the entire system will increase as stated by the second law of thermodynamics. The second law is true for a closed system, but the biosphere is an open subsystem that can exchange matter, energy and entropy with the rest of the Universe, so there is no contraction.

Theory of knowledge 1.03.01

The laws of science, such as the laws of thermodynamics, can be said to be different from the laws of other subjects such as economics. The reasons for this can be explained by examining how science is carried out, a process known as the scientific method. In most cases, scientific research leads to the development of a law only after a rigorous process which can be explained as a series of steps. First there is a problem, which leads to a hypothesis, then a prediction followed by testing, review by others, and replication of an experiment or experiments. This process may lead to a theory and, following any corrections or modifications, the formulation of a law.

Thus, science should provide a law based on impartial research backed by careful checking. Can this process be applied to other subjects such as economics? If you don't think it can be, explain why not.

1 Which aspect of our acquisition of knowledge in the sciences do you think is the most important and why?
2 Is there a role for reason in the process of gathering knowledge in the sciences?

1.03.02 Equilibria

Equilibrium is a state of balance that exists between the different parts of any system. As we have seen, natural systems are open and most are in a **steady-state equilibrium**. There are fluctuations in the system, but these are within narrow limits and the system usually returns to its original state after being disturbed.

Regulation of body temperature in mammals is an example of a steady-state equilibrium. If the temperature of a mammal rises above 37 °C, processes occur in the body to return the temperature to normal. If the temperature falls, the processes are reversed to enable the body to warm up (see Figure 1.11). Another example of a steady-state equilibrium can be seen in a

Equilibrium is a state of balance among the components of a system.

Steady-state equilibrium (also known as dynamic equilibrium) is a stable form of equilibrium that allows a system to return to its steady state after a disturbance.

1.03 Energy and equilibria

population of animals that remains approximately the same size. Some animals may be born and others may die, but if birth and death rates are equal, there is no net change in the population size.

At the level of the ecosystem, a steady state can be achieved after a disturbance either in the short term, for example as a woodland recovers after heavy rainfall, or in the long term, as new growth occurs to replace damaged plants in an area of the wood. If a tree dies or is felled, a new area is opened up in the woodland and a phase of new growth can take place. New plants will receive extra light, and young trees in the clearing can grow rapidly. After a long time, a new tree will become established to replace the one that was lost. Eventually, the system will return to its previous equilibrium.

Systems in **static equilibrium** are not living and they remain unchanged for long periods of time. We can observe static equilibrium in a rock formation where the rocks do not move their position or change their state over time. On a domestic level, objects such as a sofa or armchair or a bottle placed on a table can be said to be in static equilibrium. A graph showing static equilibrium is given in Figure 1.12.

> A **static equilibrium** is a type of equilibrium in which there are no changes over time because there are no inputs to and outputs from the system.

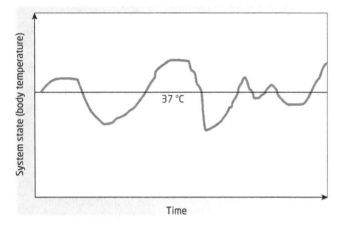

Figure 1.11 The body temperature of a human varies between 36 °C and 39 °C but remains at an average of 37 °C. Small rises and falls in temperature are corrected by processes such as sweating or shivering so that a steady state is maintained.

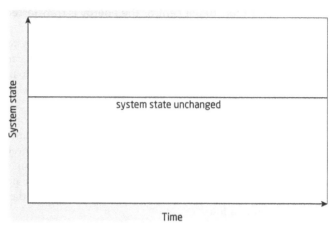

Figure 1.12 Nothing changes in a static equilibrium.

Stable and **unstable equilibria** are situations in a system where change occurs but in each case the final result is different. In a stable equilibrium, the system tends to return to the same equilibrium after a disturbance, while in an unstable equilibrium a new equilibrium is formed after the disturbance (see Figure 1.13).

A pendulum swinging from a suspended string is said to be in stable equilibrium because, if it is pushed to the side, it will return to its original position. But a ruler balanced vertically on a finger is in unstable equilibrium – if it is disturbed, it will fall and continue to fall until it hits the ground, creating a new and different equilibrium.

In natural ecosystems, which are open systems, there is normally a stable equilibrium. The stable equilibrium may be a steady-state equilibrium, like the pendulum, or the ball shown in Figure 1.13, or it may be a **developing steady-state equilibrium**. If there is a disturbance to the steady-state ecosystem, such as a natural event like a storm, the system will be disturbed but will return to its equilibrium. A developing steady state occurs in an ecosystem that is changing over time. We can see this happening as a succession takes place in a newly colonised area (see Chapter 2.04).

> A **stable equilibrium** in a system is a state in which a system that is disturbed returns to its former position.
>
> An **unstable equilibrium** is a state of equilibrium in which a small disturbance produces a large change and a new and different equilibrium.
>
> **Developing steady-state equilibrium** is a steady-state equilibrium that is developing over time (e.g. in a succession).
>
> **Feedback** is the return of part of the output from a system as input, so as to affect succeeding outputs.

Positive and negative feedback

Natural systems are able to regulate themselves through **feedback** systems. Information, which may come from inside or outside the system, starts a reaction which affects the processes of the

Equilibria

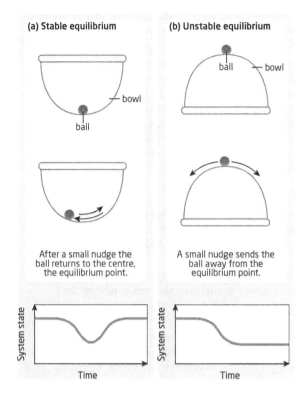

Figure 1.13 (a) The ball inside the bowl will return to its original position after a disturbance and is said to be in stable steady-state equilibrium. (b) A ball balanced on top of the bowl is in unstable equilibrium because a new equilibrium is formed if it is disturbed.

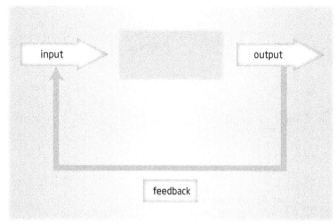

Figure 1.14 A feedback loop.

system. Changes in these processes lead to changes in the level of output, and this in turn affects (feeds back) to the level of input. This whole cycle is known as a feedback mechanism or feedback loop (see Figure 1.14). Feedback loops can be either positive or negative. Feedback can change a system to a new state or maintain a system at a steady state.

Positive feedback

A **positive feedback** loop is destabilising – it allows a system to change rapidly (see Figure 1.15). One example is the population growth of a plant, the water hyacinth, which has spread into new environments from South America. If one water hyacinth is introduced into a large, uncolonised lake, the plant will reproduce exponentially: one plant divides to become two, two become four, and so on. At first, the growth of the plants does not seem to be significant, but after about two years of unchecked growth the number of plants can reach 10^9 (1 billion). At this level, serious problems occur for other species and for navigation across the lake. The plant prevents sunlight reaching the water, which is starved of oxygen; fish die and boats cannot move across the water. The exponential growth of the water hyacinth population is a positive feedback relationship between the population size and the number of new organisms added to the population. The greater the population, the greater the number of new additional organisms, so the faster the population grows (see Topic 2).

Other examples involving positive feedback include the increase in the Earth's temperature through global warming. Higher atmospheric temperatures increase the evaporation of water from the Earth's surface; this increases water vapour in the atmosphere and, because water vapour helps to trap heat in the atmosphere, the outcome will be more heat trapped and a further increase in atmospheric temperature. In addition, if higher air temperatures cause polar ice to melt, the reflection of heat from the white surface of the ice will decrease. More of the Sun's energy will be absorbed and the temperature will increase still further.

Positive feedback must eventually come to an end, as the resources which allow the rapid change will also come to an end. But there is no guarantee that the situation will revert back to its original state. For example, lakes in many parts of the African continent remain covered with the invasive water hyacinth.

Positive feedback results in a change in the system that leads to more and greater change. It amplifies or increases change and leads to exponential deviation away from an equilibrium and thus destabilises the system.

Negative feedback tends to damp down or counteract any deviation from an equilibrium and promotes stability. It stabilises the system to eliminate any deviation from the preferred conditions.

1.03 Energy and equilibria

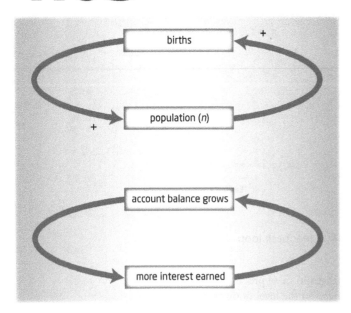

Figure 1.15 Exponential population growth is an example of positive feedback. A bank account earning interest is another.

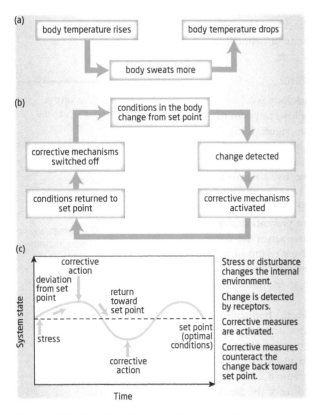

Figure 1.16 Negative feedback is a regulating mechanism in which a change in a variable results in a correction. If body temperature rises, the body activates cooling processes until the temperature returns to normal.

Not all examples of positive feedback cause an increase of a factor in a system. If a change is downward, positive feedback can make the downward change even greater. For example, if the population of an endangered species becomes very small, it is difficult for the animals to find mates. As a result, births are fewer and the population decreases. This decrease in population makes it even more difficult to find mates, and the population decreases still further. Positive feedback causes population decline, which leads to loss of biodiversity and extinction (see Topic 3). Undesirable positive feedback like this leads to a vicious cycle of events.

Negative feedback

Negative feedback works to counteract any deviation from the stable state or equilibrium. It stabilises a system and allows it to regulate itself and eliminates any deviation from the preferred conditions.

In engineering, one of the first examples of negative feedback was a device known as a governor, used by the Scottish engineer James Watt (1736–1819), who incorporated it into his steam engines in order to maintain their speed. If the speed of the engine increased, the governor cut off the supply of steam to slow it down to the correct speed. If the engine speed decreased, the governor allowed in more steam, so that the engine ran faster. In this way a constant speed was achieved. Another similar example of negative feedback is the thermostat on a heating system, which can be set to maintain a constant temperature. In the human body, negative feedback helps to maintain a constant body temperature (see Figure 1.16).

Maintenance of a steady-state equilibrium involving negative feedback is vital to keep the internal conditions of animals' bodies relatively constant. The temperature of a mammal's body must be maintained at about 37 °C so that life processes can take place in their optimum conditions. An increase in temperature leads to increased sweating and widening of the capillaries in the skin, so that heat is lost. As the body is cooled by these processes, it returns to its normal temperature. Many other body functions, including the regulation of sugar in the bloodstream by the hormone insulin and the maintenance of the correct level of water in the body, are controlled by physiological processes that involve negative feedback.

In an ecosystem, one example of negative feedback is the control of the relative numbers of species in food webs (a food web is a complex interacting set of food chains). If one species becomes too successful and its numbers rise excessively, it will use up the resources it needs or be overcome by its waste. Negative feedback ensures that its numbers fall back to sustainable levels.

On a global scale, the increase in carbon dioxide released into the air from burning fossil fuels provides more carbon dioxide for plants, which can increase their rate of photosynthesis. As plants increase their rate of photosynthesis, they remove more carbon dioxide from the atmosphere; therefore, as long as there are sufficient plants, the system can be rebalanced.

A French proverb has been used to summarise the process of negative feedback: *Plus ça change, plus c'est la même chose* (the more things change, the more they stay the same).

1.03.03 System resilience

As we have seen, feedback is the effect that change in one part of an ecosystem (or social system) has on the same part after a series of events in other parts of the system. Negative feedback provides stability, and all ecosystems have negative feedback loops that keep each part of the system stable and within the limits needed to continue functioning efficiently. But positive feedback stimulates change and is responsible for the sudden appearance of environmental problems and other rapid changes in natural systems. Positive feedback is destabilising and tends to amplify any changes that occur.

If we consider the example of water hyacinths (above), we can begin to understand how the exponential growth of a population can outstrip the **resilience** of an ecosystem so that it reaches a **tipping point** that will lead to the establishment of a new equilibrium. If the lake ecosystem where water hyacinths were introduced had been resilient and able to resist the enormous growth of the plants, it would have remained more or less unchanged. But because the numbers of water hyacinths became so great, the ecosystem reached a tipping point and the original ecosystem was destabilised. Now many species have been lost, diversity has reduced and the lake ecosystem is significantly different from its original form.

Exponential growth of the human populations in recent years (see Topic 8), plus the exponential growth in our use of natural resources and production of pollutants, may also test the limits of resilience of our ecosystems.

The resilience or ability of a system to avoid such undesirable tipping points is influenced by two key factors:

- the diversity present in the system
- the size of the storages the system contains.

A resilient system contains a wide diversity of organisms and large storages.

Humans can affect the resilience of systems by reducing either diversity or storages. An agricultural system where crops are grown in a monoculture has few species and is unable to resist changes such as drought or an attack of pests. A natural system, which has a much wider diversity of species, is more resilient and, although parts of the system might be affected by such changes, the system will usually be able to return to its stable equilibrium. If storages are depleted by human interference, such as the excessive removal of trees from a forest or overfishing in the ocean (see Chapter 4.03), the systems may be unable to recover, and a new equilibrium, without the overexploited resource, may be the result.

> The **resilience** of a system is the tendency of a system to maintain stability and resist tipping points.
>
> The **tipping point** is the minimum amount of change within a system that will destabilise it and cause it to reach a new equilibrium or stable state.

SELF-ASSESSMENT QUESTIONS 1.03.01

1. How does the first law of thermodynamics explain how energy moves through an ecosystem?
2. What is meant by 'entropy' and how does it relate to a natural system?
3. Outline the difference between a steady-state equilibrium and a static equilibrium.
4. **Research idea:** The human population is growing at an exponential rate. Research the possible consequences of this example of positive feedback.

1.03 Energy and equilibria

CASE STUDY 1.03.01

Predator–prey relationship

In an ecosystem, predation is a mechanism of population control that involves negative feedback. If we consider the relationship between a prey species such as a rabbit and a predator species such as a fox, we can predict that, when the number of foxes is low, the number of rabbits should rise (see Figure 1.17). An increase in the number of rabbits means that the foxes have more food and can produce more offspring and may be able to change their hunting habits. As the number of foxes increases, the number of rabbits declines as more are caught and eaten. This results in food scarcity for foxes, so that more starve and die or fail to reproduce. Negative feedback balances out the two populations and the cycle can begin again (see Topic 2).

Questions

1. Outline how negative feedback allows this system to maintain a steady state.
2. How could the equilibrium in this example be disturbed so that the steady state is not re-established?
3. Research idea: Investigate some strategies and adaptations of predator and prey species that enable them to either to catch prey or to avoid being eaten.

Extension

Rabbits and wolves
www.shodor.org/interactivate/activities/RabbitsAndWolves

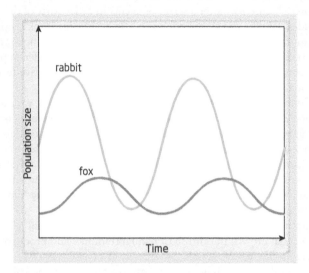

Figure 1.17 When the numbers of predators are low, the prey species is able to increase in number. As more prey animals become available, predators increase in number and the population of prey falls.

Sustainability

1.04

LEARNING OBJECTIVES

After reading this chapter you should be able to:

- explain what is meant by sustainability and how it is possible to view a system
- outline how sustainable development meets the needs of the present without compromising the future
- describe what is meant by the terms 'natural capital' and 'natural income'
- describe how environmental indicators and ecological footprints are used to assess sustainability
- outline the concept of sustainability in terms of natural capital and natural income
- explain the importance of Environmental Impact Assessments (EIAs) in sustainable development
- understand that biodiversity, pollution, population and climate can be used quantitatively as indicators of sustainability
- describe how the ecological footprint is used to assess sustainability.

KEY QUESTIONS

What is sustainable development?
How are environmental indicators used?
Where and when are Environmental Impact Assessments (EIAs) used?

1.04.01 Sustainable systems: natural capital and natural income

A sustainable system is one that remains diverse and productive. It will survive changes and return to its natural state. Some good examples of ecosystems that are naturally sustainable include wetlands and forests that have existed almost unchanged for long periods of time. Humans need healthy environments and ecosystems to survive, and it is important that new ways are found to limit the damage to ecosystems when we take resources from them, so that they can maintain their sustainability. As you study environmental systems in this book, you will discover that ecology, economics, politics and environmental values are all important for sustainability.

Resources such as water, timber, animals and plants are used by humans and are described as a system's **natural capital**. We can think of natural capital as the stock in an ecosystem that provides a flow of valuable goods and services on into the future. (Natural capital is a way of extending the idea of capital in economics to goods and services from the natural environment.) Two examples of goods might be a forest of trees or a stock of fish that each provide a flow of new trees or young fish. Mineral deposits and fertile soil are two other examples. These resources can be indefinitely sustainable if they are used and managed wisely so that the ecosystems that contain them will recover after some of the resource has been removed.

> **Natural capital** describes natural resources that produce a sustainable income of goods and services.

1.04

Sustainability

Natural income is the yield obtained from natural resources.

Sustainability is the use and management of resources so that full natural replacement of exploited resources can take place.

Ecosystem services are a form of natural income derived from natural capital.

We define the yield from sources of natural capital, such as timber, fish or plants, as **natural income**. If the amount of natural income reduces the ability of natural capital (the woodland or the ocean) to continue to provide the resources at the same rate, then this is the point when **sustainability** is no longer possible. But a supply of natural resources such as fish or timber is not the only type of natural income; we know that a forest also absorbs rainwater, which controls flooding, provides a habitat for plants and animals, produces oxygen and is important in the water cycle. From this we can see that natural capital also provides services such as water purification, waste recycling, water catchment and control of erosion.

Ecosystem services are another form of natural income derived from the same natural capital of the forest that generates timber for economic use. Ecosystem functions like these have value to us, and the 'flow' of these services depends on the fact that the system functions as a whole, so it is vital that the structure and diversity of the system is preserved.

Any society that supports itself in part by depleting essential forms of natural capital cannot do so forever, because this action is unsustainable. The rate at which natural capital is used should not exceed the rate at which is it renewed. Sustainability means living within the means of nature. We must do all we can to reduce the level of climate change, overconsumption of natural resources and the amount of damage we cause that is likely to degrade the environment, if ecosystems are to remain healthy and sustainable.

1.04.02 Sustainable development

The term 'sustainable development' was first used in 1987 and was defined as 'development that meets current needs without compromising the ability of future generations to meet their own needs' (the World Commission on Environment and Development's *Our Common Future* (1987), also known as the Brundtland report).

The issue of sustainable development has been the subject of considerable debate. Some traditional economists view sustainable development as a stable annual return on investment whatever the environmental impact. In contrast, a now common environmental view sees it as a stable return without environmental degradation. There can be little doubt that the latter view has gained ground over the former.

The UN's view on sustainable development encompasses:

- keeping population densities below the carrying capacity of a region so that humans do not overwhelm an area
- doing everything possible to ensure the renewal of renewable resources so that systems can recover
- conserving and establishing priorities for the use of non-renewable resources such as coal and oil
- keeping the environmental impact below the level required to allow affected systems to recover and continue to evolve.

Sustainability at different scales

The idea of sustainability can be applied to:

- The full range of scales, from the individual to the Earth as a whole: Increasingly, governments are reminding individuals and households about their carbon footprints and how these can be reduced at the domestic level, while tackling the problem at the national level. In many countries, adults have been asked to think about reducing their daily driving distance and to reuse plastic bags. At the largest scale, sustainability focuses on the total carrying capacity of the planet.
- Different geographical environments, such as rainforests, temperate grasslands and urban areas: Satellite photography has been a major advance in our ability to see what is happening over large land areas. It has allowed short-term changes to be recognised quickly.

- Individual economic activities, such as tourism, agriculture and forestry: Each sector has its own impact on the environment which can be modified by careful management (see Topic 8.2).

Sustainability need not require a reduction in the quality of life, but it does require a change in attitudes and values towards less consumptive lifestyles. These changes must embrace global interdependence, environmental stewardship, social responsibility and economic viability. Environmental sustainability in a country or region is difficult to achieve without economic and social sustainability, because of the strong interconnectedness between these three vital spheres of life (see Figure 1.18). Economic sustainability involves maintaining income and employment. Social sustainability means maintaining social capital, including that devoted to health, education, housing and the rule of law. You can study sustainability and economic growth in more detail in Topic 8.2.

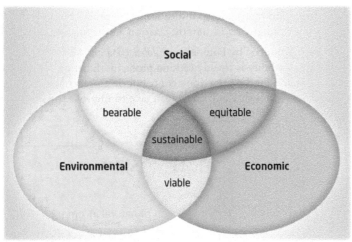

Figure 1.18 Social, economic and environmental sustainability.

Millennium Ecosystem Assessment

The Millennium Ecosystem Assessment (MA) is an international assessment of the effects of human activity on the environment, which gave a scientific appraisal of the condition of the world's ecosystems. During its production, the term 'ecosystem services' was first used to describe the benefits humans gain from ecosystems. The MA was launched in 2001, and more than 1300 individuals from 95 countries took part over a period of four years. Participants included representatives from the UN, governments, NGOs, academics, business leaders and representatives of indigenous peoples. The report concluded that nearly two-thirds of the services we derive from natural systems that support life on Earth are in decline, and that environmental degradation is already a serious barrier to reducing global poverty.

The statement below summarises ten important themes of the MA and the conclusions that were drawn from it. It draws our attention to the areas where action is needed to conserve and use ecosystems sustainably.

- Everyone in the world depends on nature and ecosystem services to provide the conditions for a decent, healthy and secure life.
- Humans have made unprecedented changes to ecosystems in recent decades to meet growing demands for food, fresh water, fibre and energy.
- These changes have helped to improve the lives of billions, but at the same time they have weakened nature's ability to deliver other key services such as purification of air and water, protection from disasters, and the provision of medicines.
- Among the outstanding problems identified by this assessment are: the dire state of many of the world's fish stocks; the intense vulnerability of the 2 billion people living in dry regions to the loss of ecosystem services, including water supply; and the growing threat to ecosystems from climate change and nutrient pollution.

Sustainability

Theory of knowledge 1.04.01

Nineteenth-century American writers such as Henry David Thoreau and Ralph Waldo Emerson did much to raise environmental awareness in the USA. Writers such as Rachel Carson and artists in other countries have also done much to raise appreciation of the natural world. More recently, writers and television presenters such as the American author Jared Diamond, who has written popular science books such as the Pulitzer prize winning *Guns, Germs and Steel* (1997), and also *Collapse* (2005) and *The World Until Yesterday* (2012), and the Canadian broadcaster David Suzuki, who presents the CBC Television science programme *The Nature of Things*, have continued to publicise environmental issues through the medium of the creative arts.

Some international organisations such as Julie's Bicycle, a global charity that aims to highlight environmental sustainability and the creative arts, and the International Federation of Arts Councils and Culture Agencies (IFACCA) promote activities and policies to integrate arts and cultural activities with environmental issues.

1 How important do you think such contributions from the world of art and culture are compared with the impact of scientific evidence?
2 Are popular writers and presenters better equipped to explain difficult concepts to the public than scientists are?

Environmental Impact Assessments (EIAs) are studies carried out before a development project is undertaken to assess the possible damage to the environment.

- Human activities have taken the planet to the edge of a massive wave of species extinctions, further threatening our own well-being.
- The loss of services derived from ecosystems is a significant barrier to the achievement of the Millennium Development Goals to reduce poverty, hunger and disease.
- The pressures on ecosystems will increase globally in coming decades unless human attitudes and actions change.
- Measures to conserve natural resources are more likely to succeed if local communities are given ownership of them, share the benefits, and are involved in decisions.
- Even today's technology and knowledge can reduce considerably the human impact on ecosystems. They are unlikely to be deployed fully, however, until ecosystem services cease to be perceived as free and limitless and their full value is taken into account.
- Better protection of natural assets will require coordinated efforts across all sections of governments, businesses and international institutions. The productivity of ecosystems depends on policy choices on investment, trade, subsidy, taxation and regulation, among others.

More recently, a report produced by Foresight – *Global Food and Farming Futures* (2011) – concluded that lack of sustainability is already causing problems. Examples the report describes include nitrogen pollution and food production's contribution to greenhouse emissions. The report proposed a range of actions that individuals can take to improve the situation, including change of diet and reduction of food waste.

SELF-ASSESSMENT QUESTIONS 1.04.01

1 Define 'sustainability'.
2 Outline two problems relating to the protection of natural income.
3 **Discussion point:** How important do you think reports such as the UN's Millennium Ecosystem Assessment are in raising awareness about sustainability?

1.04.03 Use of environmental impact assessments

An **Environmental Impact Assessment (EIA)** report will be prepared before a proposed large-scale project gets underway to assess the possible positive or negative impact that the project may have on the environment. An EIA should include not only effects on the environment but also social and economic aspects. In some countries, an EIA is a rigorous process truly linked to sustainable development, but in too many instances environmentalists claim that this is a cosmetic process where the profit motive dominates thinking. However, action by environmental groups and changing public attitudes have placed increasing pressure on politicians worldwide.

The purpose of the EIA is to ensure that decision-makers consider all the likely consequences when they decide whether or not to proceed with a project such as a dam, road, forestry scheme or tunnel. EIAs were first used in the 1960s to help to remove personal bias when decisions were made. EIAs became a legal requirement in the USA in 1969 and were included in European legislation in 1985. Similar laws are now used in many countries around the world. Figure 1.19 illustrates the stages that should be considered in assessing the sustainability of development schemes.

An EIA considers advantages and disadvantages of a project and how the abiotic and biotic environments would be changed if the project went ahead. The first step is a baseline study which looks at potential changes to biodiversity, microclimate and amenities, as well as the potential effect on people in the area and changes to their community structure, livelihood or health.

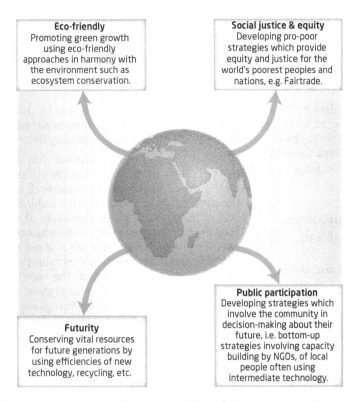

Figure 1.19 Assessing the sustainability of development schemes.

The baseline study monitors the aspects of the environment before the project so that they can be reassessed and monitored afterwards. Species types, diversity and numbers, habitats and soil characteristics are all recorded, and surveys of human populations, land use and hydrological factors are carried out.

Following the baseline study, forecasts and predictions must be made about the likely impact of the proposed project. These are difficult to do because so many variables are present in a natural system, but if potential impacts are identified, the scale of possible changes can be predicted with more certainty. Planners can then make changes to the proposal to limit the effects to a reasonable level.

To summarise, an EIA should include:

- a baseline study to record the current situation
- a survey – a report to assess the potential impact of the project
- a prediction to indicate the importance of the likely impacts
- a summary to consider how the effects can be limited to reasonable levels.

After the project has begun, changes should be monitored both during and after the development. An interesting example of an EIA is the one carried out for the Three Gorges Dam project on the Yangtze River in China (see Case study 1.04.01).

Many people argue about the benefits of an EIA. They provide valuable information about ecosystems, but a number of problems affect their effectiveness:

- Each country has different rules about the use of EIAs.
- There is no certainty that the proposals of an EIA will be implemented.
- Many socioeconomic factors influence the decision-makers, who may be influenced by local opinions or lobbying.
- There is no standard training for those who prepare the reports.
- It is difficult to define the boundaries of an individual project, which may cover a large area.
- Often indirect impacts of a project, such as the influence on other areas, are not included.

1.04 Sustainability

CASE STUDY 1.04.01

Three Gorges Dam, Yangtze River, China

The largest hydroelectric power generating dam in the world is located at Three Gorges on the Yangtze River (see Image 1.11). It was completed in 2009 and has been estimated to produce almost an eighth of China's energy needs. The EIA carried out before construction began considered the following factors:

- ecosystem destruction
- population relocation and the social consequences
- effect of sedimentation behind the dam due to reduction in river flow
- potential for landslides due to increased pressure on the land
- possibility of earthquakes.

Image 1.11 Building the Three Gorges Dam created a reservoir behind the dam almost 350 km long. Millions of people living by the Yangtze have been affected.

The enormous dam stands over 1 km wide and 200 m high. It is a physical barrier which disrupts the river ecosystem and fragments the habitats it contains. The EIA report identified 47 endangered species in the area, including the Chinese river dolphin and the Chinese sturgeon. Since the dam was built, the river dolphin has been declared officially extinct and fewer than 1000 sturgeon are thought to remain. The EIA report identified and considered the impact that the dam would have on spawning of sturgeon and other fish and balanced it against other potential benefits.

The report also highlighted social consequences. The dam flooded 100 000 acres of fertile agricultural land as well as 1600 factories and mines, 13 cities, 140 towns and 1352 villages, and almost 2 million people were forced to leave their homes and livelihoods along the river.

Another major environmental impact identified was deforestation. Large areas of forest had to be cleared both for the construction of the dam and to provide homes and farms for people who were displaced. A further impact identified was the increased risk of landslides due to the steep sides of the river, which were already unstable. The risk of landslide was made worse by deforestation.

Increased chances of earthquakes due to the huge mass of water held by the dam, which put pressure on the rock below, were studied. And, in addition, the problem of silt accumulation behind the dam and its potential to block the sluices of the dam was considered, along with water quality downstream, where effluent flow to the sea and deposition of silt are both reduced. The dam's contribution to flood control was also investigated. The reservoir's storage capacity would lessen the frequency of severe floods downstream from once every ten years to once every 100 years, but the dam could not prevent floods on downstream tributaries.

As a result of the EIA, the decision-makers concluded that the social and environmental benefits of the dam outweighed the negative impacts of the project. The energy produced by the dam does not release greenhouses gases, so air quality will improve with the increased use of electrical power. The planners also considered the positive benefits of reduced seasonal flooding and increasing economic development along the banks of the reservoir and river.

Some factors that could not be measured are those that affected lifestyles and scenery which people have traditionally enjoyed. The Chinese government took the view that people would be better off being resettled in new homes. Nevertheless, a major problem has been the lack of land: many farmers are now without land to cultivate and have been forced to adapt to new ways of life and new forms of employment. In the past, the banks of the Yangtze River were lined by spectacular cliffs and mountains which have been lost. Coffins used to hang in caves high up on the mountain cliffs, and there were ancient writings on the cave walls. Most of these have been submerged under the dam reservoir, and little was done to protect or preserve cultural relics in the area.

Questions

1. The dam has been operating since 2009, and as well as the extinction of the Chinese river dolphin other environmental impacts have resulted from it. Research some of these and try to assess the value of the EIA that was made.
2. Summarise the environmental costs and social benefits of the Three Gorges Dam project. Evaluate the overall usefulness of the EIA.
3. Using this example, highlight and discuss the strengths and weaknesses of EIAs.

SELF-ASSESSMENT QUESTIONS 1.04.02

1 What is meant by 'Environmental Impact Assessment (EIA)', and what is the purpose of the assessment?
2 Suggest why EIAs may not be able to predict all the outcomes of a new project. Give an example of a factor that was not predicted in the Three Gorges Dam assessment.
3 **Discussion point:** To what extent should environmental concerns limit development projects?

1.04.04 Ecological footprint

The **ecological footprint** is a sustainability indicator that expresses the relationship between a population and the natural environment. It considers the total use of natural resources by a country's population. The concept of ecological footprints has been used to measure our consumption of natural resources, how it varies from country to country, and how it has changed over time. The ecological footprint for a country is calculated by examining the cropland, grazing land, forest and fishing grounds needed to produce the food, fibre and timber the country consumes and to absorb the wastes it produces as it uses energy, and to provide space for homes, roads and factories. Six components are used to calculate a country's ecological footprint. You can study the details of these aspects in Chapter 8.04:

- *Built-up land:* the land area taken up by infrastructure, including housing, transportation and industrial sites
- *Fishing grounds:* the estimated primary production needed to support the fish and seafood caught (see Image 1.12)
- *Forest:* the total amount of lumber, pulp, timber products and fuelwood consumed
- *Grazing land:* the area used to raise livestock
- *Cropland:* the area used to produce food and fibre for human consumption, feed for livestock, oil crops and rubber (see Image 1.13)
- *Carbon uptake:* the amount of forest land required to absorb carbon dioxide emissions from burning fossil fuels, land-use change and chemical processes, other than the portion absorbed by oceans.

An **ecological footprint** is the area of land and water needed to sustainably provide all the resources at the rate at which they are consumed by a given population.

CONSIDER THIS

The idea of an ecological footprint was conceived in 1990 by Wackernagel and Rees at the University of British Columbia.

Image 1.12 Fish being sold in Peru – the fishing grounds' footprint is one of the six components of the ecological footprint.

Image 1.13 Market gardening in Bangladesh for both domestic consumption and for export – cropland footprint is another of the six components of the ecological footprint.

1.04 Sustainability

CONSIDER THIS

In many countries, the carbon footprint is the dominant element of the six components that comprise the ecological footprint. In others, such as Australia, Uruguay and Sweden, other aspects of the ecological footprint are more important. In Uruguay, the demand on grazing land is by far the dominant component of the ecological footprint. In Sweden, the demands on its forests comprise the country's major impact on the natural environment. In general, the relative importance of the carbon footprint declines as the total ecological footprint of countries falls. In many sub-Saharan African countries, the contribution of carbon to the total ecological footprint is extremely low. You will discover more about this in Chapter 8.04.

Knowing the extent of human pressure on the natural environment helps us to manage ecological assets more wisely, both as individuals and as societies. It is an important tool in understanding and advancing sustainable development. According to the latest reports, the global ecological footprint now exceeds the planet's regenerative capacity by about 30 per cent. This global excess is increasing, and as a result ecosystems are being harmed and waste is accumulating in the air, land and water. Unsustainable use of the Earth's resources leads to deforestation, water shortages, declining biodiversity and climate change, and puts the future development of all countries at risk.

Different countries have different ecological footprints, and an individual person's ecological footprint depends on their country of residence the quantity of goods and services they consume and the resources used and wastes produced. Nations at different income levels have considerable differences in their ecological footprints. Currently, the lowest per capita figures are attributed to Bangladesh, Congo, Haiti, Afghanistan and Malawi. All these countries have an ecological footprint of about 0.5 gha. This means that, although these countries use few resources, they may be unable to meet the basic needs of their populations for food, shelter, infrastructure and sanitation (see Chapter 8.04, Figure 8.31). Ecological footprint is influenced by the size of a country's population and its standard of living and how its people's needs are met. It includes only the resources used and waste produced that can be replaced and removed, and so it is a clear indicator of sustainably. Ecological footprint calculations provide snapshots of past resource demand and availability and are vital to understanding whether the population is living sustainably or not. You can read more about how ecological footprints are calculated for different countries in Chapter 8.04.

How big is your environmental footprint? The lifestyle choices we make contribute to our ecological footprint. You can calculate yours in just a few minutes on the WWF website.

Extension

For more information on ecological footprints, please access the website below:
Ecological footprints http://footprint.wwf.org.uk

SELF-ASSESSMENT QUESTIONS 1.04.03

1. Briefly explain why built-up land is part of the overall ecological footprint.
2. Has anything been done in your family, school or local community to reduce its ecological footprint?
3. **Research idea:** Visit an appropriate website to calculate your own ecological footprint. How does your ecological footprint differ from that of other people in your class?

Humans and pollution

1.05

LEARNING OBJECTIVES

After reading this chapter you should be able to:

- outline how pollution is defined
- describe how pollutants may be organic or inorganic substances, heat, light or invasive species and may be derived from human interference
- identify the difference between point-source and non-point-source pollution and the challenges we face to manage them
- explain the difference between primary and secondary pollutants
- describe how pollution management strategies may be applied.

KEY QUESTIONS

What types of substance cause pollution?
How are humans responsible for pollution?
How can pollution be managed?

1.05.01 Defining pollution

We can all recognise an environment that is badly polluted (see Image 1.14), but **pollution** can be defined in a number of different ways. Pollution is the contamination of air, water or soil by substances that are harmful to living organisms. It can be in the form of gases, liquids, solids or energy. Pollution can occur naturally, for example through volcanic eruptions, or as the result of human activities such as the disposal of industrial waste or an oil spillage. Pollution has been an almost inevitable side-effect of industrialisation and economic development. The processes of pollution have adverse effects on people and their environments. The amount of pollution that different organisms can tolerate varies considerably. A level of pollution that may not have any significant impact on one organism may be extremely harmful to another.

At its worst, pollution can cause serious health problems and even be fatal for humans. Figure 1.20 shows how people are exposed to chemicals and how exposure to these chemicals can affect human health. The methods of exposure to pollutants are:

- breathing in chemical vapours and dust (inhalation)
- drinking or eating the chemical (ingestion)
- absorbing the chemical through the skin (absorption).

Of all types of pollution, air pollution has the most widespread effects on human health. Air pollution affects people at a range of scales from local to global. In many LEDCs, indoor air pollution caused by the use of fires for cooking and heating homes is more severe than the pollution people experience outdoors.

Not all pollution involves inhalation, ingestion and absorption. Noise and light pollution are increasing hazards in developed societies. As people travel more widely by air, the disturbance from aircraft and airports is a major contributor to noise pollution. Another source of pollution is the heat from hot water that is discharged from factories into rivers or lakes, where it can kill or endanger aquatic life. This type of pollution is known as thermal pollution. Every type of pollution affects animal life in some way. For example, light pollution (see Image 1.15) is changing the behaviour of nocturnal animals, especially nocturnal birds whose day–night patterns of activity are upset.

> **Pollution** is the addition of any substance or form of energy (e.g. heat, sound, radioactivity) to the environment at a rate faster than the environment can accommodate it by dispersion, breakdown, recycling or storage in some harmless form.

1.05 Humans and pollution

Image 1.14 Pollution from residential coal fires in Ulaanbaatar, Mongolia.

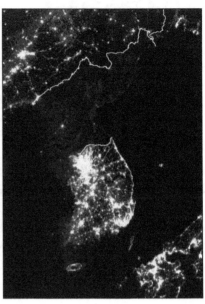

Image 1.15 North and South Korea at night from space. There are few areas in industrialised South Korea where it is truly dark.

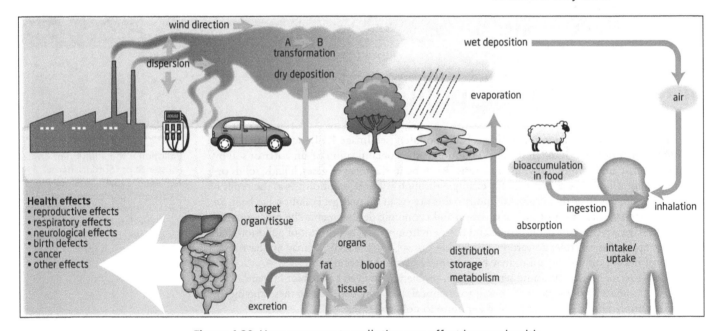

Figure 1.20 How exposure to pollution can affect human health.

1.05.02 Point-source and non-point-source pollution

Point-source pollution

Point-source pollution (sometimes called incidental pollution) can be traced to a single source such as an oil refinery, a power station or a chemical plant.

Most examples of **point-source pollution** are relatively small in scale in terms of the degree of pollution emitted, but some examples have been catastrophic, having a severe impact on both people and environment. Some of the worst examples of point-source pollution include the incidents at Chernobyl and Bhopal (see 1.01.01 *Historical influences on the modern environmental movement*) and the explosion at a large petrochemical plant in the Chinese city of Harbin in 2005. The Chinese incident released toxic pollutants including benzene, which is poisonous and cancer-causing, into a major river. Benzene levels were 108 times above national safety levels. Water supplies to the city were suspended and 10 000 residents were evacuated. Point-source pollution incidents can have extremely long-lasting consequences which are hard to predict in the earlier stages. The effects of all these accidents are still being felt decades after they occurred.

The examples of Chernobyl and the oil spill in the Gulf of Mexico (see 1.01.01 *Historical influences on the modern environmental movement*) are instances of **acute pollution**, where pollution occurred in a single isolated incident.

Non-point-source pollution

Non-point-source pollution such as exhaust from vehicles, which affects the ozone layer, or methane from cattle (see Image 1.16), which contributes to the greenhouse effect, usually takes much longer to have a substantial impact than point-source pollution, but it is likely to have a much greater affect in the long term. These and the emission of smoke or chemicals from factories are examples of long-term **chronic pollution**. It is more difficult to tackle the causes of non-point-source pollution, because the sources are widespread and common (see Image 1.17). Some governments have passed laws such as clean air acts to try to reduce non-point-source pollution from traffic.

Acute pollution is a single isolated incident such as an oil spill.

Non-point-source pollution (also called sustained pollution) is more dispersed in nature and may come from vehicle exhaust, an industrial area with many factories or domestic heating in towns.

Chronic pollution is long term, such as emissions from a factory.

Biodegradable pollution breaks down naturally in the environment; for example, certain types of waste paper will naturally decay.

Theory of knowledge 1.05.01

Defining key terminology is generally seen as the starting point for academic enquiry. How could knowledge be gained without definition of key terms?

1   Why do you think it is regarded as important to have clear definitions of key terms before examining an issue or topic in more detail?
2   What problems might arise if terms were not clearly defined?

Image 1.16 Cattle on a Russian farm – cattle create methane pollution.

It is often the poorest people in a society who are exposed to the greatest risks from point-source and non-point-source pollution. In the USA, for example, the geographic distribution of both minorities and the poor is highly correlated to the distribution of air pollution, municipal landfills and incinerators, abandoned toxic waste dumps, and lead poisoning in children.

Biodegradable pollution

Biodegradable pollution includes those things that can be broken down to simpler substances by the action of microorganisms over a period of time. Paper, wood, faeces, animal dung, bones, leather, wool and vegetable waste are all biodegradable and will naturally be removed from the environment. Some materials take longer than others to break down, but eventually they will all become harmless to the environment.

Humans and pollution

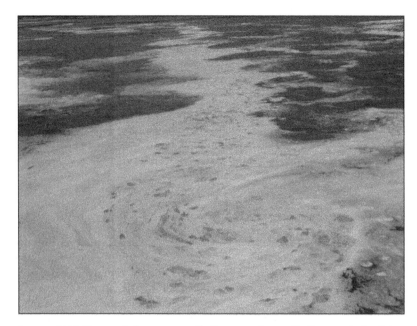

Image 1.17 Non-point-source pollution – water pollution on Lake Erie, mainly from the large number of vessels using the lake.

Persistent pollution

Materials such as plastics, mercury, lead and other heavy metals, aluminium cans, synthetic fibres, glass and persistent pesticides such as DDT (see 1.05.05 *Dealing with pollution – approaches to pollution management*) are all examples of non-biodegradable, **persistent pollution**. These pollutants cannot be broken down naturally; they will remain in the environment unless they are treated at special recycling plants. In some cases, the environment must be cleaned in other ways, such as by physically removing contaminated soil or water.

Persistent pollution is non-biodegradable, and these substances remain in the environment for a long time, often many years. An example is DDT or other organochlorine pesticides.

A **primary pollutant**: a substance that is active as soon as it is emitted.

A **secondary pollutant**: a substance that is formed from a primary pollutant that has undergone a change.

1.05.03 Major sources of pollution

Some pollutants are **primary pollutants** which are active as soon as they are emitted; others are **secondary pollutants** which become active after a primary pollutant has been physically or chemically changed. Examples of primary pollutants include smoke or carbon monoxide. Secondary pollutants include substances such as sulfuric acid, which forms when sulfur dioxide emitted from factory chimneys dissolves in water in the atmosphere. As the acid falls back to the ground it is known as acid rain.

As we have seen, pollution is created both naturally and by human activity. Natural sources of pollution include:

- volcanic eruptions (sulfur, chlorine, ash particulates)
- wild fires (smoke, carbon monoxide – primary pollutants)
- cattle and other animals (methane – primary pollutant)
- pine trees (volatile organic compounds; VOCs).

Sources of pollution from human activity that are now a major problem include the following:

- Industrial plants, power plants and vehicles with internal combustion engines producing nitrogen oxides, carbon monoxide, carbon dioxide, sulfur dioxide and particulates. In large urban areas, cars are the main source of these pollutants.
- Incinerators, stoves and farmers burning their crop waste produce carbon monoxide, carbon dioxide and particulates.

- Agricultural waste, including nitrates, pesticides and organic waste.
- Other human-made sources include aerosol sprays and leaky refrigerators, as well as fumes from paint, varnish and other solvents.

Almost every substance is toxic at a certain dosage. The most serious polluters are the large-scale processing industries, which tend to be grouped together as they have similar requirements for access and raw materials. The impact of a large industrial agglomeration may spread well beyond the immediate area and region to cross international borders. For example, prevailing winds in Europe generally carry pollution from west to east.

1.05.04 The distribution of pollutants

Pollution rarely stays in one place. Air pollution can travel vast distances, particularly if it reaches upper air currents. For example, particulate pollution examined in Alaska in the 1970s was traced back to West Germany.

Each year, about $450\,km^3$ of waste water is discharged into rivers, streams and lakes around the world. While rivers in most MEDCs have become steadily cleaner in recent decades, the reverse has been true in many LEDCs. It has been estimated that 90 per cent of sewage in LEDCs is discharged into rivers, lakes and seas without any treatment. For example, the Yamuna River, which flows through Delhi, has 200 million dm^3 of sewage draining into it each day. For many people, the only alternative to using this water is to turn to water vendors who sell water at greatly inflated prices.

Water pollution can also travel substantial distances, as it is carried by river flow and ocean currents. Pollution created near the source of a river may have a serious effect on a country further downstream. Examples of rivers that flow through a number of countries are the Nile, the Mekong, the Rhine and the Danube. Ground pollution may seep down into the water table and be moved by groundwater. Pollution of an aquifer by one country may impact on neighbouring countries if the aquifer crosses international boundaries. (You can read more about water pollution in Topic 4.)

CONSIDER THIS
China's rapid economic growth has led to widespread environmental problems. Pollution problems are so severe in some areas that the term 'cancer villages' has become commonplace. In the village of Xiditou, south-east of Beijing, the cancer rate is 30 times the national average. This has been blamed on water and air contaminated by chemical factories.

1.05.05 Dealing with pollution – approaches to pollution management

Collaboration and early intervention are key to managing and dealing with pollution in a timely and effective manner. In many cases, there is a time lag between a pollution incident and its impact on an ecosystem. When managing pollution, action may be taken at one of three stages, and this is summarised in Figure 1.21, which shows the three key stages in the process of pollution and the strategies for reducing its impacts. The three levels of intervention form a clear sequence. Action at each level of intervention has merits and limitations. The earlier pollution problems can be tackled, the better. Altering human activity to reduce pollution means that less action is required in the following two stages. Early action has less impact on the environment and frequently proves to be the most cost-effective strategy.

Altering human activity

Changing human activity to stop pollution occurring in the first place is the desirable preventative strategy. Human activity that creates pollution can be altered through:

- education
- incentives
- penalties.

1.05 Humans and pollution

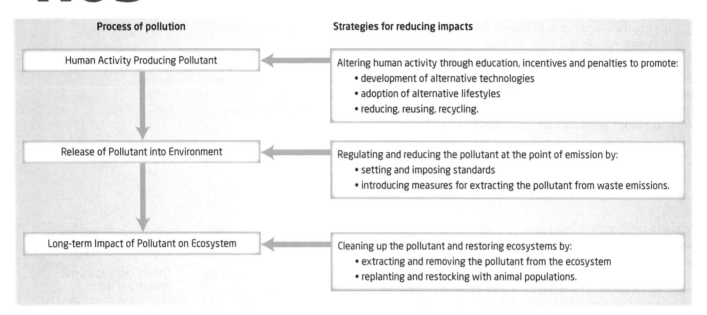

Figure 1.21 Pollution – different levels of intervention.

These processes should promote:
- development of alternative technologies
- adoption of alternative lifestyles
- reducing, reusing and recycling.

Unfortunately, efforts to alter human activity often only come into play after pollution has become a problem. Then it takes time for education to change people's perception of the situation. Offering incentives and imposing penalties can help. This is why all three strategies are often used together. Collaboration is generally essential to the effective management of pollution, because a large number of people and a range of different organisations may be involved in terms of both causing pollution and being affected by it. The pollution problem may also require action by different levels of government. In such a situation, coordinated planning and management will be essential to a successful outcome.

CASE STUDY 1.05.01

Local authorities in California try to change human behaviour

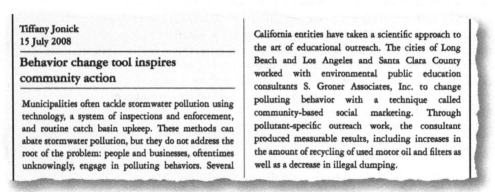

Figure 1.22
Source: 'Curbing pollution through outreach', Environmental Protection, www.eponline.com 2008

CASE STUDY 1.05.01 (continued)

This article describes a technique known as community-based social marketing, which is a way of changing people's behaviour and is rooted in social psychology. There are five general steps to this approach:

1. Identify a specific behaviour in which the target population should engage.
2. Determine the barriers and incentives to engaging in the desired behaviour.
3. Develop strategies to break down barriers or promote incentives.
4. Implement a pilot project.
5. Evaluate the pilot and refine the intervention strategies.

Questions

1. Why is the recycling of motor oil and filters important in reducing pollution?
2. Which groups of people do you think the consultants were trying to reach in this social marketing exercise?
3. Research other examples of education programmes that help reduce pollution.

As the impact of pollution has become more obvious, increasing investment has been made in developing alternative technologies. In many countries, the renewable energy sector has led the way, but there are examples of alternative technologies in many other sectors of human activity. Examples of modest lifestyle alterations are: using compost bins in the garden, installing solar panels, using energy-saving light bulbs and taking greater care to recycle a greater variety of waste and, as a result, recycle a higher percentage of waste overall. Reducing, reusing and recycling are themes which are standard in environmental education in schools in many countries. Education is a major factor in promoting environmentally friendly behaviour.

Regulating and reducing pollution

Once pollutants have been released into the environment, the main strategy is to:

- regulate and reduce the pollutant at the point of emission by setting and imposing standards, and
- introduce measures for extracting the pollutant from waste emissions.

Most countries seek to control levels of pollution by setting standards for air and water quality. In the USA, the Clean Air Act 1970 and major amendments to the Act in 1977 and 1990 serve as the backbone of efforts to control air pollution. Efforts to control air pollution in the USA date back to 1881, when Chicago and Cincinnati passed laws to control smoke and soot from factories in these cities. (You can read more about air pollution in Topic 6 and water pollution in Topic 4.)

In the European Union (EU), the emissions trading system rewards companies that reduce their carbon dioxide emissions and penalises those that exceed limits. The scheme was introduced in 2005 and covers about 12 000 factories and plants responsible for about half the EU's emissions of carbon dioxide. Under the scheme, limits are set on the amount of carbon dioxide emitted by energy-intensive industries like power generation and steel and cement makers. If companies want to emit more carbon dioxide than their quota, they have to buy spare permits from more efficient companies.

As technology has advanced, more efficient systems have been developed to extract pollutants from waste emissions. A major example is flue gas desulfurization, which is used to control emissions of sulfur dioxide (a secondary pollutant) from power plants and refineries. Sulfur dioxide is one of the elements forming acid rain. Flue gas is the air coming out of a chimney after combustion has taken place. The main methods to remove sulfur dioxide are:

- dry absorption injection systems
- spray-dry scrubbing (using sorbet slurries)
- wet scrubbing.

Cleaning up and restoring ecosystems

This is the final and least desirable approach to pollution management, because of the high cost involved and the considerable time that such an approach often takes. This is a reactive strategy that

CONSIDER THIS

In November 2008, a high court judge in the UK ruled that people in rural communities had suffered damage to their health from long-term exposure to pesticides. The judge stated that there was solid evidence that residents had suffered harm to their health from crop spraying close to their homes. He also criticised the current model for assessing the impact of crop spraying on humans.

1.05 Humans and pollution

may only achieve partial success because of the seriousness of the pollution problem. Catastrophic events such as the BP oil spill in the Gulf of Mexico in 2010 (see 1.01.01 *Historical influences on the modern environmental movement*) can incur enormous costs because of the widespread nature of the pollution. Large-scale strip (open-cast) mining can have a devastating impact on ecosystems, requiring careful and costly remediation. The same degree of pollution can have varying effects on different types of ecosystem. In cold environments such as Siberia and northern Canada, pollutants can persist for a long time because of the low level of bacterial activity.

SELF-ASSESSMENT QUESTIONS 1.05.01

1. Define 'pollution'.
2. Describe how pollution can affect human health.
3. What is the difference between point-source pollution and non-point-source pollution?
4. State two natural sources of pollution and two sources of pollution from human activity.
5. **Discussion point:** What are the main sources of pollution in the region in which you live? What is the evidence for this pollution and what is being done to reduce it?

1.05.06 The problem of DDT – agricultural insecticide and controller of disease

DDT was first synthesised in 1874, but its insecticidal properties remained undiscovered until 1939, when the Swiss scientist Paul Hermann Muller realised that DDT could be used for 'vector control' – the control of insects that carry major diseases. Muller was awarded the Nobel Prize in medicine and physiology in 1948 for this discovery.

DDT was used successfully to control the mosquitoes that spread malaria and the lice that spread typhus in the latter half of World War II (1939–1945). In the South Pacific, it was sprayed aerially to control malaria and dengue fever. After the war, DDT was used as an agricultural insecticide, and its production increased rapidly as it was cheap and effective. In 1955, the World Health Organization (WHO) began a programme to eradicate malaria. The programme relied mainly on DDT and achieved considerable success in a number of countries. However, the wide use of DDT in agriculture resulted in resistant insect populations, which eliminated some of the earlier gains.

The publication of *Silent Spring* in 1962 (see 1.1.01 *Historical influences on the modern environmental movement*) drew the public's attention to the dangers of DDT and, in particular, the fact that it is not biodegradable and accumulates in food chains. The author, Rachel Carson, set out the environmental impact of the large-scale spraying of DDT in the USA and argued that DDT was used without clearly understanding the long-term effects on the environment and human health. She suggested that DDT and other agricultural chemicals could cause cancer and be a considerable threat to wildlife, particularly birds (see Case study 1.05.02). Growing public concerns about DDT led to it being banned in the USA in 1972. In the 1970s and 1980s, the agricultural use of DDT was banned in most MEDCs. Using it to control mosquitoes was not banned, but DDT was largely replaced by less persistent alternative insecticides.

In 2004, an international treaty prohibited the use of DDT throughout the world, except for a clause allowing its manufacture and use in disease control. The chemical is still sprayed inside houses to kill malaria-carrying mosquitoes. This process is known as indoor residual spraying (see Image 1.18). India is now the only country that still manufactures DDT, after China stopped production in 2007.

The story of DDT shows us how there can be a conflict when a polluting substance has beneficial effects on one hand, but on the other hand harmful effects on the environment. It could be argued that in many parts of the world the effect of malaria is more devastating than the effect of DDT on the environment.

Image 1.18 Indoor spraying of DDT.

CASE STUDY 1.05.02

The recovery of the bald eagle

Bald eagles are found on the continent of North America and are the symbol of the USA (see Image 1.19). They live near rivers, lakes and marshes, where they find nesting sites and feed on their staple diet of fish, as well as water birds, rabbits and other small animals and carrion.

In the late 1940s, DDT was used extensively to control mosquitoes and other insects that were agricultural pests. DDT also washed off the land into waterways where plants, and the fish that fed on them, absorbed it. DDT does not decompose inside the bodies of these organisms but accumulates inside them so, further along the food chain, levels of DDT became dangerously high. Bald eagles at the top of the food chain were poisoned when they ate these contaminated animals. The chemical did not kill large adult birds, but it interfered with the ability of the birds to produce eggs with strong shells. As a result, their eggs often broke during incubation when the female eagles sat on them, or the eggs failed to hatch. DDT also affected other species such as peregrine falcons and brown pelicans.

The dangers of DDT became known, in large part due to the 1962 publication of Rachel Carson's book *Silent Spring*. In the late 18th century, there were more than 100 000 nesting pairs of bald eagles in America, but by 1963 only 487 remained. The US Environmental Protection Agency took the historic and, at the time, controversial step of banning the use of DDT. This was the first step on the road to recovery for the bald eagle, which was listed as an endangered species in 1967. Since that time, bald eagle numbers have recovered

Image 1.19 American bald eagle on nest.

substantially so that bald eagles no longer need the protection of the Endangered Species Act. In 2007, the bald eagle was removed from the list of threatened and endangered species.

Questions

1. Outline the reasons why animals at the end of a food chain suffer serious effects from DDT poisoning.
2. Why was the decision to ban DDT in 1963 described as 'controversial'?
3. Suggest why falcons and pelicans were also affected by DDT.

1.05 Humans and pollution

> **SELF-ASSESSMENT QUESTIONS 1.05.02**
>
> 1. What do you understand by the term 'accumulative' pollutant in relation to DDT? (You may like to read Case study 1.05.02 before you attempt this question.)
> 2. Why are international treaties important to deal with the use of pesticides such as DDT?
> 3. **Research idea:** Research the precautionary principle. Should this be applied in the case of persistent pesticides?

End-of-topic questions

1. **a** Describe what is meant by a model of a system. [2]
   b Outline how models are used to make predictions about:
      i changes in the climate based on carbon emissions
      ii the effect of measures taken to reduce carbon emissions. [6]

2. **a** Outline the difference between *negative feedback* and *positive feedback*. [2]
   b Draw a diagram to show how a positive feedback process involving methane may affect the rate of global warming. [4]
   c Suggest why most ecosystems are negative feedback systems. [2]

3. **a** **i** State the first law of thermodynamics. [1]
      ii Calculate the amount of output X in the system shown below. Show your working. [2]

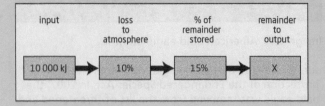

   b The diagram represents the energy (in arbitrary units) that enters and leaves a trophic level in a food chain.

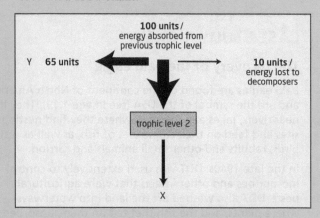

   i Calculate the value of output X and suggest where this output will go. [2]
   ii Output Y will be lost to the ecosystem. State the form of this energy. [1]

4 Complete the systems diagram below to show *three* inputs, *three* processes and *three* outputs for an intensive arable farming system that produces wheat.

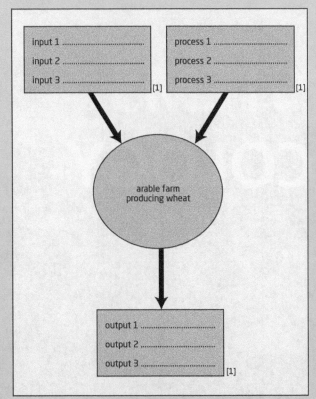

5 This cartoon is trying to convey a message about the environment.

a Suggest what message the cartoon communicates about the attitude of business (or individuals) to the environment. [2]

b State what is meant by an 'environmental value system (EVS)'. [1]

c Compare the approach of a technocentrist and an ecocentrist to the problem of carbon dioxide emissions. [4]

5 'The air, the water and the ground are free gifts to man and no one has the power to portion them out in parcels. Man must drink and breathe and walk and therefore each man has a right to his share of each.'

James Fennimore Cooper (1789–1851), *The Prairie*, **1827**

'We have forgotten how to be good guests, how to walk lightly on the Earth as its other creatures do.'

Barbara Ward, *Only One Earth*, **1972**

a Consider the two quotations and state whether they are likely to be views shared by an ecocentrist or a technocentrist. Explain your answer. [3]

b Some people claim that environmental protection is a luxury that only wealthy countries can afford.
  i Discuss this proposal using examples from MEDC and LEDC societies. [5]
  ii Explain why it is hard to value some aspects of an environment such as its educational value or the pleasure it provides. [2]

Topic 2
ECOSYSTEMS AND ECOLOGY

Species and populations

2.01

LEARNING OBJECTIVES

After reading this chapter you should be able to:

- define a species and a population
- explain the terms 'habitat' and 'niche'
- state what is meant by the abiotic and biotic components of an ecosystem
- outline how a species interacts with its abiotic and biotic environment
- describe and explain population interactions that take place between species
- describe J- and S-curves, which show how populations respond in different conditions
- describe how any system has a carrying capacity for the species it contains.

KEY QUESTIONS

How do species interact with their biotic and abiotic environment?

How do populations change and respond to interactions with their environment?

What is the significance of a system's carrying capacity?

2.01.01 Species and ecosystems

Every ecosystem contains both living and non-living components. The living organisms – plants, animals, fungi, bacteria and algae – are together known as the **biotic** components of the system. The non-living environment includes light, water, air, minerals and soil, as well as temperature and climate. Together these make up the **abiotic** or physical components. Biotic and abiotic components interact closely within an ecosystem, and the structure of the whole system relies on the relationship between them.

Examples of **species** include humans, leopards, pea plants and silver birch trees. Each species is given a scientific name which identifies it precisely and is used by scientists all over the world. The first part of the name is the genus and the second part the species (see Table 2.01).

Biotic components are the living components of an ecosystem.

Abiotic components are the non-living components of an ecosystem.

A **species** is a group of organisms that share common characteristics and can interbreed to produce fertile offspring.

Scientific name	Common name
Homo sapiens	human
Panthera pardus	leopard
Pisum sativum	pea
Betula papyrifera	silver birch

Table 2.01 Scientific names are always written in italics or underlined. The genus name has a capital letter.

2.01 Species and populations

CONSIDER THIS

Every species is placed into a larger group of similar species, called a family. There are more than 400 families of flowering plants. One of the largest was thought to be the figwort family (Scrophulariaceae), first named in 1789. Recently scientists have used new technology to investigate the genes of members of the group. Tests have revealed that some species are not closely related at all, and many have been moved into new families.

Theory of knowledge 2.01.01

The problem of defining a species

In the past, when only morphology (physical features) was used to identify species, scientists struggled to classify some organisms. Many bird species, such as the peacock, have males and females that look very different. If appearance alone was used to define a species, these birds would be placed in two separate groups. Sometimes, different species look identical but do not interbreed. One example is the reed warbler and the marsh warbler found in Britain, which can only be distinguished by their songs. Other organisms, such as tigers and lions, can interbreed but produce hybrid offspring (ligers and tigons). The hybrids are sterile and so they cannot be called species.

Today, biochemical tests and DNA analysis are used to identify and compare species and work out relationships between them. Molecular comparisons have meant that scientists have changed their view of what a species is. The relationships of some organisms have been reconsidered in the light of new evidence and some organisms have been given new scientific names.

1. Why is it important to define a species when studying an ecosystem?
2. Has our knowledge of the species concept been increased by more information about species? How can we recognise when we have made progress in the search for knowledge?

2.01.02 Niches and habitats

A **niche** is unique to each species, because it offers the exact conditions that the species needs or has adapted to.

A **habitat** is a wider area offering living space to a number of organisms, so a habitat comprises a number of niches and includes all the physical factors in the environment. An example might be a woodland habitat, which contains a range of niches from ground cover for a variety of different species living on low-growing plants to areas in the tree canopy that provide niches for birds.

Spatial habitat

Looking more closely at different habitats, we can observe that every organism has its own space in an ecosystem, which is known as its spatial habitat. The surroundings will be changed by the presence of the organism; for example, a woodpecker lives inside hollow trees, adapting them to provide nesting places and shelter. In a similar way, a rabbit that burrows under the soil affects the soil and grass.

Feeding activities

As an organism feeds within its niche, it affects the other organisms that are present. For example, an owl feeding on mice in woodland helps to keep the population of mice at a stable level, and rock limpets grazing on small algae control the degree of algal cover.

Fundamental and realised niches

We have defined niche as the special space inhabited by a particular plant or animal. This is the fundamental niche for that organism. The **fundamental niche** of a species is defined as the potential mode of existence of the species, given its adaptation.

Often the environment of a niche will change through natural phenomena, competition or through human intervention. So a species inhabiting a niche may find that its niche and place to live becomes more restricted or it begins to overlap that of another species. This more restricted niche is known as the **realised niche**. The realised niche is defined as the actual mode of existence of a species which results from its adaptation and competition from other species – that is, its actual lifestyle due to biotic interactions. A realised niche can only be the same size as or smaller than the fundamental niche.

A **niche** is the particular environment and 'lifestyle' that a species has. It includes the place where the organism lives and breeds, its food and feeding method, activity patterns and interactions with other species.

Habitat is the environment in which a species usually lives. For example, the habitat of an orang-utan is the rainforest of Borneo and Sumatra.

2.01.03 Abiotic components of an ecosystem

Looking at the larger systems that species inhabit, we can recognise three main types of ecosystem:

- marine systems – the sea, salt marshes, mangrove swamps and saline estuaries
- freshwater systems – rivers and lakes
- terrestrial (land-based) systems.

Each of these ecosystems has abiotic (non-living) factors which are crucial to the system and may vary with time. You can read about how these factors are measured or estimated in Chapter 2.05.

Marine ecosystems

The key abiotic factors in a marine ecosystem (see Image 2.01) are:

- *Salinity* – a measure of the salt content of water. Salinity affects both the buoyancy of organisms, as salinity affects the density of water, and their ability to control salt levels in their bodies.
- *Temperature* – determines the amount of oxygen that will dissolve in seawater and thus become available to marine organisms. Since many of these organisms are ectothermic (having a body temperature approximately equivalent to that of the water), temperature also affects their metabolic rate. In addition, surface water tends to be warmer than lower layers, and this can affect water currents and the mixing of water and nutrients.
- *pH* – a measure of the H^+ ions dissolved in water – which affects the solubility of inorganic ions and their availability to living things.
- *Dissolved oxygen*, which is needed by all marine organisms for respiration. The amount of dissolved oxygen in water depends mainly on water temperature but also on wave action or other disturbance to the water.
- *Wave action*, which is important because it increases the amount of dissolved oxygen by mixing air with moving water. Coastal areas and coral reefs have high levels of dissolved oxygen, and many organisms live in their surface waters.

Image 2.01 A marine ecosystem on the shores of Port Townsend in Washington, USA.

Freshwater ecosystems

These are the key abiotic factors in a freshwater ecosystem (such as that shown in Image 2.02):

- *Turbidity* – a measure of the cloudiness of water (in either a marine or freshwater ecosystem). Cloudy water has a high turbidity, and clear water low turbidity.
- *pH*, which can affect the uptake of nutrient ions by plants and the metabolism of both plants and animals.
- *Flow velocity*, which determines what organisms can survive in flowing water. If flow velocity is high, plants and animals must be firmly anchored to maintain their positions in the fast-flowing water.

Image 2.02 River Rouge basin in Michigan USA.

2.01 Species and populations

- *Temperature and dissolved oxygen.* The temperature of water not only affects the organisms that live in a lake or stream, it also influences the amount of oxygen that can remain dissolved in the water (see Table 2.02). Warm water may speed up the growth of plant life, but excessive temperatures can increase the rate of decomposition of biodegradable organic waste and reduce the water's ability to hold oxygen. A minimum of $5\,\text{g}\,\text{m}^{-3}$ of dissolved oxygen is needed to support a balanced aquatic community; if values fall below this, organisms die. Additional oxygen dissolves in water if the water is agitated, for example as it flows over rocks or waterfalls. The maximum amount of oxygen that can dissolve in water at $15\,°C$ is only $9.8\,\text{g}\,\text{m}^{-3}$.

Temperature /°C	Dissolved O_2 / $g\,m^{-3}$
4	12.7
10	10.9
15	9.8
20	8.8
25	8.1
30	7.5

Table 2.02 Effect of temperature on the solubility of oxygen in water.

Terrestrial ecosystems

These are the key abiotic factors in a terrestrial ecosystem (such as the forest habitat shown in Image 2.03):

- *Temperature* – as in other ecosystems, temperature affects the activity patterns and metabolism of the organisms present.
- *Light intensity*, which varies throughout the day and also depends on cloud cover and seasons, is crucial for photosynthesis and affects behaviour patterns and daily rhythms of animals in an ecosystem.
- *Soil moisture* content provides water for plants and affects the texture of the soil.
- *Wind speed* affects the ability of organisms to maintain their positions and survive.
- *Soil particle size* is important in determining how much water the soil can hold and how quickly the soil will drain.
- *Soil mineral content* – the ratio of mineral to organic material present – is important in determining the soil's ability to hold water, and also influences its fertility.
- *Slope* influences the water runoff and also whether or not erosion is likely to be a problem.
- *Drainage (porosity)* – too much or too little water in the soil can reduce plant growth. Poorly drained soil becomes waterlogged, so plants are unable to take up nutrients. The lack of oxygen in such soils may mean that toxic compounds build up in anaerobic conditions and denitrifying bacteria increase the loss of essential nitrogen compounds from the soil. Waterlogged soils are also slower to warm up in spring and summer, and this inhibits the germination and growth of seeds. Soil systems are discussed in Topic 5.

Image 2.03 Stone Pine (*Pinus pinea*). Forest habitat with fan palms and Halimium sp. Photographed in the Coto Donana, Southern Spain (Bob Gibbons).

You can learn about how these abiotic factors are measured in Chapter 2.05.

2.01.04 Biotic components and their interactions in an ecosystem

Within any ecosystem, all the biotic components – that is, the living things – are interdependent, so a change in the size of one population will affect all the other organisms as well. We can identify many different interactions between the **populations** in an ecosystem. The most important interactions relate to the survival of the population, and most involve feeding. They are:

- competition
- predation
- herbivory
- parasitism
- mutualism.

> A **population** is a group of organisms of the same species that live in the same area at the same time, and which are able to interbreed.

Competition

Individuals compete for resources such as food and space. Populations may compete with members of their own species (intraspecific competition) or with members of other species that use the same resources (interspecific competition).

Plants compete with one another for access to light and water. A population of fast-growing birch trees soon becomes established on an area of cleared land, but these trees require high levels of light. Slower-growing oak trees will grow alongside them for a short time, but eventually the birch trees will be overshadowed and replaced by a population of taller oaks (see Chapter 2.04.)

In a similar way, animals compete for food and homes. Many birds, for example the wren and the robin, establish a territory containing food and a nesting site that males defend with displays of aggression or by singing from points on its boundary. Birds that do not establish territories will be unable to find a mate and breed.

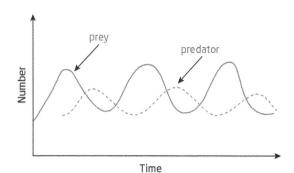

Figure 2.01 Changes in the populations of a predator and its prey over time. This is an example of negative feedback in the system.

Extension

For more information access the following webiste:
birdsong www.birds.cornell.edu/AllAboutBirds/studying/birdsongs/whysing/document_view

Predation

Predation is an interaction between species in which one species, the predator, kills and eats another, its prey. The effect of a change in the size of one population and its influence on another is shown clearly by the relationship between predator and prey populations. Predators such as bald eagles, lions and sharks are adapted for efficient hunting, because they must catch enough food to survive. Prey species such as rabbits, antelope and smaller fish, on the other hand, must be well adapted with camouflage or behaviours that allow them to escape their predators, if enough of them are to survive for the population to be maintained.

If the prey population in an ecosystem grows, predators will have more food available to them and their population will also increase. Increasing numbers of predators will eventually reduce the food supply until it can no longer sustain the predator population (see Figure 2.01).

Using the systems approach, we can also represent the interaction of predator and prey species in a diagram, as shown in Figure 2.02.

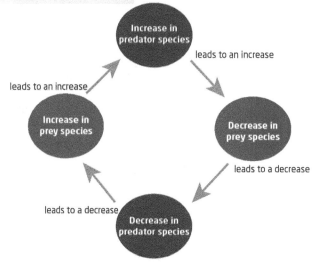

Figure 2.02 Relationship between predator and prey populations.

2.01 Species and populations

Herbivory

Herbivorous animals feed on plants. A single plant has leaves, fruits, seeds and roots that provide food for many organisms, from those that chew leaves or burrow inside leaves to those that suck sap from them (see Image 2.04).

Parasitism

Parasitism is a relationship in which one organism, the parasite, benefits from another, the host, which may suffer as a result of the parasite's presence. Tapeworms are endoparasites which live inside the gut of their host, where they feed on digested food. An animal that has many tapeworms may be weakened, as it is deprived of nutrients it needs. Ectoparasites such as fleas and ticks live on the outside of their host, piercing the host's skin to obtain a blood meal. Both may cause irritation and skin infections if the host cannot remove them.

Mutualism

Mutualism is a relationship between two organisms that gives benefit to both. Lichens form as a result of a mutualistic union between a fungus and an alga (see Image 2.05). The fungus absorbs minerals and protects the alga from intense light and desiccation (drying out), while the alga photosynthesises to provide sugars for both organisms.

Animal populations also establish mutualistic relationships. Egyptian plovers and Nile crocodiles both benefit when the birds feed on parasites and food particles in the crocodile's mouth. The crocodile's teeth are kept clean and healthy, and the birds obtain the food they need.

The range of possible interactions between two species is summarised in Table 2.03.

CONSIDER THIS

Herbivorous insects make up more than 50 per cent of the species in a forest but insect pest damage reduces the productivity of a forest by about 20 per cent per year. Researchers are concerned that damage to forests will increase because many insects will benefit from global warming and drying.

	Effects on organism 1	
	Benefit	Harm
Effects on organism 2 Benefit	mutualism	predation/parasitism
Harm	predation/parasitism	competition

Table 2.03 Ecological interactions between two species.

Image 2.04 An oak tree provides niches for many herbivores, including the gypsy moth caterpillar.

Image 2.05 Mixed lichens growing on acid rock.

Many interactions between species centre on the resources that are available. Food, space, hiding places and breeding sites are key requirements. Ecologists use the concept of a niche as a way of defining the position of a species within a community and the range of physical conditions and resources it needs. The fundamental niche is the range of conditions in which it will survive if there are no competitors, predators or other influences. When other species are present, the resources a species needs may be reduced due to competition, so that the first species cannot survive so well. In this case, the fundamental niche is reduced to what is known as the realised niche. We examine the niches of two species of barnacle in Case study 2.01.01.

CASE STUDY 2.01.01

Competition on a rocky shore

The interaction between two species of barnacle, *Balanus balanoides* and *Chthamalus stellatus*, demonstrates the effects of one species on the other and the differences between their fundamental and realised niches – that is, where they have the potential to live and where they actually live.

The two species live on rocky shores of the North Atlantic. Adult *C. stellatus* live higher in the intertidal zone (further from the sea) than adult *B. balanoides*, but many young *Chthamalus* do settle in the *Balanus* zone (see Figure 2.03). If no *Balanus* are present, these *Chthamalus* will survive well; but if *Balanus* are present, they will be smothered and die, because *Balanus* grow more rapidly and are larger. In a similar way, young *Balanus* may settle lower down the shore (further from the sea), but they will not grow well here because they are less resistant to desiccation and are outcompeted by *Chthamalus*.

When both species are present, the realised niche of *Chthamalus* is more restricted than its potential or fundamental niche because of competition with *Balanus* in the lower regions of the shore. The realised niche of adult *Balanus* is also different from the fundamental niche of its young because of the effects of drying out and competition with *Chthamalus* in the higher region.

Experiments with the barnacles have shown that, if one species is removed, the range of the other is increased.

Questions

1. Barnacles are affected by the presence of other species and by their physiology (ability to survive desiccation). Suggest two other aspects of their environment that could affect their survival.
2. Outline the difference between an ecological niche and a habitat, using *Chthamalus* as an example.
3. Suggest two factors that could influence the realised niches of the two species of barnacle.

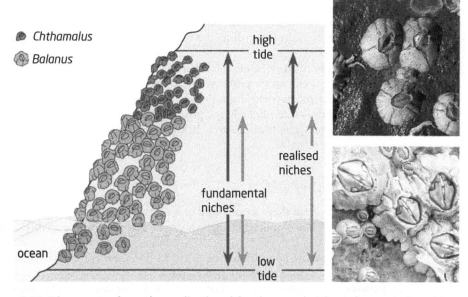

Figure 2.03 Diagram to show the realised and fundamental niches of two species of barnacle.

2.01 Species and populations

> ### SELF-ASSESSMENT QUESTIONS 2.01.01
> 1. Define a species.
> 2. Two bird species, the yellow-rumped warbler and the bay-breasted warbler, feed on similar prey and can be seen feeding in the same trees. But if the two are present together, one feeds at low levels in the tree and the other in high branches. Suggest an explanation for this behaviour.
> 3. State three abiotic factors that might affect the growth of the trees that provide food for the warblers.
> 4. **Research idea:** Find other examples of mutualistic relationships between organisms in terrestrial and aquatic habitats.

2.01.05 Limiting factors and carrying capacity

As a population grows, individuals must complete with one another for the resources they need; these may include space, light, food, nutrients and water. When the demand for a particular resource is greater than the supply, we say that the resource has become a **limiting factor**.

Competition for resources between members of the same species is known as **intraspecific competition**. This increases as population numbers increase. Once the **carrying capacity** is reached, the population growth rate will slow down, either because organisms die through lack of an essential resource, or because they fail to breed and their birth rate falls. Exactly how different organisms respond depends on the species. For example, frogs reduce their rate of growth and reach maturity at a smaller size than normal when food resources become a limiting factor. Birds that maintain nests in specific territories do not compete directly, but only those birds whose territories contain sufficient food will be able to breed successfully. Shortage of food can also mean that very few of a species survive at all. For example, if large numbers of caterpillars emerge together and compete to feed on a limited number of leaves at the same time, the result may be that none is able to get enough food and most will die.

Plant populations are most often limited by light, temperature and carbon dioxide, since these are the basic requirements for photosynthesis. Plants also compete for space to extend their roots and obtain water and nutrients from the soil. If light is a limiting factor, plants will grow faster and taller towards available light. Many plants that grow on a woodland floor have evolved to flower early to complete their life cycle before tall trees have come into leaf, so the two species do not compete directly.

S- and J-curves

Populations change as they interact with one another and compete for the resources in their environment. We can plot these changes as graphs that show how populations increase or decrease over a period.

S- and J-curves are population growth curves that show two of the different ways that populations may change over time (see Figure 2.04).

J-curves demonstrate the growth of a population that does not slow down. Populations that are becoming established in a new habitat often undergo rapid exponential growth. The habitat may contain abundant resources, so that the birth rate is high and the death rate is low. The curve is J-shaped, starting with a period of slower growth but rising sharply as more and more individuals

---

A **limiting** factor is a resource in limited supply that can affect the growth of a population.

Intraspecific competition is competition between members of the same species for a limited resource.

Carrying capacity is the number of individuals in a population that the resources in the environment can support for an extended period.

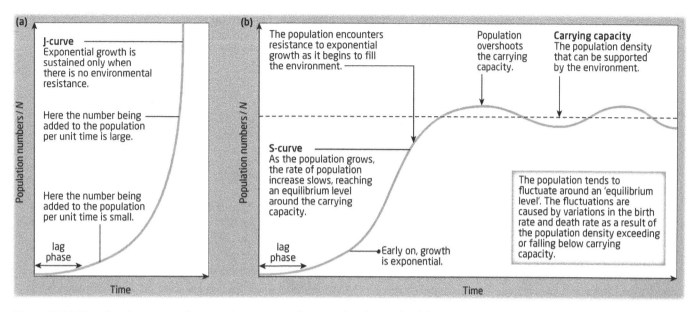

Figure 2.04 The development of a population over time can be shown by (a) a J-curve and (b) an S-curve.

join the population. Curves of this shape continue as long as resources do not become exhausted, but this seldom happens in a natural situation.

S-curves demonstrate rapid, exponential growth in a population in the initial stages of its development, but as the population grows, its increase slows down as factors in the environment become limiting. Eventually, the population stabilises at a level that the environment can support, and we say that the carrying capacity of the environment is reached.

Phases of the growth curves

The phases of the growth curves are:

- *Exponential growth phase*
  The exponential growth phase, seen in both types of curve, is the period in which there are no limiting factors and the population can double in size in set time periods, which depend on the species. For example, a slow-breeding population of dolphins might double every two years, whereas fast-breeding rabbits might double their population in just a few months. Any species growing exponentially with unlimited resources can reach an enormous population density in a short time (see Case study 2.01.02).

- *Transition phase*
  The transition phase begins as a resource starts to become limiting and the rate of growth decreases. This may occur when there are too many individuals in the area, there is an increase in the number of predators or there is an increase in disease and mortality due to overcrowding.

- *Plateau phase*
  The plateau phase shows that a population has reached stable, sustainable numbers, with births and immigration approximately equal to deaths and emigration. The plateau phase occurs as the carrying capacity of the environment is reached.

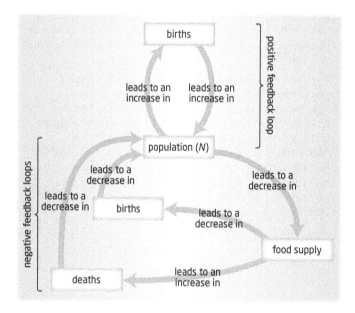

Figure 2.05 Positive feedback leads to exponential growth and a J-curve. Negative feedback stabilises the population at the carrying capacity of the ecosystem.

2.01 Species and populations

Species may show fluctuations in population numbers above and below the carrying capacity. As this occurs, negative feedback will ensure that a population that exceeds or falls below the carrying capacity of the environment soon corrects itself (see Figure 2.05).

CASE STUDY 2.01.02

Exponential growth of a rabbit population

Rabbits were introduced to Australia by European settlers in the 19th century. Rabbits were bred for food but many escaped into the wild. Some were released deliberately by settlers, one of whom wrote, 'The introduction of a few rabbits could do little harm and might provide a touch of home, in addition to a spot of hunting.' But Australia has its own unique flora and fauna and there were no natural predators of rabbits. The rabbits flourished because there were plentiful supplies of space and food, little competition from other species and their only predators were humans.

Australia had ideal conditions for a rabbit population explosion. Its mild winters enabled rabbits to breed throughout the year. As agriculture developed, wooded areas that might have halted their progress were cleared and turned into grassy habitats where rabbits could thrive, and huge populations developed (see Image 2.06).

Today, rabbits are considered by many to be the most significant factor in the reduction in the native species of Australia. Rabbits often kill young trees by chewing away their bark. They also cause serious erosion because they eat native plants and leave the soil exposed to wind and rain.

The most successful attempts to control the rabbit population have involved the deliberate release of rabbit-borne diseases such as myxomatosis. This was first tried in 1950, when the rabbit population was reduced from an estimated 600 million animals to about 100 million. But by the end of the 20th century, rabbits had become resistant to the disease and the population had recovered to an estimated 300 million. Despite efforts to control the number of rabbits in Australia, it continues to rise.

Questions

1. What type of curve could be drawn to show the population growth of the rabbit population from the 19th century to 1950?
2. What factors prevented the rabbit population from naturally reaching a plateau phase?
3. Invasive alien species are defined as those that are outside their natural distribution area and threaten biological diversity in their new habitat. The kudzu vine (*Pueraria lobata*), which was introduced into the USA from Japan in 1876, and the yellow star thistle (*Centaurea solstitialis*), which was accidentally introduced to California through contaminated seed fodder, are two examples of invasive alien species, which crowd out native species and have become major pests. Why have these plants become major pests in their non-native habitat?

Image 2.06 Rabbit populations in Australia.

SELF-ASSESSMENT QUESTIONS 2.01.02

1 Which group of organisms is an example of a population?
    A   Leopard frogs in a stream
    B   Birds in California
    C   Reptiles in the Sahara Desert
    D   Trees in a forest

2 What is the most likely outcome if two different plant species compete for the same requirements in an ecosystem?
    A   The two species will develop different requirements.
    B   One species may adapt to a different environment.
    C   One species may be eliminated from that ecosystem.
    D   The two species will change the environment so that they can both survive.

3 The behaviours of two species of iguana living on the Galapagos Islands were observed. The two species live in different habitats (see Table 2.04).

Species 1	Species 2
feeds on algae	feeds on cactus and land plants
never travels more than 10 m from the seashore	is found more than 10 m from the seashore
spends most time in the sea	spends most time on land

Table 2.04 The behaviours of two species of iguana living in different habitats.

State the reason for the survival of the two species on the islands.

4 **Research idea:** The populations of cane toads and prickly pear cacti have both followed a J-curve in their growth in Australia. Research the impact of these two species on Australian flora and fauna.

2.02 Communities and ecosystems

LEARNING OBJECTIVES

After reading this chapter you should be able to:

- define the terms 'ecosystem' and 'community'
- outline how ecosystems may vary in size and complexity
- describe the roles of producers, consumers and decomposers in an ecosystem
- explain how feeding relationships are modelled using food chains, webs and ecological pyramids.

KEY QUESTIONS

How do interactions between species result in energy flow and nutrient cycling?
How are photosynthesis and respiration important in the flow of energy?
How are models used to show feeding relationships in an ecosystem?

2.02.01 Photosynthesis and respiration

Every **ecosystem** contains **communities** of living organisms which are interlinked by their sources of food and the organisms that feed on them. Food chains and food webs are the basis of these interactions, but with very few exceptions food chains begin with the capture of light energy from the Sun in the process of photosynthesis (see Figure 2.06).

Photosynthesis is the process by which green plants make their own food using water and carbon dioxide. Light energy from the Sun is used to split water and combine it with carbon dioxide to produce the sugar glucose.

Respiration is the process in which food, often in the form of glucose, is broken down to release the energy it contains. All living things must respire to stay alive and maintain their life processes such as movement, growth and reproduction.

> An **ecosystem** is a community of interdependent organisms and their abiotic environment.
>
> A **community** is a group of populations living and interacting in the same area.

Photosynthesis

The process of photosynthesis (see Figure 2.07) can be thought of as a series of inputs, energy **transformations** and outputs. Photosynthesis involves a complex series of chemical reactions catalysed by enzymes, but it can be summarised in an equation below.

Water and carbon dioxide, together with sunlight, which is the energy source, are the inputs. The outputs are glucose and oxygen. Glucose contains chemical energy which is used by the plant and also by any organisms that feed on the plant. Oxygen is a waste product of photosynthesis and is released into the atmosphere, where it supports the **respiration** of other organisms.

> **Transformations** lead to an interaction within a system and the formation of a new end product or they may involve a change of state.
>
> **Respiration** The process in which food, often in the form of glucose, is broken down to release the energy it contains.

$$6CO_2 + 6H_2O \xrightarrow{\text{Photosynthesis}} C_6H_{12}O_6 + 6O_2$$

carbon dioxide + water $\rightarrow$ glucose + oxygen

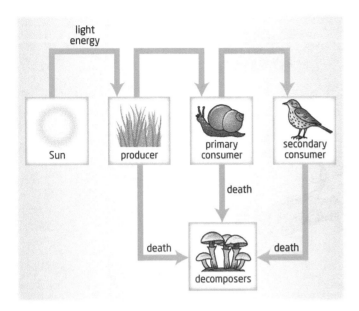

Figure 2.06 Light energy is captured by producers and passed to consumers in a food chain. Dead and waste material provide nutrients for decomposers.

As photosynthesis proceeds, light energy is captured by chlorophyll (the green pigment found in leaves and stems) and the energy is used to bond hydrogen atoms from water with carbon dioxide molecules to form glucose. Light energy is converted to stored chemical energy in the bonds of glucose molecules. Glucose is an organic compound which the plant can go on to convert into other substances such as starch or protein. In this way, photosynthesis is the process that leads to the accumulation of plant biomass.

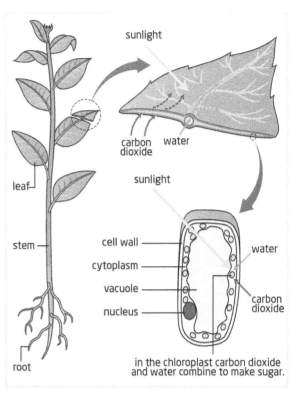

Figure 2.07 Plants absorb water from the soil, and carbon dioxide enters through stomata (pores) in their leaves. Chloroplasts inside plant cells contain chlorophyll, which captures light energy. The chemical reactions of photosynthesis occur inside chloroplasts. Glucose may either be stored or used in the life processes of the plant. Oxygen leaves via the stomata.

Respiration

Respiration can also be considered as a series of inputs, energy transformations and outputs. In a similar way to photosynthesis, respiration involves a complex series of chemical reactions controlled by enzymes, and can also be summarised in an equation. The process is shown in Figure 2.08.

$$\text{glucose} + \text{oxygen} \rightarrow \text{water} + \text{carbon dioxide} + \text{energy}$$

$$C_6H_{12}O_6 + 6O_2 \rightarrow 6H_2O + 6CO_2 + \text{energy}$$

Respiration takes place within all living cells. The inputs are glucose (or another source of energy) and oxygen. The outputs are water, carbon dioxide and energy, which is released to be used and eventually lost as heat. The stored chemical energy tied up in the bonds of the glucose molecules is transformed into a form that can be used by organisms for movement, growth and other life processes.

The equation above shows the process known as aerobic respiration. This occurs when oxygen is present. Some organisms, such as yeast, can respire anaerobically (i.e. without oxygen). Anaerobic respiration releases less energy inside body cells. Respiration releases energy from glucose, and the outputs of the process in yeast are carbon dioxide and ethanol. In the case of animals, the product of anaerobic respiration is lactic acid. Animals such as mud-dwelling worms, which live in low-oxygen environments, are most likely to respire anaerobically, but whales and seals, which dive to great depths in the ocean, can also do so. Humans respire anaerobically for short periods during vigorous exercise, but a build-up of lactic acid eventually leads to cramp.

> **CONSIDER THIS**
>
> The green pigment chlorophyll appears green to our eyes because it reflects green light from the visible spectrum. Green light is not absorbed, so a plant that is exposed to green light will not be able to photosynthesise. Chlorophyll absorbs the red and blue parts of the spectrum very well, and these colours of light produce the maximum rate of photosynthesis.

2.02 Communities and ecosystems

CONSIDER THIS

Glucose is the source of energy most often used in respiration in both plants and animals. Most organisms convert their stores of glycogen (animals) and starch (plants) into glucose when extra resources are need. However, some animal tissues, such as liver, and also some seeds are able to use fat as a source of energy during respiration without converting it to glucose first. On very rare occasions, when all other sources of energy have been used up, protein can also be used.

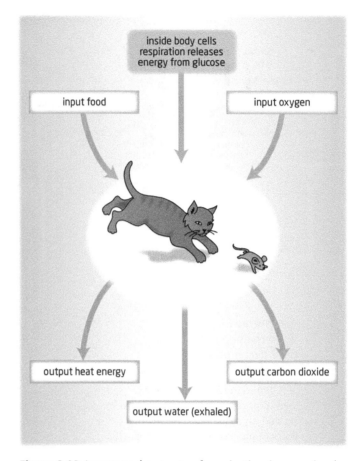

Figure 2.08 Inputs and outputs of respiration in an animal.

Respiration is essential because it provides the energy needed to build an organism's body and keep the molecules used in a well ordered (low entropy) state. But during respiration, large amounts of heat energy are lost. We can experience this in our own lives when we exercise vigorously and increase our rate of respiration and become hot. Heat is not a useful form of energy for other organisms. Heat is said to increase the entropy in the ecosystem. Entropy is a term used to explain the way a system is arranged and is a measure of the disorder of its components (See Topic 1.0).

2.02.02 Producers, consumers and decomposers

A **producer** (autotroph) is organism that converts light energy into chemical energy.

A **consumer** is an organism that feeds on other organisms.

Decomposers are organisms that feed on dead material and recycle nutrients in the ecosystem.

Producers, **consumers** and **decomposers** are the subdivision of organisms that make up the biotic component of the ecosystem, based on the way they feed or obtain the energy they need (see Figure 2.06). Producers (or autotrophs) build up their own food from simple substances and light energy, while consumers (or heterotrophs) feed on other organisms to supply their needs. Decomposers are the bacteria and fungi that feed on dead and decaying material and recycle nutrients in the ecosystem. All these organisms are essential to maintaining the flow of energy and nutrients in the system.

Producers

Producers have a key role in every ecosystem. Green plants, algae and some bacteria are the organisms that produce the biomass (living material such as leaves and fruits) that supports all the consumers in the food chains and webs that an ecosystem contains. Producers (autotrophs) do this using a process known as photosynthesis in which sunlight is a source of energy. Light energy is captured by the green pigment chlorophyll and used to build complex organic materials from the basic raw materials of water and carbon dioxide.

Consumers

Consumers cannot make their own food; they must feed on other organisms to obtain the nourishment they need to live. Consumers are also known as heterotrophs; they break down the complex organic molecules in their food in a process known as digestion. This releases minerals and the nutrients that can be respired to fuel their activities and build up their bodies. Herbivores are consumers that feed on plants; carnivores feed on other animals; and omnivores, such as humans, have a varied diet. Detritivores are a special group of heterotrophs that feed on decomposing plant and animal material and help to speed up the process of decay.

Decomposers

Decomposers break down dead organic material at the microscopic level to provide the nutrients they need, and as they do so they release inorganic materials into the soil. Decomposers in the soil feed on dead leaves and the bodies of animals, and are vital to soil fertility (see Topic 5). Part-digested material forms humus, which influences the drainage of the soil, and the inorganic minerals released by decomposers are absorbed by living plants through their roots. Decomposers are vital in the process of recycling nutrients in an ecosystem. They can provide food for other decomposer organisms and form the first link in decomposer food chains.

2.02.03 Trophic levels in food chains and webs

Organisms in any ecosystem are interlinked by their source of food and the organisms that feed on them. A hierarchy of feeding relationships influences the way that nutrients and energy pass through every ecosystem. Ecologists recognise this hierarchy and categorise it as a series of trophic or feeding levels. As we have seen, organisms that convert simple abiotic components into living matter are known as producers. Most producers are green plants, which capture light energy from the Sun and synthesise new material from water, carbon dioxide and minerals. This new biological material is known as **biomass**. Plant biomass, which may be leaves, roots, flowers or fruits, provides the starting point for the many interconnected food chains in an ecosystem. Organisms that cannot make their own food and must eat other organisms to obtain their energy and nutrients are known as consumers. Figure 2.09 shows two examples of food chains from different ecosystems.

Biomass is the amount of biological matter in a living or recently living organism, measured in units of mass per unit area (e.g. $g\,m^{-2}$)

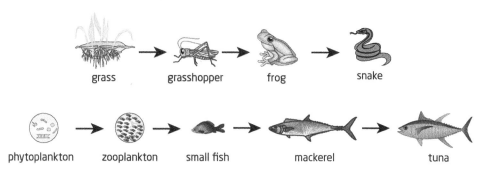

Figure 2.09 A terrestrial and an aquatic food chain.

2.02 Communities and ecosystems

Arrows in a food chain always point in the direction in which food and nutrients flow. Any food chain can be summarised as:

Producer → consumer → secondary consumer

The position an organism occupies in a food chain is known as its **trophic level**. The word *trophic* means 'feeding', and every organism in a food chain is said to be at a particular trophic level:

- Trophic level 1 is occupied by producers that make their own food.
- Trophic level 2 is occupied by consumers, the herbivores that eat plants.
- Trophic level 3 is occupied by secondary consumers – carnivorous animals that feed on herbivores.
- Trophic level 4 is occupied by tertiary consumers – carnivores that feed on secondary consumers.

Few consumers rely on only one source of food, so ecosystems contain many interlinking food chains, which together form a food web (see Figure 2.10). Organisms may change trophic level depending on their source of food, so that the fox will be a primary consumer at trophic level 2 when it eats a crab apple, or a secondary consumer at trophic level 3 when it eats a wood mouse.

Longer food chains that include more carnivores (quaternary consumers) are also possible. Food chains and webs provide a useful snapshot of the relationships in an ecosystem and can be used to predict the consequences of changes that may occur. For example, if the number of rabbits in the food web in Figure 2.10 fell as a result of disease, we could predict that there would be less food available for foxes. Foxes might decrease in numbers or they might change their feeding habits and eat more mice so that their numbers declined. But with fewer rabbits in the area, grass could grow longer, which might provide food for different consumers to use and change the food web again. A change in one species can affect many others.

> **Trophic level** is the position that an organism occupies in a food chain as a result of its feeding habits. The term also defines the position of a group of organisms that feed at the same position in their food chain.

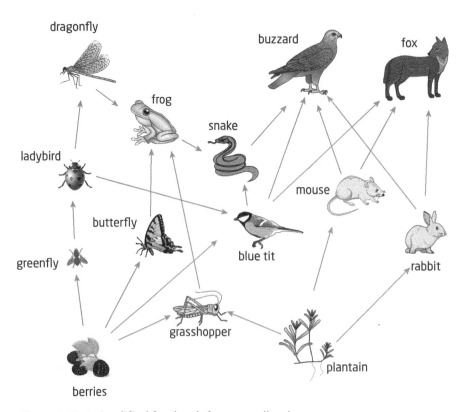

Figure 2.10 A simplified food web for a woodland.

CASE STUDY 2.02.01

Spraying with DDT

In the 1950s, the World Health Organization (WHO) sprayed parts of an island in Borneo with an insecticide called DDT to eradicate the malarial mosquito. The treatment was successful and a malaria epidemic was halted. But there were other results that no one had predicted.

The villagers' huts were roofed with thatch, which was food for a particular species of caterpillar. This caterpillar was not affected by DDT, but the wasp which was its natural predator was wiped out. Soon the thatched roofs collapsed. Insect-eating geckos and the local gecko-eating cats also died of DDT poisoning. The roofs of the huts were replaced with corrugated iron, but there was soon a big increase in the number of rats, which threatened to spread diseases in the village.

The WHO solution was to airlift new cats and drop them into the village by parachute.

Questions

Use your knowledge of food chains to explain these events on the island:

1. Why did the thatched roofs collapse?
2. DDT is an insecticide, so what caused the death of the cats?
3. Why did the rat population increase and how did the new cats help the villagers?

You can read more about the effects of DDT on food chains in Topic 1.05 and 2.02.06.

SELF-ASSESSMENT QUESTIONS 2.02.01

1. State why plants are at the beginning of aquatic and terrestrial food chains.
2. Name the trophic level of the mouse in the food web shown in Figure 2.10.
3. Which two of the following organisms are *both* essential components of an ecosystem: producers, herbivores, carnivores, decomposers?
4. **Research idea:** Find the longest aquatic and terrestrial food chains in your local area. Compare the two and think about reasons for the differences.

2.02.04 Pyramids of numbers, biomass and productivity

Ecological pyramids are diagrams that ecologists use to provide a picture of the quantities of organisms present at each trophic level in an ecosystem. Pyramids show in a visual form how the different trophic levels relate to one another in the system.

Pyramids of numbers

Pyramids of numbers are constructed by counting the number of organisms at each trophic level. The information for each level is drawn to scale about a central vertical axis (see Figure 2.11). Because counting every individual in an ecosystem is usually an impossible task, samples from measured areas are used and the results multiplied to represent the whole ecosystem.

Pyramids of numbers give a good approximate representation of an ecosystem but can be misleading because they cannot account for the size of organisms. For example, one large tree can support many hundreds of small herbivorous insects and birds, so that pyramids of numbers

Communities and ecosystems

for woodland may not be pyramid shaped (see Figure 2.11b). Similarly, a single-celled diatom in a pond would count as 1, the same as a large water lily in the same ecosystem. Factors like this must be taken into account when pyramids of numbers are used to show feeding relationships. Advantages and disadvantages of these pyramids are shown in Table 2.05.

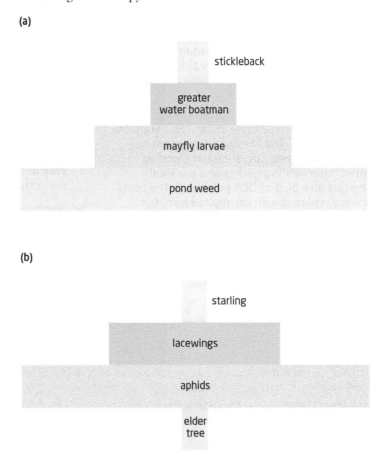

Figure 2.11 (a) Pyramids of numbers are pyramid shaped when there are a large number of producers or the producers are small. This is a typical pyramid of numbers for a pond ecosystem. (b) In a woodland, the pyramid is inverted where a single large tree supports many herbivorous and carnivorous insects.

Pyramids of biomass

Pyramids of biomass overcome some of the problems of pyramids of numbers. Biomass is the measure of mass of all the organisms at each trophic level in an ecosystem.

Ecologists estimate biomass at a particular moment in time, and it is recorded as the total (dry) mass multiplied by the number of organisms for each trophic level in a given area. Biomass is calculated in units of mass per unit area ($g\,m^{-2}$ or $kg\,m^{-2}$) or sometimes as units of energy per unit area ($J\,m^{-2}$). Water is excluded because, it has no energy value.

Pyramids of biomass are usually pyramid shaped. This is because, at each trophic level, the proportion of energy and biomass that can be transferred to the next level reduces, as some is used up by the organisms. But even biomass pyramids cannot take into account seasonal variations in populations' biomass, which varies through the year. For example, at certain times of year, the phytoplankton (small green aquatic plants) in a pond will grow quickly as temperatures increase and light levels rise. More phytoplankton biomass provides more food for the primary consumers and allows a burst of reproduction among small fish, snails and insects. Soon afterwards, as phytoplankton is eaten, its biomass falls rapidly. Sampling this ecosystem when the phytoplankton are abundant gives a pyramid-shaped diagram, but a sample taken after the consumers reproduce gives an inverted pyramid. Advantages and disadvantages of these pyramids are shown in Table 2.05.

Pyramids of productivity

Pyramids of productivity are the most accurate way to model an ecosystem (See Figure 2.13). They show the flow of energy in an ecosystem over a period, usually a whole year. Each level represents energy per unit area per unit time and is measured in mass or energy per square metre per year (e.g. $g\,m^{-2}\,yr^{-1}$ or $kJ\,m^{-2}\,yr^{-1}$). Pyramids of productivity show how quickly organisms are accumulating biomass. Each trophic level always has a shorter bar than the one below, because energy is used at each level to keep the organisms alive. Only about 10 per cent of the energy at any one trophic level is passed on to the next level, as shown in Figure 2.12. Advantages and disadvantages of these pyramids are shown in Table 2.05.

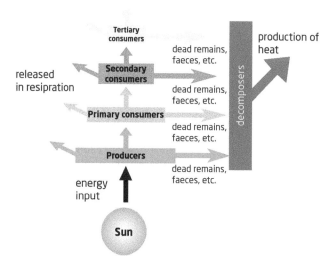

Figure 2.12 Energy losses at each trophic level of a food chain.

Type of pyramid	Advantages	Disadvantages
Numbers	• Gives a quick overview. • Useful for comparing population numbers in different seasons.	• No account is taken of the size of organisms, so pyramids that involve large producers such as trees are inverted.
Biomass	• Takes account of the size of organisms, so overcomes some of the problems of pyramids of numbers.	• Difficult to measure accurately, because sampling involves killing organisms. • Seasonal variation leads to inverted pyramids. • Some animals have a lot of bone or shell, which can distort the results.
Productivity	• Shows energy transferred over a period of time, so allows for different rates of production. • Ecosystems can be compared easily. • Pyramids are never inverted.	• Data is difficult to collect, as rate of biomass production over time must be measured. • Many species feed at more than one trophic level which can affect the results (true for all pyramids).

Table 2.05 Advantages and disadvantages of pyramids of numbers, biomass and productivity.

2.02.05 Pyramid structure and ecosystem function – energy transfers

As pyramids of productivity (Figure 2.13) show, energy and nutrients are passed up through trophic levels to the top carnivores at the end of the food chain. The lengths of food chains vary in different ecosystems and are limited by the organisms and their energy needs. On land, food chains seldom exceed four trophic levels, whereas in water there may be five or more.

2.02 Communities and ecosystems

Each time energy is transferred from one trophic level to the next, some is lost to the living organisms. Energy is used for growth, movement and life processes such as respiration; it is also lost as heat. In accordance with the second law of thermodynamics, the energy is not destroyed but is converted to a form that is not useful to sustain the lives of organisms (see Chapter 1.03). On land, energy may be used to keep animals warm and to support their bodies. In water, animals are more likely to be cold-blooded and use water to support them. Thus, in aquatic ecosystems, more energy remains at each trophic level to pass to the next, so food chains can be longer. In land-based systems, energy runs out after four transfers and there is not enough to support another consumer (see Figure 2.13).

When ecologists study the energy in living systems, they may measure the biomass or standing crop of species in the system, or they may examine the productivity of the system (see below). The units used to measure these components are shown in Table 2.06.

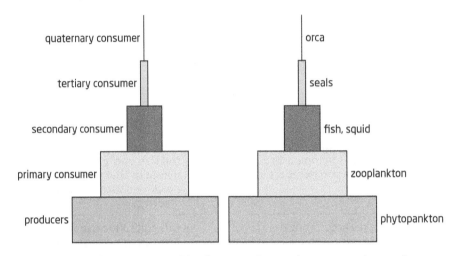

Figure 2.13 Comparing two pyramids of energy. Energy losses at each transfer are very inefficient, and this limits the number of animals that can live at the top trophic levels.

Pyramid type	Units used
biomass (standing crop)	$g\,m^{-2}$
productivity (flow of energy or biomass)	$g\,m^{-2}\,yr^{-1}$ $J\,m^{-2}\,yr^{-1}$

Table 2.06 Units used in constructing pyramids.

2.02.06 Bioaccumulation and biomagnification

Certain substances (pollutants) may accumulate in food chains as animals feed or are eaten because they are not biodegradable and do not break down easily in the environment or in organisms' bodies. At higher trophic levels, there may be an increase in the concentration of these substances because carnivores tend to feed on many smaller animals so that they take in more of the accumulated pollutant.

Pesticides are used to improve human food production but they also affect the way ecosystems function. The first pesticides made in the 1940s and 1950s were non-specific and killed many different species indiscriminately. Early insecticides killed both pests and useful pollinating insects.

The most well-documented case is that of DDT, an insecticide that was used by farmers to reduce losses and maximise crop yields. When it was first used, no one understood that DDT was **non-biodegradable** and remained poisonous in the environment for a long time. Although only low concentrations of DDT were used each time, the small amounts soon built up in the environment and in organisms' bodies – a process known as **bioaccumulation**. Primary consumers that ate many small plants containing the poison accumulated a much greater concentration of DDT, mostly in their body fat. Because primary consumers were eaten by secondary and tertiary consumers, the concentration of DDT increased many thousands of times at each link in the food chain in a process known as **biomagnification**.

The effect on top carnivores finally attracted attention to the high levels that DDT reached: people noticed what was happening when the DDT accumulated to such toxic levels in some predatory birds that their numbers plummeted (see Figure 2.14 and Case study 1.05.02). Persistent pesticides were even found stored in the fat reserves of Antarctic animals such as seals and penguins, despite the fact that they were more than 1000 km from the nearest agricultural land. Larger animals feed on many smaller ones and tend to live longer, so DDT accumulated to toxic levels in birds such as eagles, which died or failed to reproduce.

DDT was banned for use in agriculture in many countries during the 1970s and 1980s. In 2004, the Stockholm Convention permitted DDT to be used only for the control of malarial mosquitoes. This convention has been signed by more than 170 countries and is endorsed by most environmental groups.

Non-biodegradable pollutants are those that cannot be broken down within an organism or trophic level.

Bioaccumulation is the build-up of a persistent pollutant within an organism or trophic level because it is not biodegradable.

Biomagnification is the increase in concentration of persistent pollutants along a food chain.

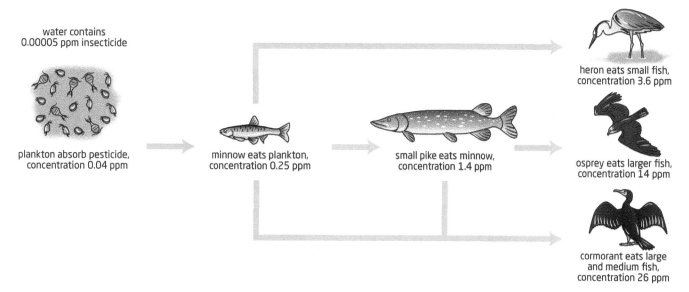

Figure 2.14 DDT sprayed on the land runs off into waterways too. Each time minnows eat plankton they take in just a small dose of DDT. The DDT remains in the minnows' bodies and builds up as they feed on more plankton. As small pike and birds complete the food chain, DDT builds up in each of them in the same way. Eventually the level of DDT in the cormorant is 500 000 times greater than in the water.

2.02 Communities and ecosystems

CASE STUDY 2.02.02

Minamata Bay and mercury poisoning

Minamata is located on the coast of Japan's westernmost island. The city and the adjacent Minamata Bay form a relatively closed ecosystem. The bay used to be a source of fish for the residents of the city, until the mid-1950s when people started to fall ill with an unknown neurological condition. They suffered increasing loss of motor control when walking or carrying out simple everyday tasks. Sometimes they became partly paralysed and were unable to see or speak properly. By the end of 1956, medical researchers had identified the cause as mercury poisoning from eating the fish and shellfish from Minamata Bay (see Image 2.07). The primary source of the problem was the town's Chisso Corporation factory, which had manufactured acetaldehyde to produce plastics since the 1930s. Mercury from the production process spilled into the bay and entered the food chain. At the time, Minamata residents relied almost exclusively on fish and shellfish from the bay as their source of protein.

Direct evidence that mercury from the factory was responsible did not emerge until 1959, by which time nearly 100 people had been identified as suffering from poisoning and more than 20 had died. Mercury is still present in the sediment of the bay, where fishing remains prohibited. Mercury is an accumulative poison. It accumulated in the food chain because animals that lived in the bay and fed on contaminated food were unable to break it down. When people in the area ate fish from the bay, they ingested mercury with each meal, until the amount in their bodies reached dangerous levels.

Questions

1. How do you think incidents such as Minamata Bay have influenced present-day thinking about environmental issues?

2. Suggest two reasons why the problem of mercury poisoning was not identified and dealt with until 1956, even though the factory had been operating since the 1930s.

3. Research methods of removing mercury and other poisonous metals from ecosystems that are polluted with them.

Image 2.07 Mercury poisoning from a plastics factory caused local residents to fall ill.

SELF-ASSESSMENT QUESTIONS 2.02.02

1. a Construct a pyramid of numbers for the following organisms in 0.1 ha of grassland in South America (represent the tertiary consumer – top carnivore – by a single line):
    - producers 15 000 000
    - primary consumers 200 000
    - secondary consumers 90 000
    - tertiary consumers 1.
   b Comment on the number of tertiary consumers in this ecosystem.
2. Why are large carnivores relatively rare in ecosystems?
3. Give reasons why a pyramid of biomass might not be pyramid shaped.
4. **Discussion point:** Consider how pyramids of productivity can be used to show how much human food is produced by different farming methods.

CONSIDER THIS

Heavy metals and pesticides are not the only substances that can accumulate in animals' bodies. Some natural toxins also do so. Vitamin A accumulates in the livers of carnivores such as polar bears, which feed on seals. Seals also have high levels of the vitamin in their livers. Indigenous people in the Arctic know that the livers of these animals are poisonous and should not be eaten, but some inexperienced explorers have been killed by eating them.

In other cases, animals use bioaccumulation to defend themselves from predators. The tobacco hornworm, which feeds on tobacco plants, builds up high levels of nicotine in its body. Nicotine does not affect the hornworm but it is poisonous to its predators and affects their muscles and ability to move. High levels of nicotine provide a good deterrent to animals that might eat the hornworms.

2.03 Flows of energy and matter

LEARNING OBJECTIVES

After reading this chapter you should be able to:

- outline how solar radiation enters and leaves the atmosphere
- explain how energy flows through ecosystems
- define productivity, net primary productivity (NPP), gross primary productivity (GPP), gross secondary productivity (GSP) and net secondary productivity (NSP)
- describe the flow of matter through ecosystems, including the carbon and nitrogen cycles
- describe the storages and flows in the nitrogen and carbon cycles
- outline the impact of human activities on energy flow and the carbon and nitrogen cycles.

KEY QUESTIONS

How are ecosystems linked together by the flow of energy and matter?
How are humans affecting the flow of energy and matter at a local and global levels?

2.03.01 Energy transfer and transformation in an ecosystem

Ecosystems are maintained by the flow of energy and matter, which also links different systems together. The primary source of energy for life on Earth is the Sun, and this energy drives ecosystems. Light from the Sun is the source of energy for life on our planet. Light that enters the Earth's atmosphere consists of a range of wavelengths known as the electromagnetic spectrum. The visible spectrum is just the part of this range of wavelengths that we can perceive with our eyes. Other wavelengths such as the short-wavelength X-rays and ultraviolet radiation (UV), and much longer infrared and radio waves, cannot be detected by the human eye. No matter what the wavelength of the electromagnetic waves, they can all be reflected and refracted (bent), and this has important consequences for ecosystems.

About 30 per cent of the energy reaching the Earth's surface is reflected back into space from clouds, ice, snow, water and land. A further 69 per cent is absorbed, as shown in Figure 2.15. Land and sea absorb some of the energy and are heated up. Less than 1 per cent of the Sun's energy is available to plants.

Of the energy that enters an ecosystem:

- Energy may be converted from light energy to chemical energy during photosynthesis.
- Energy may be transferred from one trophic level to the next as organisms feed.
- UV and visible light may eventually be converted to heat energy in the ecosystem.
- Heat energy is re-radiated into the atmosphere.

Only a very small fraction of the light from the Sun that does reach green plants is eventually converted to plant biomass. There are many reasons for this, including the following:

- *Reflection:* Some light is reflected from the surface of leaves or passes through them without being captured.
- *Wavelength:* Chlorophyll only captures certain wavelengths of light for use during photosynthesis. For example, green light is reflected and not absorbed. Red and blue wavelengths are the most effective for photosynthesis.
- *Efficiency:* Photosynthesis has inbuilt inefficiencies and is limited by factors such as temperature and carbon dioxide concentration and the rate at which enzyme catalysts can work.
- *Not being absorbed:* Even light that does enter leaves may not strike the chloroplasts.

Measurements show that less than 0.05 per cent of light energy falling on the Earth is captured by plants and converted to glucose during photosynthesis. Some of the glucose from photosynthesis is respired and the energy used to keep the plant alive, while the remainder is converted to other compounds such as starch and protein to be stored in the plant. The conversion of energy into biomass in a given period is measured as **productivity**.

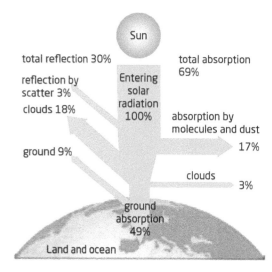

Figure 2.15 Approximate amounts of solar radiation entering the Earth's atmosphere that are absorbed and reflected. Less than 1% remains for absorption by plants.

The chemical energy in the molecules of glucose and stored compounds is available to be passed on to consumers when the plant is eaten. But not all the energy is transferred to consumers. Whenever consumers eat, energy is transferred from one trophic level to the next, but the process is not efficient and there are significant losses. Some energy is used by organisms for respiration and converted to heat, which cannot be transferred to the next trophic level. Still more energy is lost in the form of faeces and other waste products which are not eaten and so do not pass up the food chain. On average, only about 10 per cent of the energy assimilated in one trophic level is available to be passed to the next. This is known as **ecological efficiency**. One consequence of this is that the biomass at higher trophic levels is almost always less than at lower levels, as shown in pyramids of energy and biomass (see Chapter 2.02).

Energy is eventually lost from all food chains in all ecosystems in the form of heat, which is re-radiated back into the atmosphere.

Decomposer food chains are also important in energy transfer. Dead material and faeces may be used as food by bacteria and fungi to supply energy for their life processes. These organisms also break down detritus and make food available for other decomposers. Heat is released as decomposers respire, and valuable inorganic nutrients are released back into the soil.

The details of the pathway of energy flow in an ecosystem depend on the organisms and on the characteristics of the ecosystem. Most of the primary production of a tropical rainforest ecosystem will enter the decomposer food chains, but in a marine ecosystem a far larger proportion will be passed through consumer food chains. On intensively used agricultural land, more than 50 per cent of primary production can enter the grazing food chain.

Productivity is the conversion of energy into biomass in a given period of time.

Ecological efficiency is the percentage of energy assimilated in one trophic level that is available to be passed to the next.

Energy flow diagrams

Environmental scientists represent the way energy is stored and flows between trophic levels in ecosystems as diagrams, similar to the ones shown in Figures 2.16 and 2.17. These diagrams enable us to compare different ecosystems easily.

In Figure 2.17, the width of the arrows represents quantities. The arrows also show how much energy is:

- passed in and out of each trophic level
- passed to decomposers or wasted
- lost through respiration.

2.03 Flows of energy and matter

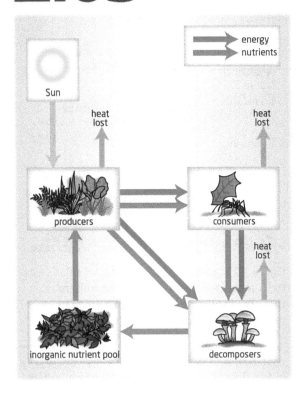

Figure 2.16 Summary of the transfer of energy and nutrients through an ecosystem.

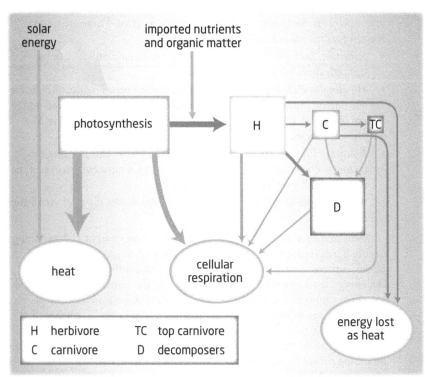

Figure 2.17 Simplified energy flow diagram showing inputs and outputs for a stream ecosystem.

CASE STUDY 2.03.01

Using diagrams to model energy flow for a herbivore and a carnivore

Diagrams like those shown here are known as Sankey diagrams. They are used by engineers as well as environmental scientists to show energy inputs and losses for a machine or an individual trophic level. They show energy taken in as food, lost in respiration or lost as waste, and the energy that is available to be passed on for different groups of organisms. The thickness of any arrow in these diagrams is proportional to the numerical value for the energy flow at that point. Comparing the diagrams shows differences between the energy balances of different types of animals.

Figure 2.18 shows diagrams for a herbivore and a carnivore. The arrows are drawn in proportion to the amount of energy flowing in each case. The input is the energy in the food that the animals eat.

The widths of the arrows in the diagrams above show that herbivores lose a larger amount of the energy from their food in faeces than the carnivores do. The arrows also show that carnivores lose more energy through respiration than herbivores do.

Questions

1. Consider the food sources of herbivores and of carnivores such as zebras and lions, and suggest a reason why herbivores lose more of the energy from their food in faeces than carnivores do.
2. Consider the activity patterns of herbivores and carnivores and suggest a reason why carnivores lose more energy through respiration than herbivores do.
3. Consider the amount of energy needed by a non-mammalian carnivore (e.g. a crocodile or jellyfish) to maintain body temperature and support its body. How and why do the outputs for such an animal differ from those of a mammalian carnivore? Try sketching a Sankey diagram for such an animal.

CASE STUDY 2.03.01 (continued)

(a)
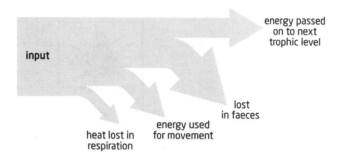

(b)
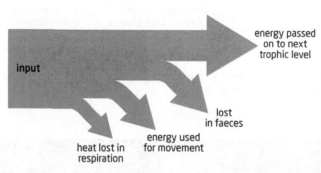

Figure 2.18 Sankey diagrams: (a) a herbivore, (b) a carnivore.

2.03.02 Productivity

Primary productivity

Primary productivity is a measure of the conversion of light energy into chemical energy in living organisms. **Gross primary productivity (GPP)** is the total energy converted by photosynthesis – that is, the rate at which photosynthesis occurs. This is not easy to measure directly, and so **net primary productivity (NPP)** is usually calculated instead from measurements of changes in biomass.

This can be represented by the equation below (also see Figure 2.19). R represents respiratory losses.

$$GPP = NPP + R$$

Measurements used to estimate NPP vary, and many involve sampling plant material and drying it to estimate the biomass. These estimates are used when pyramids of biomass and energy are drawn (see 2.02.04 *Pyramids of numbers, biomass and productivity*). It is difficult to account for the productivity of every part of a plant, such as that by roots, the amounts eaten by herbivores and leaves lost in leaf litter, so NPP estimates are often too low. Estimates of respiration are gained from the amount of carbon dioxide produced and can be obtained by measuring the concentrations of the gas in the atmosphere.

Extension

Terrestrial and marine NPP can be observed by remote sensing from satellites (see Image 2.08).

Primary productivity is the production of chemical energy in organic compounds by autotrophs (producers). It is usually measured as biomass per unit area per unit time. Primary production in rainforests is high because there are ample resources; sunlight, water, nutrients and temperature are all present in appropriate quantities (see Chapter 2, 2.04.02, p. 89). Primary production is low in a desert because the necessary resources are scarce.

Gross primary productivity (GPP) is the total gain in energy or biomass per unit area per unit time fixed by photosynthesis in green plants.

Net primary productivity (NPP) is the gain by producers in energy or biomass per unit area per unit time minus respiratory losses (R).

2.03 Flows of energy and matter

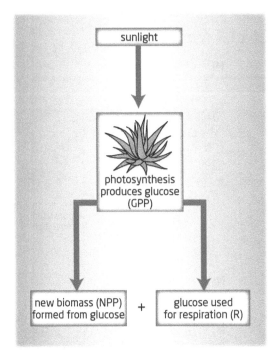

Figure 2.19 Net primary productivity is the gain in biomass from photosynthesis after respiration has been deducted.

CONSIDER THIS

The Earth's NPP has been estimated to be 105 petagrams (10.5×10^{10} metric tonnes) of carbon per year. About 54 per cent of this carbon is taken up by terrestrial ecosystems, and the remaining 46 per cent by primary producers in the oceans.

The average rate of NPP for terrestrial ecosystems is estimated to be $426\,g\,C\,m^{-2}\,yr^{-1}$, and this is higher than for oceans ($140\,g\,C\,m^{-2}\,yr^{-1}$). Tropical regions (forests and savannahs) account for about 60 per cent of terrestrial NPP.

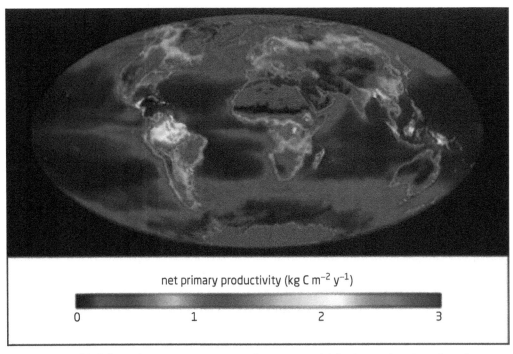

Image 2.08 This false-colour map represents the rate at which plants absorb carbon (as carbon dioxide) from the atmosphere. The map shows an annual average of the net productivity of vegetation on land and in the ocean. The scale shows kilograms of carbon taken in per square metre per year. Notice that NPP is highest in tropical regions.

Photo from NASA Goddard Space Flight Center
http://science.nasa.gov/earth-science/oceanography/ocean-earth-system/ocean-carbon-cycle

Secondary productivity

Secondary productivity measures feeding or absorption of stored energy. When the productivity of consumers is considered, two losses are significant: first, loss in faeces; and second, respiratory loss. Loss in faeces occurs because animals do not or cannot use all the biomass they eat. Faeces contain food that cannot be digested by the animal. Respiratory loss (R) is the energy that is assimilated but then used in respiration to maintain the animal's life processes. (Recall that these two losses account for the shape of pyramids of energy and biomass; see 2.02.05 *Pyramid structure and ecosystem function – energy transfers.*) **Gross secondary productivity (GSP)** is energy gained through absorption by consumers through absorption and **net secondary productivity (NSP)** is the biomass that is available to the next trophic level. This can be represented by the equations below (also see Figure 2.20).

$$GSP = \text{food eaten} - \text{faecal losses}$$
$$NSP = GSP - R$$

> **Secondary productivity** is the biomass gained by heterotrophs (consumers) as they feed. It is also usually measured as biomass per unit area per unit time.
>
> **Gross secondary productivity (GSP)** is the energy or biomass gained by consumers through absorption.
>
> **Net secondary productivity (NSP)** is the gain in energy or biomass per unit area per unit by consumers minus respiratory losses (R).
>
> **Net productivity (NP)** is the gain in energy or biomass per unit area per unit time remaining after the deduction of losses through respiration (R = respiratory loss) – that is, the biomass available to consumers at subsequent trophic levels.
>
> **Gross productivity (GP)** is the total gain in energy or biomass per unit area per unit time, which could be through photosynthesis in primary producers or absorption in consumers (including that lost to respiration).

Calculations of NPP, NSP and GSP

Table 2.07 shows the flow of biomass (as carbon) in an aquatic ecosystem.

	$g\,C\,m^{-2}\,yr^{-1}$
gross productivity of aquatic plants	129
respiratory loss by aquatic plants	32
aquatic plants eaten by primary consumers	28
loss in faeces of primary consumers	5
respiratory loss of primary consumers	11

Table 2.07 The flow of biomass (as carbon) in an aquatic ecosystem.

From these data we can calculate:
The **net productivity** of the plants:

$$GPP = NPP + R$$
$$129 = NPP + 32$$

Rearranging the equation:

$$NPP = 129 - 32 = 97\,g\,C\,m^{-2}\,yr^{-1}$$

The **gross productivity** of the primary consumers:

$$GSP = \text{food eaten} - \text{faecal losses}$$
$$GSP = 28 - 5$$
$$= 23\,g\,C\,m^{-2}\,yr^{-1}$$

Net productivity of the primary consumers:

$$NSP = GSP - R$$
$$NSP = 23 - 11$$
$$= 12\,g\,C\,m^{-2}\,yr^{-1}$$

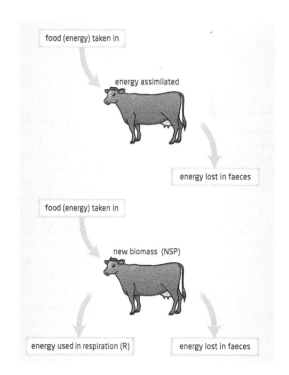

Figure 2.20 Top: gross secondary productivity. Bottom: net secondary productivity.

2.03 Flows of energy and matter

Sustainable yield

Sustainable yield (SY) is the rate of increase in biomass that can be exploited without preventing the organisms being taken from replenishing themselves.

The relationship can be expressed in the equation:

$$SY = (\text{annual growth and recruitment}) - (\text{annual death and emigration})$$

Examples of sustainable yield include:

- *Forestry:* SY is the largest amount of wood that can be harvested without reducing the productivity of the remaining trees.

  An original stock of 10 000 trees produces 1200 harvestable trees each year. If only 1200 are harvested, next year there should be another 1200 available and 1200 for the year after.

- *Fishing:* SY is the amount of fish that can be caught on a regular basis without compromising the ability of the species to reproduce and maintain its population.

Sustainable yield can vary over time with the needs of the ecosystem. For example, a forest that has recently suffered a fire will require more of its own productivity to sustain and re-establish a mature forest. While doing so, the sustainable yield may be much less than in previous years.

Calculating sustainable yield

Sustainable yield (SY) can be calculated from the rate of increase in natural capital, which means the amount of crop or number of animals that can be removed without depleting the stock so that it cannot replenish itself.

For a crop, the annual sustainable yield can be estimated as the annual gain in biomass:

$$SY = (\text{total biomass at time 2}) - (\text{total biomass at time 1})$$

Alternatively, for animals:

$$SY = (\text{growth} + \text{new individuals}) - (\text{death} + \text{emigration}) \text{ (in a given time, usually one year)}$$

Consider an area of woodland with a herd of 740 deer. In one year, 54 deer were born and 50 died. Ten male animals moved to new territories and 15 new individuals entered the woodland.

We can calculate the sustainable yield for the woodland as follows:

$$SY = (54 + 15) - (50 + 10)$$
$$SY = 69 - 60$$
$$SY = 9$$

This means that, if hunting is permitted in the woodland, no more than nine deer can be taken for the population to remain sustainable.

The related concept of **maximum sustainable yield (MSY)** is often used in terms of commercial production. Maximum sustainable yield is the largest amount of a raw material that can be taken without permanently depleting the stock. The maximum sustainable yield has been exceeded in many of the world's fishing grounds, resulting in various attempts to return to a situation of sustainability. The EU's Common Fisheries Policy (see Topic 4 and Case study 8.02.01) is perhaps the best known of such actions.

Every year or breeding season the fish population will increase due to new offspring and maybe also due to the arrival of immigrants. Population loss will occur due to death and emigration. A net increase in population will occur if the number added to the population is larger than the number leaving. If only the increase in the population of fish is harvested, the population will remain the same. This level of harvesting is the maximum sustainable yield for the population.

The sustainable yield for a fishing ground can be calculated in a similar way to the example above. If we measure:

$$R = \text{biomass of juvenile fish reaching harvestable size}$$
$$G = \text{growth in biomass of fish already at harvestable size}$$
$$M = \text{fish lost through death and emigration}$$
$$(R + G) = \text{growth and new individuals}$$
$$SY = (R + G) - M$$

Notice that the equations does not use the biomass or size of the population at the start of the year; it is only important to know the change in their biomass.

The **maximum sustainable yield (MSY)** from a system is equivalent to either the net primary or net secondary productivity of the system.

CONSIDER THIS

The term 'assimilation' is sometimes used instead of 'secondary productivity', because secondary productivity describes the increase in new stored energy.

SELF-ASSESSMENT QUESTIONS 2.03.01

1. Explain the difference between GSP and NSP.
2. Which of these statements is correct?
   - A Net primary productivity is the energy fixed per unit area.
   - B Net productivity is the gain in energy or biomass per unit are a per unit time after allowing for respiration losses.
   - C Net productivity is the biomass gained by heterotrophs per unit time.
   - D Net productivity is the total light energy stored by photosynthesis in green plants.
3. Cattle graze on ryegrass plants. The fate of one year's growth of 1 m² ryegrass is shown in Figure 2.21.

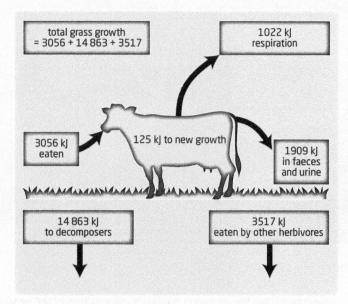

Figure 2.21 The fate of one year's growth of 1 m² ryegrass.

   - a What proportion of the net primary productivity was eaten by the cattle?
   - b What proportion of the energy in the plants eaten by the cattle went to increasing the mass of the cattle (NSP)?
4. **Research idea:** Find data on the NPP of temperate and tropical grasslands. Explain the difference in NPP of these two biomes.

2.03.03 Transfer and transformation of materials in an ecosystem

As we have seen, energy flows through an ecosystem, entering as solar radiation and eventually leaving the system as heat. Energy cannot be recycled in an ecosystem, but nutrients are different and are reused many times. Nutrients are absorbed from the soil and pass along food chains through each trophic level until eventually they return to the soil via the decomposers.

It is often said that energy flows but nutrients cycle. The most important cycles are the water cycle, the carbon cycle and the nitrogen cycle, but sulfur and phosphorus also cycle in a similar way.

2.03 Flows of energy and matter

Carbon cycle

Organic compounds are carbon-containing compounds excluding carbon dioxide and carbonates that are found in the bodies of living organisms.

Inorganic compounds are compounds of mineral origin.

Carbon is an element present in all the **organic compounds** that make up the bodies of living things. Proteins, carbohydrates and fat all contain carbon as a vital component of their molecules. Carbon is cycled in an ecosystem, so the same atoms are reused over and over again (see Figure 2.22). Storages in the carbon cycle include organisms (organic) and the atmosphere, soil and fossil fuels (**inorganic**). Flows, indicated by arrows, include feeding, death, photosynthesis and respiration.

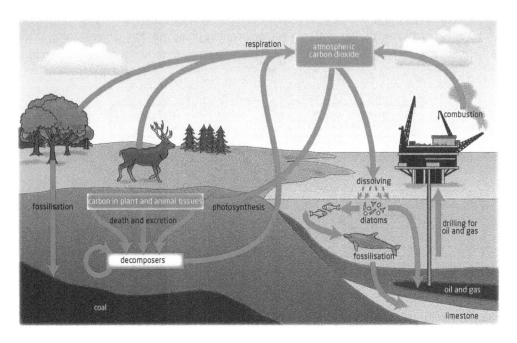

Figure 2.22 The carbon cycle.

Inorganic carbon dioxide from the air enters the biotic component of the ecosystem when it is converted to organic molecules such as glucose and starch during photosynthesis. Autotrophs use some of the glucose they produce for their own respiration and return some of the carbon to the atmosphere as carbon dioxide. The remainder forms part of their biomass, so the carbon becomes tied up or 'fixed' as organic compounds in their bodies. These compounds may either be eaten by consumers or, if the plant dies, will pass to decomposers. Decomposers also respire and return carbon dioxide to the atmosphere as they do so. Consumers, which take in carbon compounds as they eat, either respire glucose and release carbon dioxide back into the air, or assimilate the carbohydrates, proteins and fat into their body biomass. The biomass of one consumer then provides a carbon source for the next animals in the food chain which feed on it, and so on. Thus carbon is passed on through the food chain, passed to decomposers or returned to the atmosphere in respiration.

Not all the carbon in plant and animal material is recycled in this way, because conditions at certain times in the Earth's history have prevented decomposers from feeding on it. Over millions of years, some biotic material has become limestone or been fossilised to form fossil fuels. The most important fossil fuels are coal, oil and gas. The formation of coal began when dead plant material from ancient forests did not decay but became compressed in anaerobic conditions. Oil and gas were formed from the bodies of dead marine organisms which sank to the bottom of the ancient oceans where they heated and compressed deep below the surface for millions of years. Fossil fuels contain reserves of carbon which are locked in and excluded from the carbon cycle for very long periods of time. The carbon is only released when the fuels are burnt and the carbon in them combines with oxygen to form carbon dioxide.

Limestone is another reserve of carbon, which is stored in the form of calcium carbonate. Limestone was formed from the remains of the shells and bones of marine animals.

CONSIDER THIS

In many less economically developed countries (LEDCs), fertilisers are too expensive for farmers to use. Sometimes other solutions are equally effective. In Cuba, Humberto Ríos Labrada, a scientist and biodiversity researcher, encouraged local farmers to increase their crop diversity and develop low-input agricultural systems that greatly reduced the need for pesticides and fertilisers. He is recognised for encouraging Cuba's shift from agricultural chemical dependence towards more sustainable farming methods and has led similar projects in Mexico.

Nitrogen cycle

Although almost 80 per cent of the Earth's atmosphere is nitrogen gas, the gas is so stable that it cannot be used directly by living organisms. Nevertheless, nitrogen is a vital element for the formation of protein and nucleic acids in the bodies of plants and animals. It is recycled through ecosystems by the actions of many microorganisms (see Figure 2.23).

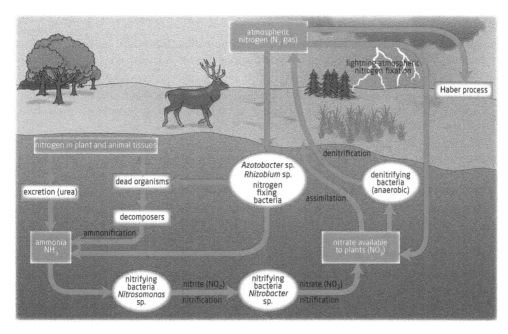

Figure 2.23 The nitrogen cycle.

Nitrogen is made available to an ecosystem by bacteria that are crucial in transferring nitrogen and nitrogen compounds from the abiotic to the biotic environment. Two types of bacteria, *Azotobacter* and *Rhizobium*, which live in the soil, are able to take in and fix nitrogen from the air and convert it into nitrates, a form that can be taken in by plants. *Rhizobium* is found in nodules on the roots of leguminous plants such as peas, beans and clover. The bacteria and plants form a symbiotic relationship in which the bacteria receive sugars from the plant, which in turn receives nitrates from the bacteria. The only other natural method of fixing nitrogen gas so that it is useful to living things is the effect of lightning, which combines the nitrogen with oxygen, forming nitrates that enter the soil. Humans also fix nitrogen in the Haber process, which is used to manufacture fertilisers.

Other important groups of bacteria in the nitrogen cycle are the nitrifying bacteria, *Nitrosomonas*, which convert ammonia from excretory material into nitrites, and *Nitrobacter*, which convert nitrites into nitrates. Nitrates are all soluble compounds that can be absorbed by plants through their roots and assimilated into their biomass. Ammonium compounds and nitrites cannot be taken in directly by plants.

Nitrogen is returned to the atmosphere by the activity of denitrifying bacteria, which convert nitrates to nitrogen gas and are found mainly in anaerobic conditions. These bacteria tend to live in waterlogged soils which are depleted of their nitrates and so not useful for cultivation.

SELF-ASSESSMENT QUESTIONS 2.03.02

1 Consider Figure 2.23 and list three storages (sinks) of nitrogen in an ecosystem.
2 For each of the cycles discussed above, outline one way in which human activity has had an effect on the natural cycle.
3 **Discussion point:** What would happen to the carbon cycle if decomposers were removed from an ecosystem?

Theory of knowledge 2.03.01

Ethics and fossil fuels

To use a 100-watt light bulb constantly for one year uses 325 kg of coal in a coal-fired power station. The efficiency of the power plant is about 40 per cent.

One litre of petrol comes from 23.5 tonnes of ancient organic material deposited on the ocean floor.

The total amount of fossil fuels used in a typical year at the turn of the century was estimated to have come from all the plant matter that grew on the ancient Earth for 400 years.

At our current rate of use, the Earth's supply of coal will last for 1500 years. But if we increase our usage by 5 per cent, the supply will only last 86 years. If supplies of other fossil fuels diminish, even greater quantities of coal may be used.

The burning of fossil fuels is responsible for environmental issues that are high on the political agendas of many countries. Examples include not only global warming, but also acidification of water, air pollution and damage to land surfaces.

1 How should an individual react to this information? What can one person do to limit his or her own impact on the environment? You might like to consider your own environmental value system (EVS) here.
2 LEDCs need electricity and transport to fuel their industries and to provide for the basic needs of their populations. How should MEDCs respond to increasing fossil fuel consumption in these countries?

2.03.04 Human influences on energy flow and nutrient cycling in an ecosystem

Human activities such as burning fossil fuels, deforestation, urbanisation and agriculture have had a profound effect on the flow of energy through natural ecosystems. Here are a few examples of how this occurs:

- Humans convert biomass such as wood into fuel, and energy is lost as heat. This means that energy that might otherwise have passed along a food chain or to a decomposer food chain is no longer available to living things.
- The cycling of carbon and nitrogen have been affected as storages are removed and flows interrupted. Carbon stored as fossil fuel is removed from storage and added to the atmosphere as carbon dioxide. In addition, the cycling of nitrogen has been modified by the use of fertilisers that add nitrates to agricultural areas in disproportionate amounts.
- Agriculture removes biomass from the land as food and adds fertiliser to the soil to boost productivity. In a natural system, biomass is used in a food chain and dead material is recycled. In an agricultural system, biomass is removed for human food and natural cycling cannot take place. (You can read more about this in Topic 5.)

In tropical regions, vast rainforests trap carbon dioxide through photosynthesis and have been important in maintaining the low level of atmospheric carbon dioxide. Humans have upset this balance by deforesting large areas for agriculture and timber production. Deforested areas cannot capture carbon dioxide from the atmosphere and create new biomass, thus the natural carbon cycle is disrupted. The destruction of rainforest has multiple effects, but the most important for the atmosphere are the loss of carbon dioxide uptake by photosynthesis, and the increase of carbon dioxide released from rotting or burnt vegetation. Burning material and the respiration of decomposers that feed on rotting material both add carbon dioxide to the atmosphere and upset the natural carbon cycle.

SELF-ASSESSMENT QUESTIONS 2.03.03

1. Outline what is meant by 'nitrogen fixation'.
2. Name three examples of sustainable natural capital.
3. Describe the concept of sustainability.
4. **Research idea:** Where is the greatest storage of nutrients held in a rainforest ecosystem? Why does this make the problem of deforestation worse?

CASE STUDY 2.03.02

Human impacts on nutrient cycles and energy flow – slash and burn

Slash and burn is a type of agriculture that involves cutting down vegetation on a plot of land, allowing it to dry, then setting fire to it (see Image 2.09). The ashes then act as a short-term fertiliser and provide nutrients to the soil so that food crops can be grown. But land can only be used like this for a short period, because the fertility of the burnt land is reduced. After a few years, it must be left alone, sometimes for up to ten or more years, probably longer than it was under cultivation, so that natural vegetation can regrow. This may allow nutrient cycles to be re-established and the land to recover its fertility. Sometimes, when vegetation has regrown, slash and burn may be repeated some years later in a practice known as shifting cultivation.

Slash and burn agriculture can provide communities with food and income. It also enables people to farm in parts of the world where agriculture is not usually possible because of dense vegetation, for example in the rainforest, or where the soil is infertile or has a low nutrient content. Slash and burn is most often used for subsistence farming, and crops are grown for the use of a family or small community. For example in the Amazon, yams, cassava and sweet potatoes can be grown. Where population densities are small and indigenous communities closely connected to the land, shifting cultivation which involves using a series of forest clearings, in turn and returning to them after a period of maybe 50 years, can be used. But in many places land may be reused too soon, before fertility has been restored, and this leads to problems. Today, it is estimated that up to 7 per cent of the world's population use slash and burn agriculture.

Slash and burn agriculture can create many environmental problems, and these become more serious as increasing numbers of people use this type of agriculture. The main problems include:

- Deforestation: temporary or permanent loss of forest cover.
- Erosion: roots and temporary water storages are removed so they cannot prevent nutrients running off the soil and away from the area.
- Loss of nutrients: fields may gradually lose the fertility they once had as crops are harvested and nutrients are not replaced. In the long term, this may cause desertification, leaving infertile land that cannot support growth of any plants.

Questions

1. Outline two ways in which slash and burn agriculture has an impact on the flow of nutrients in the cultivated areas.
2. Suggest why indigenous communities may be better able to limit these problems.
3. Loss of vegetation in a cleared area may lead to soil erosion, which in turn leads to more loss of vegetation. What type of feedback is described by this sentence?

Image 2.09 Slash and burn agriculture in Madagascar.

2.04 Biomes, zonation and succession

LEARNING OBJECTIVES

After reading this chapter you should be able to:

- define a biome and give examples of five major classes of biome
- outline the factors that influence the distribution of biomes
- explain what is meant by zonation along an environmental gradient
- describe how succession changes the appearance, energy flow and productivity of an ecosystem over time
- describe the difference between r and K- strategist species in pioneer and climax communities
- explain the general patterns of change in communities undergoing succession.

KEY QUESTIONS

What are the factors that determine the type of biome in a given area?
How does succession lead to climax communities?
How are stability, succession and biodiversity of an ecosystem linked?

2.04.01 Key factors that define a biome

A **biome** is a group of ecosystems that share similar climatic conditions and therefore similar patterns of vegetation.

Each **biome** has particular abiotic factors and contains species that distinguish it from other biomes (see Figure 2.24). The key abiotic factors that define biomes are:

- rainfall
- temperature
- insolation (sunlight).

Latitude and altitude also influence climate and so affect biome distribution. Lower latitudes and altitudes tend to be warmer, but high mountains such as Mount Kilimanjaro and some peaks in the Andes are snow covered despite being close to the equator. The location of some of the world's biomes is shown in Figure 2.25. The tricellular model of atmospheric circulation (see Topic 6) explains the distribution of rainfall and temperature, which influences the structure and productivity of biomes.

2.04.02 Distribution, structure and productivity of major biomes

Biomes are grouped into five major classes:

- aquatic
- grassland
- tundra.
- forest
- desert

Distribution, structure and productivity of major biomes

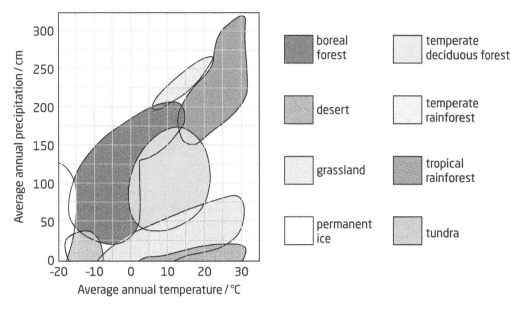

Figure 2.24 This diagram is known as a climograph. It shows the average temperature and rainfall that occur in different biomes.

CONSIDER THIS

The position of the Earth's axis varies over the course of a year, and this causes a change in the position of the Sun in the sky. This change in position affects the intensity of solar radiation. Solar radiation is mainly due to the angle at which the Sun's rays strike the Earth's surface (the angle of incidence). If the Sun is directly overhead (90° from the horizon), the incoming insolation meets the surface of the Earth at right angles and is at its most intense. But if the Sun is 45° above the horizon, the incoming insolation meets the Earth at an angle. The light rays are spread out over a larger area and the intensity of the solar radiation is reduced.

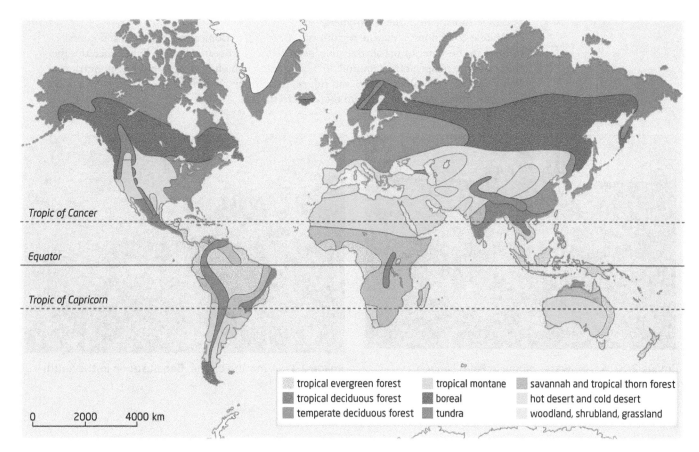

Figure 2.25 Location of various biomes around the world.

2.04 Biomes, zonation and succession

Within these classes, it is possible to recognise other divisions. Each of the five major classes has its own characteristic productivity, biodiversity and factors that limit its development.

Aquatic biomes

Water covers nearly 75 per cent of the Earth's surface. Aquatic biomes are important to thousands of living things as well as sustaining organisms that inhabit terrestrial biomes. There are two types of aquatic biome: freshwater (ponds and rivers) and marine (oceans and estuaries).

Freshwater biomes

Freshwater regions have a low salt concentration (usually less than 1 per cent). Examples include lakes, rivers, ponds and swamps. Ponds and lakes vary in size from a few square metres to thousands of square kilometres (see Image 2.10). Their temperatures vary with the seasons, and the ability of light to penetrate the water determines the level of photosynthesis that can take place. Streams and rivers contain flowing water. The temperature of the water is cooler at the source of a river and, as water flows to the mouth of the river, it becomes turbid with sediment and debris so that light penetration and species diversity are reduced.

Marine biomes

Marine regions include oceans, coral reefs (see Image 2.11) and estuaries. Oceans are the largest of all, and some scientists say that the ocean has the highest diversity of species, even though there are fewer species than in terrestrial ecosystems (see 2.05.05 *Diversity and the Simpson diversity index*). Salt water evaporates and contributes to the water cycle (see Chapter 4.01). Marine algae produce much of the Earth's oxygen as they photosynthesise, and they also absorb large amounts of carbon dioxide from the atmosphere.

Abiotic conditions in marine regions vary from warm tropical temperatures and rainfall to those in the water of the abyssal zone, deep below the ocean. Here no light penetrates the water and the temperature is around 30 °C; the water is high in oxygen but low in nutrients. Hydrothermal vents along mid-ocean ridges are home to chemosynthetic bacteria, which use hydrogen sulfide and minerals from the vents to survive.

Image 2.10 A freshwater biome in Palm Lagoon, within the Gilbert–Smithburn River Delta, Queensland, Australia.

Image 2.11 A marine biome, Beqa Lagoon in the South Pacific.

Forest biomes

Forest biomes include temperate deciduous forest and tropical rainforest.

Temperate forest

Temperate deciduous forests are found in areas with mild, short winters; they have a growing season of at least 120 days free of frost.

Temperate forests have a mild climate with rainfall of between 500 and 1500 mm per year and temperatures in the range −30 °C to +30 °C, depending on the location of the forest (see Figure 2.25). Forests of Western Europe are warmer in winter due to the influence of the Gulf Stream, but the forests of north-eastern USA are very cold at this time.

Unlike rainforests, temperate forests are found in seasonal areas, and most plant growth occurs in warmer summer months. In some regions, trees such as maple, elm and oak are deciduous and lose their leaves in the cold winter months, while in others, particularly those at higher latitudes or altitudes, trees are evergreen and have leaves or needles all year round.

The rate of photosynthesis and NPP of temperate forest is lower than that of a rainforest, because of the lower temperature and rainfall. Temperate forests also contain fewer species of tree. A forest may be dominated by one species such as oak or beech which forms a canopy with a shrub layer beneath. The fully developed canopy shades the lower layers of the forest during the summer months, but, if the trees are deciduous, the light levels are sufficient for other species to grow during the spring and autumn months. In spring, before deciduous trees come into leaf, small flowering plants may grow and complete their life cycles before they are overshadowed by the canopy and deprived of light. The floor of the forest may contain species such as bramble, ferns, grasses, bracken and other species that can grow in lower levels of light. Soil is enriched by leaf litter in autumn, and nutrients are recycled rapidly.

Many species of small herbivore are found on the forest plants. One oak tree may support up to 250 invertebrate species which provide food for other animals. Food chains are interlinked to form many food webs which include larger mammals such as deer and foxes.

Tropical rainforests

Tropical rainforest is the most ecologically rich of the Earth's biomes. It is characterised by high average temperatures of about 27 °C and high rainfall which can be as much as 5000 mm / yr-1. These remain fairly constant throughout the year.

Rainforests lie between the Tropic of Capricorn and the Tropic of Cancer, 5° to the north and south of the Equator (see Figure 2.25), where they receive high levels of light, 12 hours each day, throughout the year.

The combination of high temperature, abundant rainfall and high light levels at the canopy means that the rate of photosynthesis of rainforest plants is rapid because there are few limiting factors. The result is that rainforests grow vigorously and are highly productive, on average producing the greatest amount of organic matter of all the biomes (see Figure 2.26).

There is high NPP, which is estimated to account for 40 per cent of the productivity of terrestrial ecosystems (see Chapter 2.05).

Rainforests are hot and wet: rainfall can be as much as 5000 mm yr^{-1}, and the temperature averages around 27 °C throughout the year.

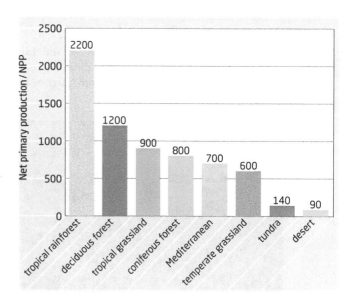

Figure 2.26 Graph comparing the NPP of eight biomes.

2.04 Biomes, zonation and succession

CONSIDER THIS

Much natural temperate forest has been cleared over the centuries to provide building materials and agricultural land. Large predators such as bears and wolves, which once roamed these forests, are now rare or absent in many places other than protected areas or areas where the species have been reintroduced.

Extension
Red list
www.iucnredlist.org/details/41687/0

Rainforests are very diverse and contain a large number of different plant and animal species. These species are also present in very large numbers. Plants compete for the available light and grow to great heights. The tallest trees may be up to 80 m high in some forests. Beneath them are layers of canopy and understorey trees, which shade the developing and shrub layers below so that the majority of NPP occurs in the highest levels of the forest. It is estimated that only about 1 per cent of the light that falls on the topmost layers reaches the ground. Epiphytes such as vines, orchids, ferns and bromeliads climb up tall trees to reach the light and they, in turn, provide habitats for many animal species.

Rainforest plants tend to have shallow roots, because most nutrients are close to the surface of the soil. Many of them have buttress roots to help support themselves. Although the rate of decay of materials is high because of the high temperature, nutrient levels in the soil are usually quite low and nutrients are easily washed away by heavy rain. In primary rainforest, the tree cover prevents this from happening, but if trees are cut down, the soil and nutrients are exposed and quickly washed away leaving the land unsuitable for agriculture.

Grassland

Grasslands are generally semi-arid areas with few trees, and they are inhabited by grazing mammals, ground-nesting birds, insects and a few species of reptiles. Grassland covers about 21 per cent of the Earth's land surface and is located in several areas of the world. The main categories of grassland are prairies, steppes and savannah.

Prairies

Prairies are usually humid and covered with tall grass (see Image 2.12). If trees are present, they are found on slopes or close to small rivers. Prairie soil is rich in nutrients, and NPP here is about one-third of that in a tropical rainforest. Grazing animals such as bison and oxen feed on the prairies, but many areas are now used for agriculture.

Steppes

Steppes have lower rainfall than prairie grassland, and grass in these regions tends to be short (see Image 2.13). Grazing animals such as antelope feed on these grasses. Some regions of steppes are threatened by overgrazing and may become semi-arid desert.

Savannah

Savannahs are warmer and drier than the other types of grassland (see Image 2.14). They may experience drought at certain times of year. Plants in a savannah are adapted to a hot, dry climate, and many can store water or become dormant during dry periods.

Image 2.12 Prairie grassland in the USA.

Image 2.13 Eurasian Steppe landscape.

Desert

Deserts are defined as very dry areas where rainfall is less than 250 mm per year. Desert regions are subdivided into four categories; arid, semi-arid, coastal and cold.

Rain often falls at irregular intervals; for example, the Atacama desert in Chile, the driest place on Earth, may have no rain for periods of up to 20 years.

Most arid desert biomes have hot days and cold nights, while cold deserts, such as the Gobi desert in China, may experience frost and extremes of temperature. The average temperature in the Gobi desert is below 0 °C from November to March, and night-time temperatures of −40 °C are common, while in summer the temperature may soar to +40 °C. The world's largest hot, arid desert is the Sahara, which covers most of northern Africa, an area of $9.4 \times 10^6 \, km^2$. Daytime temperatures here are 45–50 °C, while at night the temperature falls to below 10 °C.

Lack of rainfall means that photosynthesis and NPP in desert biomes are very low despite the relatively high light levels.

Fluctuations in temperature mean that plants have to be specially adapted to survive the extreme conditions. Plants that do grow are often xerophytes such as cacti and succulents, which have adaptations that include very small leaves, thick cuticles to reduce water loss and the ability to store water in their stems. Some have deep roots which extend far below the ground to reach water, while others have spreading surface roots to absorb any rainfall before it evaporates.

Plant biomass is low, so there are few consumers in desert biomes and food chains are limited. Reptiles such as snakes and lizards (see Image 2.15) can survive by adapting their behaviour so that they are active during the cooler parts of the day. Only a few mammals, such as the jerboa, are found in hot deserts. These animals are nocturnal, only emerging from their underground burrows at night.

Tundra

Two types of tundra are identified by their location.

Arctic tundra

The Arctic tundra is found to the south of the Arctic ice cap (see Figure 2.25).

The tundra is cold with very little rain and long, dark winters when the temperature may be as low as −50 °C. The tundra soil remains permanently frozen at this time, forming a layer known as the permafrost.

Lack of sunlight, low rainfall and cold temperatures mean that the rate of photosynthesis and NPP are very low. Because of low temperatures, the recycling of nutrients is also slow, and this can lead to the formation of peat.

In spring and summer, days lengthen and plants are able to grow. For about six weeks between May and August there is constant daylight, and ice begins to melt, releasing water so that mosses, grasses and low shrubs are able to develop.

Many of the plants have adaptations such as thick leaves and underground storage organs to prevent them drying out. No trees are found in the tundra, because there is insufficient soil and the ground remains frozen just a few centimetres below the surface, even in summer. Lemmings, hares and other small mammals feed on the tundra plants, and herds of elk and bison graze on the grasses (see Image 2.16).

Image 2.14 Oryx running in savannah grassland in Awash National Park in Ethiopia.

Image 2.15 The thorny devil is a lizard that lives in the desert of central Australia. It keeps cool by raising itself off the hot ground on its long legs. The thorny devil's ridged body allows the animal to collect water from any part of its body and channel it to its mouth.

Image 2.16 Bison give birth to their young during the summer months when there is grazing available and the calves have the best chance of survival.

2.04 Biomes, zonation and succession

These animals, in turn, are prey for arctic foxes, lynxes, wolves and owls, so simple food chains are formed. All the animals of the tundra have thick fur or feathers to keep them warm and small ears to reduce heat loss. These features distinguish them from closely related species from warmer areas. Many of the tundra species hibernate during the winter.

Alpine tundra

The alpine tundra biome exists on rocky mountaintops and is very similar to the Arctic tundra. Most of the primary production in alpine tundra is small shrubs and small leafy plants such as alpine bluegrass, which provide food for a few grazing animals such as bighorn sheep and mountain goats. Other alpine tundra animals include small mammals such as the pika and marmots, and birds such as ptarmigan and grouse.

2.04.03 Comparing four biomes

Table 2.08 shows data on temperature, rainfall, light, location and NPP for four different biomes. Use this information to compare the conditions in each one with those in a different part of the world or with a different NPP.

Biomes	Temperature range	Annual rainfall	Solar radiation (approx.) / $W m^{-2} y^{-1}$	Current % of land area on Earth	Approximate location	NPP / $g m^{-2} y^{-1}$
tropical rainforest	average 27 °C constant throughout year	2000–5000 mm rainfall exceeds evaporation	180	15	5° north and south of the equator	2200
Arctic tundra	−50 °C in winter, over 20 °C in summer	<250 mm rainfall exceeds evaporation	90	Less than 10	around the margins of the Arctic ice cap	140
hot desert	10 °C at night, up to 50 °C at mid-day, little seasonal variation	<250 mm evaporation exceeds rainfall	175	20–30	30° north and south of the equator	90
temperate forest	−30 °C to +30 °C depending on location	500–1500 mm rainfall exceeds evaporation	130	10–15	40–60° north and south of the equator	1200

Table 2.08 Biomes are identified by their abiotic factors and NPP.

CONSIDER THIS

Tundra also occurs in the southern hemisphere, where the seasons are reversed. Alpine tundra is found on high mountains in many parts of the world.

Extension
What's a paramo?
www.youtube.com/watch?v=Hf6liaUdRLA

Theory of knowledge 2.04.01

Rainforests in less economically developed countries

The Congo basin rainforest is an enormous natural resource, stretching across six countries in the centre of Africa. It provides shelter, food, income and fuel for millions of local people. However, like most of the world's rainforests, it is being destroyed at an unsustainable rate. It lies in LEDCs that are in great need of the income the forest provides, so this irreplaceable biome is being destroyed for what is a small amount to conservationists in richer parts of the world. Several countries, including the UK and Norway, have launched a fund to fight deforestation and preserve the forest. The plan is to offer funds for projects that can prevent forest destruction, for example by providing alternate sources of income or of energy.

1. Is this approach an effective piece of internationalist policy or should other methods be used to preserve precious ecosystems?
2. Would the money be better spent on trying to reduce the consumption of resources by people in more economically developed countries (MEDCs)?

SELF-ASSESSMENT QUESTIONS 2.04.01

1 What are the most important abiotic factors which determine the features used to identify biomes?
2 Why is the rate of NPP in a tropical rainforest high throughout the year?
3 What effect may global warming have on: **a** the distribution of tundra, **b** the percentage of the Earth covered by desert? Outline reasons for your answers.
4 **Discussion point:** Both trophic levels and biomes are described and defined by humans. We are observers of nature, but what we observe are systems that are relative to one another. Is it ever possible to produce an absolute definition of such a system? How useful is it to try to do so?

2.04.04 The processes of succession

Succession is the long-term process by which communities in a particular area change over a period so that the appearance of the whole area evolves and changes. Succession involves interactions between both the biotic and abiotic components of the area. A succession begins as early **pioneer communities** of simple plants such as grasses and moss arrive on bare ground and begin to modify the physical environment, which, in turn, leads to changes in the biotic community. The changes enable a wider variety of species to move in as intermediate communities. They modify the physical environment still more, until a stable situation is reached.

In the early stages of succession, there is a low density of producers, so gross productivity is low. But only small amounts of energy are lost through respiration, so net productivity is high. As the system grows and succession proceeds, more consumers arrive so that food chains and feeding relationships become more complex. As this happens, gross productivity increases. However, more consumers and a complex community mean that losses through respiration increase, so net productivity becomes almost zero. Put another way, the ratio of productivity to respiration (P:R) is close to 1 as a climax community is reached.

The different stages of succession are known as **seral stages**, and the final stable community is called a **climax community**.

> **Succession** is a long-term process in which communities in an area change over time.
>
> A **pioneer community** is the first group of organisms to colonise a bare area of land.
>
> **Seral stages** are stages in a succession.
>
> The **climax community** is a stable community that is formed at the end of a succession.

Primary succession

Primary succession begins when an area of bare ground or rock is colonised for the first time. Two well-studied examples are the area on the Indonesian island of Krakatau, which was left bare after a massive volcanic eruption in 1883, and the newly formed volcanic island of Surtsey off the coast of Iceland, which formed in 1963. Not all volcanic eruptions are followed by the development of a primary succession. In some cases, the damaged area is not completely devoid of life; species can sometimes survive because they are protected by snow, mud or water.

A typical **primary succession** in the northern hemisphere might develop as follows. The first organisms to colonise bare rock are lichens and mosses, which can settle on the rock surface. Lichens gradually erode the rocks and use dissolved minerals for growth. As lichens die they decompose and leave debris which starts to form humus and soil (after about ten years); the low-growing lichens and mosses eventually modify the environment sufficiently for seeds and grasses and small shrubs to start growing. These plants modify the structure of the ground still further. A deeper layer of soil develops as plants die and decompose, and this soil can hold more moisture and contain more organic matter. Later, fast-growing trees such as rowan and birch may begin to grow and extend their roots into the soil, which they bind together and protect from erosion. Eventually these trees will be replaced by slower-growing species which form a climax community, usually after about 100–200 years.

> **Primary succession** begins when an area of bare ground or rock is colonised for the first time.

2.04 Biomes, zonation and succession

Each stage in the succession is known as a sere. Seres of particular environments tend to follow similar patterns and are named according to the environment. For example, a hydrosere develops in water and a halosere in a salt marsh.

The sequence described above is typical of one that occurs on a lithosphere (on bare rock), but succession will be different on different substrates. If the succession takes place where volcanic ash is present, grasses may be the pioneer species because their seeds can germinate among the ash particles.

A moraine is the name given to the new ground that is exposed as a glacier melts and retreats. The newly exposed land is a mixture of bare rock, boulders and debris, and moraines can be used to study the events of primary succession.

Secondary succession

Secondary succession takes place after an area of previously established land has been cleared, for example by a fire or a landslide. In this case, soil is already present so a secondary succession proceeds more rapidly than a primary succession. Annual grasses may be present after one to two years, followed by low-growing shrubs after three to five years, then scrub of small trees and shrubs, broadleaved trees and finally a climax community of mature woodland (see Table 2.09 and Figure 2.27).

A climax community contains mature trees which deprive the shrubs beneath of light. In an established climax community, the original shrub layer is likely to be replaced by shade-tolerant species, which can survive in low light levels. Many species of plants from earlier stages of the succession die and their remains build up leaf litter and humus, which are broken down by microorganisms. Animals that come to feed on the plants also release organic matter in their

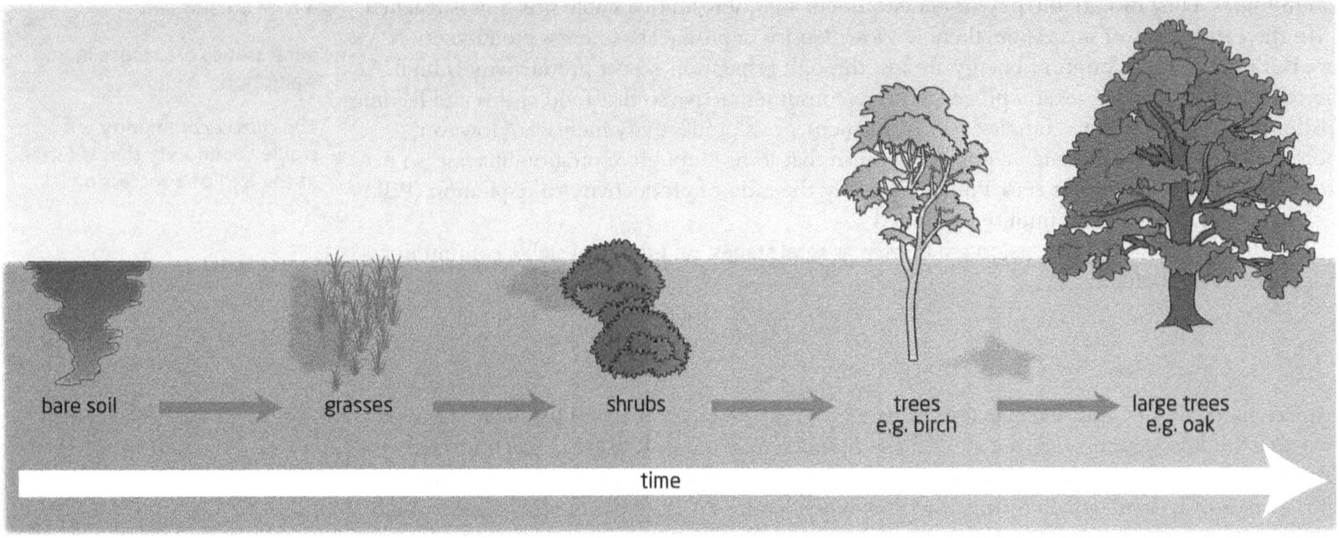

> **Secondary succession** takes place after an area of land has been cleared (e.g. by a fire or a landslide) and soil is already present.

Figure 2.27 Undisturbed bare land will gradually change from grassland to shrubland. Eventually it will become home to small, short-lived trees such as birch then to larger, slow-growing trees such as oak, which form mature woodland.

Bare earth	Pioneer grasses	Low shrubs	Scrub	Young trees	Mature woodland
Approx. years	1–2	3–5	15–30	30–150	>150

Table 2.09 Secondary succession.

faeces, which adds to the nutrients available to plants. Soil depth continues to increase as organic matter accumulates and is enriched by water and mineral retention in humus. Mineral nutrients are recycled and remain in the developing ecosystem, and are not leached away. Finally, the climax community comes to contain complex communities with a large number of plant and animal species, many niches and high species diversity.

2.04.05 Zonation

The concept of succession over time must not be confused with zonation, which refers to a spatial pattern of organisms over a particular area.

Examples of zonation patterns can be seen on rocky shores, where bands of organisms with different tolerances to environmental conditions such as immersion and desiccation can be seen at different distances from the sea (you can see an example of this in the distribution of barnacles shown in Case study 2.01.02). Different species of mangrove trees also have defined zonation patterns. In this case, elevation and salinity are key factors that determine each one can grow (see Figure 2.28).

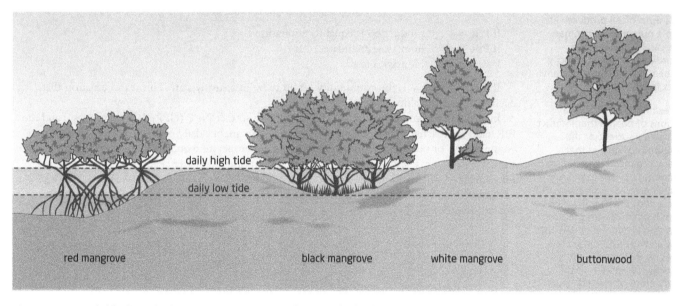

Figure 2.28 Red, black and white mangrove trees, along with the buttonwood, all grow along the same shoreline. When these species are found together, each is limited to different areas within the tidal zone. This zonation is determined by tidal changes, elevation of the land and salinity of the soil.

SELF-ASSESSMENT QUESTIONS 2.04.02

1. How does the concept of succession differ from zonation that can be observed on the seashore?
2. What are pioneer species and what is the importance of these species to succession?
3. Suggest reasons why shrubs and grasses become established before broadleaved trees as succession proceeds.
4. **Discussion point:** Surtsey has been protected from visitors since its formation in 1963. Why is this of great importance to the scientists studying the island?

2.04.06 Energy flow and productivity at different stages of a succession

As a succession develops, not only the species present in an ecosystem change. The productivity of the system also changes (see Figure 2.29).

In the first stages succession, when there are few producers present, gross productivity is low but the proportion of energy lost in respiration by these organisms is also low. This means that net productivity is high, because the ecosystem is growing and accumulating biomass.

In time, as the succession progresses, there are more consumers present and gross productivity may be high. With more consumers, there are more complex feeding interactions and food webs, so net productivity also increases. GPP and NPP stabilise as a climax community is reached.

Ecologists not only consider productivity but also the ratio of production to respiration, known as the P:R ratio. Successions may not lead to a maximum efficiency in the conversion of energy (GPP) but they do lead to a maximum accumulation of biomass. An upper limit of biomass is reached when respiratory losses (R) from the system almost equal GPP, that is when P:R is almost equal to 1. These trends are shown in Table 2.10.

If P:R = 1   production is equal to respiration
If P:R > 1   biomass accumulates
If P:R < 1   biomass is lost

If the P:R ratio = 1, the community is said to be in a steady state. This is the situation that occurs in a mature climax community.

In a climax community, GPP and R may be high so that NPP (GPP−R) approaches 0 and the P:R ratio approaches 1. This balance may be an instantaneous daily one (e.g. in tropical rainforest communities) or occur over a longer period (e.g. in temperate woodlands). But, as we have seen, if P:R is persistently greater or less than 1, then organic matter accumulates or is depleted, respectively.

Greater habitat diversity also leads to an increase in the diversity of species present and their genetic diversity.

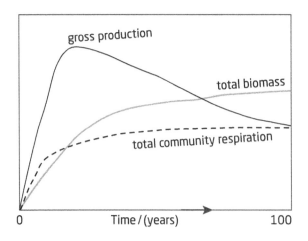

Figure 2.29 Graph to show changes in GPP, biomass and respiration during a typical succession.

Theory of knowledge 2.04.02

Complexity theory and climax communities

Complexity theory suggests that complex systems and organisms often produce very simple outcomes, outcomes that can be explained in accordance with the capacity of their individual constituent parts for 'self-organisation'.

For example, consider the millions of people that interact to create a society, or the thousands of species that interact to make up a climax community. If you examine a single person or a single plant, you cannot tell that it would be able to operate with others to create the community in which they all exist. A community takes on a life and identity of its own by means of the interactions of the individuals that live within it. Without the fact that individuals exist within a community, there could not be a community identity. We can't predict what a complex system will evolve into, and this unpredictability underlines the complexity of life.

1. To what extent do you think a climax community provides an example to support the complexity theory described above?
2. Is it ever possible to predict the effect of changes to the climate or other abiotic factors on a climax community?

Early stages of succession	Late stages of succession
GPP low	GPP high
NPP high	NPP decreasing
P:R ratio >1	P:R ratio approaches 1
food chains, simple and linear	large food webs established and decomposer food webs present
simple community structure	complex communities with many microhabitats
many *r*-strategists present	*K*-strategists predominate

Table 2.10 Comparison of early and late stages of succession.

K-strategists are organisms that produce few offspring but care for them for a long time.

r-strategists are organisms that produce large numbers of offspring but do not spend time or energy caring for them.

2.04.07 Survivorship strategies

Organisms can be divided into two groups: **K-strategists** and **r-strategists**. These groups are defined by the amount of energy and time they invest in rearing their offspring. These divisions are identified by ecologists so not all species fit perfectly into the two categories. Nevertheless, the divisions allow useful comparisons to be made.

K- and *r*-strategies

K-strategists are usually found in areas where a climax community is present. They are those organisms that tend to have few offspring but invest a large amount of time in caring for them so that most of them survive. Examples include the great apes (orang-utans and gorillas; see Image 2.17) and elephants. These species take a long time to mature and may reproduce several times during their adult life. The population size of a *K*-strategist is usually close to the carrying capacity of the environment and, in a stable ecosystem, organisms that are *K*-strategists are the predominant species.

Species that follow an *r*-strategy are ideally suited to being members of a pioneer community. They have a relatively short lifespan during which they reproduce once and produce large numbers of offspring. They are unlikely to care for their offspring, which mature quickly and are usually small. Examples include insects such as blowflies, fish and frogs, and plant species such as dandelions (see Image 2.18), which produce hundreds of offspring. Adults cannot invest time and energy in caring for such large numbers of young, so only small numbers survive to produce the next generation. Because *r*-strategists reproduce quickly, they are able to colonise new environments easily and make use of short-lived resources. They are most likely to be pioneer species in a succession. The weeds that appear on a newly cleared patch of land are a good example of this. *r*-strategists are most common in unstable ecosystems. The features of *K*- and *r*-strategists are compared in Table 2.11.

K-strategies and r-strategies are the extreme ends of a spectrum of patterns of reproduction, and many organisms fall between the extremes. Many birds and reptiles produce and care for more eggs than would be expected for a strict *K*-strategist, but this allows for the loss of young through predation, starvation and disease.

Image 2.17 Female gorillas give birth when they are about 10 years old. A baby depends on its mother for up to six years.

Image 2.18 Dandelion seeds are produced in huge numbers and carried on the wind.

2.04 Biomes, zonation and succession

K-strategists	r-strategists
tend to be dominant species	first colonisers of an area
include a varied range of species	include large numbers of a few species
development of individuals is slow	grow rapidly and develop quickly
reproduction occurs over a long time period; few young are produced but most survive	reproduce early; many young are produced but they are short-lived
individuals are large and tend to live a long time	individuals are small and have short lifespans
low rate of productivity	high rate of productivity

Table 2.11 The biological features of K- and r-strategists.

Survivorship curves

The percentage of individuals that die before they reach reproductive age is one of the main factors to affect population size; for any given population, it is far more variable than fertility. To maintain a constant population size, an average of two offspring from each pair must survive to reproduce. Different species have different-shaped survivorship curves (see Figure 2.30) based in part on their ability to reach reproductive age. Most plants and animals begin to age once they have passed maturity. Their strength declines and they die as they reach their natural life expectancy.

A survivorship curve traces the survival of a group of individuals of a species. Figure 2.30 shows three theoretical curves.

- Line 1 shows a curve for K-strategists. When members of the species are young, mortality is low because offspring are cared for by their parents. Most individuals reach reproductive age and achieve their expected lifespan. Examples include humans, big cats and eagles.
- Line 2 shows an example of a species that has an equal likelihood of dying at any age. Death could be due to factors such as hunting or diseases or simply due to chance. Examples of species with this type of survivorship curve are mice, coral and many reptiles.
- Line 3 shows a curve for an r-strategist. Most members of the species die at a young age but those that survive are likely to live for their expected lifespan. Examples include most invertebrates and plants, as well as frogs and organisms with free-swimming juvenile stages such as barnacles.

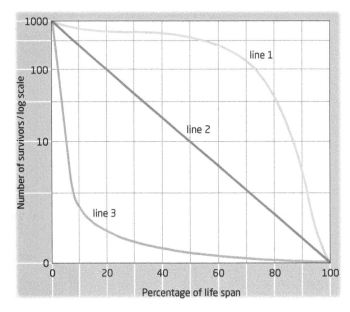

Figure 2.30 Survivorship curves for a K-strategist, an r-strategist and an intermediate organism. The scale on the vertical axis is a logarithmic scale that increases in powers of 10. This enables large numbers of individuals to be shown on the same axes as small ones.

SELF-ASSESSMENT QUESTIONS 2.04.03

1. Compare *K*- and *r*-strategies. Why are *K*-strategists likely to survive better and be more abundant in a climax community than in the early stages of a succession?
2. Why is GPP low at the start of a succession?
3. **Research idea:** If the P:R ratio is less than 1, biomass is lost from a system. Suggest a reason for this.

CASE STUDY 2.04.01

Succession on Surtsey

Surtsey is a volcanic island and a United Nations Educational, Scientific and Cultural Organization (UNESCO) World Heritage Site, approximately 32 km from the south coast of Iceland. It is a new island formed by volcanic eruptions that took place between 1963 and 1967. Surtsey has provided unique, long-term information on the colonisation of new land by plants and animals and is one of only a few sites in the world where studies of primary succession have been undertaken. The island has been protected since its formation, and only certain visitors are permitted on the island.

Scientists have studied Surtsey since 1964 and recorded the appearance of different species over the years. They have noted how these different species have arrived and where they have come from. Among the first colonisers of Surtsey's coastal environment were diatoms (tiny, mainly unicellular, algae), which floated onto a sandy beach in August 1964. Other marine organisms such as the floating spores of algae and swimming larvae of marine invertebrates soon arrived, after being carried long distances by ocean currents. Among the first terrestrial species to arrive were mosses, liverworts, lichens, fungi and algae. These species all reproduce by tiny spores that are spun up into the atmosphere by air currents and were carried to the island on the wind.

In the first spring after the formation of Surtsey, seeds and other parts of terrestrial plants were found washed up on the newly formed shore. Scientists collected samples and germinated the plants in the laboratory to confirm that seeds carried by ocean currents could be likely first pioneers. In 1965, the first vascular plant, sea rocket (*Cakile arctica*), was found growing on the shore; other coastal species, the sea sandwort (*Honckenya peploides*), sea lyme grass (*Leymus arenarius*) and oyster plant (*Mertensia maritima*), soon followed. The seeds of these species germinated in the uppermost part of the shore, where sand had been washed up. Researchers noted that all of these pioneer species have fairly large seeds that are able to float and tolerate salty seawater.

Some seeds and fruits from flowering plants are dispersed over long distances by the wind, but not all seeds that land are able to grow. Just three species of willow and the common dandelion have become established since the island formed. Other seeds have been carried by birds that feed on berries and fruits many kilometres away and expel the hard indigestible seeds with their waste. Seeds may also arrive on feathers or on twigs and plants that birds use for nest building.

The pattern of succession on Surtsey began with bacteria and fungi that were found growing in 1964, followed by lichens, mosses and liverworts and later sea rocket, which was seen in 1965. After ten years, when some humus and soil had formed, there were ten plant species growing (see Image 2.19), and by 2004 this had risen to 60. This was in addition to 75 mosses and liverworts, 71 lichens and 24 fungi. To date, 89 species of bird have been recorded on the island and 12 species are regularly found there. The island is also home to 335 species of invertebrate.

Image 2.19 Seeds of the oyster plant probably arrived in Surtsey from Iceland. It was one of the earliest plants to survive on the island.

Biomes, zonation and succession

CASE STUDY 2.04.01 (continued)

The succession began with few species and low gross productivity, but, as the consumer community has increased, the variety of nutrient and energy pathways has contributed to the stability of the developing ecosystem on the island. The number of species and the complexity of interactions between them will continue to increase as the succession proceeds.

Questions

1 Table 2.12 shows the number of moss species present on Surtsey over a period of 40 years.

Year	1967	1969	1971	1972	1980	2004
No. of moss species	1	8	40	63	70	75

Table 2.12 The number of moss species on Surtsey over 40 years.

Plot the data in a graph and account for the shape of the curve.

2 For each of these groups, suggest two ways in which they might have reached the island:
 a spiders
 b mosses
 c vascular (higher) plants.

3 Suggest three ways in which seabirds might affect the composition of plant life on the island.

4 Discussion point: Use the information above and your knowledge of succession to comment on and predict the type and size of the species that have colonised Surtsey in the time from 1964 to the present day.

2.04.08 Factors affecting climax communities

Stable ecosystems are regulated by feedback mechanisms which enable the system to rebalance itself if changes occur (see Topic 1). Negative feedback is very important in returning a system to its steady-state position when deviations occur. As a climax community is reached, such as the temperate forest shown in Image 2.20, the complexity of food webs and interactions between the diverse species present ensure that consumers can use alternative food sources in times of shortage, and also that plants have a good reservoir of nutrients and organic matter and are not dependent on inputs from outside the system.

The exact nature of a climax community is determined by climatic and **edaphic factors**, the physical, chemical and biological properties of soil, such as water content, organic content, texture and pH.

In most cases, a climax community that has reached a steady state can be distinguished from earlier (seral) stages by its higher biomass and species diversity, richer soil structure and organic content. It will also contain larger, longer-lived species and a greater proportion of K-strategists, as well as greater complexity and habitat diversity.

Human interference in climax communities

People often interrupt or deflect the course of a natural succession by their activities. We clear land for agriculture, mow away grass or use land for grazing animals, as shown in Image 2.21. Natural events such as fires or landslides can cause similar interruptions in the progress of a succession. The climax community which results after this interference will be different from the naturally occurring one, and we say that a **deflected succession** has taken place and given rise to a **plagioclimax**.

> **Edaphic factors** are physical, chemical and biological properties of soil, such as water content, organic content, texture and pH.
>
> **Deflected succession** is interruption of a succession so that it does not become a climax community.
>
> A **plagioclimax** is the community that is produced when a succession is deflected.

Factors affecting climax communities

Image 2.20 A temperate forest or woodland is a climax community but a very different one from tropical rainforest because of the different climate and soil structure and the different species they contain.

Image 2.21 Long grass is cut for hay and interrupts the natural succession in the meadow. The grassland is known as a plagioclimax.

Clearing rainforest

Burning and cutting down areas of rainforest (see Image 2.22) has disturbed these ecosystems in South America and Asia. The cleared land is planted with crops or used for grazing, which prevents the re-establishment of a climax community. Even when grazing and clearance have ended, the area may be unable to revert to a climax community because nutrients have been lost and topsoil eroded. Despite the fact that rainforest still covers about 6 per cent of the Earth's land surface, it is estimated that an area equivalent to that of a small country is lost every year as a result of clearance.

Changing the face of the prairies

The temperate grassland on the American prairies is a plagioclimax that has replaced the climax community that once existed in the interior of the USA to the south of the coniferous belt. Over many years, fire and human use have left buffalo grass and blue grama in the place of trees. These tufted grasses serve to limit tree growth with their tight clumps and deep roots, so a succession does not continue (see Image 2.23). Grass roots and rhizomes store nutrients and reduce erosion, and grassland is maintained by many grazing species including sheep and cattle. In different parts of the world, different grazing animals, including llamas, horses and goats, maintain similar grassland plagioclimax systems.

Moorland

Today, large areas of the uplands in northern parts of the UK that were once forested are treeless and covered by a plagioclimax of heather moorland.

Heather probably began to dominate the area about 7000 years ago when people first cut down trees. Evidence suggests that this coincided with a wetter period in the climate so that peat bogs of moss and rushes became established in poorly drained areas, with heather and grasses dominating drier places. Over hundreds of years, woodland clearance for fuel, construction and farming continued. Heather survives in the poorer, acidic soils which are left and now dominates the moorlands.

Grazing animals eat small plants and prevent the regrowth of young trees (see Image 2.24), and people regularly burn heather to

Image 2.22 Loss of rainforest not only affects the climax community but, because of the huge areas that have been destroyed, it also has an effect on weather patterns and biodiversity worldwide.

2.04 Biomes, zonation and succession

Image 2.23 Species of tufted grasses on the American prairies prevent the growth of trees, with their tight roots, and lead to a plagioclimax.

Image 2.24 Grazing animals feed on young shrub and tree shoots and prevent the re-establishment of woodland in this plagioclimax of heather moorland.

maintain the moors. Controlled burning is carried out in different areas in rotation so that each one is burnt at intervals of about 10–20 years. Potash left from the fires enriches the soil, and healthy new heather plants become re-established, so the moorland is made up of a patchwork of mature and developing vegetation.

SELF-ASSESSMENT QUESTIONS 2.04.04

1. What type of community would you expect to have a productivity : respiration (P : R) ratio close to 1?
2. Why does a developing succession lead to an increase in biodiversity?
3. Why are grazing animals important in maintaining grassland areas?
4. Give three examples of deflected successions that have become established plagioclimax (interrupted climax) communities as a result of human activities.
5. **Discussion point:** Discuss the different factors that cause humans to interfere in an ecosystem and lead to different climax communities.

Investigating ecosystems

2.05

LEARNING OBJECTIVES

After reading this chapter you should be able to:

- describe appropriate methods of identifying organisms in an ecosystem, including keys, reference collections and technology
- outline a range of techniques used to sample and measure biotic and abiotic factors along an environmental gradient
- describe methods of estimating biomass
- calculate abundance of organisms using the Lincoln index and the diversity of organisms using the Simpson index
- explain the difference between species richness and species diversity in an ecosystem.

KEY QUESTIONS

How does investigation of ecosystems allow us to compare and monitor them over time?

How can natural change and human impact on an ecosystem be measured?

Can quantifying the components of an ecosystem help our understanding?

2.05.01 Identifying organisms

We can come to understand and compare ecosystems more easily by investigating and quantifying the biotic and abiotic components that make up the system. We can monitor and model different systems and measure both natural changes and the impact of human activity. Before any system is studied, it must be identified by its name, location and type; for example, an open, temperate grassland. As you carry out your own practical work, make sure that you include details of the area, country and type of ecosystem you are studying in your report. You will select sampling techniques that are appropriate for your study; some of these are described below.

Using keys to identify species

In ecological studies, the different species must be identified so that results are accurate. A **key** is used for reference so that identifications can be made quickly and easily without expert knowledge of each organism. Keys are drawn so that identification is made in a series of steps, each involving just one decision (see Figure 2.31). At each step, another choice is given, so keys are usually known as 'dichotomous', which means 'divided into two'. Certain features are not used when drawing up a key. Size may be related to the age of an organism that grows, so although proportions may be used in a key, simple measures of size are not.

A key is useful for species that are known and have been identified previously. But if a new species is discovered, it may be useful to compare it with other specimens of similar organisms in a herbarium or museum. In reference collections held in museums and herbaria, accurately identified 'type' specimens are catalogued so that scientists can compare any specimen they discover with those that are already known. They can also consult with expert taxonomists who know the genera well. The most up-to-date techniques for identifying and classifying organisms involve using samples of DNA or protein from an organism and comparing the sample with

A **key** is a chart that is used to identify organisms and deduce their correct species.

2.05 Investigating ecosystems

others from known similar species. DNA and proteins are molecules that are specific to an organism and can act a 'fingerprint' for it. Closely related species will have similar DNA and protein profiles. These methods of identification require specialised equipment and can only be carried out in laboratories with the correct facilities.

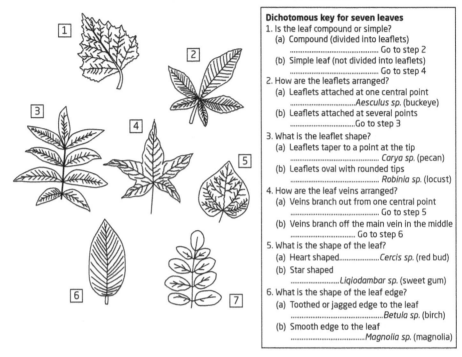

Figure 2.31 The key shown here enables ecologists to identify tree species from the shapes of their leaves. Choose one leaf and follow the steps given in the key to identify its species.

SELF-ASSESSMENT QUESTIONS 2.05.01

1   Construct your own key to separate the insects in Figure 2.32 into their different species.

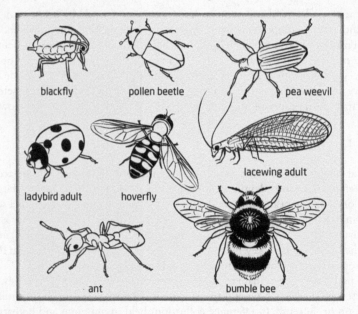

Figure 2.32 Insects.

2   Suggest which parts of a plant are most useful in making keys to identify plants. Explain your answer.
3   **Discussion point:** More than 2000 years ago, the first key made by Aristotle divided living things into only two groups: plants and animals. This classification is not used today. Discuss the reasons why keys and classifications change.

2.05.02 Measuring the biotic components of an ecosystem

Estimating the abundance of organisms

Biotic components of an ecosystem are the living organisms it contains. Ecologists study both the abiotic (non-living) and biotic factors of an ecosystem to gain an understanding of how the system works. The abundance and biomass of living things can be estimated to give a measure of the distribution and energy content of organisms. It is also important that each organism is carefully identified so that the data collected and the conclusions drawn are accurate. Ecologists may require information on the population density, percentage frequency or percentage cover of different species.

In most ecosystems, populations of organisms are too large to be counted directly, so they have to be sampled and identified to obtain information about their abundance. Different techniques are used depending on the organisms being studied and the information that is needed.

Organisms that do not move can sometimes be counted directly if there are only a few of them, but usually a percentage cover measurement is used. This can be obtained using **quadrats** to sample the area and estimate the abundance of plants and slow-moving animals.

To obtain information on mobile organisms such as insects or small mammals that live in a particular area, the capture–mark–release–recapture technique is used. Estimates of populations are calculated from the data that is collected, using the Lincoln index.

A **quadrat** is a square (or circular) frame used to sample organisms in an area.

Using the Lincoln index

The Lincoln index provides an estimate of the numbers of a mobile species living in a chosen area. It is suitable for organisms such as beetles, snails and mice, which can be trapped easily without harming them. Samples of the population are collected using methods such as the Longworth trap (see Image 2.25) for small mammals or pitfall traps for crawling insects (see Figure 2.33).

The stages in estimating a population using the Lincoln index are as follows:

1   As many organisms as possible are captured using traps or other suitable methods. The number captured is recorded (n_1).
2   Each animal is marked in a way that makes it easy to identify if it is captured again. Marking must not harm the animal or make it more vulnerable to predators. For example, a mark of non-toxic paint may be used on the underside of a snail's shell, or a small patch of fur may be clipped from a small mammal.
3   Marked animals are returned to their habitat and allowed to mix freely with the other members of the population.
4   After a suitable time, the population is sampled again and the number of animals captured in the second sample (n_2) and the number of marked animals (n_m) in this sample are counted and recorded. The length time between the two sampling sessions depends on the animals. Twenty-four hours is suitable for small animals such as woodlice, but a few days or a week is more suitable for fish in a lake.
5   The Lincoln index formula is used to estimate the size of the total population.

2.05 Investigating ecosystems

Image 2.25 The Longworth trap is baited with food. Small mammals enter and the trap door closes, keeping them inside. Traps must be visited regularly so that animals can be released quickly.

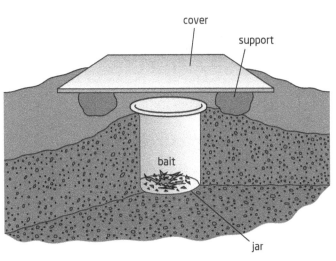

Figure 2.33 Pitfall traps are buried in the ground and covered so that animals that fall in are protected from rain. Bait is used to provide food for the captured species.

$$\text{total population} = \frac{\text{number in sample 1} \times \text{total number in sample 2}}{\text{number of marked animals in sample 2}}$$

The Lincoln index formula is often shown as follows:

$$N = \frac{n_1 \times n_2}{n_m}$$

Where N = total population, n_1 = number of animals marked in sample 1, n_2 = number of animals captured in sample 2, n_m = number of marked animals recaptured.

Calculating a population using the Lincoln index

A population of one species of fish in a lake was sampled. In the first sample, 110 fish were caught and marked. The following week, 175 fish were caught in a second sample; 53 of these were already marked.

Using the formula:
Size of first sample, $n_1 = 110$
Size of second sample, $n_2 = 175$
Number marked in second sample, $n_m = 53$

$$N = \frac{(110 \times 175)}{53}$$

$N = 363$. The estimate of the total populations (N) of fish in the lake is 363.

Quadrats

A quadrat is a sampling device, usually a square of fixed size, which may be $0.25\,m^2$ ($0.5\,m \times 0.5\,m$) or $1\,m^2$, depending on the area being sampled. Quadrat sampling is a method in which a proportion of organisms in an ecosystem are counted directly. All the individuals in a fixed number of quadrats are counted and the data used to calculate the abundance or percentage cover for the whole area (see Figure 2.34).

Quadrats may be placed randomly or in a set pattern in the area that is being sampled. Plants or slow-moving animals that are inside the quadrat can easily be counted and recorded. In most cases a $0.25\,m^2$ quadrat is suitable for sampling low-growing plants such as grasses, herbs or seaweed, while larger quadrats are used for woodlands where there are trees and shrubs. The number of quadrats to sample is determined by the type of ecosystem. If there are many different species present, a large number should be used, but in a homogeneous ecosystem with few species, fewer quadrats can be used to ensure all species are sampled.

The stages of a quadrat sampling study for an area that is similar throughout are as follows:

1. Divide the area to be sampled into a grid, marked out by tapes which are at 90° to one another.
2. Use random number tables to select the sample areas in the grid. If the area has varied vegetation with many different species present, more quadrats should be used.
3. Choose a quadrat of a suitable size for the type of ecosystem.
4. Use the coordinates from the random number tables to position the quadrats and count and record the organisms in the quadrats at these points.

If an estimate of a population in relation to an abiotic factor such as soil moisture is required, the quadrat method can be adapted with the use of a **transect** (see Image 2.26). A transect is a tape or rope which is stretched across a sample area. At suitable intervals, populations can be sampled and measurements made of abiotic factors along the transect line to assess the changing distribution of a plant or animal species.

Measures of the abundance of species

Data from quadrats can provide different measures of species abundance:
- population density
- percentage cover
- percentage frequency.

Population density is defined as the number of a species per unit area. To calculate the population density, the data totals for all the sampled quadrats are used to estimate an average density for each species:

$$\text{density} = \frac{\text{total number in all quadrats}}{\text{number of quadrats} \times \text{area of one quadrat}}$$

Percentage cover is the best method to estimate the abundance of a plant species. It is defined as the percentage of the total area within a quadrat which is covered by the species of interest. Often quadrats are divided into 100 small squares to make it easy to work out percentage cover.

Percentage cover can be given as a percentage (e.g. 5 per cent or 25 per cent) or expressed on a scale of values such as the ACFOR scale:

- A = abundant
- C = common
- F = frequent
- O = occasional
- R = rare.

Scales like this are quick and easy to use but are subjective and depend on the observations of individuals.

Percentage frequency is the percentage of the total number of quadrats sampled that a particular species is found in. So if four quadrats out of ten in a rocky shore sample contain limpets, we say that limpets occur with a frequency of 40 per cent. Since this result depends on the size of the quadrat, it is best to include a note of the quadrat size with the data. So we could say, for example, that limpets occur with a frequency of 40 per cent in a sample of $20 \times 0.25\,\text{m}^2$ quadrats.

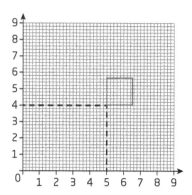

Figure 2.34 To select a part of an area to sample, divide the area into a grid of squares and use random number tables to select column and row numbers to determine where to place the quadrats. This diagram shows a quadrat placed at random coordinates 5,4.

A **transect** is a line or rope stretched across a habitat so that organisms can be sampled along its length.

Population density is the number of a species per unit area.

Image 2.26 These students have placed quadrats at measured intervals along the transect line to record the plants at a series of locations. Differences in species along the length of the transect can be related to abiotic factors such as slope, nutrients present or water content.

107

2.05 Investigating ecosystems

Percentage cover is the percentage of the area within a quadrat that is covered by the species of interest.

Percentage frequency is the percentage of the total number of quadrats sampled that contain a particular species.

SELF-ASSESSMENT QUESTIONS 2.05.02

1. In a study which sampled woodlice in an area of woodland, the following data were collected. In the first sample, 85 woodlice were collected and marked. These animals were released, and 24 hours later 99 woodlice were captured of which 21 were already marked. Estimate the size of the woodlice population in the area using the Lincoln index.
2. In a study to estimate a mouse population, a student found that the same animals were returning to the traps day after day because they had learnt that they would find food there. What effect would this have on the student's estimate of the number of mice in the population?
3. Why would it be inappropriate to mark the top of a snail's shell in a capture–mark–release–recapture study? What effect might doing so have on the estimate of population size?
4. Outline how you would use a quadrat sampling method to compare the populations of daisy plants in two different fields. What factor should you consider when deciding how many quadrats are needed?
5. What size quadrat would you use to investigate the distribution of lichen on a tree trunk: $0.5\,m^2$, $0.25\,m^2$ or $0.01\,m^2$? Suggest a reason for your answer.
6. **Research idea:** Find out how sampling techniques are used to estimate the size of very large mobile populations of animals such as wildebeest.

Theory of knowledge 2.05.01

The importance of precision

Ecology studies living organisms in dynamic systems. It tackles the big questions in environmental science. Gathering of precise, consistent data is very difficult. Biological research restricted to a laboratory can be used to determine cause-and-effect relationships, but usually disregards the broader and more complex environmental implications. Environmental measurements are constantly changing, so they cannot be accurate and reliable in the same way that laboratory studies are.

1. Does the lack of precise measurements in environmental systems make the data collected less valuable?
2. Should we doubt the validity of the knowledge obtained from sampling techniques that may not be precise?

2.05.03 Measuring the abiotic components of an ecosystem

As we have seen in Chapter 2.01, the most important abiotic components of different ecosystems are:

- Marine systems – salinity, temperature, pH, dissolved oxygen and wave action
- Freshwater systems – turbidity, flow velocity, pH, temperature and dissolved oxygen
- Terrestrial systems – temperature, light intensity, wind speed, particle size, slope, moisture content, drainage and mineral content.

Abiotic components are measured directly or estimated using equipment that can be taken out into an ecosystem. In some cases, samples of water or soil may be taken back to a laboratory for analysis.

Salinity

Salinity is a measure of the salt content of water. Seawater has a salinity of around 35 parts per thousand (‰) of water. Brackish water found in estuaries and areas such as the Baltic Sea has a salinity of less than 10 parts per thousand. Salinity is measured either from electrical conductivity readings taken with a data logger, or from density measurements, since increased salinity increases the density of water. A number of individual samples of water can be collected from different locations in bottles and returned to the laboratory for readings to be taken.

pH

pH is a measure of H^+ ions dissolved in water. The pH of seawater is usually greater than pH 7 (basic) and is measured using a pH meter or pH probe and data logger. For quick approximate readings, pH paper can also be used. pH is important because it affects the solubility of the inorganic substances that living things require.

Temperature

Temperature is a very important abiotic factor. Temperature determines the amount of oxygen that will dissolve in water and thus become available to aquatic organisms. Since many of these organisms are ectothermic (having a body temperature approximately equivalent to that of the water), temperature also affects their metabolic rate. Surface water tends to be warmer than lower layers, and this can affect water currents and the mixing of water and nutrients. In a terrestrial system, temperature can be measured using a simple thermometer. The type of thermometer used may depend on the information that is needed. A maximum and minimum thermometer gives readings of the range of temperature, and a temperature probe and data logger can be used to provide temperature readings over a long period.

Dissolved oxygen

Dissolved oxygen is needed by all aquatic organisms for respiration. The amount of dissolved oxygen in water depends mainly on water temperature but also on wave action (see below). Warmer water contains less dissolved oxygen than cold water. At 20 °C freshwater usually contains about $9 g m^{-3}$ of dissolved oxygen. A minimum of $5 g m^{-3}$ is needed to support a balanced aquatic community. Dissolved oxygen is measured using oxygen electrodes and a data logger, or by titration. Titration is usually used to check the calibration of oxygen electrodes.

Wave action

Wave action is measured using a dynamometer which assesses the force in waves. Wave action is important because it increases the amount of dissolved oxygen by mixing air with moving water. Coastal areas and coral reefs have high levels of dissolved oxygen, and many organisms live in their surface waters.

Turbidity

Turbidity is a measure of the cloudiness of water. Cloudy water has a high turbidity and clear water low turbidity. A Secchi disc (see Image 2.27) is used to measure turbidity. The disc is lowered into the water on a graduated pole or line until it disappears from view. A depth reading is taken from the pole and the measurement repeated for accuracy. To ensure results are reliable, the readings should always be taken in the same light conditions and with the measurer in the same position – that is, either seated or standing. Turbidity affects the penetration of sunlight through water and can influence the rate of photosynthesis of aquatic plants.

Flow velocity

Flow velocity of surface water can be measured very simply by recording the time for any floating object (such as an orange or a dog biscuit) to pass a fixed distance between two marked points. Flow velocity varies at different depths in a stream or river, and more accurate measurements are made using a flowmeter, which is a calibrated propeller attached to a rod (see Figure 2.35). Water velocity determines which organisms can survive in flowing water. If flow velocity is high, plants and animals must be firmly anchored to maintain their positions in the fast-flowing water.

Image 2.27 A Secchi disc in use.

2.05 Investigating ecosystems

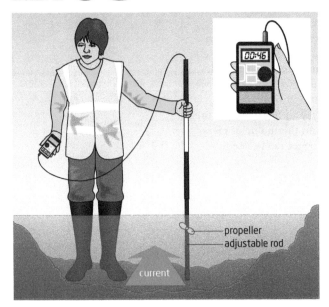

Figure 2.35 An adjustable flowmeter can be placed at any depth in a stream to measure water velocity.

Image 2.28 This is an anemometer. The cups catch the wind, making the instrument spin; the stronger the wind, the faster the spin speed.

Light intensity

Light intensity can be measured with a light meter. Light intensity varies throughout the day and also depends on cloud cover and season, so a series of readings may be needed to gain a full understanding of the ecosystem.

Soil moisture content

Soil moisture content is measured by weighing soil samples before and after gentle drying. Soil contains both moisture and organic matter, so soil samples are heated gently to remove water – but not to very high temperatures, which would burn away organic material. Samples are weighed at intervals during the drying process until a constant dry weight is reached and water content can be calculated by subtraction.

Wind speed

Wind speed is measured with a digital anemometer (see Image 2.28). Cups on the device revolve in the wind, and rotations per unit time are displayed. If only approximate wind speed is needed, the effect of the wind on objects such as trees can be observed and compared with descriptions in the Beaufort scale (see Table 2.13).

Soil particle size

Soil particle size is important in determining how much water soil can hold and how quickly the soil will drain. Sometimes, very large particles can be measured individually, but usually soil is passed through a series of graduated sieves with different mesh sizes. Very small particles of clay and silt can be measured using sedimentation, since large particles sink faster in water than smaller ones.

Soil mineral content

Soil mineral content is the ratio of mineral to organic material present. It is important in determining the soil's ability to hold water and also its fertility. Mineral content is measured by the loss-on-ignition (LOI) method. Weighed soil samples must be heated to very high temperatures for several hours in an oven so that the organic content of the soil is burnt off. The loss in mass is calculated once the sample has reached constant mass and there is no further change. The percentage weight lost gives a crude measure of the organic content of the soil.

Slope

The slope of the land in an ecosystem influences the water runoff and also whether erosion is likely to be a problem. It can be estimated using field-levelling poles or calculated using a clinometer (see Figure 2.36).

Wind description	Mean wind speed / km per hour	Force	Environmental effect
calm	0	0	smoke rises vertically
light air	4 or less	1	smoke drifts slightly
light breeze	5–9	2	leaves rustle
gentle breeze	10–19	3	leaves in constant motion
moderate breeze	20–29	4	small branches stir
fresh breeze	30–39	5	small trees sway
strong breeze	40–50	6	large branches move
near gale	51–62	7	whole trees in motion
gale	63–75	8	twigs broken off trees
severe gale	76–87	9	slight structural damage occurs
storm	88–102	10	severe structural damage occurs
violent storm	103–117	11	widespread damage
hurricane	118 or more	12	devastation

Table 2.13 The Beaufort scale.

2.05.04 Estimating the biomass and energy of trophic levels

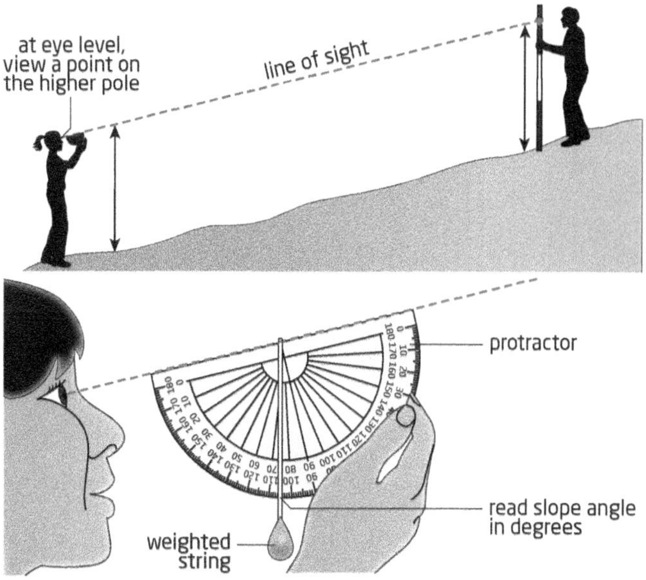

Biomass is biological material that can be used as an energy source. It is estimated from the dry mass of organisms, because water does not contain energy and water content varies considerably in different organisms.

Pyramids of biomass are diagrams that are drawn to show the distribution of biomass at each trophic level in an ecosystem (see Chapter 2.03). To estimate the total biomass at each link in a food chain, the mass of a few sampled organisms is obtained and the result multiplied to represent the total biomass.

Measuring biomass

Biomass measurements are obtained by sampling organisms, drying them until all the water has been removed and then weighing the samples on an accurate electronic balance. Representative individuals from each trophic level are collected using random sampling techniques and the number of organisms at each level is recorded. This means that biomass measurements are very time-consuming and laborious to collect. The sample organisms are dried in an oven, usually for a period of 24 hours, carefully weighed, and then dried for a further period until a constant mass is obtained. For the biomass of plant material, all parts of the plant, such as leaves, stems and wood, must be sampled and dried. Biomass estimates for animals are made in a similar way from the sampled organisms.

Direct measurements of biomass involve the destruction of wildlife and potential harm to the ecosystem. For this reason, tables of biomass drawn up from previous studies are often used in

Figure 2.36 Poles and a protractor make a simple clinometer to measure the slope of an incline. The slope angle is read off the protractor using a weighted string.

Investigating ecosystems

preference to direct measurements. These tables show the biomass for particular species from its dimensions, so that a sample organism's biomass can be recorded without harming it. Tables of biomass minimise damage to the ecosystem, and any animals that are sampled can be returned to their habitat after measurement.

Units of biomass are mass per unit area, for example kilograms per square metre ($kg\,m^{-2}$). Dry-weight measurements for the sampled material can be used to calculate total biomass for a whole trophic level and compare it with that at any other trophic levels.

Biomass can be estimated in this way, but precise measurements are not usually possible because of errors in sampling procedure and in the accuracy of weighing and measuring organisms. This is a particular problem for very small organisms such as algae and small invertebrates. Despite the difficulties involved in the procedure, biomass provides a good way of studying the energy distribution in an ecosystem.

Measuring energy content

The energy content of organic material can be measured directly using calorimetry. In this technique, samples of plant or animal material are taken and dried to constant mass to remove water (which has no energy value). A known mass of each sample is then burnt in oxygen in a device called a calorimeter (see Figure 2.37). The energy contained in the sample is released as heat, which raises the temperature of water in the calorimeter. In this way, a value of the energy that is released can be obtained from the change in temperature of a known volume of water. A calorimeter is well insulated so that heat losses are minimised. The first law of thermodynamics states that energy cannot be created or destroyed, so during the burning process it is assumed that energy is transferred from the sample to the water.

The units of energy measured are joules or kilojoules. A familiar unit used by nutritionists is the calorie. A calorie is defined as the energy needed to raise the temperature of 1 g of water by 1 °C, but the SI unit of energy is the joule or kilojoule. One calorie is equivalent to 4.2 joules.

This calorimeter method is most often used to estimate the amount of energy in foods and provide nutritional data for the labelling of food packaging (shown in kilocalories and kilojoules). In ecological investigations, it involves the destruction of living organisms so that measurements can be made, so only tiny samples are used and the results for a whole organism are extrapolated from these by multiplication. In most cases researchers will use tables of data that has already been collected for the organisms being studied rather than destroying the organisms themselves. Energy measurements for the sampled material can be used to calculate total energy for a whole area or trophic level and compare it with that of other areas or trophic levels.

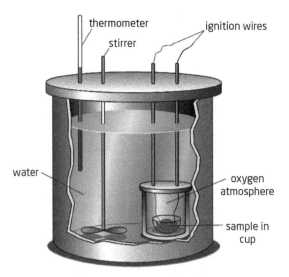

Figure 2.37 A calorimeter contains a sample of dried organic material which is burnt in oxygen. The energy released raises the temperature of the water.

SELF-ASSESSMENT QUESTIONS 2.05.03

1. **a** Draw a pyramid of biomass using the following data from an area of grassland:

   dry mass of plants = 475 g m⁻²
   herbivores = 6.0 g m⁻²
   carnivores = 0.1 g m⁻².

   b What is the trophic level of the carnivores?
2. Why are measurements of dry mass used when drawing pyramids of biomass?
3. Figure 2.38 represents an energy pyramid.

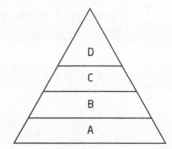

Figure 2.38 An energy pyramid.

State which of the following organisms – birds, worms, mammals and algae – would be most likely to be found at level A, and which at level C.

4. **Discussion point:** Why is it important to take measurements at regular intervals to determine the interactions and composition of an ecosystem?

2.05.05 Diversity and the Simpson diversity index

The **Simpson diversity index** allows us to quantify the diversity of a habitat or ecosystem. It takes into account both the number of different species present (the species richness) and the abundance of each species. If the habitat has similar population sizes for each species present, the habitat is said to have 'evenness'.

This index gives us a measure of both richness and evenness. It is calculated with the formula:

$$D = \frac{N(N-1)}{\sum n(n-1)}$$

in which D is the diversity index, N is the total number of organisms in the habitat, and n is the number of individuals of each species. The value of D varies between 1 and infinity. The higher the value, the greater the variety of organisms in the ecosystem.

You don't need to memorise the equation for the Simpson diversity index, but it would be useful if you can remember what the letters N, n and D represent.

An advantage of using this index is that it is not necessary to know the names of the different species, simply that they are different. If the index is calculated at intervals over a period, it can give an indication of the health of an ecosystem and whether conservation measures might be necessary.

When the Simpson diversity index is used, comparisons can only be made between habitats containing the same type of organisms, not between different habitats in different ecosystems. The index gives a comparative, not an absolute, measure of **diversity**.

> The **Simpson diversity index** is a measure used to quantify the biodiversity of a habitat.

> **Diversity** is usually explained as a function of two components: the number of different species and the relative numbers of individuals of each species.

2.05 Investigating ecosystems

CASE STUDY 2.05.01

Calculating a diversity index

Invertebrate animals found in two ponds (see Image 2.29) were sampled in similar ways using nets. The species were identified and the numbers in each pond recorded. Table 2.14 shows the results. The Simpson diversity index was calculated and used to compare the number of insect species in two separate ponds.

	Species collected					Total number of organisms
	Water boatmen	Water measurers	Pond skaters	Whirligig beetles	Water spiders	
Number of organisms in pond 1	43	18	38	3	1	103
Number of organisms in pond 2	26	18	29	11	5	89

Table 2.14 Numbers of invertebrates found in each of two ponds.

Image 2.29 Pond skater.

The Simpson diversity index for each pond was calculated as follows, using the formula:

$$D = \frac{N(N-1)}{\sum n(n-1)}$$

Pond 1:

$$D = \frac{(103 \times 102)}{43(43-1) + 18(18-1) + 38(38-1) + 3(3-1) + 1(1-0)}$$

$$= \frac{10\,506}{3525}$$

$$= 2.98$$

For pond 1, Simpson diversity index = 2.98

Pond 2:

$$D = \frac{(89 \times 88)}{26(26-1) + 18(18-1) + 29(29-1) + 11(11-1) + 5(5-1)}$$

$$= \frac{7832}{1898}$$

$$= 4.13$$

For pond 2, Simpson diversity index = 4.13

Questions

1. There are fewer organisms in pond 2, but the Simpson diversity index is higher. What does this tell us about the pond?
2. Comment on the diversity of pond 1 as shown by the data collected and the diversity index that was calculated.
3. What is the most appropriate sampling method that could be used to collect these pond organisms for diversity studies?

SELF-ASSESSMENT QUESTIONS 2.05.04

Six species of invertebrate were found in the same area of grassland. The numbers of organisms recorded for each one were: 8, 9, 12, 1, 4, 3.

1. Calculate the Simpson diversity index for this community.
2. Comment on the level of diversity in the community.
3. **Discussion point:** Discuss the advantages of using a comparative index to study diversity.

2.05.06 Choosing and evaluating field techniques

The choice of the most appropriate technique to measure or estimate abiotic components of an ecosystem depends on the level of accuracy needed and the type of ecosystem being studied. For example, a visual assessment of a field that has a thick growth of nettles around its border can provide evidence that nitrate is abundant in these areas, because nettles can only grow in soils that have a high nitrate content. However, if a study was being conducted into the nitrate content of the soil of the field so that it could be compared with another, it would be important to sample the soils and use chemical tests to measure nitrates accurately. When quadrat sampling is used, it is important that the method of doing so is appropriate. For a general overview of plants in a field, or to compare two similar fields, random quadrats could be used. On the other hand, if an investigation was being designed to assess how the slope of the land affects plant growth, a series of quadrats placed in a systematic way across the area would be more suitable.

Reliability

In most cases, measurements of abiotic factors have to be made on a single occasion in the field. This limits their usefulness, because abiotic factors vary from day to day and with the seasons. Data loggers have the advantage that they can be used to take continuous readings over much longer periods, so that the results become more reliable.

Reliable data can be obtained by taking a series of measurements and calculating an average of the results. This ensures that any errors in taking readings from equipment such as data loggers or in recording the data are noticed. Taking averages also minimises the effect of one error on the final set of data.

Accuracy

The accuracy of data can be improved by selecting equipment or a method that improves the precision of the data obtained. This usually means using a more accurate measuring device. For example, in order to compare the temperatures of two very different ecosystems, such as desert and tundra, a thermometer that can measure to an accuracy of 1 °C would be acceptable. However, within a pond ecosystem, fluctuations in temperature across the system may be very small, and a thermometer that measures fractions of a degree could be a better choice.

CONSIDER THIS

'Margin of error' is a term that is used to describe the likelihood and extent of error in quantities that we measure in experiments. Every time we repeat a reading with an instrument such as a thermometer or oxygen meter, we may get a slightly different result. To reduce errors like this, it is important to take a series of readings and calculate an average.

If we use statistics in calculations, errors are said to be of two types:

- Systematic error or bias, which always occurs, with the same value, when we use an instrument in the same way. This can be minimised by using the correct measuring instrument and the same technique each time we take a reading, for example holding the thermometer in the water for exactly the same period.
- Random errors, which may vary from reading to reading and are due to factors that we cannot control.

When you carry out practical work, you will be expected to show the margin of error in your readings, either using error bars on a graph or by ± in your table of data. This shows you understand the degree of error your data may contain.

2.05 Investigating ecosystems

Theory of knowledge 2.05.02

Removing the risk of subjective bias

Objectivity is important whenever data is collected and recorded in scientific investigations, because each individual may interpret what they see in a different way. As environmental scientists investigate the natural world, it is important to eliminate any personal bias in the data they collect. Standardised equipment such as stopwatches, thermometers and rulers helps to remove personal judgements and variation between the observers. Standard units and scales ensure that data is communicated in terms that others can understand easily. Standardised measurements help to make scientific data objective, because the accuracy of measurement can be checked and tested independently. It is also important in defining how scientific activities are carried out and how conclusions are drawn.

1. Is it possible to be entirely objective in science when human perception is involved in collecting and interpreting information?
2. How has scientific equipment helped us to investigate factors such as the presence of chemicals or microbes, which would otherwise be beyond our perception?

End-of-topic questions

1 Figure 2.39 shows a food web.

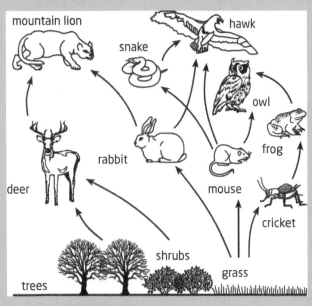

Figure 2.39 A food web.

a Identify *three* organisms that are herbivores. [3]

b State which species would be most affected by a long-term decrease in the population of rabbits, and outline the reasons for your choice. [2]

c Sketch a graph to show the population sizes of a predator and prey species over a period of several years. [2]

d How many animals in the food web are feeding at trophic level 4? [1]

2 One new source of fuel that is being investigated is plant biomass. The productivity of three different types of forest has been studied by measuring the dry mass of components of the ecosystems. The results are shown in Table 2.15.

Forest type	Total dry mass / kg m^{-2}			
	Living plant biomass	New plant material per year	Litter (fallen leaves and dead material) per year	Decaying material in soil
Tropical rainforest	52.5	3.3	2.5	0.2
Deciduous	40.7	0.9	0.7	1.5
Coniferous	26.6	0.7	0.5	4.5

Table 2.15 The dry mass of different components of three forest ecosystems.

a Compare the living plant biomass in each forest. [1]

b Compare the production of new plant material per year in relation to the total living biomass in each ecosystem. [2]

c The mass of litter produced each year is less than the mass of new material produced per year. Suggest *two* reasons for this. [2]

d Discuss the suitability of each of these three ecosystems as a source of biomass for fuel. [3]

3 a Explain what is meant by the term 'ecosystem'. [2]

b Outline what is meant by each of the following:
  i pyramid of biomass [2]
  ii abiotic factors [2]
  iii succession. [2]

4 Identical sample areas of land were cleared of weeds before pea seeds were planted in them. Different sample areas were kept clear of weeds for different periods of time after the seeds were planted. After nine weeks, the plants were harvested and weighed; the results are shown in the graph (see Figure 2.40).

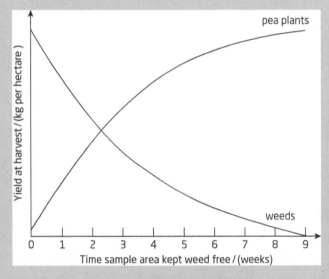

Figure 2.40 Graph of results.

a Describe the effect of competition between the pea plants and the weeds. [3]

b Evaluate the importance of weeding on yields for farmers growing peas or other crops. [3]

c Describe how you would estimate the total mass of weeds growing in a large field. [3]

5 State, with a reason, whether each of the following will increase, decrease or remain the same as a succession proceeds:

a nutrient content of the soil or water [2]

b species diversity [2]

c biomass [2]

d productivity. [2]

6 a **i** Distinguish between r-strategies and K-strategies. [2]
  ii State an example of a species that follows an r-strategy and suggest two advantages that the organism gains from this strategy. [2]
  iii Suggest why K-strategists are often vulnerable to extinction. [1]

b **i** Bacteria in a culture vessel divide to produce two new organisms every 20 minutes. Complete this table to show the number of bacteria present over a period of 160 minutes. [1]

Time / min	Population numbers
0	1
20	2
40	4
60	8
80	16
100	
120	
140	
160	

  ii Construct a graph of the data and describe the shape of the curve. [2]
  iii If no additional nutrients were added to the culture of bacteria, predict what would happen to the shape of the curve, and explain your answer. [3]

Topic 3
BIODIVERSITY AND CONSERVATION

An introduction to biodiversity

3.01

LEARNING OBJECTIVES

After reading this chapter you should be able to:

- *outline* the concept of biodiversity, which includes diversity of species, habitats and genetics
- explain how diversity is a product of the number of species (richness) and their relative proportions (evenness)
- discuss how quantification of biodiversity is important to conservation efforts
- describe how the assessment of changes to biodiversity is important in assessing human impact in a community.

KEY QUESTIONS

What are the three ways in which biodiversity can be identified?

Why is the ability to understand and quantify biodiversity important for conservation of species?

3.01.01 The importance of diversity

The United Nations (UN) has designated 2011–2020 the United Nations Decade on Biodiversity, because of the importance of **biodiversity** to all life on Earth. As biodiversity gives a measure of the amount of biological variation in an ecosystem or biome, it can be used to assess the health of a system. As we have seen (see Topic 2), biodiversity is influenced by climate, so tropical regions with high productivity are more diverse than polar regions, which contain fewer species. Biodiversity also varies considerably within regions and depends on temperature, precipitation, altitude, soil, geography and the presence of other species.

Decade on Biodiversity
www.youtube.com/watch?v=zpM-nkhZCgk
Factsheet: What is Biodiversity?
www.unep.org/wed/2010/english/PDF/BIODIVERSITY_FACTSHEET.pdf

Extension

A different kind of country by R. F. Dasmann (1968)

Habitat diversity

The range of different habits in an ecosystem is one of the most important factors to consider in a study of the conservation of biodiversity. To assess **habitat diversity**, ecologists study the variety of niches that a habitat contains. A rainforest has a high diversity of habitats, which include the canopy, the soil and pools of standing water, so there are many ecological niches present (see Figure 3.01). On the other hand, a desert has little habitat diversity, simply a sandy terrain and a few plants, so it provides few ecological niches. An increase in habitat diversity is always likely to lead to an increase in both **species diversity** and **genetic diversity**.

Biodiversity is a measure of the quantity of living (biological) diversity per unit area. It includes species diversity, habitat diversity and genetic diversity.

Theory of knowledge 3.01.01

A paradigm shift is described as 'a change in the basic assumptions within a ruling theory of science'. Today, the term 'biodiversity' is used when in previous times the word 'nature' would have been used instead.

1. Do you think that this represents a paradigm shift?
2. How important is the use of a term such as 'biodiversity' in extending our knowledge of a biological system?
3. Discussion point: Can you think of other examples in science where a change of terminology has occurred? Do they help us understand how a system functions?

3.01 An introduction to biodiversity

Habitat diversity is the number of ecological niches or range of different habitats that are present per unit area of a biome, ecosystem or community. If habitat diversity is conserved, this usually leads to the conservation of both species and genetic diversity.

Species diversity is the variety of species per unit area. It includes both the number of species that are present and their relative abundance.

Genetic diversity is the range of genetic material present in the population of a species or its gene pool.

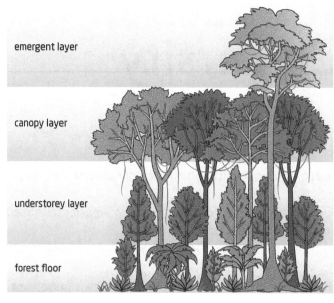

Figure 3.01 The different layers of a rainforest provide many niches and a wide range of different habitats for organisms.

CONSIDER THIS

Rainforests contain 170 000 of the 250 000 known plant species; only 80 species of frog are found in the whole of the USA, but Madagascar has 500 species; the continent of Europe has 570 species of butterfly, but a single rainforest in Peru has 1300.

Species diversity

Species diversity is a measure of the variety of species in a given area; it is quantified by measuring both the number of species present (species richness) and their relative abundance. You read about the Simpson diversity index, which is one of the ways used to quantify diversity, in Topic 2. A complex ecosystem like a rainforest contains a wide variety of species, which are likely to be abundant, so the species diversity is high. A hectare of rainforest may contain more than 400 different species of tree.

Genetic diversity

Genetic diversity is the range of genetic material present in a population. It follows that, where more species and more individuals of a species are present, the range of genetic material will also be greater. Genetic diversity is high if a population has a large gene pool. The gene pool is the number of variations of the same gene that are present in the DNA of a particular species. A large gene pool is the sign of a healthy population with high genetic variability. A small gene pool will indicate low genetic diversity.

In some cases, genetic diversity is artificially reduced, for example by inbreeding animals or cloning plants for agricultural use. Cloned plants or highly selected farm animals or seeds have a smaller gene pool than their natural relatives. A small gene pool can make them less adaptable to changes in the environment and more susceptible to diseases than traditional breeds and varieties.

3.01.02 Comparing communities

A **community** is a group of populations which live in the same habitat and interact with one another.

Communities can be compared and described using diversity indices such as the Simpson diversity index (see 2.05.05 *Diversity and the Simpson diversity index*). If two similar communities are compared and one has a much lower diversity index than the other, this may be a sign that there is pollution in the area or that it is under stress for other reasons. For example, a study of the biodiversity in the 'dead zone' in the Gulf of Mexico revealed the extent of the effect of water pollution in that area (see Chapter 4.04). In rivers and other waterways, biotic indices (see Chapter 4.04) are used to monitor key indicator species, which by their presence or absence can indicate that a habitat is under threat. Understanding the biodiversity of an area is important when conservation measures are being put in place. But it is also important to remember that undisturbed, unpolluted sites, such as Arctic ecosystems, may naturally have a low diversity due to their location and climatic conditions, and that this low diversity is not a sign of problems.

Diversity indices can be used to compare two similar habitats, but they can also be used to monitor changes to biodiversity over time. Measurements can be taken at suitable intervals, for example before and after large building projects, and used to assess the importance of the impact of new housing developments, industry or other human activities in the area over a long timescale. Quantifying biodiversity is important in conservation so that areas of high biodiversity or particular interest because of their unique species can be identified. Only after this is it possible to put in the appropriate measures to conserve species and their habitats.

Biodiversity hotspots

A biodiversity hotspot is a biogeographic region that is both a significant reservoir of biodiversity and is under threat. Ecologists have identified 34 biodiversity hotspots that cover about 15 per cent of the surface of the Earth. They include the West African rainforest, Japan, California and the Mediterranean coastline. Twenty-five of these areas have lost at least 70 per cent of their natural habitat but nevertheless support 60 per cent of the Earth's species. Hotspot biodiversity is assessed in one of three ways:

- the number of total species (species richness)
- the number of endemic species (unique to the region)
- the number of species at risk.

Two very different biodiversity hotspots are regions with a Mediterranean-climate and tropical rainforests. When these two were compared, it was been found that, despite the lack of high rainfall and productivity in Mediterranean hotspots, (features that are usually a sign of high plant diversity), some of the most species-rich Mediterranean communities occur in semi-arid regions on poor soils. They are also equally as rich as the communities in the rainforest. Further comparison of the habitat diversity, species diversity and genetic diversity in regions like these may help us understand how diversity arises and is maintained. It will also help to assess where conservation efforts should be focused.

CONSIDER THIS

In the late 1980s, biodiversity hotspots were chosen as places where conservation efforts should be prioritised, and were soon adopted by many environmental groups. But more recently, Gerardo Ceballos, Professor of Ecology in Mexico City, and Paul R. Ehrlich, Professor of Population Studies at Stanford, have suggested that, if species richness is used as the only criterion for conservation, it will not be possible to protect the endemic or endangered species in other locations as well. They propose that it is also important to study and compare communities outside these 'hot spot' areas.

CASE STUDY 3.01.01

The confusing issue of species diversity and richness – why use a diversity index?

A diversity index is a mathematical measure of species diversity in a community. Diversity indices provide more information about community composition than species richness (the number of species present) alone; an index also takes account of the relative abundances of different species.

Let's look at two hypothetical communities that both contain 100 individuals:

- Each one contains ten different species.
- The first community has ten individuals of each species.
- The second community has one individual of each of nine species, and 91 individuals of the tenth species.

Which community is more diverse? It is clear that the first community is, but both communities have the same species richness.

A diversity index depends on both species richness and also the evenness (the way in which the individuals are distributed among the different species). Thus the index can take relative abundances into account and increase our understanding of the two communities.

Questions

Check the information in Topic 2 about the Simpson diversity index (see 2.05.05 *Diversity and the Simpson diversity index*).

1 Name the two factors that affect diversity indices.
2 What effect does each one have on diversity?
3 Research idea: Investigate species richness in a rainforest, a temperate forest and a coniferous forest. Which forest contains the greatest richness of species?

3.01 An introduction to biodiversity

CONSIDER THIS

In your course, you will use the Simpson diversity index to compare communities, but in some books you will also see two other indices used, the Simpson's index and Simpson's reciprocal index. They are all calculated in a similar way, and all three represent the same biodiversity even though the values calculated from the formulae are different. The Simpson's diversity index you will use gives more importance to the more abundant species in a sample. If you look up diversity in reference books, check that you are using the correct formula each time.

SELF-ASSESSMENT QUESTIONS 3.01.01

1. Define 'biodiversity' and outline the three types of biodiversity that are recognised.
2. State two reasons why it is useful to give a numerical value to species diversity.
3. What does a high value of the Simpson diversity index tell you about an area that is being studied?
4. **Research idea:** Find out about other indices of diversity and how they are used.

Theory of knowledge 3.01.02

The importance of numbers

Diversity indices are not true measurements, like numbers, length and temperature. Diversity indices provide a way of comparing measures of proportion and richness in a numerical way, in a similar way to a ratio. A diversity index involves a subjective judgement on the combination of two measures. Philosophers justify the subjective use of numbers rather than true measurements as a means of measuring the 'degree of belief' in a certain set of data.

1. How useful do you think indices are in a scientific situation?
2. When is quantitative data more useful than qualitative data in helping us know the world?

Origins of biodiversity

3.02

LEARNING OBJECTIVES

After reading this chapter you should be able to:

- explain that biodiversity arises from evolutionary processes
- describe how evolution is a gradual change in the genetics of a population, achieved through natural selection
- outline the mechanism of natural selection
- explain how environmental change produces new challenges to species that can lead to the survival of some but not others
- describe how new species can be formed when populations are isolated by barriers that form, such as mountains, and evolve differently
- understand that mass extinctions have occurred because of factors such as tectonic plate movement and meteorite impact.

KEY QUESTIONS

What is evolution and how is it achieved?
How is environmental change important in the development of diversity?
Which events have caused mass extinctions in the geological past?

3.02.01 Natural selection and speciation

Evolutionary change happens over many generations, largely as a result of **natural selection**. As changes occur in the environment, species must adapt and change if they are to survive. As **evolution** occurs, the diversity of species increases.

Biologists define a species as a group of organisms that can interbreed to produce fertile offspring. As we have seen (see Topic 2), lions and tigers are separate species because lions breed with lions to produce more lions, and tigers breed with tigers to produce more tigers. If a lion and a tiger do breed, their hybrid offspring (ligers or tigons) (see Image 3.01) are infertile and therefore not a species.

The process of **speciation** establishes a new species which is fertile but which can no longer breed with the original species. The new population is said to be reproductively isolated.

Over the course of millions of years, speciation has led to the great variety of life on Earth. Species have evolved and changed and become separated from their original populations. Some species have died out altogether (become extinct) and are only known from their fossil remains found in rocks. The process of evolution has increased biodiversity on Earth.

> **Natural selection** is the proposed mechanism that can lead to the formation of new species.
>
> **Evolution** is a gradual change in the genetic characteristics of a population.
>
> **Speciation** is the formation of new species from an original population so that the new species cannot interbreed with the original species.

Extension

Great website about natural selection and fossils
www.ucmp.berkeley.edu/education/explorations/tours/stories/middle/B1.html

Natural selection is proposed as the key mechanism which leads to the formation of new species. The theory of evolution by means of natural selection was first suggested by Charles

3.02 Origins of biodiversity

Darwin and Alfred Wallace. Darwin gathered evidence to support his ideas and published his conclusions in 1859 in his book *On the Origin of Species*.

Natural selection explains how variation between individuals can lead to new species. By observing any species, we can see that all living things produce far more offspring than are needed to replace them when they die, and yet in most natural situations we do not see vast overpopulation with any one species. Trees can produce thousands of seeds, and fish produce hundreds of eggs, yet we rarely see population explosions. The reason is that not all the offspring survive. Several factors explain this:

- There is genetic variation between the offspring. Not all members of a species are the same. For example, some birds may have brighter plumage than their siblings, some howler monkeys may be able to call more loudly than others, and some deer may be stronger than others.
- Individuals of a species compete with one another for the resources, such as food, water, mates and the nesting sites they need. Not every individual gets the resources they need. This is sometimes referred to as the 'struggle for survival' or 'survival of the fittest'. Fitness does not only mean physical strength, it also includes behaviour and appearance. For example, the loudest monkey may be able to establish a larger territory, a stronger male deer will be able to fight off its rivals and mate with more females, and a well-camouflaged mouse may be able to hide from danger.
- So, individuals that survive are those with the best adaptations to their environment. These individuals go on to produce the next generation of offspring.
- The offspring inherit the genes which give them their parents' successful characteristics.

Overproduction of young, together with variation, competition and reproduction of individuals, can gradually change a population or a species over a long period of time. The characteristics that lead to survival are passed on and, over many generations, they accumulate so that more and more individuals in the population have the favourable characteristics. Individuals with characteristics or variations that are unfavourable tend to die out. Some neutral variations make no difference to an individual's survival at all. When Darwin first proposed his theory of evolution by means of natural selection, he did not know about DNA or exactly how characteristics are passed from parents to offspring. We now know that characteristics are passed from one generation to the next in the genes that offspring inherit from their parents. Generation after generation, the proportion of the favourable genes in the whole population will increase. The gene pool of the population is changed, and the changes may ultimately lead to the formation of a new species.

Darwin coined the term 'natural selection' to explain how a species could change, based on his studies of 'artificial selection'. Darwin knew from animal breeding and agriculture that farmers selected their best animals with the characteristics that they required to produce the next generation (see Image 3.02). People have domesticated and bred plants and animals for thousands of years, and we can see the changes that have occurred. Modern varieties of wheat are shorter and stronger than the varieties of 100 years ago, and these had many differences from the wild grasses that were their forerunners. Likewise, modern breeds of cattle and sheep produce more milk and better quality wool than older breeds were able to do.

Image 3.01 Liger cubs, born in a Russian zoo in 2013.

Image 3.02 Artificial selection: this prize specimen has been bred from animals that had the genes for long wool. Lambs bred from this sheep will also have long wool.

CASE STUDY 3.02.01

Evidence from fossils – evolution of the horse

Fossils provide evidence that can help us understand how species have changed over millions of years. The oldest fossil horses from 55 million years ago were small animals about the size of a dog, and they had several toes. These animals were forest dwellers and their feet were adapted to walking on soft, moist soil. Later, as grass species began to evolve, ancestors of the modern horse changed their diet from leaves to grass and evolved larger teeth to cope with this diet. Later still, around 24 million years ago, early horses began to diverge and to live on drier, open grassland steppes, where they evolved longer limbs and toes that enabled them to run faster and escape potential predators. Modern horses carry the weight of their body on the end of their third single toe, which is protected by a strong hoof (see Figure 3.02)

Questions

1. Which were the two most important environmental features that influenced the evolution of the modern horse from its early ancestors?
2. What was the consequence of the evolution of grassland for the lifestyle of the forerunners of the modern horse?
3. Research idea: DNA from a 600 000-year-old ancient horse species was sequenced in 2013. Find out what this revealed about relationships between horses, donkeys, zebras and Przewalski's horse alive today.

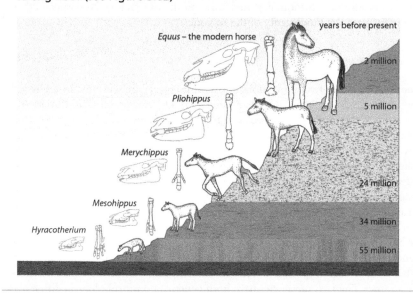

Figure 3.02 Some of the many species of fossil horses and the modern horse *Equus*. The fossil sequence shows that over time horses have developed single toes, longer legs and larger teeth.

Theory of knowledge 3.02.01

Scientific theories

A theory is a well-established principle that has been developed to explain some aspect of the natural world. A scientific theory arises from repeated observations and testing and incorporates predictions and hypotheses that are widely accepted.

The theory of evolution by means of natural selection is a good example of a scientific theory. Since Darwin's time, scientists have gathered evidence which is used to test his original explanations of natural selection and evolution. As in all science, there is a constant updating process rather like a feedback loop of questioning, experimentation, theory and revision.

Darwin's theory is still accepted widely because it remains the best explanation for the observed phenomena in the natural world. Evolutionary theory has been vital to the workings of biology, geography and history, but it has changed and adapted to new evidence in many branches of scientific knowledge.

1. How do you think discoveries in biochemistry and genetics have influenced evolutionary theory?
2. Why do scientists not describe the alternative view of life held by creationists as a scientific theory?

3.02 Origins of biodiversity

> **SELF-ASSESSMENT QUESTIONS 3.02.01**
>
> 1. What are the four key arguments that support Darwin's proposal for natural selection?
> 2. What is the difference between 'natural selection' and 'artificial selection'?
> 3. **Research idea:** Use the internet to find out about the various views on Darwin's theory of evolution in different religious faiths. Consider your own philosophy together with the evidence for the theory, and draw your personal conclusions.

3.02.02 Isolation and the formation of new species

Isolation of a natural population can lead to a new species being formed, because the isolated individuals may change so much over a period of time that they become unable to breed with the original population to produce fertile offspring. A population may become isolated either by a geographic barrier or because of behavioural, genetic or reproductive factors. New species that are formed in this way increase the biodiversity of the region.

Geographical isolation

Geographical isolation occurs when two populations of the same species are separated by a physical barrier, such as a mountain range or a large lake (see Figure 3.03). The two populations can no longer interbreed, because of the distance or difficulty in reaching one another. As time passes, each group develops slightly different physical or behavioural variations in response to the selection pressures where they live. Eventually, the differences are so great that the two populations cannot interbreed even if the barrier is removed and they do come into contact with one another.

The freshwater fish the char (*Salvelinus*) lives in lakes in Switzerland, Scandinavia and the UK. Almost every lake contains a different species, because the populations are separated by the land which forms a barrier between them. At one time, the species of fish must have been able to interbreed, but movements of the land and formation of barriers means this has not been possible for millions of years. Another example is the spotted owl species found on the west coast of North America. Two subspecies, the northern spotted owl and the Mexican spotted owl live in different regions and are separated from one another. The two subspecies have both genetic and morphological differences. The observations of the char and owls, and many other examples, provide evidence that geographic separation can lead to the formation of new species (see Figure 3.04).

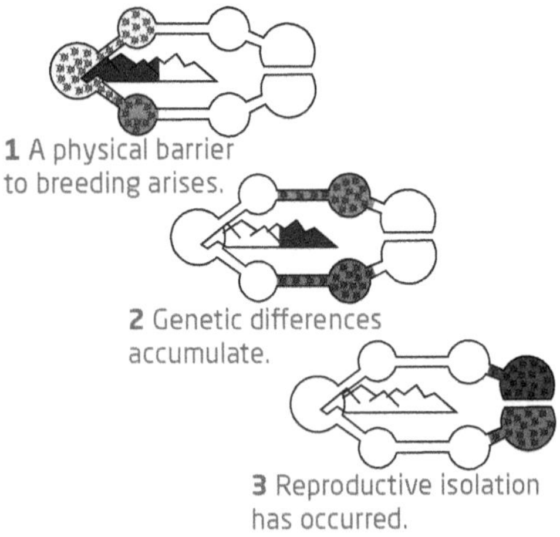

1 A physical barrier to breeding arises.

2 Genetic differences accumulate.

3 Reproductive isolation has occurred.

Figure 3.03 Allopatric speciation model initiated by a physical barrier: these two populations of beetles became separated by mountains and developed independently until they could no longer interbreed.

Reproductive isolation

Reproductive isolation can also lead to speciation. Populations do not have to be geographically separated for speciation to happen. Sometimes, species may be separated by a geographic barrier at first and later, if they are subjected to different selection pressures, behavioural differences become important in leading to reproductive isolation and the formation of new species.

Timing

Timing is a crucial factor for successful reproduction and, if the time of breeding of two groups is not synchronised, speciation may occur. For example, if one population of plants matures earlier than another, pollen of one group is extremely unlikely to come into contact with the flowers of the other, so there

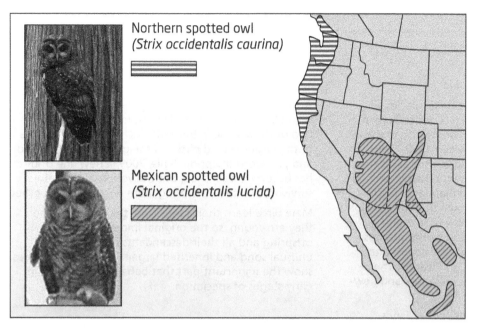

Figure 3.04 Spotted owl subspecies separated spatially already show differences in their genetics and appearance.

will be little chance of the two populations producing offspring together. Over time, they could become reproductively isolated and eventually develop into new species. A good example is that of the pine trees *Pinus radiata* and *P. muricata*, which grow together in California. *P. radiata* sheds its pollen in early February, while *P. muricata* does not produce its pollen until April. The two *Pinus* species have become reproductively isolated and no longer interbreed. In the animal world, an example is seen in two closely related crickets found in the north-eastern USA. These animals also differ in the timing of their breeding cycles: *Gryllus pennsylvanicus* becomes mature in autumn, but *Gryllus veletis* reaches reproductive age in spring, so the two species are reproductively isolated and are very unlikely to interbreed.

Behaviour

In other cases, male and female animals may fail to mate because of differences in their behaviour. Many species have very elaborate courtship behaviours and calls and, if the females of one population do not respond to the courtship displays of males from another population, mating will not take place. Genes will not be exchanged between the two groups and, after a period of time, new species may develop. An example of this can be found in the species of eastern tarsiers that live in the same geographical region of Tangkoko in Japan but have very different mating calls. Males and females only respond to calls from members of their own species.

Behavioural differences produce barriers to gene flow and, once the gene flow between two species is stopped or severely reduced, larger genetic differences between the species develop and accumulate. In this way, behavioural differences lead to reproductive isolation, and new species may develop.

Other causes

There are two other causes of speciation due to reproductive isolation:

Differences in anatomy, so that males and females cannot physically mate: for example, among domesticated dogs, Chihuahuas and Great Pyreneean hounds are so different in size that mating is impossible. Domesticated animals like these have been bred by artificial selection, but, if similar differences developed in wild animals, they could eventually lead to speciation.

Failure of offspring to survive: sometimes two species do interbreed, but the offspring fail to survive long enough to reproduce. For example, the two species of *Gryllus* cricket mentioned above do occasionally interbreed, but the offspring are weak and quickly die.

3.02 Origins of biodiversity

CASE STUDY 3.02.02

Galapagos Island finches

Darwin's finches are a group of about 15 species all found only on the Galapagos Islands. Charles Darwin collected specimens of the birds when he visited the islands on his voyage on HMS *Beagle* in 1830s. The most important difference between the species is their beaks. The original birds to colonise the islands arrived long ago from the South American mainland. As the original population grew larger, it avoided competition for food and space by colonising new islands and exploiting different sources of food. Over thousands of generations, the birds' beaks became adapted to different foods such as seeds, nuts and insects. The diversity of the birds on the islands increased and new species formed.

Darwin spent five weeks in the Galapagos Islands studying the birds, but a more recent study has been taking place for more than 30 years. Peter and Rosemary Grant have studied behavioural separation of Galapagos finches on the island Daphne Mayor since 1973 (see Image 3.03). At first, the island was occupied by two finch species, the medium ground finch and the cactus finch. In 1981, a large bird that was a hybrid between the ground finch and the cactus finch was seen on the island. It had a large beak and unusual 'strange' song. The new bird mated with a female ground finch and their offspring have been studied ever since. The female that chose the incomer as her mate selected him despite his strange song. We cannot be sure of the reason, but it might be that she found his large beak attractive and ignored his unusual singing.

After four generations, a drought on Daphne Mayor island killed many of its birds. Following the drought, two of the survivors, both of which were descendants of the large-beaked bird and the ground finch, mated and produced offspring. Since 2009, a new line of birds has become established, but the young of these two survivors and their offspring mate only with each other.

Male birds learn their songs from their fathers when they are young, so the original immigrant male's offspring and all their descendants have learnt his unusual song and inherited larger beaks. These finches show the important part that behaviour plays in the early stages of speciation.

Image 3.03 Finches on the island Daphne Mayor: new immigrant finch (left), cactus finch (top), ground finch (right).

SELF-ASSESSMENT QUESTIONS 3.02.02

1. Explain what is meant by the 'struggle for survival'.
2. How does geographic isolation differ from reproductive isolation due to behavioural differences? How does each contribute to speciation?
3. **Discussion point:** Consider the work of Francis Galton, who founded the science of eugenics and tried to apply principles of natural selection to human populations. Consider why eugenics came to be considered as unethical science.

3.02.03 Plate tectonics

The arrangement and distribution of the continents has changed as the tectonic plates of the Earth have moved over geological time. Movement of the plates has caused physical barriers as well as land bridges that have separated species, led to climatic change and contributed to evolution.

The Earth is made up of four distinct layers (see Figure 3.05):

- The **inner core** is the solid, hot centre made of iron and nickel.
- The **outer core** surrounds this layer and is made up of liquid iron and nickel, which are at temperatures in excess of 5000 °C.
- The **mantle** surrounds the outer core and is semi-molten rock called magma. In the upper areas, this rock is hard, but deeper down it is softer.
- The **crust** is the outermost layer and is relatively thin, with a thickness of up to 60 km. The land we live on, known as the continental crust, is formed of this solid rock layer. Below the sea, the crust is known as the oceanic crust and it forms the bed of the ocean.

The crust and the mantle are together known as the lithosphere, and this floats and moves on the semi-molten layer beneath it. The lithosphere is divided into eight major and several minor tectonic plates (see Figure 3.06). The plates vary in size but are able to move relative to one another between 50–100 mm per year – a phenomenon known as continental drift.

The study of the movement of these plates is called **plate tectonics**. Over millions of years, the plates have collided, slid against one another and overlapped each other. At the junctions of the plates, mountains, earthquakes and volcanoes provide evidence of their movements.

Where plates slide past one another, fault lines such as the San Andreas fault in California develop. If plates are diverging, a ridge such as the mid-Atlantic ridge develops.

The mid-Atlantic Ridge is a boundary between the Eurasian plate and the North American plate; it runs along the floor of the Atlantic Ocean forming part of the longest mountain range in the world. Volcanic eruptions occur along the ridge and can lead to the formation of new islands. Both Surtsey and Iceland owe their existence to the magma that wells up through rifts along the ridge. New islands like these provide completely new habitats for colonising species.

In other parts of the world, plates push together and collide (converge), so the continental crust is squashed and forced upwards. This process creates fold mountains as the converging plates

CONSIDER THIS

DNA is the genetic material that builds the chromosomes in the cells of all living things. Chromosomes contain the genes that give each organism its unique characteristics. Plants and animals have chromosomes arranged in pairs, so there are two copies of every gene present in a cell. Slight variations between the genes of different organisms give the variation in appearance and characteristics that we observe.

Plate tectonics is the theory that the Earth's outer covering is divided into several plates that move over the rocky inner layer above the core.

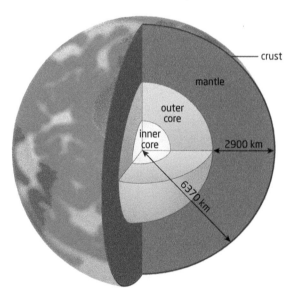

Figure 3.05 The Earth's outer layer or crust which we live on is very thin in comparison with the core and the mantle.

3.02 Origins of biodiversity

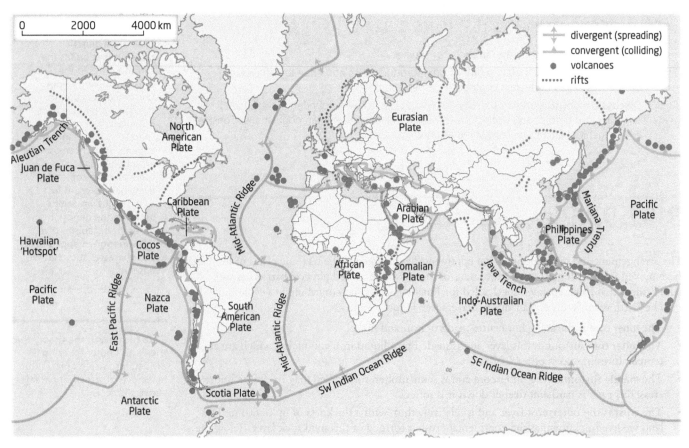

Figure 3.06 The Earth's crust (the lithosphere) is divided into eight major and several minor tectonic plates. Volcanic activity is common where the plates meet.

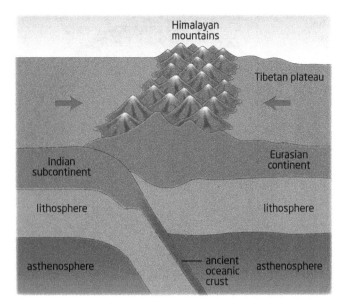

Figure 3.07 Mountains like the Himalayas form a barrier between groups of species that are separated by them. They also create new habitats with different climates and different altitudes, which can lead to speciation and an increase in biodiversity.

are forced into each other, crumpling and folding under enormous pressure. Great mountain ranges such as the Himalayas and the Alps (see Figure 3.06) were formed this way.

In some cases, as tectonic plates move, one slides under the other in a process called subduction. Subduction zones are known for their volcanoes, earthquakes and mountain building. Oceanic plates are denser than continental plates, so if the two move together, the oceanic plate is forced underneath the continental plate. As this happens, the oceanic plate begins to melt to form magma, and earthquakes are triggered. Magma then rises up through cracks in the crust as pressure builds up (see Figure 3.08).

In some subduction zones, one oceanic plate will slide under another; along the area above the subduction zone, volcanoes exist in long chains called volcanic arcs. These volcanoes tend to be extremely explosive; Krakatau, Nevado del Ruiz and Mount Vesuvius are all examples of arc volcanoes. Arcs of islands are also created in a similar way; the Aleutians, Japan and the Philippines all formed around the northern and western borders of the Pacific Plate. These new islands became habitats for pioneer species which have settled and evolved on them.

Formation of the modern continents

Until about 250 million years ago, the land mass of the Earth was formed of just one supercontinent known as Pangaea. Around 175 million years ago, this continent split into two new continents called Laurasia and Gondwanaland. Laurasia went on to become North America, Europe, Asia and Greenland, while the southern landmass Gondwanaland contained the modern-day continents of India, Africa, Australia, Antarctica and South America. These continents began to break up and move towards their present-day locations around 130 million years ago.

The displacement and rearrangement of land masses has helped to create biological diversity on our planet by separating different groups of organisms from one another. Our modern understanding of continental drift, and the structures that it has produced, have helped us understand the fossil record and provided valuable information about speciation.

Long ago, Gondwanaland was home to many types of mammal, including both the marsupials and placental groups. As Gondwanaland split in two to create South America and Australia, two very different types of mammal developed. The pouched marsupial mammals thrived in Australia, but most died out in South America. In Australia, separated from other mammals, they evolved into modern-day kangaroos, koalas, wallabies and wombats. In South America, the only type of marsupial species to survive was the opossum, because the placental mammals outcompeted the others and came to dominate the new continent.

The physical separation of the marsupials led to genetic and anatomical differences and speciation. Australia also contains unique plants such as the eucalypts, as well as unusual egg-laying monotreme mammals such as the duck-billed platypus and echidna, which are found nowhere else on Earth.

Likewise, the island of Madagascar, which was separated from the African mainland millions of years ago, is the only place on Earth where lemurs live. Many species of lemur have evolved, and they fill the niches occupied by different species in other parts of the world.

Another example of the importance of continental movement to speciation can be seen in large flightless birds (see Figure 3.09). Emus, rheas, ostriches and cassowaries are only found on the continents that were once part of Gondwanaland. Although each group of birds is now found on different continents, the groups are not closely related. The groups were separated as the continents separated and have evolved to become completely different species. The birds cannot fly, so they have never been able to interbreed.

Movement of the continents has separated organisms and led to speciation, but in some areas land bridges have formed and now link places that were previously separated. Land bridges give mobile animals the chance to colonise new habitats, but they can have other consequences. For example,

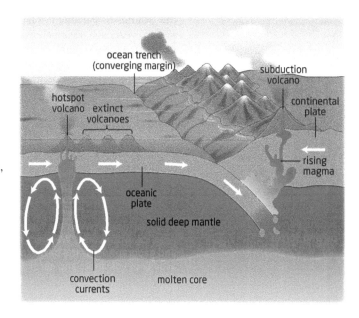

Figure 3.08 Rising magma from beneath the ocean causes new volcanic islands to form. These islands are colonised by species which become adapted to the new conditions. The Galapagos Islands and Hawaii were formed in this way.

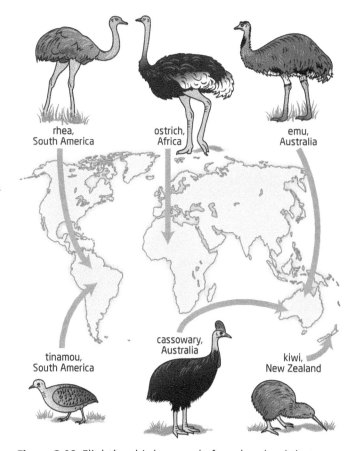

Figure 3.09 Flightless birds are only found on land that was once part of the ancient continent of Gondwanaland.

3.02 Origins of biodiversity

CONSIDER THIS

Scientists have different theories about what causes movement of the Earth's plates. One suggests that the forces of the Earth's rotation together with the tidal forces of the Sun and Moon are important. Another suggests that movement of the seabed together with drag and downward pull at subduction zones are the driving forces.

the Isthmus of Panama is a land bridge that formed between North and South America just 4 million years ago, relatively recently in geological time. This allowed bears to move from North to South America, but cut off the gene flow between groups of aquatic animals which became separated in the Pacific and Atlantic Oceans. Two other important land bridges are the Bering land bridge, which has connected Asia and North America as sea levels fell during the last Ice Age, and the Sinai Peninsula, which links Africa to Europe and Asia.

Theory of knowledge 3.02.02

Plate tectonics – a paradigm shift

Plate tectonics is a theory that scientists use to describe the large, slow movements of the Earth's crust. The theory has been built on the ideas of continental drift, which were developed during the early 20th century. The theory was first proposed by Alfred Wegner in 1912 based on his observations of similarities between the fossils of reptiles found in western Africa and on the east coast of South America. At the time, Wegner's theory was ridiculed, but the theory is now accepted by most geologists and geographers as the best explanation for the formation of continents and mountains and the activity of volcanoes and earthquakes.

The acceptance of continental drift is an example of a paradigm shift. The term describes a change in assumptions within a ruling theory when a significant number of anomalies are found to counter the accepted paradigm. In this case, the original belief was that the continents did not move, but gradually more and more evidence accumulated to support the belief that they did. As in all cases when a paradigm shift takes place, the accepted theory is put into a state of crisis until a new paradigm is formed and gains its own supporters. Today, most scientists accept the ideas of plate tectonics and continental drift, so we say that a paradigm shift has occurred.

1. Why is plate tectonics described as a scientific theory?
2. Suggest reasons why the theory proposed by Wegner was not accepted in 1912.

SELF-ASSESSMENT QUESTIONS 3.02.03

1. Give reasons why marsupial mammals are only found in Australia.
2. What types of movement of the Earth's plates can lead to speciation?
3. How do land bridges affect species and speciation?
4. **Discussion point:** Would it ever be possible to disprove Wegner's theory?

3.02.04 Mass extinctions

Mass extinctions are periods in Earth's history when very large numbers of species die out simultaneously or within a short time.

Mass extinctions are periods in Earth's history when very large numbers of species die out simultaneously or within a very short period. The most severe occurred at the end of the Permian period, 250 million years ago, when 96 per cent of all species were wiped out. There are about 12 million different species present on Earth today, but this represents only about 1 per cent of the total number of species that have lived. Since the beginning of life on Earth there have been several mass extinctions (Figure 3.10).

Movements of the continents, huge volcanic eruptions, drought and ice ages, and also the impact of huge meteorites on the surface of the Earth, have caused mass extinctions. Events like these cause such massive changes in climate or physical features of the Earth that, as well as destroying species, they present new challenges to the survivors and lead to new evolutionary paths. An increase in biodiversity may be the long-term consequence of mass extinctions.

Causes of mass extinctions

The average time between mass extinctions has been about 100 million years. Palaeontologists have recorded five mass extinction events throughout Earth's history (see Figure 3.09).

Extension

Mass extinctions
www.to14.com/game.php?id=4d486a4d35a14

Drop in sea level

The first great mass extinction event took place at the end of the Ordovician period, about 430 million years ago (mya). According to the fossil record, about 25 per cent of all families (60 per cent of all genera) of both terrestrial and marine life were exterminated. Extinction was caused by a drop in sea levels as glaciers formed.

Cause unknown

About 360 mya, in the Late Devonian period, the Earth experienced the second mass extinction event. The cause is not known, but around 19 per cent of all families (50 per cent of all genera) became extinct.

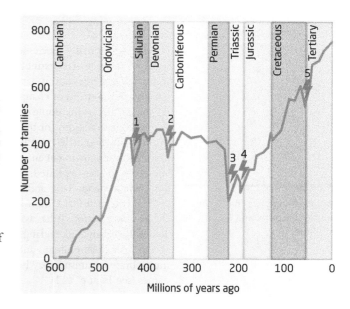

Figure 3.10 The five known mass extinctions on Earth are indicated by lightning strikes on the curve showing the general increase in the number of families over time.

Asteroid or volcano

At the end of the Permian era, 250 mya, a third mass extinction led to the loss of around 54 per cent of all families (95 per cent of all marine species) from the warm shallow seas. The cause is uncertain but could have been either an asteroid colliding with the Earth or a flood of volcanic material escaping from an area known as the Siberian Traps in what is now eastern Russia. Plants and plankton were destroyed by huge releases of carbon dioxide, and oxygen levels in the sea were reduced. Only a few molluscs survived in the black, deoxygenated mud that was left, and it took 20 or 30 million years for coral reefs to be re-established and for the forests to regrow.

Undersea volcanic eruption

The End Triassic extinction occurred about 200 mya and is estimated to have claimed about 23 per cent of all families (50 per cent of all marine invertebrates and around 80 per cent of all land quadrupeds). Oddly, plant species were not so badly affected. The probable cause was flood-like lava escaping from a volcano in the Atlantic Ocean.

Asteroid or comet collision

The Cretaceous–Tertiary mass extinction of 65 million years ago is famously associated with the extinction of the dinosaurs. Virtually no large land animals survived, plants were greatly affected and nearly 50 per cent of marine life was wiped out; total loss of families was about 17 per cent. Global temperature was 6–14 °C warmer than at present, with sea levels over 300 m higher than those of today. At this time, oceans flooded up to 40 per cent of the continents. An enormous asteroid or comet colliding with the Earth in the sea near the Yucatan peninsula in Mexico is currently accepted as the most likely cause.

The five extinctions are summarised in Table 3.01.

Extinction	Time/mya	Geological era	Loss of species (estimate)
1	439	Ordovician	25% families
2	364	Devonian	19% families
3	251	Permian–Triassic	54% families
4	199–214	End Triassic	23% families
5	65	Cretaceous–Tertiary	17% families (all dinosaurs)

Table 3.01 The five recorded mass extinctions.

3.02 Origins of biodiversity

Human activities – the sixth mass extinction

Populations of many species are in decline, and some scientists have warned that Earth is on the brink of another mass extinction. Unlike the five previous mass extinctions, the sixth is related to human activities and is happening at a much faster rate. Species have been unable to adapt because of the speed of changes on Earth largely caused by humans. Humans have already wiped out many species including large mammals and flightless birds such as the dodo (Image 3.13). The two main reasons for this are the way that humans have spread and occupied territories throughout the Earth, and the development of agriculture in the last 10 000 years. The human population is growing at an exponential rate and, with more people, more land has been taken and more species exploited. Species do not have the chance to move to new areas, and pollution and **climate change** have added to the destruction of their ecosystems.

The International Union for Conservation of Nature (IUCN) is an organisation dedicated to nature conservation. It has assessed those species that are most at risk, based on data relating to critically endangered, endangered and vulnerable species. The IUCN estimations are that 30 per cent of amphibians and 28 per cent of reptiles are seriously under threat and many smaller invertebrate organisms may become extinct before they have even been discovered and given names (see Figure 3.11).

Climate change is a change in weather patterns that lasts for a long time.

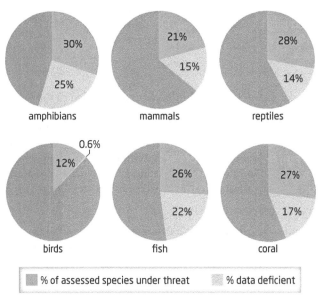

Figure 3.11 Amphibians and reptiles have a high percentage of species at risk, but coral and fish are also severely in danger.

CONSIDER THIS

Up to one in six of the species alive today may become extinct by the end of the 21st century. Some studies have suggested that, if greenhouse gases were limited and climate change was reduced, the number of extinctions could be halved.

Read more: www.smithsonianmag.com/science-nature/climate-change-will-accelerate-earths-sixth-mass-extinction-180955138/

SELF-ASSESSMENT QUESTIONS 3.02.04

1. How many mass extinctions have occurred in the past?
2. How does the current mass extinction compare with previous extinctions: **a** in its cause, **b** in possible consequences?
3. **Research idea:** Why do scientists not know exactly how many species are

Threats to biodiversity

3.03

LEARNING OBJECTIVES

After reading this chapter you should be able to:

- understand that estimates of numbers of species on Earth vary considerably
- describe how the current rate of species loss is greater than in the past, mainly due to human influence
- explain how the conservation status of a species is categorised by the IUCN
- discuss the conflict between exploitation, development and conservation of species in rainforest biomes in less economically developed countries (LEDCs).

KEY QUESTIONS

How is global biodiversity declining in response to human activity?

How is knowing the conservation status of a species useful in the conservation of biodiversity?

3.03.01 Species estimates and rates of extinction

There is no accurate figure for the number of species alive on Earth today. Organisms that are found are described by scientists and recorded or stored in institutions such as the Natural History Museum in London and the collections of other research organisations (see Figure 3.12). Estimates of the number of species that have never been found or named vary widely. An accepted view is that over 1.5 million species have been described but that there may be more than 10 million actually alive (see Table 3.02). Most estimates of the number of species are based on mathematical models. These models depend on the amount of data which is available to be put into them. Many habitats and groups of species are under-recorded because it is difficult to reach them. There may also be insufficient funding for expeditions and scientific research and, in some cases, there is disagreement about the classification of certain groups.

Some scientists think that the figure for living species may be substantially higher than 10 million.

Over geological time, the number of species on Earth has changed many times. Species become extinct and new species evolve. **Extinction** is defined as the point when a species ceases to exist or the last known individual of the species dies. Palaeontologists characterise mass extinctions as times when the Earth loses more

Extinction is the point when a species ceases to exist or the last known individual of the species dies.

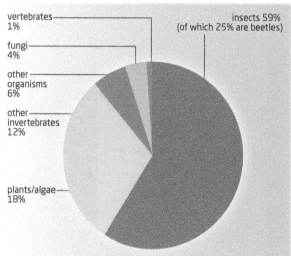

Figure 3.12 The largest group of organisms that have been identified is the insects, and 25 per cent of all named insect species are beetles. Adapted from: UNEP–WCMC (2000). Global Biodiversity: Earth's living resources in the 21st century. Cambridge, World Conservation Press.

3.03 Threats to biodiversity

Kingdom	Number of species
bacteria	4 000
fungi	72 000
protoctists (algae and protozoa)	80 000
plants	270 000
animals (vertebrates)	52 000
animals (invertebrates)	1 272 000
total number described	1 750 000
possible number of unknown species	14 000 000

Table 3.02 Estimates of total numbers of species.

Background extinction rate is the natural extinction rate of all species.

than three-quarters of its species in a geologically short interval. This has happened five times in the past 540 million years or so and has always been caused by major natural abiotic events. The five major extinction events are known from studies of the fossil record, with the most recent, the Cretaceous mass extinction, ending some 65 million years ago (see Chapter 3.02). Given the many species known to have disappeared in the past few thousand years, some biologists are convinced that a sixth such event is now underway.

The **background extinction rate** is the natural extinction rate of all species. Scientists estimate that it should be about one species per million per year, or up to 100 species per year. Just over 5000 mammal species are known to be alive today, so the background extinction rate should be one per 200 years, but in fact about 90 mammals have become extinct in the last 400 years and 170 are listed as critically endangered. This evidence suggests that the current extinction rate is far higher than it should be due to natural causes. In many cases, there is ample evidence that humans are causing the extinctions. But since the total number of species known to science is only a small fraction of the estimated total, estimates of extinction rates also vary. The current rates are probably between 100 and 10 000 times greater than the background rates.

Species loss and human activities

The key factors responsible for the current mass extinction are:

- loss of habitat to agriculture, cities, roads and industry
- overexploitation of resources such as timber and fish, and also reduction of them due to hunting and agriculture
- pollution of waterways, oceans and soil so that they become uninhabitable
- introduction of alien species as humans move species from one continent to another.

Image 3.04 shows a habitat before and after destruction. Some of the factors that influence current extinctions are summarised in Table 3.03.

Image 3.04 A habitat before and after destruction.

Factor	Influence on risk of extinction
Numbers of a species	Small populations have less chance of survival if conditions change. They may have a small gene pool and be unable to adapt (e.g. cheetah). They may be isolated in small areas or islands, like the tortoises on the Galapagos Islands. Common species that have a wide distribution are less likely to be endangered.
Degree of specialisation	Some organisms are entirely dependent on one resource for their survival. For example, some orchids rely on a single insect species to pollinate them, and giant pandas depend almost exclusively on bamboo for food. If the insect or the plant becomes scarce, the dependent organism is at risk.
Reproductive potential	Species which reproduce slowly are at greater risk. Species such as elephants live a long time and have a low reproductive rate, and so are more vulnerable. If species are hunted or overfished, those that reproduce more slowly are unable to restore their populations to sustainable numbers.
Behaviour	Flightless birds such as the dodo and the elephant bird evolved in places without humans. They were not afraid of humans and had poor defences. This made them vulnerable and easily killed, so the birds soon became extinct. Individuals of some species, such as elephants and whales, have such close bonds with one another that they remain with dying members of their group and put themselves in danger. Animals that live in herds or flocks rely on one another for protection and become vulnerable if they are isolated in small groups.
Trophic level	Top predators are more sensitive and at risk than species at other trophic levels in a food chain. The numbers of species at the end of a food chain are lower than those of other species, and reductions in prey species lower down the food chain can have serious risks. Large predators such as tigers are also at risk from hunting.
Distribution	When a species has a limited range, it is more at risk than widely distributed species. The spoon-billed sandpiper is limited to a small area in the far east of Russia, and there are thought to be fewer than 200 birds left in the wild.
Valued resources	Species that are valued for parts of their bodies are vulnerable to hunting and overexploitation. Ivory from elephants, bush meat from large primates, oil from sperm whales and desirable feathers from birds such as the emu have all added to the vulnerability of these species.

Table 3.03 Factors that influence the risk of extinction.

3.03.02 Threatened species

The Red List and determining conservation status

The International Union of Conservation of Nature (IUCN) helps to develop environmental policy and laws, working with governments to develop economic policies that take into account the preservation of biodiversity. Each year since 1963, the IUCN has published the Red List (or Red Data Book), which draws together information from many international organisations about species that are threatened. This list is the world's most accurate record of the conservation status of vulnerable species. It is compiled using precise, objective criteria to evaluate the risks to a species. The hope is that conservation issues can be quickly communicated to the public and policy-makers.

3.03 Threats to biodiversity

Image 3.05 The Roman or edible snail (*Helix pomatia*) (LC).

Image 3.06 The hog badger (*Arctonyx collaris*) (NT).

Image 3.07 The Eastern box turtle (*Terrapene carolina carolina*) (V).

The Red List places species in one of seven categories, based on their conservation status. Factors used to assign a species to a category are criteria with quantitative thresholds for population size and population trend, geographic range and range size, numbers of mature organisms, quality and size of habitat and likelihood of extinction. The criteria are designed to be objective, quantitative and repeatable, and to handle uncertainty. The seven categories are:

- least concern (LC) – species not qualifying for the other categories, including widespread and abundant species
- near threatened (NT) – close to qualifying for vulnerable status
- vulnerable (V) – facing a high risk of extinction in the wild
- endangered (EN) – facing a very high risk of extinction in the wild
- critically endangered (CR) – facing an extremely high risk of extinction in the wild
- extinct in the wild (EW)
- extinct (E).

Information on the reasons for threats and extinctions is also included in the Red List. Some examples from each category are described below.

Least concern

The Roman or edible snail (*Helix pomatia*) (see Image 3.05) is abundant in south-eastern Europe in forests and open habitats, gardens and vineyards, and especially along rivers. The snail is threatened by drainage, intensive farming and habitat destruction. In some European countries there are restrictions on collecting the snails, but its numbers are sufficient to maintain the species at present, so it is listed as least concern.

Near threatened

The hog badger (*Arctonyx collaris*) (see Image 3.06) is listed as near threatened because it is undergoing population decline. It has been hunted out of many areas of Lao, Vietnam and south-east China where it was once widespread. The badger is active in the day time and usually found in forests where it feeds on roots, worms and insects. It is not very wary of humans and it is often hunted with dogs.

Vulnerable

The eastern box turtle (*Terrapene carolina carolina*) (see Image 3.07) is a hinge-shelled turtle native to the eastern USA. It is the state reptile of North Carolina. It crawls slowly and is extremely long-lived, slow to mature, and has a low rate of reproduction with few offspring produced per year. The turtles are also injured and killed in accidents with vehicles and agricultural machinery, which makes them a particularly vulnerable species, classified as 'vulnerable'.

Endangered

The Ecuadorian anole, *Anolis proboscis* (see Image 3.08), is a small forest lizard found in Ecuador. It is listed as endangered because it has a restricted range and occurs in an area of only 200 km², with all known individuals in only five locations. Its habitat is under threat from logging, grazing, human settlement and agriculture, all of which are expected to cause a further decline in this species. As it lives in such a small area, it is more vulnerable to habitat alteration than species with a wider distribution.

Image 3.08 The Ecuadorian anole (*Anolis proboscis*) (E).

Image 3.09 The gorgeted puffleg hummingbird (*Eriocnemis isabellae*) (E).

Critically endangered

The gorgeted puffleg hummingbird (*Eriocnemis isabellae*) (see Image 3.9) is a recently discovered species from Colombia, which appeared for the first time on the IUCN Red List in 2009 and was classified as 'critically endangered'. This brightly coloured bird has only 1200 ha of habitat remaining in the cloud forests of the Pinche mountain range in south-west Colombia where it is found. Large areas of this forest are damaged every year as local people remove trees to grow coca.

Extinct in the wild

The northern white rhinoceros (*Ceratotherium simum cottoni*) (see Image 3.10) was once found grazing on the grasslands in central Africa. The last surviving population lived in Garamba National Park in the Democratic Republic of the Congo, but the animal is now extinct in the wild. Just seven animals remain in captivity in zoos and nature reserves. Only four of these are able to reproduce, and they now live in a Kenyan nature reserve where scientists hope they will successfully breed.

The main cause of their decline was poaching by hunters, who wanted only their horns to sell for their alleged aphrodisiac properties. Hunting drove the species to near extinction as these slow-breeding animals could not maintain their population. The remaining animals have had their horns humanely removed so that they are of no value to hunters. They have also been fitted with radio transmitters so that they can be monitored when they are finally released back into the wild.

Image 3.10 The northern white rhinoceros (*Ceratotherium simum cottoni*) (EW).

Extinct

The Mount Glorious torrent frog (*Taudactylus diurnus*) (see Image 3.11) was an endemic Australian species which was found in separated populations in three sub-coastal mountain ranges in south-east Queensland. In the early 1970s it was considered to be relatively common, but it has not been seen in the wild since 1979. The frog disappeared over a period of about four years. The reasons for its extinction are not precisely known. Logging in its habitat, as well as invasion of weed species (which affect water flow and quality) and activities of feral pigs are all thought to have contributed. Diseases have also caused the deaths of similar species in other parts of the world.

Extension

Closing a deadly gateway
www.youtube.com/watch?v=OC9CATzZCO4

Image 3.11 Mount Glorious torrent frog (*Taudactylus diurnus*) (E).

3.03 Threats to biodiversity

CONSIDER THIS

The most recent Red List was released by the IUCN at the 2012 Earth Summit in Rio. Four more species were added to the extinct category. The IUCN assessed 63 837 species and listed 3947 as critically endangered and 5766 as endangered, with more than 10 000 in the vulnerable category. They estimate that 40 per cent of amphibian species, 34 per cent of reef-building corals, 30 per cent of conifers, 25 per cent of mammals, and 13 per cent of birds are at risk.

SELF-ASSESSMENT QUESTIONS 3.03.01

1. Suggest four factors that make some species more vulnerable to extinction than others.
2. What are the factors used to decide whether an organism is listed in the Red List?
3. Outline the factors that have led to the extinction of a species that has been lost as a result of human activity.
4. **Research idea:** Use the internet to discover other species of invertebrate that are endangered or critically endangered. Discuss the importance of the species you have chosen to the food webs they are part of.

CASE STUDY 3.03.01

Extinct: the dodo (*Raphus cucullatus*)

The dodo (see Image 3.12) was a flightless bird that lived on the island of Mauritius, east of Madagascar. The earliest written descriptions of the bird were made by Dutch travellers to the island in 1601.

Image 3.12 Dodo.

Ecology

Dodos were about the size of a turkey, 1 m tall and weighing up to 20 kg. Their breast bone was not strong enough to support flight muscles, and the birds had short, stubby wings. They were unable to fly or swim. Specimens of the bird were brought back to Europe for study in the 17th century, and they were described as having downy plumage.

Dodos lived on the ground and evolved to take advantage of their island ecosystem, where there were no predators and they could grow to a large size. They fed on fruit and nested on the ground, laying a single egg. Mauritius has separate dry and wet seasons, so the dodo probably fed on ripe fruits at the end of the wet season to build up reserves for the dry season when food was scarce. Some reports say that the birds were greedy and others say that the birds used stones in their gizzard to help digestion.

Socioeconomic pressures

When humans first visited Mauritius, they were able to capture dodos easily, because the birds had no fear of people. This, combined with their inability to fly, made them easy prey. There are some accounts of large numbers of birds being caught and killed as food for passing ships.

Many sailors' journals reported that the meat of the dodo was tough, so dodos were probably not regularly hunted.

As people began to settle on Mauritius, they brought animals including dogs, pigs, cats, rats and macaques, which preyed on the dodos and their eggs. The population of dodos was probably reduced more by these animals than by hunting. The plight of the birds was made worse by people who destroyed the forests where the birds lived. A scientific expedition in

CASE STUDY 3.03.01 (continued)

2005 also found that, when the dodo was most in danger of becoming extinct, a flash flood on Mauritius reduced the few remaining birds' chances of survival.

Consequence of disappearance

The dodo became extinct just 80 years after it was first discovered. It is the first example of a bird whose disappearance was primarily due to human activities. Today, the dodo is used as a symbol by environmental organisations that promote conservation and the protection of endangered species. It also appears on the coat of arms of Mauritius.

Questions

1. Do you think that modern conservation measures would have saved the dodo?
2. Why was it so difficult for scientists to be sure about the behaviour and appearance of the dodo?
3. Research idea: Investigate other flightless birds and evaluate why they have not become extinct like the dodo.

CASE STUDY 3.03.02

Critically endangered: Beluga sturgeon (*Huso huso*)

Ecology

The Beluga sturgeon (see Image 3.13) is an Asian and European species found in the Caspian Sea and the Black Sea. Although it used to live in the Adriatic and Azov Seas, it is no longer found there due to overfishing and the construction of dams. Beluga sturgeon in the Caspian Sea are listed as critically endangered. They are the main producers of wild caviar. The fish spend part of their life cycle in salt water and return to rivers to breed. As they are a migratory species, the damming of rivers across Europe, Asia and North America over the past century has led to many sturgeon species losing access to vast areas of their traditional spawning grounds.

Image 3.13 Beluga sturgeon (*Huso huso*).

The last wild population in the Black Sea basin migrates up the Danube River. A second population that lives in the Volga depends on humans restocking rivers with their eggs, since the construction of Volgograd dam in 1955 led to the loss of almost all of the species' spawning sites in the Volga River.

At sea, this species is found in the pelagic zone, following food organisms. It migrates further upriver to spawn than any other sturgeon; however, this migration has now been disrupted due to river regulation (in the Danube drainage up to the Morava River). It spawns in strong-current habitats in the main course of large and deep rivers on stone or gravel bottoms.

Sturgeon can live for up to 100 years and do not reproduce every year, which means their populations take many years to recover from any decrease in numbers. Males do not reproduce until they are 10–15 years old, females when they are 15–18 years. They spawn every three to four years between April and June.

Socioeconomic pressures

The main reason for the decline in Beluga sturgeon populations, is overfishing for black caviar – the sturgeon's unfertilised eggs. This caviar is considered the finest in the world and can cost US$10 000/kg. In addition, the skin of the fish is used for leather, and its intestines used in food sauces and to produce gelatine and glue.

Overfishing occurs at sea, while poaching in estuaries and rivers for meat and caviar is another major threat. The largest and most mature fish are removed from the population, so natural reproduction is almost zero. In the Ural River, current fishing rates are estimated to be four to five times sustainable levels. As well as the Volgograd dam, the Don River dam also destroyed huge areas of spawning ground, and flow regulation in the Kuban River led to the loss of still more.

3.03 Threats to biodiversity

CASE STUDY 3.03.02 (continued)

Beluga sturgeon live for up to 100 years so pesticide accumulation may also be leading to further problems, including reduced reproductive success.

Consequence of disappearance

The decline in fish has brought problems of overfishing and habitat loss to international notice, because of the popularity of black caviar. Alternative sources of caviar have been investigated to assist in the conservation of fish. As the Beluga sturgeon lives for a long time, a few fish can still be found in areas where they are cut off from their spawning sites. One hopeful sign for the species is its ability to produce millions of eggs. If proper protection is put in place, harvested eggs may be used to replenish fish stocks. Since 2000 in Russia, fishing has only been permitted for the purposes of reproduction (for hatcheries) and science. But protective measures at the feeding grounds are also needed to maintain the population of beluga. Most of the few wild females that have been monitored are in their first year of maturation. Based on catch data and the number of recorded spawning individuals, it is estimated that the species in the wild has seen a population decline of over 90 per cent in the past three generations (60 years), and unless action is taken the remaining natural wild fish will soon be extinct.

Questions

1. How does the longevity of the sturgeon contribute to its problems of survival?
2. Suggest some other factors, in addition to those mentioned, that may contribute to the decline of migratory fish like the sturgeon.
3. Research idea: Compare the plight of the sturgeon with that of the European eel. What are the similarities and differences between the survival of the two species?

CASE STUDY 3.03.03

Improved by intervention: the Arabian oryx (*Oryx leucoryx*)

In 1986, the Arabian oryx (see Image 3.14) was classified as endangered on the IUCN Red List, and in 2011 it was the first animal to receive vulnerable status again after having been listed as extinct in the wild.

Image 3.14 The Arabian oryx has been saved from extinction by intervention and captive-breeding programmes.

Ecology

The oryx is a grazing antelope that is adapted to survive in the extreme conditions of hot, dry deserts. The animals live in small herds of 10–30 animals with a hierarchy of dominance among both males and females. They defend their territories using their horns, and they have keen eyesight to maintain the group. They also use their horns to dig shallow pits to rest from the heat of the day. The oryx lives across the Arabian and Sinai Peninsulas and has been reintroduced into Oman, Israel, Saudi Arabia and Jordan.

Socioeconomic pressures

By the early 1970s, the Arabian oryx was extinct in the wild as a result of hunting by poachers who chased them across the desert in four-wheel-drive vehicles. A new population was established by breeding animals that were held in zoos in different parts of the world. A captive-breeding programme began in 1962 at the Phoenix Zoo and was supported by the Fauna and Flora Preservation Society of London and the World Wildlife Fund (now known as the World Wide Fund for Nature; WWF). It began with nine animals, and soon oryx were sent to other zoos and parks to start new herds. The pedigree of the captive-bred animals was monitored to ensure that a sufficiently large gene pool was maintained.

CASE STUDY 3.03.03 (continued)

Animals were reintroduced into the wild in 1982, and the population thrived for about 15 years until poaching began again in 1996. However, this time laws were changed to put a stop to poaching, and a second reintroduction took place in Saudi Arabia. So far this population has survived successfully. The total reintroduced population now stands at about 1000 animals and is well over the threshold number of 250 mature individuals needed to qualify for endangered status.

The IUCN estimates there are also 6000–7000 Arabian oryx held in captivity worldwide in zoos, reserves and private collections.

Questions

1. The oryx had survived hunting for many years before it became endangered. Suggest why the situation changed in the mid-20th century.
2. Summarise the importance of zoos and nature parks in protecting endangered species.
3. Research idea: Investigate other species that survive in desert conditions. Check their conservation status using the IUCN website.

3.03.03 Vulnerability of tropical biomes

Tropical rainforest biomes are located only between the Tropics of Capricorn and Cancer and so are restricted to the land between the latitudes 23.5° north and 23.5° south of the equator; despite this they are among the most important biomes on the planet. Between the tropics, they receive constant sunlight with little seasonal variation, so they have fairly constant temperatures. Rainfall is high. The majority of the Earth's rainforest is in the Amazon Basin, which contains about one-third of all rainforests. A further one-fifth is found in Indonesia and the remainder in countries of central Africa, mainly the Democratic Republic of the Congo, but also large areas of Gabon, Cameroon, Equatorial Guinea, the Central African Republic and the Republic of Congo (see Chapter 2.04).

Rainforest as an ecosystem

Rainforests are multi-layered ecosystems that contain a wide variety of habitats in the emergent layers, the canopy layer, the understorey and on the forest floor (see Figure 3.01).

This complex layering of the tropical trees and plants means that there are many niches in the forest; it is estimated that more than 50 per cent of all species on Earth today live in rainforests. There may be as many as 300 species per hectare. The rainforest has a high species diversity and high habitat diversity, and many of the species found in forests are endemic, which means that they are not found anywhere else. Rainforests also have a high rate of photosynthesis and release almost 40 per cent of the oxygen that is required by other organisms. They also contain huge reserves of carbon in the form of timber, which is in great demand by the world's growing population. In addition, tropical rainforests have a vital role in regulating world weather patterns, as they help to maintain regular rainfall and to buffer countries against floods, droughts and erosion.

Loss of rainforest

Over recent years, rainforests have been lost at a rapid rate. In the middle of the 20th century, about 15 per cent of the Earth was covered in forest, but now the figure is less than 6 per cent

3.03 Threats to biodiversity

of the land surface. Much of the remaining forest that once covered vast expanses of land is now fragmented into small areas, separated from one another. The most pessimistic estimates predict that rainforests will have disappeared within 50 years due to human activities and interference.

Forest loss has mainly been caused by agriculture and logging. Both these activities are carried out on a commercial scale to supply the world's need for timber, cattle, palm oil and soya. But another serious threat to the forests has come not from large organisations, but from subsistence farmers and small groups. As the population of the world has increased, pressure on rainforests has also increased. Small areas of forest are cleared for crops or animals, but because of the poor soil quality the nutrients are exhausted after two or three years, so farmers move on to another area and repeat the process. Most of the nutrients in the forest are held in plant biomass rather than the soil or leaf litter. This means that abandoned areas of previously cultivated land do not recover well and may take up to 100 years to return to their original diversity. In some cases, selective harvesting of trees for timber can be carried out in a managed way so that trees are able to regrow. But if too many trees are removed, the balance between the different layers in the forest cannot recover and the climax vegetation cannot become re-established.

Unsustainable exploitation of these biomes leads to huge losses in biodiversity and reduces the ability of the forests to capture carbon dioxide and maintain the quality of the soil.

Difficulties in controlling forest loss

Controlling the loss of biodiversity requires international legislation. The willingness to participate in conservation initiatives varies from country to country and is very dependent on economic, social and political issues. Most tropical biomes are located in LEDCs, and in these countries there is a conflict between exploitation of resources, sustainable development and conservation. One LEDC where there is such a conflict of interest is Madagascar, which is trying to provide for its people and conserve its wild habitats. Some of the issues are described in Section 2.03.04.

CONSIDER THIS

Water vapour produced by rainforests carries a huge amount of heat, which is released when the water condenses and falls as rain. Air masses tend to circulate away from the equator to higher latitudes, and so the heat energy in tropical clouds is carried to cooler regions of the Earth. Rainforests spread out solar radiation to temperate zones quickly and efficiently.

Theory of knowledge 3.03.01

Extinction – a moral and a material issue

'The Red List may be less about true extinction than impoverishment: the closing down of personal experiences, the vanishing of living landmarks that help define places and communities, the eclipse of the fundamental elements of those intricate, mutually dependent networks of animals and plants that make up ecosystems. It suggests that alongside the sometimes crude index of biodiversity, we need another: bioluxuriance, a measure of the spread of organisms, of their living where they belong, not herded into biological ghettoes and token nature reserves.

Yet if we're beginning to acknowledge and document this creeping erosion of the living systems we inhabit, why aren't we more bothered? Why isn't local extinction – the first, inexorable step towards a more comprehensive termination, after all – a source of horror, of self-interested panic, of public response? Perhaps, ironically, the acceptance of the process of evolution has credited it with almost supernatural regenerative power: the great engine of life will cope, somehow. Or perhaps we believe we will eventually be able to do the job ourselves, putting back into the wild species we've nurtured in captivity, even recreating vanished plants from strips of DNA.

Extinction is a fact of life, and doesn't have the power to shock any more. Stephen Jay Gould's extraordinary book, *Wonderful Life*, put paid to any lingering beliefs in the sanctity of species, or the inexorable "upward" progress of evolution. Five hundred million years ago there was an explosion of biological creativity that generated a variety of life far surpassing that existing on the Earth today. All of it has vanished, been discarded, without a single human lifting a destructive finger.

Knowledge makes extinction a moral problem as well as a material one. Can we accept being conscious witnesses – or worse, accomplices – in the erasure of the unique outcomes of hundreds of millions of years of natural trial and error? And do we think we know enough to draw ethical lines, decide population sizes, judge which species are of most importance – not for us, but for the biosphere?'

Richard Mabey, *The Guardian*, 18 June 2005

1. Why does the writer suggest that extinction is a moral as well as a material problem?
2. Could it ever be right for humans to decide which species are of most importance?

3.03.04 Degradation of an area of biological significance by human activities

Madagascar – a biodiversity hotspot

The island of Madagascar has a unique and irreplaceable biodiversity, with 80 per cent of its species found nowhere else on Earth. These include endemic lemurs, six of the eight species of baobab tree, and *Uroplatus* geckos, species that evolved as a result of the island's isolation from neighbouring continents. But Madagascar has suffered environmental degradation over large parts of its land area, and many species have been lost or endangered. Forests that once covered the eastern third of the island have been cut down, fragmented and converted to scrub land. Spiny forests in the south are being replaced by cactus scrub as indigenous vegetation is taken and burnt for charcoal production. Soil is being eroded from the central highlands and washed away, and it is estimated that as much as a third of the country is burnt and 1 per cent of its remaining forests are felled each year. The area of natural forest is less now than at any time since Madagascar was first inhabited by humans 2000 years ago.

Human activities

Madagascar is among the world's poorest countries, where people depend on the land and its resources for their survival. Madagascar's population has increased from 5 million in 1960 to 20 million in 2010, and 85 per cent of local people live on the poverty line. Population increase and poverty are significant factors contributing to the loss of the island's biodiversity.

Many factors have caused damage to the environment, but the most important are:

- deforestation and destruction of habitat
- agricultural fires
- erosion and soil degradation
- hunting and collection of wild species
- introduction of alien species
- mining for natural resources.

Deforestation

Most deforestation has taken place as people convert rainforest into rice fields in a traditional practice known as 'tavy'. Small areas of forest are cut and burnt and planted with rice for subsistence farming. After a few years, the field is left fallow before the process is repeated. After two or three cycles of this type of agriculture, the nutrients in the soil are exhausted and the land is left to be colonised by scrub vegetation or alien grasses. Fires that are started for land clearance often spread to other areas and increase the amount of damage.

Logging also causes serious loss of forests and is a particular problem in eastern Madagascar, where the valuable hardwood ebony and rosewood trees grow. Some areas are protected, but even here illegal logging is a significant problem.

Following deforestation, erosion becomes a serious issue. Astronauts have commented that, when viewed from space, Madagascar's rivers make the country look as though it is bleeding, as red soil runs into them and out into the Indian Ocean. Deforestation in the highlands, added to natural losses from weathering, mean that up to 400 tonnes per hectare per year of top soil are being lost (see Image 3.16).

Exploitation of living resources

Because they are so unusual, the indigenous animals of Madagascar have been hunted and trapped for collectors and as pets. Since 1964, it has been illegal to kill or keep lemurs, but they are still hunted in some areas. Many reptiles and amphibians are in demand as pets in other countries, so

3.03 Threats to biodiversity

Image 3.15 Onibe River, Madagascar carries precious topsoil eroded from the land down into the sea.

local people collect large numbers of geckos, snakes, tortoises and chameleons. Tenrecs (small insectivores) and other carnivores are also sometimes killed for food.

The seas around Madagascar are rich in fish, but local laws are not powerful enough to keep away the many foreign fishing boats which take large catches from the area. Sharks, lobster and sea cucumbers are often harvested in unsustainable numbers. In recent times, permits have been issued for oil exploration in Madagascan waters, and this may add further to the damage to species in the ocean.

Threat from alien species

Alien species are those that are not native to Madagascar but which have been introduced by people. One of the best examples is the tilapia, an aggressive fish which was introduced for food but which survives well on the island and soon displaces the native cichlid species in rivers and lakes. The snakehead murrel is another species of carnivorous fish that was introduced into eastern Madagascar where the highly vulnerable Alaotra grebe used to live. The grebe was already endangered by habitat loss and the increase in fishing, and since the arrival of the murrel, grebe numbers have fallen so much that it is now officially extinct.

Consequences and conservation

Madagascar has 24 critically endangered and endangered species and a further 26 that are listed as vulnerable. Loss of biodiversity not only has consequences for local people and the beauty of the countryside; it also deprives future generations of valuable resources that threatened species might hold. Many medicines have been developed from natural sources, and if organisms are destroyed before they have been studied, many potential new sources may disappear. In recent years, efforts have been made to help save what remains of the island's special biodiversity. The country's government began a programme to extend environmental protection, and in 2007 the number of national parks was increased to 60. The National Association for the Management of Protected Areas in Madagascar has introduced a new park management system to conserve wildlife using sustainable development programmes that can also provide direct benefits to the local people.

International conservation organisations have also helped in conservation efforts. The World Bank and the WWF purchased $5 million of the country's foreign debt at a discounted rate in exchange for government support for local conservation projects. Ecotourism, agriculture, expansion of international trade as well as investment in education and health are key elements of a policy to develop Madagascar's economy. But the country is faced with the political challenge of balancing growth and development with conservation.

SELF-ASSESSMENT QUESTIONS 3.03.02

1. The environment of Madagascar has been damaged by human activities. What are the key economic and ecological pressures that have led to its degradation?
2. What are the four most important human activities that threaten species on Madagascar and in other similar regions?
3. **Research idea:** Magadascar is described as a biodiversity hotspot because of the large number of indigenous species that live there. Investigate other hotspots and the species they contain.

Conservation of biodiversity

3.04

LEARNING OBJECTIVES

After reading this chapter you should be able to:

- describe how arguments about conservation can be based on aesthetic, ecological, ethical or economic grounds
- explain the criteria used to manage protected areas
- describe how the loss of biodiversity increases conservation efforts
- understand how local community support is vital to the success of conservation efforts
- explain various approaches to conservation and evaluate their strengths and weaknesses.

KEY QUESTIONS

How does the loss of biodiversity drive conservation efforts?

How is a society's environmental value system (EVS) linked to arguments about conservation?

What are the strengths and weaknesses of different approaches to the conservation of biodiversity?

3.04.01 Arguments for preserving species and habitats

It is difficult to quantify the value of a species or a habitat, and conservation arguments struggle to balance ethical and aesthetic values with commercial considerations:

- Economic arguments cite the valuation of ecotourism, genetic resources and natural capital.
- Ecological arguments centre on preservation of the ecosystem.
- Ethical arguments include the intrinsic value of a species.

So-called direct value is easier to measure because it can be calculated in terms of economics. Indirect value is assessed in terms of the 'services' that an ecosystem provides for the local or global community. These values include scientific and educational value, potential sources of medical products and value to the world climate and weather patterns.

Values of biodiversity that can be measured directly

Food comes from both plant and animal sources, and its value can be measured. There are many strains of common food plants such as rice and maize and species of farmed animals such as sheep and cattle. It is important to maintain a gene pool of these species so that there is sufficient diversity available if changes occur in the future. For example, older or wild varieties of plants may have genes that could be introduced or reintroduced into domesticated strains to promote resistance to a pest. Rare or traditional breeds of animal may retain genes that confer protection against new diseases, which modern varieties may have lost.

3.04 Conservation of biodiversity

All the cultivated bananas that we eat have been produced from one original variety. The bananas are clones that are produced asexually and so are genetically identical. This means they cannot evolve resistance to disease. In Honduras, an agricultural research organisation is cross-breeding cultivated bananas; wild ones to create new varieties that are resistant to Panama disease. Panama disease is a serious threat to bananas, and if it spreads throughout the world, it has the potential to wipe out the worldwide banana crop.

Natural products including timber and pharmaceuticals are obtained from plants and animals. Many modern medicines are synthesised using ingredients from plants; examples include digoxin from digitalis, and aspirin, which is a chemically synthesised salicylic acid that resembles compounds in white willow bark. Morphine and other opiates are derived from the opium poppy, and other plants provide humans with rubber, palm oil and cotton. Animals are sources of silk, honey and milk as well as meat and fish.

Values of biodiversity that are difficult to measure directly

Many aspects of biodiversity cannot be evaluated or measured directly. It is impossible to quantify the aesthetic value of an unspoilt ecosystem or the pleasure it provides for recreation and tourism, and furthermore an ecosystem with natural biodiversity is also a valuable source of education. New species are being discovered all the time, and information about them will be vital to educate people everywhere.

Ecosystem productivity is not simply a measure of biomass; it also includes aspects of the system that are vital to its function and to the function of other systems. Plants remove carbon dioxide from the air and release oxygen, which is essential for environmental stability. As we have seen, climate is regulated by rainforest plants. Within any ecosystem, decomposers also have an essential role in recycling waste to provide nutrients. Insects and other pollinating animals are needed by crop plants to ensure their productivity, and many plants rely on animals to disperse their seeds. Preservation of biodiversity keeps natural systems in their stable, balanced state. Stable systems are less likely to be disturbed by **external factors** such as disease or abiotic events.

A diverse ecosystem such as a stream contains a variety of organisms which have different tolerances to pollutants or changes in other factors. These are known as indicator species. Changes to an indicator species can show that changes are occurring in an ecosystem long before the change has become measurable. For instance, lichens are key indicator species of changes in the atmosphere. Many are unable to survive if there is a small increase in sulfur dioxide in the air. Their death is an indication of poor air quality.

Genetic diversity is another factor that is preserved in complex ecosystems but is hard to quantify. A larger gene pool is important in enabling populations to adapt to changes in their environments, and wild varieties are a source of new genes for cultivated plants and domesticated animals.

Human rights and ethical considerations are also important. Many different indigenous peoples live in rainforests and other areas. An estimated 50 million of the world's 300 million or more indigenous people live in rainforests. By preserving forests and native lands, these tribes can remain in their traditional homelands. Many modern philosophies accept that, as well as humans, *every* species has a right to exist. From this point of view, it is important that people accept their responsibility to preserve biodiversity for the sake of all species.

External factors Factors regulating populations from outside the population (e.g. predation or disease).

Theory of knowledge 3.04.01

Who has the right to decide?

'Miners going into the mines often used to carry small birds, such as canaries, which were highly sensitive to the buildup of toxic gases. If the birds died, the miners quickly fled. Today, the world's 500 million indigenous peoples are the miners' canary: and the Earth – particularly the rainforest – is the mine. That the canary is dying is a warning that the dominant cultures of the world have become toxic to the Earth. In this case, however, we cannot flee the mine.'

Jason W. Clay, anthropologist

Theory of knowledge 3.04.01 (*continued*)

'*Any discussion about the forests should start by looking at the remaining tribal people for whom the tropical forest has been home for many generations. Their story is one for which we must all be profoundly ashamed. The Yanomami of Brazil (see Image 3.16) are being driven to extinction by measles, venereal disease and mercury poisoning following illegal invasion of their lands by gold prospectors, even now the collective genocide continues.*'

HRH the Prince of Wales

Image 3.16 The Yanomami are a group of indigenous people who depend on the Amazon rainforest, where they live, hunt and grow food.

1. What are the ethical issues involved in conservation efforts in relation to indigenous peoples?
2. How might a member of the Yanomami tribe value rainforest conservation, and how might this opinion differ from that of an outsider?

3.04.02 Intergovernmental and non-governmental conservation organisations

The key objective of conservation organisations is to preserve species and their habitats throughout the world. Some organisations work at a local level, while others are global. Different organisations are categorised according to the way they are set up and funded. **Governmental organisations** follow the policies of one or more governments and are funded by them. **Non-governmental organisations (NGOs)** are funded by individuals or independent groups. The effectiveness of the different organisations varies due to the different strategies they adopt in their work. Some of the differences are summarised in Table 3.04.

> **Governmental organisations** are groups that follow the policies of one or more governments and are funded by them.
>
> **Non-governmental organisations (NGOs)** are groups that are funded by individuals or independent groups.

UNEP and WWF

The United Nations Environment Programme (UNEP) is a governmental organisation that coordinates UN work on the environment and helps LEDCs to implement environmentally sound policies. UNEP was founded in 1972 and has its headquarters in Kenya. Its stated objectives are: 'to provide leadership and encourage partnership in caring for the environment

3.04 Conservation of biodiversity

by inspiring, informing and enabling nations and peoples to improve their quality of life without compromising that of future generations'.

UNEP gathers, collates and verifies data on biodiversity and ecosystems from many sources. This can be used as a reliable source of information. UNEP also promotes global and regional cooperation and develops environmental laws covering a range of issues from the atmosphere to marine and terrestrial ecosystems and the green economy. Like many governmental organisations, it also works with NGOs to implement its policies.

The WWF is an NGO and one of the best-known international conservation organisations. Since 1961 is has campaigned for the natural world and worked to ease pressure on the world's natural resources. The WWF is an independent organisation, but around the world it works with businesses, governments and local communities to create sustainable solutions that take account of the needs of both people and nature.

The organisation's conservation work focuses on safeguarding wildlife and places it considers to be of global importance. It also lobbies governments and runs campaigns to change legislation and policy to protect the environment and biodiversity. Major campaigns have focused on climate change, energy, housing and the protection of the marine environment. The WWF states that its ultimate goal has always been 'people living in harmony with nature' and finding ways to share the Earth's resources fairly.

Although both UNEP and WWF have similar ideals and objectives, the ways in which they operate can be quite different. These differences are summarised in Table 3.04.

	Government organisation, e.g. (UNEP)	Non-governmental organisation (NGO), e.g. (WWF)
Use of media	Professionals produce statements and communiqués. Organisations have good communication with media outlets such as TV news.	Uses the internet and social media, advertisements, membership drives and direct action for communication and publicity. May produce press packs and leaflets. Links can be worldwide via web links.
Speed of response	Slow and bureaucratic; often many countries are involved in negotiations. Each country may have its own view or legal position.	Generally faster; because NGOs are independent, they are able to take decisions quickly.
Diplomatic constraints	Often held back by political arguments between different countries. Must respect legal requirements of each nation.	Not affected by diplomatic constraints.
Political influence	Organisation has direct links to governments of many countries.	Influence is indirect and depends on lobbying and pressure groups and public protests.
Legal powers	Can pass laws on environmental issues.	Public opinion and pressure used rather than legal powers.

Table 3.04 Comparing UNEP and WWF.

Often, there is conflict between what is good for the environment and what is required for a country's economic development and the needs of its population. But conservation organisations work to encourage countries and multinational companies to consider development that is sustainable as present-day needs are met, so that resources and biodiversity can be conserved for the future.

3.04.03 Important international conventions on biodiversity

Meetings, conventions and global summits on issues of world importance such as global warming, whale conservation and carbon trading all raise the profile of conservation with people all over the world. Some produce legally binding agreements, while others simply put pressure on governments to act on conservation issues as a result of publicity and public opinion. The effectiveness of such international conferences and the agreements they produce depends on the cooperation of all signatories and the support of one country by others. The purpose of all these meetings is to encourage collaboration between nations so that biodiversity can be preserved and conservation efforts are effective.

Some key dates and meetings are:

- 1948 – IUCN founded
- 1961 – WWF set up
- 1966 – Red Data Book published
- 1973 – First Convention on the International Trade of Endangered Species of Wild Fauna and Flora (CITES convention) held
- 1980 – World Conservation Strategy announced – highlighting the need to preserve genetic diversity and to ensure sustainable use of species and ecosystems
- 1982 – UN World Charter for Nature agreed
- 1991 – IUCN Caring for the Earth conference
- 1992 – United Nations Conference on Environment and Development (UNCED) – the Earth Summit, in Rio de Janeiro
- 2000 – UN Millennium Summit
- 2002 – UN World Summit on Sustainable Development
- 2005 – UN World Summit, New York
- 2012 – UN Earth Summit Rio +20

3.04.04 Approaches to conservation

When a programme of conservation of a species is being considered, three approaches may be taken:

- conservation of a habitat so that species can survive
- conservation based on protection of a species
- a combined approach to conserve both the habitat and the species within it.

Criteria used to design areas of protected habitat

One effective way to conserve biodiversity and protect endangered species is to protect their habitats from interference. When governments set aside land for new protected nature reserves, many factors must be taken into account to ensure they are successful in promoting conservation of diversity. The shape, size and interconnections between protected areas are very important in achieving the maximum conservation of species and habitats (see Figure 3.13).

3.04 Conservation of biodiversity

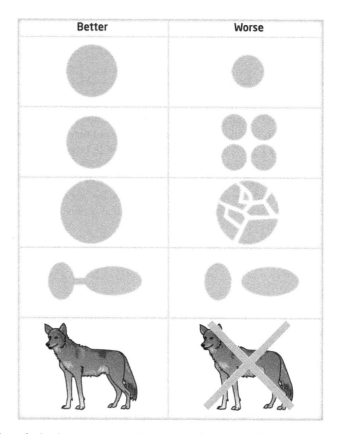

Figure 3.13 When designing a conservation area, a large area is better than a small one; a single large area is better than several small areas of the same total size; an intact area is better than a fragmented or disturbed one; areas connected by corridors are better than separate isolated areas; and it is better to have large native carnivores present in the protected area than not.

SELF-ASSESSMENT QUESTIONS 3.04.01

1. List three important reasons to conserve species and habitats.
2. Why is international cooperation important in conservation and the preservation of biodiversity?
3. Give two similarities between the roles of governmental organisations and NGOs, and two differences between them.
4. **Discussion point:** How much influence do NGOs have on the work of governmental organisations?

Size

Large reserves usually work better than small ones; they can support more species and a greater diversity of any one species, and so provide more complex interactions. Small reserves can only support small population numbers, so there is a risk that inbreeding will occur and the genetic diversity of species will diminish. In a small reserve, the risk of a natural disaster such as a flood or forest fire wiping out all the individuals of a species is more likely than if the reserve is large. Edge effects are also less significant in larger reserves than in small ones.

Edge effects

The centre of any nature reserve is likely to have different features from the areas around the edges. A woodland reserve has more light and wind but less moisture at the edge than at the centre.

Organisms that live in the centre are protected from the influence of other organisms such as farm animals or humans living outside the reserve. Organisms at the edge may be disturbed by or even have to compete with organisms from outside the reserve. Small reserves have more edge per hectare than large ones, so edge effects have a greater impact on the whole ecosystem of a small reserve. Shape is also a factor in determining edge effects: a long, thin reserve will have a larger edge than a rounder shape that covers the same area, so the shape as well as the size of a reserve is important.

Wildlife corridors

Sometimes it is impossible to create a large nature reserve, but good planning can make it possible to link two smaller areas through a corridor (see Image 3.17), which is a strip of land used to link two or more separated reserves. Corridors may be built under busy roads or railways lines so that organisms have a larger area to move about in and colonise.

Image 3.17 This bridge forms a wildlife corridor between two nature reserves divided by a road.

A corridor is not ideal, because animals using it may be exposed to dangers as they travel and may come into closer contact with humans. Corridors can also act as conduits for the spread of disease and may make certain species easy targets for poachers or hunters. On the other hand, the benefits of corridors include that gene flow between two otherwise isolated areas can take place and promote diversity. Seasonal movements and even large migrations can happen safely and there is less chance that animals will be killed by traffic once they are accustomed to using the corridor.

Buffer zones

Semi-protected areas surrounding a nature reserve are known as buffer zones. Most successful reserves are protected from outside disturbance by these zones. Buffer zone habitats may be managed or left wild, but the objective of the zone is to minimise outside influences of people in nearby towns or the impact of local agriculture, as well as limiting the spread of pests and disease into the reserve.

Management

Newly created nature reserves must be managed, usually by government or government-funded organisations, so that local species are encouraged and areas damaged by humans can be restored. Without active management, some species can come to dominate and threaten the survival of other important organisms. A well-planned policy for protected areas should conserve the local ecosystem and allow for scientific study and education, as well as permitting access to visitors in manageable numbers that will not cause disturbance or damage.

Many reserves contain iconic species which receive much attention. Examples include orang-utans in Borneo, and giant pandas in China. If well-known species are present, it may be easier to encourage visitors and to raise funding for conservation. Although the key species may not be the only important organism in the area, their presence encourages visitors to come and to support conservation projects.

Species-based conservation strategies

These strategies focus on individual species or groups of species that are threatened, and aim to protect them and increase their numbers.

The role of CITES

CITES is an international agreement that aims to limit international trade in wild plants and animals and their body parts (such as elephants' ivory and rhino horns), so that trade in specimens does not endanger the survival of a species.

The agreement came into force in 1975 and membership of CITES is voluntary, with each government producing its own national laws to support its overall aims. Within the agreement, species are grouped according to the degree of threat they face. Those threatened with extinction cannot be traded at all; less threatened species may be traded within regulations that make trade sustainable, and a third group includes those species included

CONSIDER THIS

It is not always easy for local people to maintain their traditional life and rights in conservation areas or areas managed by the authorities. Thuli Brilliance Makama is a lawyer who deals with environmental issues in Swaziland. In 2009, she won a landmark case to include environmental representation in conservation decisions made by the country's Environmental Authority. She has also challenged the forced evictions of poor communities that live on the edges of conservation areas and who depend on the environment for basic needs such as building materials, medicinal plants, food and firewood. Makama has helped to preserve traditional knowledge and skills among people by ensuring that their rights to their homes are taken into consideration.

3.04 Conservation of biodiversity

at the request of a specific country because cooperation is needed to prevent illegal trade. Species listed include all primates, cetaceans (whales and dolphins), turtles and tortoises, and plants such as orchids and mahogany. Importing or exporting specimens, body parts or derivatives of the species must be authorised and licensed, and so far CITES has been responsible for a significant reduction in the trade in endangered species and material from them, such as ivory.

Strengths of CITES:

- Trade in endangered plants and animals has been significantly reduced.
- Permits and licences are required to trade in listed species.
- The agreement has raised awareness of trade in endangered species.

Weaknesses of CITES:

- The agreement is voluntary and countries can withdraw.
- Penalties may be less than the profit to be made from trade or smuggling.
- Some countries may be unable to enforce the laws effectively due to lack of resources, long unmanned borders or corruption among politicians or law enforcement agencies.

Captive breeding programmes, reintroduction programmes and zoos

The best modern zoos are no longer simply places to view animals; most are actively involved in captive breeding and conservation programmes that aim to increase the numbers of endangered animals and work to reintroduce them into their native habitats. This may be the best hope of saving species that are severely endangered.

Zoos provide a location for *ex-situ conservation* – that is, preserving a species whose numbers are very low, by selectively breeding the animals away from their natural habitat. Difficulties arise because animals behave very differently in zoos and breeding may be problematic, particularly if animals have complex breeding behaviours or require special environmental conditions to breed. On the other hand, scientific knowledge can be used to select animals to breed based on their genetic profiles, so that a large gene pool is maintained to provide as much diversity as possible. Zoos in different parts of the world keep pedigree charts of their animals so that those with the greatest genetic variation can be selected and exchanged with other zoos for breeding programmes. Techniques such as artificial insemination and embryo transfer can be used if animals fail to breed naturally. Difficult pregnancies can be monitored and the young can be cared for by zoo staff after they are born.

Plants are more straightforward to maintain in an ex-situ programme. Botanic gardens can supply the correct environmental conditions for plant growth, and computer-controlled greenhouses maintain the correct temperature and humidity for each one. Many countries keep national collections of native plants, exotic genera and important food plants. Seed banks are an additional resource; seeds can be stored in cool, dark conditions that prevent germination for many decades. The Millennium Seed Bank at Wakehurst Place in England and the Svalbard Seed Bank in Norway both provide a safety net against the loss of valuable species.

Reintroducing species into the wild is not easy. If habitat has been lost, there may be nowhere for a reintroduced species to live. Released animals may not have the behaviours they need to survive, and humans may need to intervene to feed or protect them. For example, young orang-utans have to be taught to socialise, climb and forage for themselves before they can be fully independent. Some notable successes in the reintroduction of captive-bred species include the Arabian oryx (see Case study 3.03.03) in Oman and Saudi Arabia, the condor in USA, Przewalski's horse in Mongolia (see Image 3.18), and the golden lion tamarind in Brazil.

Image 3.18 Przewalski's horse, a critically endangered wild horse, has been reintroduced to its natural habitat.

The strengths of captive-breeding and reintroduction programmes are that:

- Numbers of rare species can be increased in captivity to boost numbers in the wild.
- Genetic diversity can be maintained by selecting animals to breed.
- Artificial insemination or embryo transfer can help when animals fail to breed naturally.
- Offspring may have a better chance of survival.
- Zoos have a valuable role to play in education and public awareness.
- Plant species can be held in seed banks for many years and provide a source of genetic variability.

The weaknesses are that:

- Captive breeding programmes are expensive.
- Reintroduction of species is difficult, and poorly supported programmes may leave vulnerable animals at risk.
- Reintroduced species may be targeted by hunters or poachers if local people do not support the programme.
- Ethical issues (such as whether humans should be interfering with nature and keeping animals in zoos) must be considered.

Selection of 'charismatic' species to help protect others in an area

Any conservation programme must select which species are to be protected. Should endangered animals be given priority over other species, and on what basis should one species be chosen over another? When captive breeding programmes are carried out in zoos, the zoo must select animals that are likely to increase visitor numbers and provide financial support. This may mean that decisions have to be taken for aesthetic or economic rather than ecological reasons. Primates and rhinos are more likely to attract visitors than beetles or worms. Different zoos have different levels of experience and expertise with different species, and they tend to develop those species that they know best.

Conservation programmes also base decisions on which species to conserve on the level of threat to the species. For example, as we have seen, countries can ask for certain species to be added to the CITES list if there is a need to conserve them.

Choosing species for aesthetic reasons can:

- raise public awareness
- increase funding and support
- engage local populations, who may benefit from increased tourism
- be important in preserving the overall beauty of an area.

Choosing species for ecological reasons can:

- be more likely to benefit the whole ecosystem
- fail due to lack of support and funding
- result in species which the public consider unattractive being conserved.

Selection of keystone species to protect food webs

Keystone species include specific predators or grazers. For example, limpets are a keystone species on a rocky shore because they control the level of algae as they graze.

Lobsters are a predatory keystone species. We can see how important they are from observations of overfishing. When fishermen removed too many lobsters from the Atlantic Ocean, there were not enough left to feed on and control the sea urchin population. Numbers of sea urchins increased and destroyed large areas of kelp, a species of seaweed. As the kelp disappeared, the complex community of molluscs and other small organisms that lived in it was also destroyed.

The diversity of species and the complexity of food webs are much reduced if a keystone species is in decline. Specific efforts to conserve these species will not only protect the species, but also protect the whole community (The grey wolf provides another example of this effect.).

CONSIDER THIS

The term 'keystone species' was first used by Robert Paine in 1969. It is now a widely used concept in conservation ecology.

CONSIDER THIS

Grey wolves (*Canis lupus*) are native to Arctic tundra and were present across the USA until the 1900s. But by the 1970s they had been hunted to such an extent that they became an endangered species. Wolves were a keystone species in Yellowstone Park, where they had helped to control elk numbers. When wolves were gone, elk increased in numbers causing substantial damage to the park by overgrazing. In the 1990s, wild wolves were reintroduced to the national park. Today, the wolf population in the park has grown and the elk numbers have declined. Willows, cottonwoods and aspens along the fringes of heavily timbered areas are recovering from overgrazing. Wolves kill some elk but have modified the behaviour of the elk population by their presence. Elk now live in more marginal habitats where their nutrition is poorer, and because they live with greater stress their birth rate has fallen. This shows the importance of keystone species in maintaining key features of a whole ecosystem.

A **keystone species** is one that has a disproportionate effect on the structure of a community.

3.04 Conservation of biodiversity

CASE STUDY 3.04.01

Keystone species

On exposed rocky shores between low and high tide, marine organisms live in delicate balance with one another (see Figure 3.14). One organism, the starfish *Pisaster ochraceus*, is the keystone species on which the stability of the ecosystem depends.

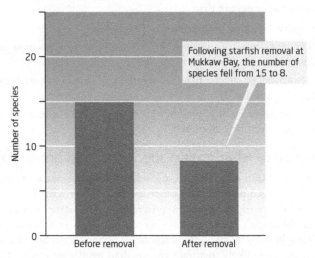

crowded out other species. With starfish present (to prey on mussels), barnacles, echinoderms and other marine invertebrates are able to maintain a presence in the ecosystem (see Figure 3.15).

Following starfish removal at Mukkaw Bay, the number of species fell from 15 to 8.

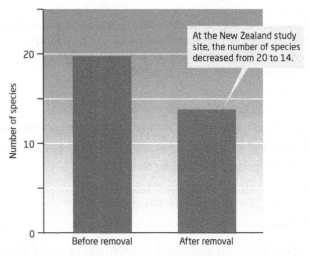

At the New Zealand study site, the number of species decreased from 20 to 14.

Figure 3.14 (a) Food web on a rocky intertidal shore, showing two predators (*Thais* sp. and *Pisaster* sp.); (b) *Pisaster* sp. feeding on a bed of mussels.

An experiment was carried out on the shore of Washington, USA, by Robert Paine in the 1960s. He removed starfish from selected areas and compared these areas with others that were left undisturbed. He discovered that, if starfish were removed, mussels

Figure 3.15 Graphs showing the effect of removal of starfish in intertidal food webs. Removing a starfish acting as top predator in intertidal food webs reduced the number of species both studies carried out in Mukkaw Bay, Washington, and New Zealand.

Questions

1. Does a higher proportion of predators in a food web indicate that predators contribute to higher diversity?
2. Is a keystone species the dominant species in this food web? Explain your answer.
3. Why was it necessary to carry out experiments in the USA and New Zealand?

3.04.05 Evaluating the success of a protected area

Despite the protected status of Chitwan National Park (see Case study 3.04.02) and other similar protected areas, there is no guarantee that a project will be successful unless it has government support, adequate funding, and research and education programmes. The support of the local community is also vital, and it is important that the needs of local people are met as part of the project.

Other important factors include the location of conservation areas. They must be situated in regions that are a suitable distance from urban centres, so that surrounding land can act as a barrier to human interference. But they must be close enough to allow access for local people and, if tourism is used to fund the conservation project, it is vital that visitors can reach the area easily and safely.

SELF-ASSESSMENT QUESTIONS 3.04.02

1. Why are NGOs often able to respond more quickly to environmental needs than government organisations are?
2. How are size, shape and edge effects important in the design of a protected area for wildlife?
3. Why are wildlife corridors important to species in protected areas?
4. List four factors that are important to the success of a protected area.
5. **Research idea:** Find a conservation area in your locality or country and research its shape and size. Check whether they are optimised for the conservation of species in the area.

Theory of knowledge 3.04.02

Can knowledge help us decide?

As we have seen, there are a number of different approaches to conservation of biodiversity. Much information is collected about different species that are in danger because their habitat is under threat and their population is very low. When is it right to take action?

1. Do all species have the same rights to be conserved?
2. To what extent are decisions on which species to conserve based on emotion rather than reason?

3.04 Conservation of biodiversity

CASE STUDY 3.04.02

Chitwan National Park

Chitwan National Park in the south of Nepal was declared a national park in 1973. At that time, it covered an area of 932 km², and laws controlling fishing, hunting and logging within the park were passed to protect it. The park was enlarged in 1977 and an adjacent wildlife reserve was established in 1984. The habitat of Chitwan had been well protected when it was a royal hunting reserve from 1846 to 1951, but after this time human population increase and farming had begun to cause damage and reduce its wildlife. Nevertheless, the area has a long conservation history. As early as 1958, part of the area was proposed as a rhinoceros sanctuary after a survey showed numbers were dangerously low. Chitwan National Park became a World Heritage Site in November 1984.

Funding

Support for the park at the local level has come from the King Mahendra Trust for Nature Conservation (KMTNC), the WWF and Biodiversity Conservation Network (BCN). These organisations monitor populations of important species, carry out surveys and work on animal breeding programmes.

UNESCO has provided support and funding for public awareness campaigns and the development of an educational and interpretative centre. It has also helped the park with consultancy services for long-term planning and for the purchase of equipment. Such international support also helps involve the local community and improves the effectiveness of conservation initiatives.

Community support

In 1997, a buffer zone of 766.1 km² of forest and private land was added to the park and provided a means of isolating the park from commercial activities in the surrounding area (see Figure 3.16). In the early stages, there were some problems because communities living in the buffer zone did not gain any benefit from the park. The park's valuable resources were out of bounds to local people, whose crops were sometimes eaten or trampled by wildlife from the protected area.

A Nepalese proverb says, 'If you want to control stealing by any household member, then give him or her the keys to the treasury.'

This policy has been applied in the buffer zone. Local people are now included in projects and given responsibilities for them, so their attitudes have changed. The park and the local people jointly carry out community projects and manage natural resources in the buffer zone. The government has agreed to return up to 50 per cent of the park's revenue for community

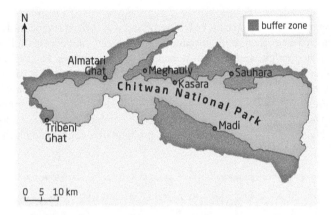

Figure 3.16 A buffer zone around the edge of the national park reduces the disturbance from outside activities such as agriculture.

development and to provide a stimulus to the local economy.

Benefit of local involvement

When poachers attempt to kill rhinos, they dig pits to trap the animals. Community members have not only informed the authorities but also filled in the pits so that wildlife are not harmed. Today, if a domestic animal is killed by a tiger, people are more likely to accept the situation, because they understand that tourists come expressly to see tigers. As local people have benefited from tourism, they have become more aware of the importance of protecting and living alongside local wildlife.

Another key objective of the project was to decrease the pressure on the park for firewood and fodder. Recent data suggests that the number of people entering the park to collect wood has decreased by almost 30 per cent after a scheme to provide firewood and fodder in the buffer zone was introduced.

Ecotourism

Chitwan National Park is one of Nepal's most popular destinations for ecotourism, and it provides revenue and work for local people. In 1989 more than 31 000 people visited the park, and in just ten years the number had more than doubled. Today, it is even more popular. Lodges and hotels in and around the park have enhanced the local economy and supported conservation work, as well as educating and informing visitors about the park and its biodiversity.

Although most visitors come to see two important charismatic endangered species, the Asian single-horn

CASE STUDY 3.04.02 (continued)

rhinoceros and the Bengal tiger (see Image 3.19), the National Park provides habitats for more than 700 species of animal and many invertebrates which have not been fully surveyed. It has ensured the survival of many species, including the king cobra and rock python as well as the starred tortoise and monitor lizard. The river system is also home to 113 recorded species of fish and rare mugger crocodiles.

Conservation successes and education projects

The protected area contains a rich flora and fauna typical of the lowland forest which once stretched across the foothills of the Himalayas through India and Nepal. On the hills are pines and scattered palms, and moister slopes support bamboos. This is recognised as prime habitat for the Bengal tiger. Since the establishment of the protected area, the Bengal tiger population has increased from about 25 individuals to more than 70 in 1980. Although in some years the population declined due to poaching and floods, a study carried out between 1995 and 2002 concluded that the population had risen to 82 breeding tigers with a density of six females per 100 km^2.

In 2011, the WWF released the results of its rhinoceros census in Nepal, reporting a 23 per cent increase in rhino numbers since 2008. Numbers increased from 408 to 503. These numbers reflect the success of conservation efforts for the species and are a result of improved rhino-protection measures and management of habitat. To help the survival of the rhinos, animals are carefully managed and transferred between Chitwan and other national parks to ensure genetic diversity and prevent disease. But despite all the successes, poaching is still a threat: in 2002 alone, poachers killed 37 animals to saw off and sell their valuable horns.

Bengal tigers and rhinoceroses are the most famous inhabitants of the park, but many other endangered and rare species now also have protection. The fishing cat, leopard cats and rare marbled cats are all found in the park, and Chitwan has one of the highest populations of sloth bears. It is also home to the world's largest wild cattle species, the guar, which grazes on the grassland.

Study and research projects are vitally important. Breeding programmes are in place for two critically endangered species of vulture and for the critically endangered gharial crocodile. Gharial eggs are collected and hatched in a breeding centre, where animals are reared for between six and nine years. Young gharials are reintroduced into river systems, but unfortunately few animals survive, despite the best efforts of the conservationists.

Image 3.19 Bengal tiger stalking prey.

Although several species in the Chitwan National Park are still under threat, good management, local involvement, scientific study and conservation programmes, together with the support of the government, have ensured that the protected area has had considerable success in achieving its objectives.

Questions

1. Why does granting protected status to an area alone not guarantee that that area will be conserved and the species it contains survive?
2. How and why are Bengal tigers important to the park?
3. What is the role of ecotourism in conservation?

End-of-topic questions

1. **a**
   i Outline *two* reasons why Australia has unique flora and fauna. [2]
   ii Explain how the characteristics of an isolated population can become changed over time. [3]
   iii Outline the mechanism of natural selection as a driving force for speciation. [4]

   b Define:
   i species diversity [1]
   ii habitat diversity. [1]

2. **a** Discuss the strengths and weaknesses of a species-based approach to conservation. [3]

   b Outline the factors leading to a loss of biodiversity; use an example you have studied. [5]

   c Explain the importance of shape and size in the design of a nature reserve. [4]

   d Outline *two* advantages and two disadvantages of wildlife corridors. [4]

3. **a**
   i List *three* factors that make a species more prone to extinction. [3]
   ii Outline *two* differences between the current (sixth) mass extinction and the five previous mass extinctions. [2]

   b Western grey whales are a critically endangered species on the Red List.
   i Outline *three* factors used to establish the conservation status of a species on the Red List. [3]
   ii Describe the human factors that have led to the conservation status of the grey whale or other endangered species you have studied. [3]

4. The Kluane Boreal Forest Ecosystem Project operated from 1986 to 1996 in the south-western Yukon. Food and predators of the Arctic ground squirrel population were controlled and modified.

   Three trials were carried out in three areas, as follows:

   Area 1: Squirrels were given extra food

   Area 2: Predators were excluded from the area

   Area 3: Extra food was provided and predators were excluded

   Figure 3.17 shows the average spring body mass of females in the area from 1990 to 1996.

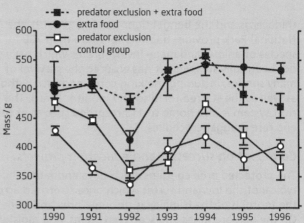

Figure 3.17 The average spring body mass of females from 1990 to 1996.

Figure 3.18 shows the number of squirrels per hectare after 1996, when feeding and exclusion of predators was stopped.

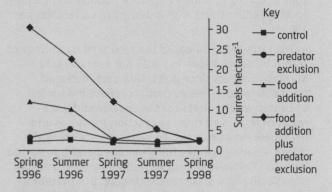

Figure 3.18 The number of squirrels per hectare after 1996, when feeding and exclusion of predators was stopped.

a Which group of female squirrels had the highest average body mass in 1996? [1]

b Suggest a reason for this result. [1]

c Suggest why the researchers were interested in female body mass rather than the average mass of the whole population. [1]

d State the number of squirrels in the 'food addition + predator exclusion' area in summer 1996. [1]

e Describe the effect that stopping the feeding programme in spring 1996 had on the squirrel population. [2]

f State the sampling technique that the researcher are likely to have used to monitor the populations. [1]

Topic 4

WATER AND AQUATIC FOOD PRODUCTION SYSTEMS AND SOCIETIES

4.01 Introduction to water systems

LEARNING OBJECTIVES

After reading this chapter you should be able to:

- understand that the hydrological cycle is a system of water flows and storages that may be disrupted by human activity
- appreciate that the ocean circulatory system (ocean conveyor belt) influences the climate and global distribution of water (matter and energy).

KEY QUESTIONS

How does water flow and how is it stored in the hydrological cycle?

How do human activities impact on surface runoff and infiltration?

What impact does the ocean circulatory system have on the global distribution of water?

4.01.01 The distribution and storage of fresh water

Although water covers 70 per cent of the Earth's surface, less than 3 per cent of global water is fresh water (see Figure 4.01). The definition of fresh water is water containing less than 500 parts per million (ppm) dissolved salts.

Water exists in three states – liquid, solid and vapour – and the three states are constantly interchanging. Almost 70 per cent of the Earth's fresh water is in the form of ice caps and glaciers (see Image 4.01). By far the largest ice stores are in Antarctica and Greenland. Antarctica covers an area of almost 14 million km^2 and contains 30 million km^3 of ice. This equates to around 61 per cent of all fresh water on Earth. The Antarctic ice sheet holds an amount of water such that if it were to melt, the sea level would rise by 70 m. The Greenland ice sheet covers 1.7 million km^2, which is about 70 per cent of the surface of Greenland.

About 30 per cent of the Earth's fresh water is held as groundwater. This is water contained in or moving through rock strata. Rocks that are permeable and store underground water are known as aquifers. Aquifers are located at various depths. Those found close to the surface are more likely to be used for water supply and irrigation because of their relatively easy access. They are also more likely to be topped up by local precipitation. At over 1.7 million km^3, the Great Artesian Basin in Australia underlies 22 per cent of the country and is arguably the largest groundwater aquifer in the world. It plays a vital role in water supplies for Queensland and remote parts of South Australia.

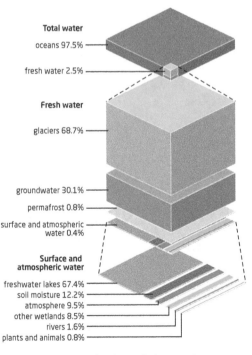

Figure 4.01 Distribution of the Earth's water.

The rest of the world's fresh water, less than 1 per cent of the total, is found in lakes (see Image 4.02), soil water, atmospheric water vapour, rivers and biota, in decreasing order of storage size. Lake Baikal in eastern Russia is the largest volume freshwater lake in the world. It is also the world's deepest lake. It covers an area of 31 500 km² with a maximum depth of 1637 m. The largest lake in terms of surface area is Lake Superior in North America.

Image 4.01 Meltwater stream emerging from the snout of a glacier. Evidence of condensation/sublimation in the form of cloud can be seen high on the mountain side.

Image 4.02 Water tower in a reservoir, Lake Vyrnwy, Wales.

4.01.02 The hydrological cycle

Hydrology is the study of water. The atmosphere holds less than 0.001 per cent of the Earth's total volume of water. This amounts to about ten days' supply of average precipitation around the world. However, the Earth's annual precipitation is more than 30 times the atmosphere's total capacity to hold water, indicating the constant recycling of water between the Earth and the atmosphere. NASA has estimated that every drop of fresh water has been consumed at least once before. Water flows through a closed hydrological system (see Figure 4.02), which means that the volume of water in the hydrosphere today is the same as has always been present in the Earth's atmosphere system. The different parts of the water budget are replenished at greatly varying time intervals. The time for a water molecule to enter and leave part of the system is known as the **turnover time**. This varies from about 12 days for atmospheric moisture to about 10 000 years for polar ice caps.

Extension

Good interactive site on water cycle
https://water.usgs.gov/edu/watercycle.html

Solar energy drives the hydrological cycle

Advection is an important part of the **hydrological cycle**. Without advection, water could not be transported from the oceans to land masses. Evaporation from water surfaces on land would not be enough to keep rivers and lakes full and provide the human population with drinking water.

Energy from the Sun drives the hydrological cycle by evaporating water (to form water vapour) from oceans, rivers, lakes and other water stores. The higher the temperature, the greater the potential for **evaporation**. Look how quickly water evaporates from a concrete or tarmac surface on a hot day compared with on a cooler day! Evaporation is also faster on a windy day compared with on a calm day.

> **Turnover time** is the time taken for a water molecule to enter and leave part of the hydrological system.
>
> **Advection** is the horizontal movement of water in the atmosphere, in vapour, liquid or solid states in air masses – that is, wind-blown movement.
>
> The **hydrological cycle** is the natural sequence through which water passes into the atmosphere as water vapour, precipitates to Earth in liquid or solid form, and ultimately returns to the atmosphere through evaporation.
>
> **Evaporation** is the process of water in a liquid state changing to a gaseous state (water vapour) due to an increase in temperature.

4.01 Introduction to water systems

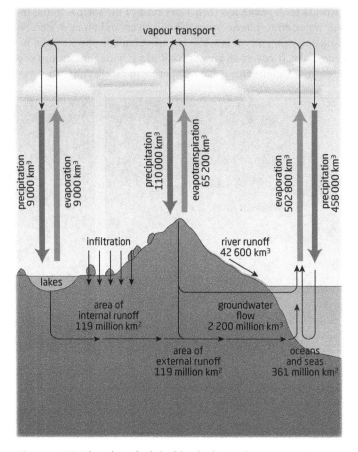

Figure 4.02 The closed global hydrological system.

Of course, a source of water is required, so there is little evaporation from hot deserts. The moisture-holding capacity of the atmosphere also increases with temperature. For every 1 °C increase in global temperatures there is a 7 per cent increase in the moisture-holding capacity of the atmosphere.

Water also moves from vegetation to the atmosphere through the process of **transpiration**. The combination of evaporation and transpiration is called **evapotranspiration**. Where ice masses are concerned, the change in state from a solid (ice) to water vapour is known as **sublimation**.

Upward movements of air carry water vapour to higher levels in the atmosphere. In the troposphere, the lowest layer of the atmosphere, temperature declines with altitude. When air rises, it expands due to the decrease in atmospheric pressure. As air molecules move apart in the process of expansion, energy is used up and the air cools. Cold air can hold less water vapour than warm air, so if the air continues to rise and cool, **condensation** will eventually occur and clouds will form. Condensation occurs when water vapour is cooled to a level known as the dew point.

If the water droplets and ice particles in a cloud become large and heavy enough, they will fall towards the Earth's surface as **precipitation**. Thus, the three major processes in the hydrological cycle are evaporation, condensation and precipitation. Water molecules continuously move from location to location in this cycle.

Climate graphs

The climate graph is the standard way of summarising variations in precipitation and temperature over time for a particular location. The graph illustrates the average climate over a long period, usually 30 years. Precipitation for each month is shown by bars, while temperature is illustrated by a line graph.

Figure 4.03 shows the climate graphs for four climatic regions (polar, temperate, desert, equatorial) north of the equator and close to the 30 degrees east line of longitude. Kisangani, deep in the Congo Basin, has by far the highest precipitation, but there are clear variations throughout the year.

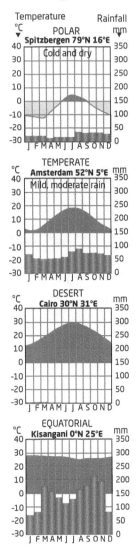

Figure 4.03 Four climate graphs.

Transpiration is the process by which plants absorb water through the roots and transport it to the leaves from where it is lost as water vapour.

Evapotranspiration is the sum of evaporation, sublimation and transpiration from land and ocean surfaces to the atmosphere.

Sublimation is the process of changing from a solid to a gas without a liquid phase.

Condensation is the process by which water vapour in the air is changed into liquid water.

Precipitation is water that falls to Earth from the atmosphere (e.g. rain, snow, sleet, hail, dew and frost).

SELF-ASSESSMENT QUESTIONS 4.01.01

1 Define the hydrological cycle.
2 What is evapotranspiration?
3 Why is advection a vital part of the hydrological cycle?
4 **Research idea:** Find out the average annual precipitation where you live. How evenly is this precipitation spread over the year?

CONSIDER THIS

The highest precipitation in a 24-hour period was 1825mm at Foc-Foc, Réunion, 7–8 January 1966, during tropical cyclone Denise. The highest precipitation in one year was 26470mm Cherrapunji, Meghalaya, India, 1860–1861.

The drainage basin system

When precipitation occurs, the movement of water over and through the Earth's surface takes various forms, as Figure 4.04 illustrates. However, 61 per cent of the total precipitation that falls on land is never available for capture or storage because it evaporates from the ground or transpires from plants; this fraction is called **green water**. The remaining 39 per cent channels into **blue water** sources – lakes, rivers (see Image 4.03), wetlands and aquifers that people can tap directly. This is available for withdrawal before it evaporates or reaches the ocean.

When precipitation reaches the surface it can follow a number of different pathways (see Figure 4.04). A small amount falls directly into rivers as direct channel precipitation. The rest falls onto vegetation or the ground. If heavy rain has fallen previously and all the air pockets in the soil are full of water, the soil is said to be saturated. Because the soil is unable to take in any more water, the rain flows on the surface under the influence of gravity. This is called surface runoff or overland flow.

If the soil is not saturated, rainwater will soak into it. If the rock below the soil is permeable (allows water into it), the rainwater will continue to soak down deeper into the rock. This water will eventually come to impermeable rock that does not allow water into it. The underground water level will build up towards the surface from here. It does not remain stationary but flows downslope under gravity. The upper level of underground water is the water table. Water contained in rocks is known as groundwater and water on the move in rocks is called groundwater flow.

Water flowing through the soil is called **throughflow**. **Infiltration** is the passage of water into the soil. **Percolation** is the downward vertical movement of water within a soil or rock.

Rainwater can be intercepted by vegetation. Interception is greatest in summer when trees and plants have most leaves. Some rainwater will be stored on leaves and then evaporated directly into the atmosphere. The remaining intercepted water will either drip to the ground from leaves and branches or it will trickle down tree trunks or plant stems (stemflow) to reach the ground.

Stores are places where water is held (see Figure 4.04). Transfers are where water is flowing through the drainage basin system. Inputs are where water enters the system. Outputs are where water is lost to the system.

In some countries precipitation is fairly even during the year. However, in other countries there may be distinct wet and dry seasons. Here, rivers may dry up completely for a number of months. In deserts, small river channels may be dry for most of the time.

Green water is the proportion of total precipitation absorbed by soil and plants, then released back into the air.

Blue water is the proportion of precipitation that collects in water courses (in rivers, lakes, wetlands and as groundwater) and is available for human consumption.

Shared hydrological cycles

Many hydrological cycles (in terms of drainage basins) are shared by different nations. This is not generally a problem when there is enough water for everyone. However, in some cases shared drainage basins have led to international disputes. As the demand for water has increased, the number of disputes has risen. This issue will be discussed in more detail in Chapter 4.02.

Image 4.03 River with a high discharge as a result of a recent precipitation event.

4.01 Introduction to water systems

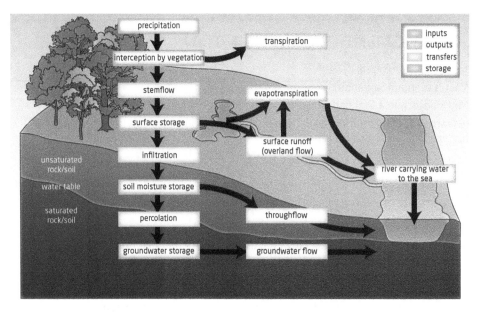

Figure 4.04 Diagram of the drainage basin system.

SELF-ASSESSMENT QUESTIONS 4.01.02

1. What is the difference between blue water and green water?
2. What happens to precipitation reaching the surface when the soil is saturated?
3. What are stores? Give two examples.
4. **Research idea:** In which drainage basin is your school/college located?

4.01.03 Infiltration and surface runoff

The flow of water on and below the surface is affected by a range of factors. These can be divided into meteorological factors and surface characteristics (see Table 4.01).

Meteorological factors affecting runoff	Surface characteristics affecting runoff
• type of precipitation (rain, snow, sleet, etc.) • rainfall intensity • rainfall amount • rainfall duration • distribution of rainfall over the drainage basin • direction of storm movement • precipitation that occurred earlier and resulting soil moisture • other meteorological and climatic conditions that affect evapotranspiration, such as temperature, wind, relative humidity, and season	• land use • vegetation • soil type • drainage area • shape of the drainage basin • elevation • topography, especially the slope of the land • drainage network patterns • ponds, lakes, reservoirs, sinks, etc., in the basin, which prevent or delay runoff from continuing downstream

Table 4.01 Meteorological and surface characteristics affecting agriculture.

Different forms of human activity can also have a considerable impact on infiltration and surface runoff. The impacts of agriculture, deforestation and urbanisation are considered below.

Agriculture

The impact of agriculture on runoff processes is considerable. The greatest impact is usually on infiltration rates. The soil surface acts as a filter that lets water pass through (infiltrate) at a rate known as the infiltration rate or infiltration capacity. Compaction of soils reduces the size of pore spaces and the infiltration rate decreases.

Undisturbed soils have much higher infiltration capacity than soils in agricultural use. Once cultivated, soils become easily compacted. Heavy machinery used in agriculture increases soil compaction. Infiltration can fall to levels where surface runoff can occur even in rainfall of modest intensity.

Successive rainfall events compact the bare soil further, and impermeable crusts often form. Raindrops falling on bare soil can also compact the soil surface in ploughed fields, leading to increased runoff and erosion of farmland.

Some management techniques can help alleviate the situation. Grassed filter strips in farm fields help reduce runoff and erosion by slowing water velocities in the vegetated areas. No-till farming is an even more effective technique, which will be discussed in Topic 5.

Deforestation

Areas that have been deforested often experience:

- reduced infiltration
- increased surface runoff
- lower groundwater recharge
- increased soil erosion
- greater flood risk.

A major reason for these changes is the greatly reduced rate of interception by trees and other vegetation when deforestation has occurred. Interception slows down the movement of water through the system and is a vital part of the natural balance of water movement on the Earth's surface before human intervention takes place. Populated areas at risk from flooding often look to reforestation to reverse the process that results from deforestation.

A **hydrograph** is a graph showing the rate of flow (discharge) over a certain period past a specific point in a river. The rate of flow is expressed in cubic metres per second.

CASE STUDY 4.01.01

The hydrology of the Rivers Severn and Wye

Figure 4.05 shows the storm hydrographs of two rivers in the same region of the UK. A **hydrograph** measures the discharge from a river. A storm hydrograph shows the discharge when a precipitation event occurs. The horizontal axis of the graph shows the period under consideration, while the vertical axis shows the discharge in cubic metres per second.

Although the sources of both rivers are close together, the land use in the two river basins varies significantly. The River Wye flows over moors and grassland, while the River Severn flows through an area of coniferous forest. The geology, soils, topography and precipitation are similar in both river basins.

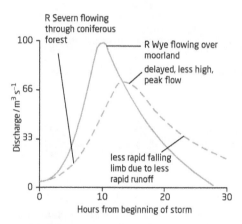

Figure 4.05 Storm hydrographs for the Rivers Severn and Wye.

4.01 Introduction to water systems

CASE STUDY 4.01.01 (continued)

The two catchments lie in the upland massif of mid-Wales and are characterised by rolling hills. The geology comprises slates, mudstones and sandstone rocks, which are generally classified as impervious. The climate is wet with up to 2500 mm of precipitation falling on the highest ground.

Figure 4.05 shows that the peak flow on the River Severn is significantly less than that of the River Wye. The former is about 70 m³/s⁻¹, while the latter is around 95 m³/s⁻¹. Both the rising and falling limbs of the hydrograph for the River Severn are more gentle than those for the River Wye. This is almost entirely due to the much higher rate of interception in the upper part of the drainage basin of the River Severn because of the presence of coniferous forest. This results in lower surface runoff in the drainage basin of the River Severn.

Questions

1. Describe the location and characteristics of the two river basins.
2. Describe and explain the difference between the two storm hydrographs shown in Figure 4.05.
3. How might building a new urban area in the drainage basin of the River Severn affect the river's discharge?

Urbanisation

Urbanisation can considerably increase the proportion of precipitation that is converted to surface runoff. Urbanisation replaces permeable vegetated surfaces with impermeable surfaces of concrete, tarmac, brick and tiles. Where open spaces exist in urban areas in terms of parks, commons and other open land, the soil is often heavily compacted due to high recreational use.

Urban systems are designed to move water off buildings, roads and other surfaces as quickly as possible. Think of the way buildings are designed to move water away rapidly – pitched roofs, gutters, water downpipes and drains. Most roads have a camber, or curvature to the surface, which helps water drain off into drains at the sides of the road rather than pooling in the centre of the road and making driving difficult.

All these actions drastically reduce infiltration and increase surface runoff. As a result, water moves into river systems much more quickly than before, increasing the risk of flooding. Figure 4.06 shows the impact of increasing the proportion of impervious surfaces in an area on infiltration, surface runoff and evapotranspiration.

The increasing size of many urban areas has increased the potential for **flash floods**. Flash flooding happens when precipitation falls so fast that the underlying ground cannot cope, or drain it away, fast enough. With flash flooding there is often little time between the precipitation falling and flash flooding occurring. Flash floods are capable of inundating roads, undermining buildings and bridges, tearing out trees, and scouring new channels.

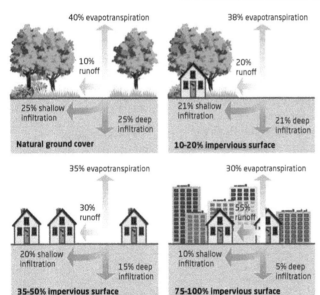

Figure 4.06 Urbanisation and hydrological responses.

A **flash flood** is a flood caused by heavy or excessive rainfall in a short period, generally less than six hours. Raging torrents of water can rip through river beds, urban streets or mountain canyons, sweeping everything before them. They can occur after minutes or a few hours of excessive rainfall.

SELF-ASSESSMENT QUESTIONS 4.01.03

1. How can deforestation affect the movement of water on and below the surface?
2. Explain the changes shown in Figure 4.06.
3. **Research idea:** Find a recent example of a flash flood affecting an urban area.

4.01.04 Ocean circulation systems

Ocean circulation systems are driven by differences in temperature and salinity. The resulting difference in water density drives the ocean conveyor belt, which distributes heat around the world and thus affects precipitation and other aspects of climate.

The motion of the oceans is due to a combination of thermohaline currents (thermo = temperature; haline = salinity) in the deep ocean and wind-driven currents on the surface.

According to NASA, the ocean conveyor belt occurs as follows:

1 The conveyor belt system (see Figure 4.07) can be thought of as beginning near Greenland and Iceland in the North Atlantic, where dry, cold winds blowing from northern Canada chill ocean surfaces.
2 The combination of chilling of surface waters, evaporation and formation of sea-ice produces cold, salty North Atlantic deep water (NADW).
3 The cold, dense NADW sinks and flows south along the continental slope of North and South America towards Antarctica, where the water mass flows east around the continent of Antarctica. This latter circulation is known as the Antarctic Circumpolar Current.
4 The NADW mixes with Antarctic waters. The resulting common water (Antarctic Circumpolar Water) flows northward at depth into the three ocean basins (Pacific, Indian, Atlantic).
5 The deep waters gradually warm and mix with overlying waters as they flow northwards into the northern hemisphere, continuing the ocean conveyor belt that circles the globe. These deep waters move to the surface at a rate of only a few metres per year.

It takes about 500 years for the conveyor belt to turn over the oceans' waters and make one complete journey around the Earth.

Importance of the ocean conveyor belt

This major circulation system redistributes vast amounts of the Sun's heat around the globe. It supplies heat to the polar regions and regulates the amount of sea ice. The warm surface waters of the North Atlantic Drift give north-west Europe a more moderate climate than most regions

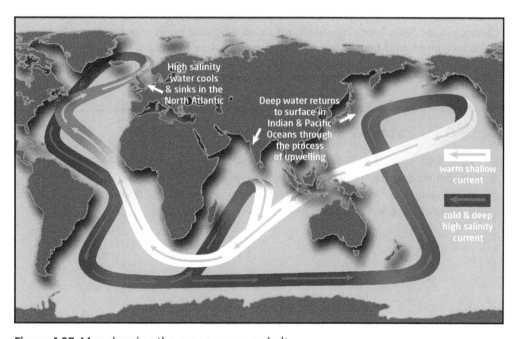

Figure 4.07 Map showing the ocean conveyor belt.

4.01 Introduction to water systems

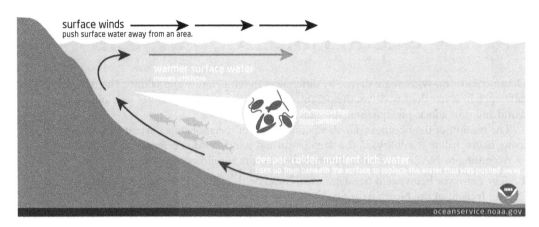

Figure 4.08 Diagram showing upwelling of cold, nutrient-rich water.
Source: www.oceanservice.noaa.gov

Winds blowing across the ocean surface often push water away from an area. When this occurs, water rises up from beneath the surface to replace the diverging surface water. This process is known as **upwelling**.

Theory of knowledge 4.01.01

Because of population growth and social trends such as the decline in average household size in many countries, urban areas have increased in size and new urban areas have been built. Due to the resultant decreased infiltration and increased surface flow, discharge downstream of such urban areas may cause damage to human environments and natural capital.

1. Should urban residents be financially responsible for the increased discharge downstream?
2. What could be done to rebalance the hydrological situation in a drainage basin where such changes were occurring?

of similar latitude. Without the warming effect of the North Atlantic Drift, north-west Europe would be at a temperature like that in south-east Canada, which suffers harsh winters. The North Atlantic Drift is the most important ocean current in the northern hemisphere. Oceanographers have estimated that it moves at approximately 8.5 million m^3 per second.

The predominant winds in north-west Europe are from the south-west. These winds pick up heat and moisture from the North Atlantic Drift, both of which benefit the maritime areas of north-west Europe. If the waters of the North Atlantic Drift were colder, less moisture would be picked up by the south-westerly winds, as colder air can hold less moisture than warmer air. This would result in the maritime areas having lower precipitation.

The **upwelling** of deep waters when they become warm enough to rise brings vital nutrients to the surface of the ocean (see Figure 4.08). Upwelling occurs in the open ocean and along coastlines. The nutrient-rich upwelled water stimulates the growth and reproduction of primary producers such as phytoplankton, which in turn results in high levels of fishery production. Upwelling effectively fertilises surface waters.

Good fishing grounds are invariably found where upwelling occurs, for example, the rich fishing grounds along the west coasts of South America and Africa. Approximately 25 per cent of the total global marine fish catches come from five upwellings that occupy only 5 per cent of the total ocean area.

Climate change could have a significant impact on the global ocean conveyor belt, which could lead to considerable temperature changes. For example, a weakening of the North Atlantic Drift would bring colder winters and lower precipitation to north-west Europe. This would impact significantly on people and the environment in the region.

SELF-ASSESSMENT QUESTIONS 4.01.04

1. Which two factors drive ocean circulation systems?
2. What effect does the North Atlantic Drift have on the climate of north-west Europe?
3. **Research idea:** Compare climatic graphs for locations in southern England and Newfoundland, Canada. Both regions are at a similar latitude.

Access to fresh water

4.02

LEARNING OBJECTIVES

After reading this chapter you should be able to:

- understand that supplies of freshwater resources are inequitably available and unevenly distributed, which can lead to conflict and concerns over water security
- appreciate that freshwater resources can be sustainably managed using a variety of different approaches.

KEY QUESTIONS

What are the reasons for the global water crisis?

How can water supplies be enhanced?

How serious is the prospect of international conflict over water resources?

4.02.01 The global water crisis

The longest a person can survive without water is about ten days. All life and virtually every human activity needs water; it is the world's most essential resource and a pivotal element in poverty reduction. But for about 80 countries, with 40 per cent of the world's population, lack of water is a constant threat. And the situation is getting worse, with demand for water doubling every 20 years. In those parts of the world where there is enough water, it is being wasted, mismanaged and polluted (see Image 4.04) on a large scale. In the poorest nations, it is not just a question of lack of water; the paltry supplies available are often polluted. **Water security** (see Figure 4.09) has become a major issue in an increasing number of countries.

Securing access to clean water is a vital aspect of development.

- More than 840 000 people die each year from a water-related disease. While deaths associated with dirty water have been virtually eliminated from more economically developed countries (MEDCs), in less economically developed countries (LEDCs) most deaths still result from water-borne disease.
- At any one time, half of the world's hospital beds are occupied by patients suffering from water-borne diseases.
- Around the world, 750 million people – approximately one in nine people – lack access to safe water. Of those who lack access to improved water, 82 per cent live in rural areas, while just 18 per cent live in urban areas.
- Women and children spend 140 million hours a day collecting water.

Water scarcity has been presented as the 'sleeping tiger' of the world's environmental problems, threatening to put world food supplies in jeopardy, limit economic and social development, and create serious conflicts between neighbouring drainage basin countries. In the 20th century, global water consumption grew six fold, twice the rate of population growth. Much of this increased consumption was made possible by significant investment in water infrastructure, particularly dams and reservoirs, affecting nearly 60 per cent of the world's major river basins.

Water security is the capacity of a population to safeguard sustainable access to adequate quantities of acceptable quality water for sustaining livelihoods, human well-being and socio-economic development, for ensuring protection against water-borne pollution and water-related disasters, and for preserving ecosystems in a climate of peace and political stability (UN-Water, 2013).

4.02 Access to fresh water

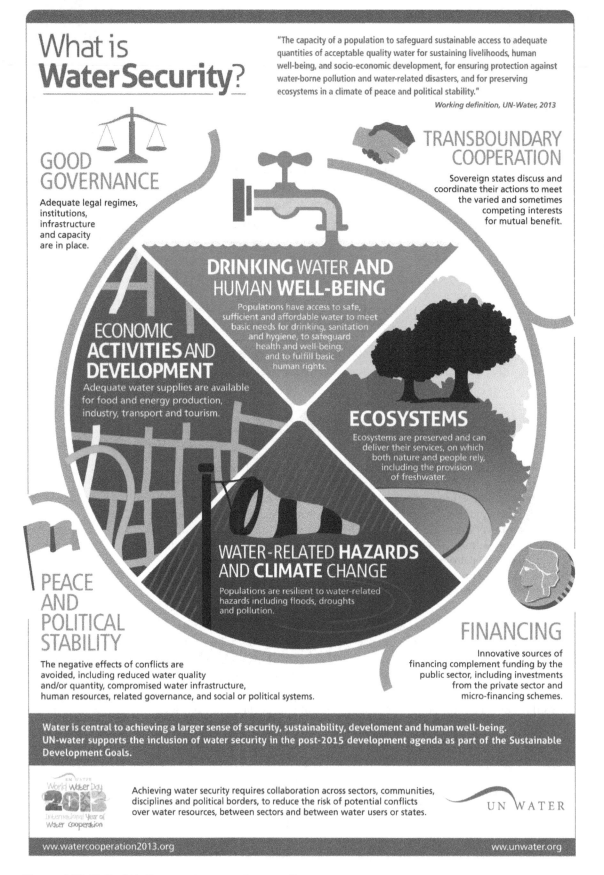

Figure 4.09 United Nations poster on water security.

Image 4.04 Polluted water – Christchurch, New Zealand.

Image 4.05 Water distribution in central Asia by donkey.

The United Nations (UN) estimates that two-thirds of the world's population will be affected by 'severe water stress' by 2025. The situation will be particularly severe in Africa, the Middle East and South Asia. The UN notes that already a number of the world's great rivers, such as the Colorado in the USA, are running dry, and that groundwater is also being drained faster than it can be replenished. Many major aquifers have been seriously depleted, which will present serious consequences in the future. In an effort to add impetus to global water advancement, the UN proclaimed the period 2005–2015 as the International Decade for Action, 'Water for Life'.

The link between poverty and water resources is clear, with those living on less than $1.25 a day roughly equal to the number without access to safe **potable water**. Improving access to safe water can be among the most cost-effective means of reducing illness and mortality. In LEDCs, it is common for water collectors, usually women and girls, to have to walk several kilometres every day to fetch water. Once filled, pots and jerry cans can weigh as much as 20 kg. In urban areas in LEDCs, water is still often distributed by donkey (see Image 4.05).

Potable water is water that is free from impurities, pollution and bacteria, and is thus safe to drink.

Theory of knowledge 4.02.01

Expanding access to water and sanitation is an ethical and moral imperative rooted in the cultural and religious traditions of many communities around the world. The UN sees the right to water as 'indispensable for leading a life in human dignity'. This is an issue central to life itself, crossing a whole range of academic disciplines.

1 How important is the right to water compared with other human rights, such as education or freedom?
2 Is it appropriate to charge households for water, and if so should everyone pay the same for household water?

SELF-ASSESSMENT QUESTIONS 4.02.01

1 State three facts that illustrate the global water crisis.
2 According to the UN, what proportion of the world's population will be affected by severe water stress by 2025?
3 **Discussion point:** How far do you think you could walk carrying water weighing 20 kg?

4.02.02 Increasing demand and unequal access

Since 1930, the global population has increased from 2 billion to over 7 billion, putting ever-increasing pressure on the world's water supplies. However, not just the increase in population is influencing the demand for water, but also rising per-capita usage in many countries. As households become more affluent, they use more water in an increasing number of different ways.

4.02 Access to fresh water

Increasing demand for water is only half the story – the other half is unequal access. Every year, 110 000 km³ of precipitation falls onto the Earth's land surface. This would be more than adequate for the global population's needs, but much of it cannot be captured and the rest is unevenly distributed.

- Over 60 per cent of the world's population live in areas receiving only 25 per cent of global annual precipitation.
- The arid regions of the world cover 40 per cent of the world's land area but receive only 2 per cent of global precipitation.
- The Congo River and its tributaries account for 30 per cent of Africa's annual runoff in an area containing 10 per cent of Africa's population.

The Middle East and Africa face the most serious problems. Since 1972, the Middle East has withdrawn more water from its rivers and aquifers each year than is being replenished. Yemen and Jordan are withdrawing 30 per cent more from groundwater resources annually than is being naturally replenished. Israel's annual demand exceeds its renewable supply by 15 per cent.

Total world blue-water withdrawals are estimated at 3390 km³, with 74 per cent for agriculture, mostly for irrigation (see Figure 4.10). About 20 per cent of this total comes from groundwater. Although agriculture is the dominant water user, industrial and domestic uses are growing at faster rates. Demand for industrial use has expanded particularly rapidly.

Figure 4.11 contrasts water use in more and less economically developed countries. In LEDCs, agriculture accounts for over 80 per cent of total water use, with industry using more of the remainder than the domestic allocation. In the developed world, agriculture accounts for slightly more than 40 per cent of total water use. This is lower than the amount allocated to industry. As in the developing world, domestic use is in third place.

Extension

Try out a fun survey to calculate water footprint using the link below:
http://environment.nationalgeographic.com/environment/freshwater/change-the-course/water-footprint-calculator

As LEDCs industrialise and urban–industrial complexes expand, the demand for water grows rapidly in the industrial and domestic sectors. As a result, the competition with agriculture for water has intensified in many countries and regions. This scenario has already played itself out in many economically developed countries, where more and more difficult decisions are having to be made about how to allocate water.

Large variations in water allocation can also exist within countries. For example, irrigation accounts for over 80 per cent of water demand in the west of the USA, but only about 6 per cent in the east.

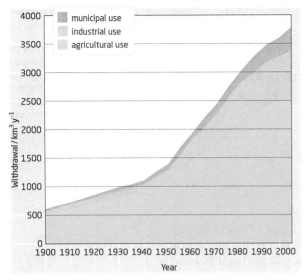

Figure 4.10 Global water use in 1900–2000, agriculture, industry and domestic.

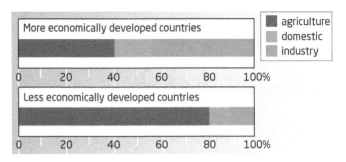

Figure 4.11 Water used for agriculture, industry and domestic for more and less economically developed countries.

The highest proportion of **irrigation water** use is in the Middle East and North Africa and South Asia. Irrigated farming accounts for 70 per cent of global annual water consumption. This rises to over 90 per cent in some countries such as India. Conserving irrigation water would have more impact than any other measure. Most irrigation is extremely inefficient, wasting half or more of the water used. A 10 per cent increase in irrigation efficiency would free up more water than is evaporated by all other users. The most modern drip-irrigation systems are up to 95 per cent efficient (see Table 4.02 and Image 4.06) but require significant investment.

Style of irrigation	Irrigation method	Efficiency
Surface: used in 80% of irrigated fields worldwide	*Furrow*: traditional method; cheap to install; labour intensive; high water loss; susceptible to erosion and salinisation	20–60%
	Basin: cheap to install and run; needs a lot of water; susceptible to salinisation and waterlogging	50–75%
Aerial: used in 10–15% of irrigated fields worldwide	*Sprinklers*: expensive to install and run; low-pressure sprinklers preferable	60–80%
Subsurface: used in 1% of irrigated fields worldwide	*Drip*: expensive to install; has sophisticated monitoring	75–97%

Table 4.02 Main types of irrigation.

Figure 4.12 shows improved drinking-water coverage in 2006 according to the UN's *Global Annual Assessment of Sanitation and Drinking-Water* report, 2008. From 1990 to 2006, approximately 1.56 billion people gained access to improved drinking water sources. Currently, 87 per cent of the world uses drinking water from improved sources (see Image 4.07) compared with 77 per cent in 1990. Although the world is on track to meet the Millennium Development Goal (MDG) drinking-water supply target by 2015, in general many countries in Sub-Saharan Africa and in Oceania are currently projected to miss MDG country targets.

Extension

Find out more about clear water programmes using these links:
www.youtube.com/watch?v=c321dj-1IYo
www.youtube.com/watch?v=M41fhqpTDwE

Access to safe drinking water is influenced by both physical and human factors (see Table 4.03).

Image 4.06 Sprinkler irrigation system in northern Spain.

Image 4.07 An improved drinking water source in a rural area in Nepal.

4.02 Access to fresh water

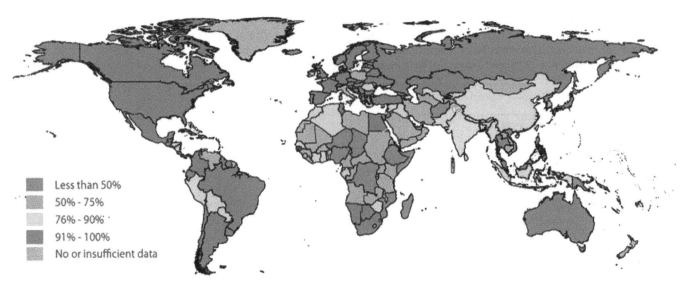

Figure 4.12 World map showing improved drinking-water coverage.

Physical factors	Human factors
• amount of precipitation • seasonal distribution of precipitation • physical ability of the surface area to store water • rate of evapotranspiration • density of surface access points to water • ease of access to groundwater supplies if they exist	• wealth of a country in terms of its ability to afford water infrastructure • distribution of population between urban and rural areas with the concentration of investment in water infrastructure in urban areas • socio-economic differences in urban areas: affluent urban districts invariably have better access to safe water than poor districts • degree of contamination of urban and rural water supplies • civil war and international conflict

Table 4.03 Physical and human factors influencing access to safe drinking water.

The amount of water used by a population depends not only on water availability but also on levels of urbanisation and economic development. As global urbanisation continues, the demand for potable water in cities and towns will rise rapidly. In many cases, demand will outstrip supply.

SELF-ASSESSMENT QUESTIONS 4.02.02

1 Describe the differences in irrigation efficiency shown in Table 4.02.
2 State three physical and three human factors that influence access to safe drinking water.
3 **Research idea:** How is water utilised in the country in which you live? Has this changed to any significant extent over time?

4.02.03 Water scarcity

The world's population is increasing by about 80 million a year. This converts to an increased demand for fresh water of around 64 billion m³ per year, which equates to the total annual flow rate of the River Rhine.

The Pilot Analysis of Global Ecosystems (PAGE), undertaken by the World Resources Institute, calculated water availability and demand by river basin. This analysis estimated that, at present, 2.3 billion people live in **water-stressed areas**, with 1.7 billion resident in **water-scarce areas**. The PAGE analysis forecasts that these figures will rise to 3.5 billion and 2.4 billion people respectively by 2025.

A country is judged to experience water stress when water supply is below 1700 m³ per person per year. When water supply falls below 1000 m³ per person a year, a country faces water scarcity for all or part of the year. These concepts were developed by the Swedish hydrologist Malin Falkenmark.

Water scarcity is to do with the availability of potable water. **Physical water scarcity** is when physical access to water is limited. This is when demand outstrips a region's ability to provide the water needed by the population. The **arid and semi-arid** regions of the world are most associated with physical water scarcity. Here temperatures and evapotranspiration rates are high and precipitation low. In the worst affected areas, points of access to safe drinking water are few and far between.

However, annual precipitation figures fail to tell the whole story. Much of the freshwater supply comes in the form of seasonal rainfall, as exemplified by the monsoon rains of Asia (see Image 4.08). India gets 90 per cent of its annual rainfall during the summer monsoon season from June to September. Also, national figures can mask significant regional differences. Analysis of the supply and demand situation by river basin can reveal the true extent of such variations. For example, the USA has a relatively high average water-sufficiency figure of 8838 m³ per person per year. However, the Colorado River Basin has a much lower figure of 2000 m³ per person per year, while the Rio Grande River Basin is lower still at 621 m³ per person per year.

However, in increasing areas of the world, physical water scarcity is a human-made condition. This is largely due to overuse. Examples of physical water scarcity include:

- Egypt, which has to import more than half of its food because it does not have enough water to grow it domestically
- the Murray-Darling Basin in Australia, which has diverted large quantities of water to agriculture
- the Colorado River Basin in the USA, where once-abundant resources have been heavily overused leading to serious physical water scarcity downstream.

Figure 4.13 shows these regions and the other parts of the world suffering from physical water scarcity.

Economic water scarcity exists when a population does not have the necessary monetary means to utilise an adequate source of water. The unequal distribution of resources is central to economic water scarcity, where the crux of the problem is lack of investment. This occurs for a number of reasons, including political and ethnic conflict. Figure 4.13 shows that much of sub-Saharan Africa is affected by this type of water scarcity.

Scientists expect water scarcity to become more severe, largely because:

- the world's population continues to increase significantly
- increasing affluence is inflating per-capita demand for water
- biofuel production is creating increasing demands and biofuel crops are heavy users of water
- climate change is increasing aridity and reducing supply in many regions
- many water sources are threatened by various forms of pollution.

A **water-stressed area** is one in which water supply is below 1700 m³ per person per year.

A **water-scarce area** is one in which water supply falls below 1000 m³ per person a year.

Physical water scarcity is when physical access to water is limited.

Arid and semi-arid conditions are those where precipitation is less than 250 mm and 500 mm per year respectively.

Economic water scarcity is when a population does not have the necessary monetary means to access an adequate source of water.

Image 4.08 Monsoon rain in South East Asia.

4.02 Access to fresh water

Figure 4.13 World map showing physical water scarcity and economic water scarcity.

The Stockholm International Water Institute has estimated that each person on the Earth needs a minimum of 1000 m^3 of water per year for drinking, hygiene and growing food for sustenance. Whether this water is available depends largely on where people live on the planet, as water supply is extremely inequitable. For example, major rivers such as the Yangtze, Ganges and Nile are severely overused, and the levels of underground aquifers beneath major cities such as Beijing and New Delhi are falling.

In many parts of the world, the allocation of water is largely down to the ability to pay. A recent article in *Scientific American* entitled 'Facing the freshwater crisis' quotes an old saying from the American west: 'Water usually runs downhill, but it always runs uphill to money.' Thus, poorer people and non-human consumers of water, the fauna and flora of nearby ecosystems, usually lose out when water is scarce.

Figure 4.14 illustrates the huge extent of the global water gap, using selected groups of more developed (see Image 4.09) and less developed countries. The daily usage figures for the USA and Australia are particularly high. Many of the less developed countries illustrated have water use figures below the water poverty threshold. Water scarcity is playing a significant role in putting the brakes on economic development in a number of countries.

Depleted aquifers

Aquifers provide approximately half of the world's drinking water, 40 per cent of the water used by industry and up to 30 per cent of irrigation water. Falling water tables can bring severe ecological, economic and social consequences. However, detailed knowledge of the state of many major aquifers is limited. Nevertheless, there is no doubt that the water table in many aquifers has fallen significantly. For example, the Ogallala aquifer in the American mid-west has fallen by more than 35 m in 50 years. In China, the overexploitation of aquifers has been a major factor in the decline in rice production from 140 million tonnes in 1997 to 127 million tonnes in 2005.

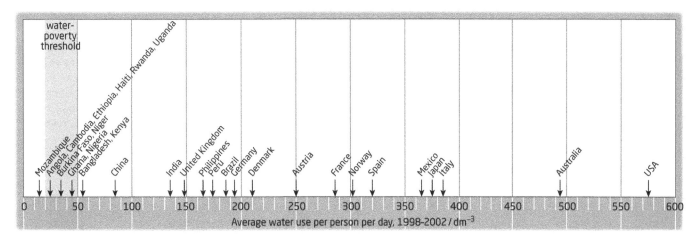

Figure 4.14 The global water gap.

Apart from falling water tables due to water withdrawals exceeding natural replenishment:

- in some areas, for example around the Mediterranean, seawater has begun seeping into depleted aquifers, making the water unusable for most purposes
- in other areas, sewage is beginning to contaminate some aquifers.

Where aquifers cross international borders, more detailed knowledge of their potential and problems will be required for neighbouring countries to reach agreement about allocating withdrawals of water. An atlas of underground water sources published by UNESCO identifies 273 transborder aquifers, some crossing as many as four nations. Some of the world's largest untapped aquifers are in North Africa and could have significant future development potential. Egypt, Libya, Sudan and Chad share the Nubian sandstone aquifer, which contains 10 000-year-old 'fossil water'. The four countries have formed a joint organisation to manage the aquifer.

Image 4.09 Public drinking water point: Dubrovnik, Croatia.

Water pollution

Water pollution has had a massive adverse effect on potential water supplies in many parts of the world. Water pollution comes from a variety of sources including:

- contamination by agricultural runoff, particularly from factory farming
- industrial pollution – about 70 per cent of the industrial waste is dumped into the water bodies, polluting the usable water supply
- urban runoff carrying pollutants from cars, factories and other sources into water courses
- untreated sewage, which has increased in many areas as population has increased.

Water pollution is contamination of a source of water making it unsuitable for use.

Each year, about 450 km^3 of waste water is discharged into rivers, streams and lakes around the world. While rivers in more affluent countries have become steadily cleaner in recent decades, the reverse has been true in much of the developing world. It has been estimated that 90 per cent of sewage in developing countries is discharged into rivers, lakes and seas without any treatment.

- According to the World Health Organization (WHO), around 2.5 billion people do not have access to improved sanitation.
- The UN estimates that almost half the population in many cities in the developing world do not have access to safe drinking water.
- Eighty per cent of the water pollution is caused by domestic sewage such as throwing rubbish on open ground and into water bodies.

Rivers in Asia are the most polluted. For example, the Yamuna River that flows through Delhi has 200 million litres of sewage drained into it each day. For many people, the only alternative to using this water for drinking and cooking is to turn to water vendors, who sell tap water at greatly inflated prices.

4.02 Access to fresh water

Although most people in developed countries think that their water supplies are clean and healthy, there is growing concern in some quarters about traces of potentially dangerous medicines that may be contaminating tap water and putting unborn babies at risk, according to a report published in the UK in September 2008. One newspaper headline read, 'Is our water being poisoned with a cocktail of drugs?' Scientists are worried that powerful and toxic anti-cancer drugs are passing unhindered through sewage works and making their way back into the water supply. The greatest concern is about cytotoxic, or cell-killing, cancer drugs, which are taken by about 250 000 people in the UK. Easily dissolved in water, they remain highly toxic when leaving the body and are hard to destroy in water treatment plants.

SELF-ASSESSMENT QUESTIONS 4.02.03

1. What is the difference between water stress and water scarcity?
2. Give three reasons why scientists expect water scarcity to become more severe.
3. **Research idea:** Find out about the degree of water scarcity in an arid country of your choice.

Climate change

Many countries, particularly those that are less developed economically, are concerned about the effects that climate change may have on both the quantity and quality of drinking-water resources. Potential adverse effects include:

- Reduced precipitation in many areas.
- Higher evapotranspiration, causing increasing risk of drought.
- Increased pollution of water supplies resulting from more flooding as extreme weather events become more common.
- Reduced water supplies and increased costs due to silting caused by lower stream flows and higher evaporation rates. A 2008 UN report quotes a recent water resources inventory in Mongolia which found that, compared with previous knowledge, 22 per cent of rivers and springs and 32 per cent of lakes and ponds have dried up or disappeared.
- Shrinking glaciers will lead to long-term loss of fresh water to rivers supplied by the glaciers. More than 50 per cent of the world's fresh water comes from mountain runoff and snow melt.
- As temperatures rise, people and animals require more water to maintain their health.

4.02.04 Enhancing water supplies

Water supplies can be enhanced through reservoirs, redistribution, desalination, artificial recharge of aquifers and rainwater harvesting schemes. **Water supply** is the provision of water by public utilities, commercial organisations or community efforts. The objective in all cases is to supply water from its source to the point of usage. However, as water supply is of vital importance, each method of water supply has disadvantages as well as the obvious benefits.

Water supply is the provision of water by public utilities, commercial organisations or community endeavours.

Dams and reservoirs

Much of the increase in water consumption over the past century has been made possible by significant investment in water infrastructure, particularly dams and reservoirs affecting nearly 60 per cent of the world's major river basins. Figure 4.15 shows water supply and management methods in the large Canadian province of Alberta, where water supply is a concern in many parts of the region. Not all reservoirs are held behind dams, but the largest ones usually are. In contrast 'off-channel' reservoirs (see Figure 4.15) usually use depressions in the existing landscape or human-dug depressions to store water.

Globally, the construction of dams has declined since the height of the dam-building era in the 1960s and 1970s. This is because most of the best sites for dams are already in use or such sites are strongly protected by environmental legislation and therefore off limits for construction. An alternative to building new dams and reservoirs is to increase the capacity of existing reservoirs by extending the heights of the dams.

A number of environmental problems are caused by the construction of dams. Dams trap sediments, which are vital for maintaining physical processes and habitats downstream of the dam. Large dams have led to the extinction of many aquatic species. They have also resulted in the disappearance of birds in flood plains, losses of forest, wetland and farmland, and erosion of coastal deltas.

Water can also be stored underground (see Figure 4.15) as well as on the surface, thus reducing losses from evaporation. Underground storage usually uses existing chambers such as abandoned mines.

> **CONSIDER THIS**
>
> The San Vicente Dam Raise project in southern California is adding 36 m to the existing 67 m structure. At a cost of $530 million, it will more than double the current capacity of the reservoir.

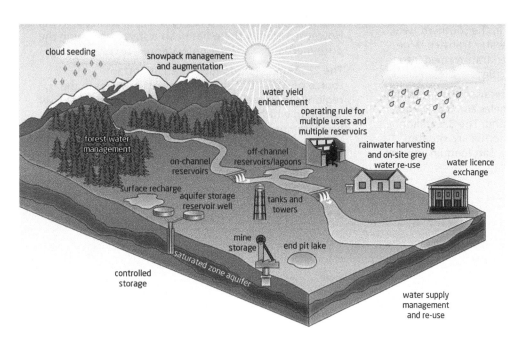

Figure 4.15 Alternative water supply and management methods in Alberta, Canada.

Wells and bore holes

A **well** or **bore hole** is a means of tapping into aquifers to gain access to groundwater. Wells and bore holes are sunk directly to the water table. Aquifers provide approximately half of the world's drinking water, 40 per cent of the water used by industry and up to 30 per cent of irrigation water.

Groundwater is most important in arid and semi-arid areas. This is the main source of water in oasis settlements such as those in the Sahara Desert in North Africa. However, groundwater supplies can be contaminated in various ways. For example, nearly 70 million people living in Bangladesh are exposed to groundwater contaminated with arsenic beyond WHO recommended limits. The arsenic occurs naturally at deep groundwater locations.

A **well** is relatively large in diameter and often sunk by hand, although machinery may be used.

A **bore hole** is typically drilled by machine and is relatively small in diameter.

Redistribution

In general, water is redistributed through water networks and grids over longer distances in more economically developed countries (MEDCs) than in LEDCs, because of the substantial cost of

4.02 Access to fresh water

such infrastructure schemes. However, water grids are nowhere near as extensive as power grids. Thus, in many countries there is limited ability to move water from areas of water surplus to areas of water deficit.

CASE STUDY 4.02.01

Water redistribution in the south-western USA

Southern California, in the south-western USA, has long benefited from major water redistribution (transfer) schemes, with large volumes of water brought to the region from Northern California, the Sierra Nevada mountains and the Colorado River. A great imbalance exists between the distributions of precipitation and population in California. Seventy per cent of runoff originates in the northern third of the state, but 80 per cent of the demand for water is in the southern two-thirds. As a result of large-scale water transfer, large areas of this dry region have been transformed into fertile farmlands and sprawling cities.

While irrigation is the prime water user, the sprawling urban areas, particularly Los Angeles and San Diego, have also greatly increased demand. The 3.5 million hectares of irrigated land in California are situated mainly in the Imperial, Coachella, San Joaquin and the lower Sacramento valleys. Figure 4.16 shows the major component parts of water transfer and storage in the state.

Agriculture uses more than 80 per cent of the state's water, although it accounts for less than a tenth of the economy. Water development, largely financed by the federal government, has been a huge subsidy to California in general and to big water users in particular. However, recently there has been a start at bringing the price mechanism to bear on water resources.

The 2333 km Colorado River is an important source of water to California and other states in the region (see Figure 4.16). The Colorado was the first river system in which the concept of multiple use of water was attempted by the US Bureau of Reclamation. In 1922, the Colorado River Compact divided the seven states of the basin into two groups: Upper Basin and

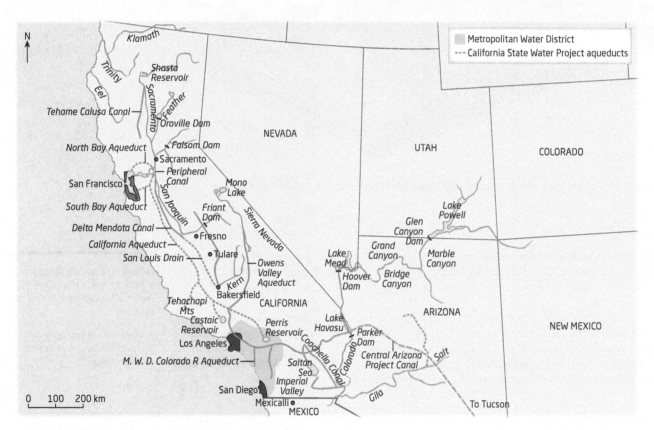

Figure 4.16 Water management schemes in California.

CASE STUDY 4.02.01 (continued)

Lower Basin. Each group was allocated 9.25 trillion litres of water annually, while a 1944 treaty guaranteed a further 1.85 trillion litres to Mexico. Completed in 1936, the Hoover Dam and Lake Mead marked the beginning of the era of artificial control of the Colorado.

Despite the interstate and international agreements, major problems over the river's resources have arisen:

- Although the river was committed to deliver 20.35 trillion litres every year, its annual flow has averaged only 17.25 trillion litres since 1930.
- Demand has escalated with population growth and rising living standards. The river now sustains around 25 million people and 820 000 ha of irrigated farmland in the USA and Mexico.

The $4 billion Central Arizona Project (CAP) is the latest, and probably the last, big-money scheme to divert water from the Colorado River (see Figure 4.16). Before CAP, Arizona had taken much less than its legal entitlement from the Colorado; it could not afford to build a water transfer system from the Colorado to its main cities, and at the time the federal government did not feel that national funding was justified. Most of the state's water came from aquifers, but it was overdrawing this supply by about 2 million acre-feet a year. If thirsty Phoenix and Tucson were to remain prosperous, something had to be done. The answer was CAP, which the federal government agreed to part-fund. Since CAP was completed in 1992, 1.85 trillion litres of water a year have been distributed to farms, Indian reservations, industries and fast-growing towns and cities along the river's 570 km route between Lake Havasu and Tucson. However, providing more water for Arizona has meant that less is available for California. In 1997, the federal government told California that the state would have to learn to live with the 4.4 million acre-feet of water from the Colorado it is entitled to under the 1922 compact, instead of taking 5.2 million acre-feet a year.

Resource management strategies and possible future options

Implementation of the following strategies would conserve considerable quantities of water:

- Reduce leakage and evaporation losses
- Recycle more water in industry
- Recycle municipal sewage for watering lawns, gardens and golf courses
- Introduce more efficient toilet systems
- Charge more realistic prices for irrigation water
- Extend the use of drip irrigation systems, changing from highly water-dependent crops to those needing less water
- Change the law to permit farmers to sell surplus water to the highest bidders
- Require development to identify the source of water to be used before construction can commence.

Other future options have also been discussed. These include developing new groundwater resources. Although groundwater has been heavily depleted in many areas, in regions of water surplus such as northern California they remain virtually untapped. However, the transfer of even more water from such areas would probably prove politically unacceptable. It has been claimed that various techniques of weather modification, especially cloud seeding (already operating on a modest scale), can provide water at a reasonable cost. However, environmental and political considerations cannot be ignored here.

In early 2015, California's first desalination plant was under construction. The Carlsbad Desalination Project will convert as much as 56 million gallons of seawater each day into drinking water for San Diego County residents. The $1 billion project has taken nearly 15 years to move from concept to construction, surviving 14 legal challenges along the way.

There is now general agreement that planning for the future water supply of the south west should embrace all practicable options. Sensible management of this vital resource should rule out no feasible strategy if this important region is to sustain its economic viability and growing population.

Questions

1. How great is the imbalance between the distributions of precipitation and population in California?
2. From which source regions is water transferred to southern California?
3. Why can the Colorado be described as a river under pressure?

Desalination: the answer to water shortages?

Desalination plants are in widespread use in the Middle East, North Africa and the Caribbean, where other forms of water supply are scarce. The journal *Scientific American* noted that there were over 13 000 desalination plants worldwide in 2009. Most of these plants distil water by boiling,

Desalination is the conversion of salt water into fresh water by the extraction of dissolved solids.

4.02 Access to fresh water

generally using waste gases produced by oil wells. Without the availability of waste energy, the process would be expensive. This is the main reason why desalination plants are few and far between outside the Middle East.

However, another method of desalination exists. Originally developed in California in the mid-1960s for industrial use, the reverse osmosis technique is now being applied to drinking water. Recent advances have substantially reduced the cost of reverse-osmosis systems. Large-scale systems using this new technology have been built in Singapore and Florida.

The seawater will still have to undergo conventional filter treatment to rid it of impurities such as microbes pumped into the sea from sewage plants. Thus it is likely that, even when the technology has been refined, desalinated water will always be more expensive than obtaining water from conventional sources. However, desalination has advantages:

- Desalination does not affect water level in rivers.
- It could mean that controversial plans for new reservoirs could be shelved.

Desalination is generally viewed as a last resort in terms of water supply, because:

- The plants are expensive and do not offer a viable solution to the poorest countries unless costs can be drastically reduced.
- Desalination plants are energy intensive with a high ecological footprint.
- Salt is returned to oceans and seas as waste, which can be harmful to local marine ecosystems.

Marine biologists warn that widespread desalinisation around the world could have a big impact on ocean biodiversity. The intake pipes of desalination plants essentially vacuum up large quantities of plankton and other microbial organisms that make up the base layer of the marine food chain.

Replenishing aquifers

Aquifers can be replenished artificially in two main ways:

- By spreading water over the land in pits, furrows or ditches, or by erecting small dams in stream channels to detain and deflect surface runoff, thus allowing it to infiltrate to the aquifer below. For example, large volumes of reclaimed water that have undergone advanced secondary treatment are reused by pumping the water into rapid infiltration basins in a 40 square mile area near Orlando, Florida, USA. The water put into these basins recharges the shallow aquifer and is used to irrigate local citrus crop fields.
- By constructing recharge wells and injecting water directly into an aquifer. For example, large volumes of groundwater used for air conditioning are returned to aquifers through recharge wells on Long Island, New York.

The use of recharge wells is the more expensive of the two methods. While some artificial-recharge projects have been successful, others have proved disappointing. There is still much to be learned about different groundwater environments and their receptivity to artificial-recharge practices.

Cloud seeding

Cloud seeding is any technique of adding material to a cloud to alter its natural development, usually to increase or obtain precipitation.

Cloud seeding (see Figure 4.17) is a technique used to increase rainfall (or snowfall) in an area. It can be used directly over an agricultural area where rainfall is required immediately, or mountain or 'orogenic' cloud seeding can be used for snowpack augmentation, particularly in snowmelt-dominated basins like those originating in the Rocky Mountains in the USA and Canada. The more snow that falls in winter, the more water from snowmelt in spring.

Cloud seeding raises a number of concerns:

- It is an expensive method of water supply.
- The placement of chemicals in the atmosphere, mainly silver iodine, will impact on people and the environment below. Little is known about this impact.
- Management of cloud seeding in not an exact science. The amount of precipitation created may be more than that desired, thus causing flooding.

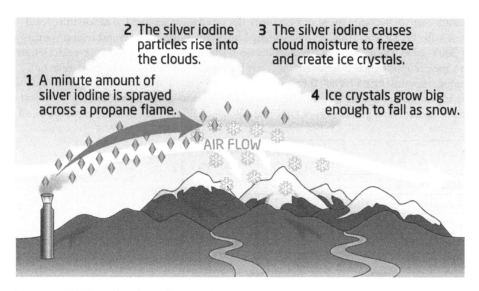

Figure 4.17 How cloud seeding works.

Forest water management

Land management activities can affect water flow and degrade the quality of water. Many countries rely on 'protection forests' to preserve the quality of drinking water supplies, alleviate flooding and to guarantee against erosion, landslides and the loss of soil.

Extension

For more information on filtering/reusing sewage/grey water, San Diego, visit the webistes below:
www.utsandiego.com/news/2014/nov/18/water-recycling-sewer-tap-council-approves/
www.theguardian.com/sustainable-business/2015/jan/20/turning-human-waste-into-drinking-water

4.02.05 Conservation measures and investment

The general opinion in the global water industry is that in the past the cost of water in the developed world has been too low to encourage users to save water. Higher prices would make individuals and organisations, both public and private, think more carefully about how much water they use. Higher prices would:

- encourage the systematic reuse of used water, or **grey water**
- encourage the installation of rainwater-harvesting systems
- spur investment in recycling and reclamation systems
- lead to greater investment in the reduction of water losses.

However, many consumers still see water as a free or low-cost resource, and campaign groups are concerned that higher prices would impact unfairly on people on low incomes. Water pricing for both domestic and commercial users is a sensitive issue. It has also become much more of a political issue as more and more countries have privatised their water resources.

Although some industries have significantly reduced their use of water per unit of production, most water analysts believe that much more can be done. For example, production of 1 kg of aluminium can require up to 1500 litres of water. Other industries such as paper production are also water intensive. Some countries such as Japan and Germany have made considerable improvements in industrial water use. For example, Japanese industry recycles more than 75 per cent of process water.

> **Grey water** is water that has already been used for one purpose, but can possibly be reused for another purpose.

4.02 Access to fresh water

As water scarcity becomes more of a problem, the investment required to tackle this global challenge will rise. Table 4.04 shows the estimated investment needed by world region for the period 2005–2030. There are large contrasts between the different regions of the world. Delivering water to the points where it is required is a costly business in terms of both constructing and maintaining infrastructure. Overall, the sums of money illustrated in Table 4.04 are huge, and funding may need to be diverted from other sectors of national government funding. However, investment in water as a proportion of GDP has fallen by half in most countries since the late 1990s.

Area	Investment needed / trillions of US dollars
Asia and Oceania	9.0
South America	5.0
Europe	4.5
USA and Canada	3.6
Middle East	0.3
Africa	0.2

Table 4.04 Water investment needs by area, 2005–2030.

Urban sanitation services are heavy users of water. Demand could be reduced considerably by adopting dry, or low water use, systems such as dry-composting toilets with urine-separation systems. A number of pilot projects are in operation, such as the Gebers Housing Project in Stockholm. Hong Kong has nearly 90 per cent of its toilets flushing with salt water.

Rainwater-harvesting systems (RHS) harvest the rain water that has fallen typically onto the roof of a household or business. In contrast to a water butt, which typically captures about 200 litres of rain water, a rainwater-harvesting tank can easily filter and store up to 6500 litres of clean water. New technology means that an RHS, which was traditionally used to water gardens, can now be plumbed into existing household pipework and the rain water used to flush toilets and wash clothes. According to the Rainwater Harvesting Association, households could reduce water consumption by up 40 per cent. For those with water meters, this could result in a considerably reduced water bill.

More regions and communities have promoted local conservation measures. Educational materials have made people more aware of how water is wasted and how it can be used more efficiently. In some countries, people have been encouraged to install water meters. If people become more aware of the cost of water, they are more likely to be careful about its use. A lot of water can be lost through leaking pipes. The replacement of old piping systems, although expensive, can save significant amounts of water.

> **Rainwater harvesting** is the accumulation and storage of rainwater for reuse on site, rather than allowing it to run off.

SELF-ASSESSMENT QUESTIONS 4.02.04

1. Define water supply.
2. How important are aquifers to global water supply?
3. **Discussion point:** Why is desalination not more widely used to enhance water supply?

4.02.06 External development assistance

Individual countries, multilateral organisations, non-governmental organisations (NGOs) and private foundations all provide assistance to the drinking water and sanitation sectors of developing countries. Such aid is provided in various ways and can account for the majority of spending on drinking water and sanitation in some countries. For some developing countries, this figure rises to nearly 90 per cent. For example, the NGO WaterAid is currently working in 30 different countries in Africa, Asia and the Pacific region.

Mali, in West Africa, is one of the world's poorest nations. The natural environment is harsh and deteriorating. Rainfall levels, which are already low, are falling further, and desertification is spreading. Currently, 65 per cent of the country is desert or semi-desert. Eleven million people still lack access to safe water. WaterAid has been active in the country since 2001. Its main concern is that the fully privatised water industry frequently fails to provide services to the poorest urban and rural areas. The NGO is running a pilot scheme in the slums surrounding Mali's capital, Bamako, providing clean water and sanitation services to the poorest people. Its objective is to demonstrate to both government and other donors that projects in slums can be successful, both socially and economically.

WaterAid has financed the construction of the area's water network. It is training local people to manage and maintain the system and to raise the money needed to keep it operational. Encouraging the community to invest in its own infrastructure is an important part of the philosophy of the project. According to Idrissa Doucoure, WaterAid's West Africa Regional Manager, 'We are now putting our energy into education programmes and empowering the communities to continue their own development into the future. This will allow WaterAid to move on and help others.' Already, significant improvements in the general health of the community have occurred. The general view is that it takes a generation for health and sanitation to be properly embedded into people's day-to-day lives.

In 2006, the grant and loan aid commitments of bilateral and multilateral external support agencies to the drinking water and sanitation sectors totalled $6.4 billion. The total for the period 2002–2006 was $18.3.

CONSIDER THIS

The concept of virtual water is becoming increasingly recognised. Virtual water is the amount of water that is used to produce food or any other product, and is thus essentially embedded in the item. For example, a kilogram of wheat takes around 1000 dm^3 of water to produce, so importing a kilogram of wheat into a dry country saves the country 1000 dm^3 of water (compared with the cost of growing the wheat itself). The size of global trade in virtual water is more than 800 billion m^3 of water a year. This is equivalent to the flow of ten Nile Rivers. Greater liberalisation of trade in agricultural products would further increase virtual water flows.

4.02.07 Water resources: the potential for international conflict

More than 300 major river basins as well as many groundwater aquifers cross national boundaries. However, as yet there are no enforceable laws that govern the allocation and use of international waters. The UN Watercourses Convention, approved by many countries in 1997, still lacks enough signatories to come into effect. An increasing number of countries are becoming concerned about water insecurity. The countries along the River Nile are a case in point.

The Nile Water Agreement of 1929 gave Egypt by far the largest share of Nile water. This agreement forbids any projects that could threaten the volume of water reaching Egypt. However, this has been criticised by upstream countries such as Kenya as a relic of the colonial era. Kenya and other countries now want to take more Nile water than they have done in the past as their domestic demands increase. In 2004, the Egyptian water minister described Kenya's intention to withdraw from the 1929 agreement as an act of war.

Denying access to water has become a familiar characteristic of conflict in recent years. In the Bosnian war of the early 1990s, one of the first acts of the Serbs besieging Sarajevo was to shut off the electricity and with it Sarajevo's water pumps. People then had no option but to gather at wells around the city, making them easy targets for Serb snipers and mortar shells. At about the same time, water terror was also a potent weapon in the Somalian civil war. People retreating from the fighting filled wells with rocks and dismantled all water infrastructure. These two conflicts could be previews of 'water wars' that some environmentalists warn will eventually engulf the world. For example, in the Middle East, King Hussein of Jordan has said that only a dispute over water could break the peace his country has established with Israel.

Corruption at various levels impacts on the efficient development of water resources. The Fifth World Water Forum, held in Istanbul in 2009, heard that some 30 per cent of water-related budgets are being siphoned off by corrupt deals, including falsifying meter readings, nepotism and favouritism in the awarding of contracts.

4.03 Aquatic food production systems

LEARNING OBJECTIVES

After reading this chapter you should be able to:

- explain that aquatic systems provide a means of food production
- describe how unsustainable use of aquatic ecosystems can lead to environmental degradation and collapse of wild fish stocks
- define the term 'maximum sustainable yield' (MSY) as applied to fish stocks
- evaluate strategies used to avoid unsustainable fishing
- explain the potential value of aquaculture for future food production
- discuss the impact of aquaculture.

KEY QUESTIONS

How do aquatic systems provide us with food?

How is the environment degraded and how do wild fisheries collapse if aquatic systems are used unsustainably?

In what ways does aquaculture provide the potential to increase food production?

4.03.01 Food from aquatic systems

As the human population continues to grow (see Topic 8), there is an ever-increasing demand for food and food-production systems. In aquatic systems, most food is harvested from higher trophic levels where the total storages are much smaller. Where human populations have a choice of fish products as a result of purchasing power and availability, they generally choose fish at a high trophic level such as cod, herring or bass (see Figure 4.18). The fish at this level are generally larger and more palatable to humans. As energy is lost at each trophic level, the energy efficiency of aquatic food production is lower in comparison with the efficiency of terrestrial systems.

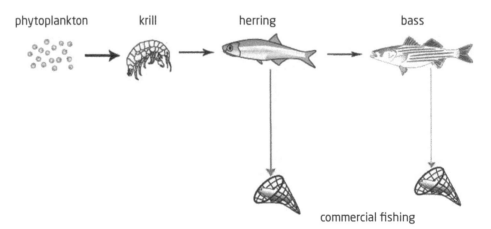

Figure 4.18 A simple marine food chain. Many aquatic food chains have five or more links.

Just as clothes and music go in and out of fashion, so too do societies' tastes in fish. Changes in attitudes to foods that are affected by fashion or influenced by healthy eating campaigns drive the demand for food resources. In Europe today, expensive white-fleshed fish such as cod and haddock is a regular feature on the menus of the most fashionable restaurants, whereas once these fish were common and cheap. Oysters, once a food for working people and equivalent to present day hamburgers, are considered a delicacy today. In London 200 years ago, apprentices even went on strike in protest against being fed salmon for more than five meals a week, but modern diners now consider smoked salmon a luxury food. Other fish that have seen a rise in popularity are swordfish and tuna, even though swordfish is on the list of endangered species. In the 1990s, environmental campaigns in the USA tried to persuade people to stop buying the fish, and new legislation was needed to protect the stocks of the fish. Tuna is another endangered fish. It is eaten in expensive restaurants and is one of the most popular types of tinned fish. In 2000, a large frozen tuna was sold for £55 000 ($40 000) in Japan, but in previous times it might have been used for animal feed.

Aquatic primary production

Phytoplankton are primary producers and the starting point of aquatic food chains and webs. Phytoplankton contain chlorophyll and photosynthesise in the sea and fresh water. In the right conditions, the population of phytoplankton can grow exponentially and produce a phenomenon known as a bloom. Such blooms may last for several weeks as individuals reproduce. Blooms may extend over many kilometres and can be seen in photographs taken from satellites (see Image 4.10). Phytoplankton are a source of food for an enormous number of organisms, which range in size from tiny zooplankton to whales.

Phytoplankton are also responsible for capturing carbon dioxide from the atmosphere and storing it in the oceans as they photosynthesise. Some of the carbon returns to surface waters when the organisms are eaten, but some descends to the depths of the ocean. Phytoplankton are responsible for the biological carbon pump which transfers up to 10 gigatonnes per year of carbon from the atmosphere. Small changes in phytoplankton affect the amount of carbon dioxide in the air. Scientists consider that phytoplankton are a vital contributor to reducing global warming (see Topic 7).

> **Phytoplankton** are microscopic organisms that live in aquatic environments (either freshwater or marine) and start aquatic food chains.

> **CONSIDER THIS**
> The term 'phytoplankton' is derived from the Greek words *phyto*, which means plant, and *plankton*, which means to make wander or float.

Harvesting aquatic production

Humans have always used aquatic systems as a source of both food and materials. Aquatic systems provide fish and shellfish, which include molluscs (such as mussels and whelks), crustaceans (crabs and lobsters) and also echinoderms (sea cucumbers and sea urchins). Sea mammals such as whales and dolphins have also been hunted for food, oil and other resources, although the numbers taken today are declining. Seaweeds and algae of various kinds are part of the human diet in many parts of the world. Not all the products taken from the sea are used as food for humans; some are used as food in fish farms, others such as large seaweeds (kelp) are used in fertilisers and thus produce different foods for humans.

Highest rates of productivity

The highest rates of productivity in aquatic systems are found near coastlines or in shallow seas. The phytoplankton, which are vital to all food chains, grow best along coastlines and continental shelves, where winds drive currents that bring deep water containing an abundance of nutrients to the surface. These regions are known as upwelling zones (see 4.1.04 *Ocean circulation systems*) and are the most productive ecosystems in the oceans. Phytoplankton grow

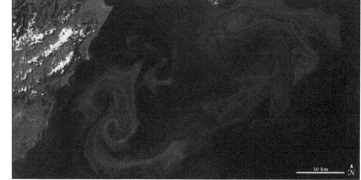

Image 4.10 Blooms of phytoplankton can be seen floating in the ocean as this image from space shows.
Source: NASA – Robert Simmon and Jesse Allen

seasonally and tend to bloom in spring and summer in high latitudes when the light intensity and temperature increase. At lower latitudes, growth is related to the monsoon seasons that change the distribution and availability of nutrients.

4.03.02 Fishing and fisheries

The commercial fishing industry is of enormous importance worldwide. Fish provide what should be a renewable source of food, but catching fish is no longer a straightforward process involving fishing lines and simple nets. It has become an industrial process, involving factory ships that use technology such as sonar (sound waves) to track shoals of fish, huge trawling nets and large-scale machinery. Many species are in danger of being over-fished, as their populations are reducing to unsustainable levels. In some species, the numbers of adult fish available to breed is too low to replace the animals removed by fishing. There is a pressing need to conserve certain fish populations so that the industry can survive.

Fish and shellfish can be caught in a variety of ways, from traditional rods and lines to traps and huge trawl nets. Table 4.05 and Figure 4.19 summarise some methods used and their environmental impacts.

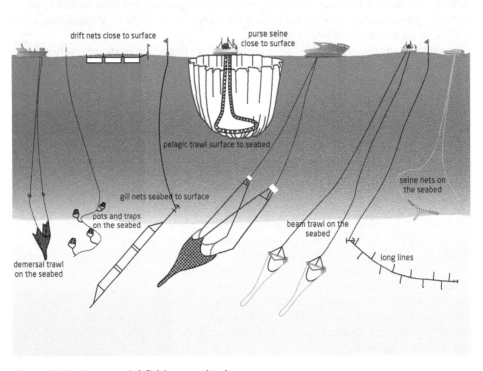

Figure 4.19 Commercial fishing methods.

Maximum sustainable yield

Maximum sustainable yield (MSY) in the context of fisheries is the largest proportion of fish that can be caught without endangering the population.

The ideal is to fish at **maximum sustainable yield (MSY)** – that is, at a level that maintains the maximum yield by allowing fish stocks to replenish at the optimum rate (see Chapter 1.4). Fish are a renewable resource and can always be available for food if they are only taken in a way that allows them to survive and reproduce their numbers. At extremes, if the fish population is small, there will be few adults to produce young, and if the population is large, competition for food will slow growth. How the MSY figure is arrived at is hotly debated, and many countries have different views on the issue, often clouded by traditions or local interests (see Chapter 8.02).

Fishing method	Description	Environmental problems caused
Beam trawls	Net is held open by a beam and, mounted at each end, guides travel along the seabed. Tickler chains (for sand or mud) or heavy chain matting (for rough, rocky ground) drag along the seabed ahead of the net to disturb fish so that they swim up from the seabed into the net.	Chains and matting cause extensive damage to seabed communities.
Bottom trawls	Cone-shaped nets towed across the seabed. The front of the net is kept open by boards. Fish are herded between the boards and swept into the mouth of the nets, where they swim until they are exhausted. Fish drift back through the funnel of the net to the end where they are held.	Many immature fish are taken. In some places, laws set net sizes so that prawns and shrimps are not removed from the sea. The mesh size of two parts of the net can be adjusted to the size of the fish being gathered.
Dredging	A metal-framed dredging basket with a base of iron rings or netting. The frame has a rake, which lifts oysters, clams and scallops from the seabed and directs them back into the basket.	One ship may carry ten or more dredges at once. This may damage the seabed and lead to overharvesting.
Purse seining	A purse net encloses a shoal of fish with a wall of netting. The net is drawn together underneath the fish like a purse, so that they are completely surrounded. This method of fishing aims to catch large, dense shoals of fish such as tuna, mackerel and herring.	In parts of the Pacific Ocean, dolphins used to be enclosed to trap yellowfin tuna that swam underneath them. Large numbers of dolphins were killed, so legislation and an environmental campaign were needed to protect them. Today numbers of yellowfin tuna taken in purse nets account for less than 3% of tuna on sale. Dolphin deaths have reduced by 98%.
Trammel nets	Trammel nets have three layers of netting. The innermost is fine and is set in between two nets with a larger mesh. The whole net is fixed at the bottom and hangs vertically in the water. Fish are caught in the inner looser net.	Trammel nets are not towed along, so do less damage to the environment, but they do trap dolphins and porpoises. In the EU, acoustic devices must be attached to them so that mammals are not caught by accident.
Long-lining	Long-lining involves stretching out a fishing line, up to 100 km long, either vertically or horizontally in the water. Baited hooks are attached to the line at intervals. It is used to fish for swordfish and tuna, but the size of fish and the species caught depend on the type the fishermen use.	A great threat to seabirds. Albatross, petrels, shearwaters and fulmars scavenge on the baited hooks near the water surface, become entangled and are dragged under water, where they drown.
Industrial fishing	Fisheries targeting small fish species that are not eaten but used for the manufacture of fish oil and meal. This is usually done with a purse seine or trawling net.	Removes large numbers of species from the base of food chains. For example, fishing for sand eel in the North Sea has been blamed for the breeding failure of seabirds, and a lack of food for marine mammals and fish such as cod and haddock.

Table 4.05 Fishing methods and their impacts.

Mitigating unsustainable exploitation of aquatic systems

In recent years, there have been several alarming reports on declining fish populations worldwide (see Figure 4.20). In 2003, 29 per cent of open-sea fisheries were in a state of collapse, defined as a decline to less than 10 per cent of their original yield. Bigger vessels, bigger nets and new technology for locating fish are not improving catches, simply because there are fewer fish to catch.

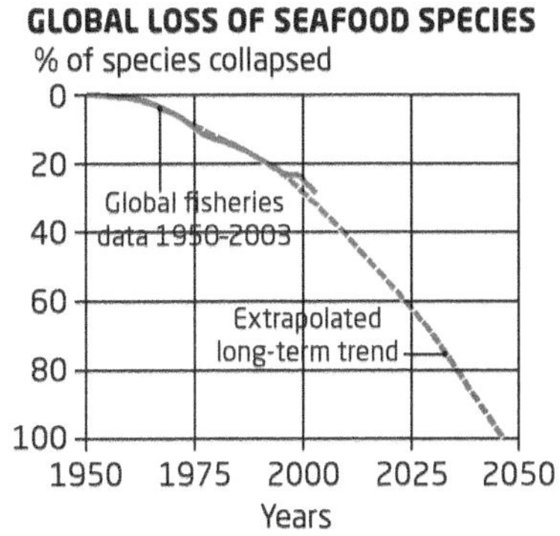

GLOBAL LOSS OF SEAFOOD SPECIES
% of species collapsed

[Graph showing Global fisheries data 1950–2003 and Extrapolated long-term trend from 1950 to 2050, with % of species collapsed rising from 0 to near 100%]

Figure 4.20 Global loss of fish.

4.03 Aquatic food production systems

To conserve populations, several measures can be put in place at the international, national, local and individual level:

- Populations of fish must be monitored to assess whether they are in danger.
- Quotas or closed seasons can be put in place to reduce fishing during the breeding seasons.
- Net sizes must be monitored and controlled, so that smaller immature fish can be left in the water to breed. Where fishing is banned or regulated, biodiversity can improve and fish populations may be restored relatively rapidly. This is shown in Figure 4.21 for the population of North Sea cod.
- Laws must be enforced by individual countries, many of which establish exclusion zones around their coastlines.

These methods not only protect fish populations but also other organisms within a marine ecosystem. This, in turn, has a beneficial effect on the ecosystem biodiversity and fish stocks.

International and national policy

In most cases, international cooperation is essential if conservation measures like these are to be successful, because fish travel long distances and are seldom within the international waters of just one nation. The case of North Sea cod is a success story that is a well-documented example of how fish can be protected using the methods listed above (see Figure 4.21). Similar methods can be used for most species.

The International Council for Exploration of the Sea (ICES) monitors harvests in the North Atlantic. Fish are not easy to count, because they move over long distances. The usual method to estimate a population involves collecting data from landings at fish markets, from the numbers of fish discarded from fishing boats, and from targeted surveys with research vessels.

The numbers of fish of different ages are recorded to give an idea of the age distribution in the population. The age of individual fish is a useful indicator of fish stocks. Too few young ones indicate that fish are not spawning sufficiently to replace caught fish, and too few large fish indicates that overfishing is occurring.

The age of fish can be estimated by the length and weight of individuals or the rings in the ear bones. As fish grow, the number of rings increases, and this can be measured using a microscope. The data collected from catches and age estimation can be used to deduce spawning rates and survival of different species.

ICES offers advice on over 130 species of fish and shellfish. Using the advice from this and other similar organisations, scientists can work out the health of a particular fish population and whether it is being overfished. Fishing boats are monitored using satellite technology to detect fishing grounds and vessels.

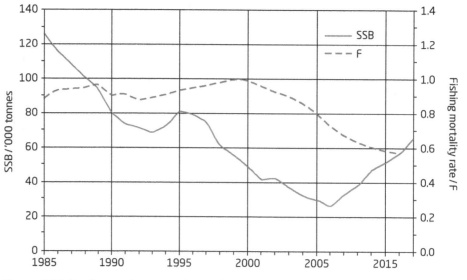

Figure 4.21 Decline and recovery in North Sea cod population (SSB = spawning stock biomass, F = fishing mortality rate).
Source: Napier (2013) *Trends in Scottish Fish Stocks*.

The International Union for Conservation of Nature (IUCN) also supports scientific research and helps governments to develop and implement policies, laws and best practice. It currently works with 140 different countries on a variety of issues, including marine biodiversity conservation. With the IUCN's support, the EU enacted several measures in the early 21st century to try to ensure a sustainable fishing industry. These include:

- restricting the sizes of nets so that only adult fish are taken
- a ban on drift nets (which catch many different species together)
- imposing quotas for different fish
- creating exclusion zones where marine life is protected.

If a trawler netted more than the quota, it had to return the excess to the sea – a process known as discarding. While discarding sounds reasonable and sensible, in practice it did not work because, when bony fish are brought to the surface, their swim bladders are destroyed and the fish die. Forcing fishermen to return them to the sea did not help the fish populations, so the rules have been amended to allow some species of fish to be brought ashore rather than discarded.

The results of these measures can be seen in Figure 4.20. In the period after legislation was introduced North Sea cod populations have begun to recover to a more sustainable level.

New technology and the fight against illegal fishing

The total amount of fish taken illegally by poachers is estimated to be about 20 per cent worldwide, but as high as 40 per cent in some areas that are difficult to police. New satellite tracking systems are being used to help the fight against pirate fishing. Computers can watch satellite feeds from parts of the world that are difficult to monitor in other ways, and gather data about country of registration and ownership of suspicious ships. Once specific information is available, action can be taken by local governments to stop the illegal activity.

Local and individual action

Consumers can also help conserve endangered species. One example is the 'Check the fish on your dish' campaign, which encourages consumers to select sustainable seafood from their local market or shop and select species that are not endangered in preference to those whose numbers are in decline. Organisations such as the Marine Stewardship Council (MSC) and Sustainable Fisheries Partnership check and label products from sustainable sources (see Figure 4.22).

Other campaigns have drawn attention to the plights of the tuna and the swordfish.

Figure 4.22 The logos of the Marine Stewardship Council and Sustainable Fisheries Partnership.

SELF-ASSESSMENT QUESTIONS 4.03.01

1. List three important sources of food that are obtained from aquatic systems.
2. Outline the reasons for the abundance of phytoplankton in coastal waters.
3. Suggest how consumers may be persuaded to reduce their purchases of endangered fish and buy alternative sustainable species.
4. **Research idea:** How did the numbers of certain species of tuna become depleted? How successful were conservation efforts?

4.03 Aquatic food production systems

CASE STUDY 4.03.01

Two fishery-management strategies

Atlantic cod (*Gadus morhua*) is an important food fish that is caught in waters around both Newfoundland and Iceland.

Newfoundland

Cod has been fished in Newfoundland for centuries, and the sea around the country was one of the richest fishing grounds in the world. Traditionally, fishing was carried out at a subsistence level until around 1500 when the first Europeans arrived in North America. By the 20th century, overfishing was a serious problem. In the 1950s, factory fishing began with super-trawlers that took so many fish that in 15 years 8 million tonnes of fish were caught. This amount equalled the total fish taken in 100 years between 1650 and 1750 and was unsustainable, leaving too few fish to replenish the cod population. In the 1960s 800 000 tonnes of fish were taken, but by the 1990s the fisheries had collapsed due to overfishing and a failure of the authorities to recognise or deal with the problem.

In 1993 a moratorium on cod fishing was declared, but even ten years later the cod had still not returned and the population remained depleted. The reasons are not fully understood, but the problem was thought to be due to changes in the ecosystem: it seems that capelin (a fish that had been eaten by cod) was eating young cod that were present. In the early 2000s shrimp and crab dominated the ocean around Newfoundland and it was not until 2011, almost 20 years after the moratorium, that fish numbers gradually began to recover.

Iceland

Over the last three decades the numbers of cod caught in Icelandic waters have varied between approximately 200 000 and 500 000 tonnes per year. But since the turn of the century catches have been decreasing, and now quotas are set to conserve stocks. In 2009, the quota was 130 000 tonnes. Iceland has adopted two strategies to conserve its fish stocks. The first involves management of traditional fisheries, and the second farming of cod in sea cages.

The fisheries management system adopted within Iceland controls not only cod but also fishing of several other species. There are controls on the access to fishing, with permits being used within designated areas and a control on the total annual harvest. The **total allowable catch** varies from year to year depending on the results of monitoring that takes place in these waters. In 2011, the total catch was 182 000 tonnes and the quota for 2013 was 194 000 tonnes. At present the Icelandic cod stock is no longer considered to be overfished to a level that is unsustainable. Nor do the authorities think that overfishing is occurring.

Fish farming began in the 1990s with the production of juvenile fish on an experimental scale. The scheme continued in the 1990s, and in the autumn months from 2003 to 2008 about 1 million young fish were captured and reared in an inshore nursery (see Image 4.11) during the winter to be placed in sea cages (see Image 4.12) later in the year.

Image 4.11 Indoor nursery for cod.

Today, sea cages are no longer stocked in this way, but instead eggs from natural spawning of captured fish are used to raise more fish. The hatcheries produce up to 200 000 fish per year and these are delivered to 11 farms in Iceland and the Shetland Islands. When juveniles reach about 5 g, they are transferred from hatcheries to nurseries, and when they weigh 50–250 g, they are placed in sea cages (see Image 4.12) for about two and a half years, where they grow large enough for market at 3–4 kg. In an alternative farming method, wild cod of about 1 kg in size are captured and reared in sea cages for about one year. This seems to be the more efficient way of producing fish, and a large increase in production is expected after 2015. The amount of farmed cod produced by the two methods increased from 10 tonnes in 2000 to 1450 tonnes in 2007.

Image 4.12 Feeding cod in a sea cage in Iceland.

CASE STUDY 4.03.01 (continued)

Questions

1. Suggest two reasons why the strategy adopted in Newfoundland took so long to be effective.
2. Why are factory ships so damaging to fisheries worldwide?
3. Outline what is meant by maximum sustainable yield (MSY) for a fishery such as those in Iceland and Newfoundland.

4.03.03 Aquaculture

Aquaculture is a method of providing additional food resources. Farming of fish and shellfish has been increasing over recent years and is likely to continue to increase in the future. Aquaculture includes the production of seafood from fish and shellfish that are hatched in ponds then grown to market size in large containers such as tanks or cages. Marine aquaculture can take place in the ocean in cages on the seabed or suspended in open water. Species that have been successfully farmed include sea bass and salmon, turbot, halibut, cod and Arctic char, as well as prawns, mussels, oysters, scallops and clams (see Case study 4.03.01). In freshwater systems, aquaculture can be used to produce species such as carp, trout, salmon, tilapia and catfish. Freshwater aquaculture usually takes place in ponds or lakes and artificial systems such as tanks that can recycle water.

Aquaculture not only improves food supply but it can also help LEDCs support their economic and social development. Fish is a vital source of dietary protein in sub-Saharan Africa, where it provides up to 22 per cent of the population's protein intake. Marine fisheries are overexploited, and aquaculture has the potential to increase supplies of food for poorer consumers. Growth is taking place in countries including Ghana, Nigeria, Uganda, Kenya and Namibia. This is partly due to initiatives such as the Food and Agriculture Organization of the United Nations' Special Programme for Aquaculture Development in Africa (SPADA), which provides policy and technical support to new aquaculture enterprises. Farming of prawns on a large scale takes place not only in South East Asia, including Thailand, Vietnam and Indonesia, but also in Latin America, including Ecuador, Honduras, Guatemala and Mexico. These farms export their produce to Europe, the USA and Japan, and the farms contribute a considerable amount to the economies of the producing countries.

Problems with aquaculture

It is important that the health of fish or other species being farmed is kept under review. Figure 4.23 shows some of the factors that influence the health of fish. External factors are considered when the sites for fish farms in lakes or the ocean are selected, but they cannot be controlled by fish farmers thereafter. Internal factors must be closely monitored. Sites may be selected where there is a good current of water which maintains a healthy flow and high levels of oxygen. In indoor nurseries, tanks must be flushed through regularly to prevent the build-up of bacteria or fouling and remove the waste produced by fish or by the build-up of excess food. Fish are usually separated by size and the density of individuals is monitored to ensure there is sufficient space for swimming, so that the fish build up strong muscle tissue. Food is carefully managed, so that fish receive the correct balance of nutrients to maximise their growth but minimise losses.

Fish farms have an impact on the ecosystems in which they are found. Figure 4.24 shows a fish farm used to rear salmon in a large loch (sea inlet) in Scotland. Food supplied to the fish contains 57 per cent protein, 15 per cent oil, 5 per cent coloured pigment and 23 per cent other ingredients.

> **Total allowable catch** The maximum quantity of a particular type of fish that may be caught each year in total or by the fishing vessels of individual nations.
>
> **Aquaculture** is the farming of aquatic organisms such as fish, crustaceans, molluscs and aquatic plants.

4.03 Aquatic food production systems

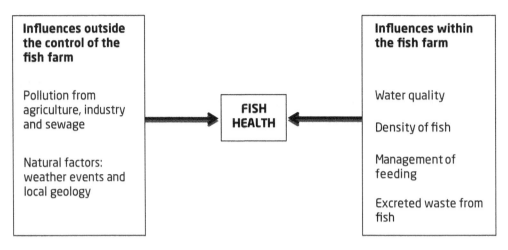

Figure 4.23 Some factors that affect the health of fish reared in aquaculture.

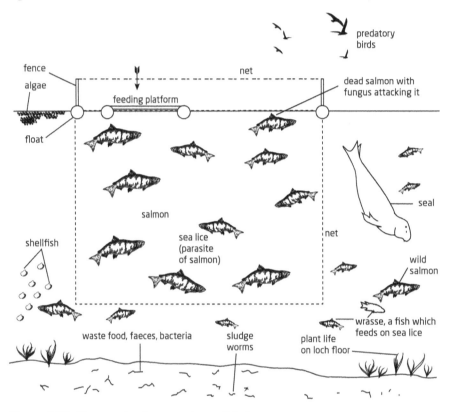

Figure 4.24 Diagram to show a salmon fish farm.

The main problems caused by fish farms include:
- loss of habitat for the local species when cages are placed in lakes or oceans
- disruption of natural food chains and webs
- spread of disease from the farmed species to the wild population
- escape of species from the farms, which may be a problem if genetically modified species are farmed
- pollution of the local environment with excess food, antibiotics and medicines used to treat the fish, and with waste from the farmed organisms.

Extension

For more information on indoor urban aquaculture, visit the link below:
www.youtube.com/watch?v=K5zP4WPgcqY

SELF-ASSESSMENT QUESTIONS 4.03.02

1. Suggest two reasons why the level of food provided to the salmon must be closely monitored.
2. What effect does a fish farm have on the population of **a** plant life on the bed of the loch and **b** wrasse in the ecosystem?
3. Suggest two ways in which the number of parasitic sea lice inside the cage could be controlled.
4. **Research idea:** Why does the food for these farmed fish include coloured pigment?

Refer to Figure 4.23 to answer these questions.

> **CONSIDER THIS**
>
> There have been many criticisms of the prawn-farming industry and the damage it causes, particularly to mangrove habitats in Asia. Madagascar is one nation that is working towards making all its prawn fisheries sustainable and is increasing organic production. Some people regard prawns from Madagascar as a better choice than those from other countries.

4.03.04 Harvesting marine mammals and ethical issues

The harvesting of some species such as seals, dolphins and whales has taken place for centuries and is part of the traditional life of some indigenous cultures. Until the 20th century the hunting methods used involved hand-thrown harpoons and small boats, but as fast boats and high-powered weapons came into use the numbers of mammals that were killed became unsustainable. Today, countries that want to hunt whales may be issued with a special permit to take animals for scientific research. But governments that are members of the International Whaling Commission (IWC) do not agree on special-permit whaling. In 2010, the government of Australia brought a case against the government of Japan at the International Court of Justice regarding Japan's special permit programme in the Antarctic (Chapter 1.1). The IWC also monitors factors such as entanglements, ship strikes, mass strandings, whale watching and pollution that affect the welfare of whales and dolphins.

Campaigns such as the 'Save the Whales' and the Greenpeace campaign against the hunting of Canadian seal pups for their skins (see Image 4.13) in the 1970s turned public opinion against the hunters. Europe banned the import of white-coat harp seal pups in 1983, and whale hunting is now controlled by international treaties (see Topic 1, where ethical issues are also examined).

These developments left some peoples like those of the Nunavut Inuit community, who live on the Clyde River in Canada, without their traditional source of food and income. Although the Inuit could still hunt, the European ban destroyed the market for seal skins. In some Northern Canadian communities, annual income from seal hunting revenue reportedly dropped from $50 000 to as low as $1000. Thirty-five years after Greenpeace's initial anti-sealing campaign, Nunavut Inuit were found to be suffering high rates of malnutrition and poverty. As a result, Greenpeace has changed its view and since 2014 respects the rights of some indigenous communities to carry out sustainable, traditional hunting and fishing.

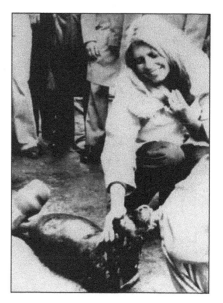

Image 4.13 French actress Brigitte Bardot supported the campaign against hunting of seal pups in the 1970s.

> **Theory of knowledge 4.03.01**
>
> The Inuit people have a historical tradition of whaling. To what extent does our culture determine or shape ethical judgements about issues like this? How would your view of whaling differ from that of an Inuit student?
>
> 1. To what extent are the ethics of a community influenced by the campaigns of environmental groups?
> 2. How can the rights of indigenous cultures, international regulations and the rights of species to be protected be reconciled?

4.04 Water pollution

LEARNING OBJECTIVES

After reading this chapter you should be able to:

- explain that there are many types of fresh and marine water pollution
- outline types of aquatic pollutant, including debris, organic and inorganic material, toxic metals, suspended solids and synthetic compounds
- describe the parameters used to test water quality
- explain the use of indicator species in monitoring water quality
- define the term 'biochemical oxygen demand' and outline how the measure is used
- describe the process of eutrophication
- outline water pollution management strategies.

KEY QUESTIONS

What problems are caused by water pollution?
How does water pollution influence human and other biological systems?
How is water pollution managed?

4.04.01 Types of aquatic pollutants

Pollution is the addition of any substance or form of energy (e.g. heat, sound, radioactivity) to the environment at a rate faster than the environment can accommodate it by dispersion, breakdown, recycling or storage in some harmless form.

Many substances that pollute air and soil (see Topics 1, 5 and 6) also cause **pollution** problems in aquatic systems. Examples of pollutants that affect water are shown in Table 4.06. Pollutants such as sewage and industrial discharges that affect freshwater systems may also cause problems in marine systems, because rivers and pipelines discharge into the sea. Human activities at sea, such as fishing, drilling for oil and shipping goods across the oceans, can also lead to accidental pollution or may themselves discharge pollutants into the sea.

4.04.02 Monitoring water pollution

Water quality indicators include physical, chemical and biological measurements. Most involve the collection of samples (see Figure 4.25) followed by analytical tests. However, some data such as temperature may be collected in situ, without sampling. Direct measurements of water quality are made using measurements of pH, temperature and suspended solids (turbidity), as well as the presence of metals and nitrates and phosphates. You can review how these measurements are made in Topic 2.

Indicators of water quality, the methods used to gather data and interpretation of the results are shown in Table 4.07, which summarises the chemical indicators of water quality. The usual method of measuring water quality is to take samples to assess the concentrations of different chemicals in the water. If chemical concentrations are beyond certain limits, the water is regarded as polluted. Combinations of the measurement of dissolved solids (turbidity) and conductivity may also be used to establish the level of pollution.

Pollutant	Examples
floating debris	plastic waste, bags, nets, microplastic fragments, litter, woody debris, dead plants and animals
organic material	runoff from fields, sewage, industrial discharge from factories and solid waste from domestic sources
inorganic plant nutrients	nitrates, phosphates and other ions from fertilisers
toxic metals	spillages from ships, discharge from factories
synthetic compounds	by-products of industrial processes
suspended solids	effluent from waste water treatment
hot water	discharge from factories and power stations
oil	accidental discharge from ships into the environment, deliberate pollution of fresh water
radioactive material	accidental discharge from power stations
pathogens	in water contaminated with human or animal waste, or in leakage from sewage treatment plants
biological pollutants	invasive species such as water hyacinth (see Chapter 3.03)

Table 4.06 Types of aquatic pollutant and examples of their sources.

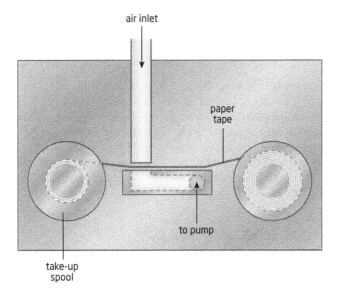

Figure 4.25 Paper-tape sampler.

Water quality data are used to:
- characterise waters
- identify emerging problems
- identify trends over time
- determine whether pollution control programmes are working
- help direct pollution-control efforts to where they are most needed
- respond to emergencies such as floods and spills.

4.04 Water pollution

Indicator	Method	What the results mean
dissolved oxygen	test kit/meter/sensor; oxygen is usually measured as percentage saturation	• >75% oxygen saturation = clean healthy water • 10–50% oxygen saturation = polluted water • <10% oxygen saturation = raw sewage present
pH	indicator paper; compare the colour of the paper with the given colour chart	• pH 1–6 = water is acidic • pH 7 = water is neutral • pH 8–11 = water is alkaline
phosphate	test kit; phosphate is measured in $mg\,dm^{-3}$	• >5 $mg\,dm^{-3}$ = clean water • 15–20 $mg\,dm^{-3}$ = polluted water
nitrate	test kit; nitrate is measured in $mg\,dm^{-3}$	• 4–5 $mg\,dm^{-3}$ = clean water • 5–15 $mg\,dm^{-3}$ = polluted water
chloride (salinity)	test kit; chloride is measured in $mg\,dm^{-3}$	• 20 000 $mg\,dm^{-3}$ = seawater • 100–20 000 $mg\,dm^{-3}$ = tidal or brackish water
ammonia	test kit; ammonia is measured in $mg\,dm^{-3}$	• 0.05–1.00 $mg\,dm^{-3}$ = clean water • 1–10 $mg\,dm^{-3}$ = polluted water • 40 $mg\,dm^{-3}$ = raw sewage

Table 4.07 Chemical indicators of water quality.

4.04.03 Biochemical oxygen demand

Organic material such as dead plants and animals, sewage and other waste contains nutrients which provide a source of food and energy for other organisms. Bacteria and fungi, which feed on these materials, break them down and recycle nutrients, but, at the same time, these decomposers require oxygen. If water contains large amounts of organic waste and many bacteria, the level of oxygen in the water will fall and give rise to **anoxic** conditions. In these circumstances decomposition will proceed anaerobically and lead to the formation of methane, hydrogen sulfide and ammonia. These gases are toxic, and water that is polluted in this way is likely to have a characteristic smell of rotten eggs.

> **Anoxic** water is an area of seawater or fresh water that is depleted of dissolved oxygen.
>
> **Biochemical oxygen demand** is the amount of oxygen required by aerobic microorganisms to decompose the organic matter in a sample of water such as that polluted by sewage. It is an indirect indicator used as a measure of the degree of water pollution.

Defining biochemical oxygen demand

A river system produces and consumes oxygen. Oxygen is gained from the atmosphere and from plants through photosynthesis. Oxygen is consumed by the respiration of aquatic mammals, decomposition and various chemical reactions.

Biochemical oxygen demand (BOD) is determined by the number of aerobic microorganisms and their rate of respiration at any particular location in a river. The greater the BOD, the more rapidly oxygen is depleted in a river, resulting in less oxygen being available to higher forms of aquatic life. When this happens, aquatic organisms can become stressed, suffocate and die. BOD is used as an indirect measure of the amount of organic matter within a water sample.

One of the most common pollution categories is organic pollution caused by oxygen-demanding wastes. Sources of BOD include:

- leaves and woody debris
- dead plants and animals
- animal manure
- effluents from wastewater treatment plants, animal feed, food-processing plants and pulp and paper mills
- failing septic systems
- urban storm-water runoff.

Measuring biochemical oxygen demand

BOD is measured as follows:

1. Take two samples of water at each site (see Figure 4.26). The bottles should be black to prevent photosynthesis. Place the cap on the bottle underwater so that additional air is not introduced.
2. Test one sample immediately for dissolved oxygen.
3. Incubate the second sample in the dark at 20 °C for five days. (The lack of light prevents photosynthesis which would release oxygen and affect the BOD figure.)
4. Test the second sample to see how much dissolved oxygen is left.

The difference in oxygen levels between the two tests, in milligrams per dm^3 (mg dm^{-3}) is the amount of BOD. This is the amount of oxygen consumed by microorganisms to break down the organic matter in the second sample over the five-day period.

Thus:

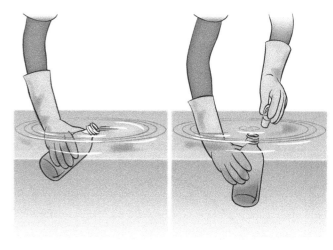

Point the bottle downstream and fill gradually.   Cap underwater when full.

Figure 4.26 Taking a water sample for dissolved oxygen analysis.

dissolved oxygen (mg dm^{-3}) of bottle 1 − dissolved oxygen (mg dm^{-3}) of bottle 2 = BOD (mg dm^{-3})

4.04.04 Effects of organic pollution

In rivers with much organic pollution, the dissolved oxygen level is zero after the five-day period. Some species such as *Tubifex tubifex* (the sludge worm), rat-tailed maggot and bloodworm (see Image 4.14) are tolerant to the low oxygen levels associated with significant organic pollution. The population densities of these species will be high where an organic pollution incident occurs. Other species such as mayflies and stoneflies are unable to tolerate such conditions and will move away if possible. Figure 4.27 shows the effects of organic pollution on a stream ecosystem. Notice that there are significant changes downstream of the point of waste discharge as the aquatic system gradually deals with pollution and reduces its effects. Where pollutants enter the aquatic system, biochemical oxygen demand increases, dissolved oxygen decreases and pollution-tolerant **macro-invertebrates** (see Image 4.14) increase, while the diversity of the system declines.

Macro-invertebrates are animals without backbones that can be seen with the naked eye. Examples include water fleas, dragonfly nymphs, mayflies and worms.

Dissolved oxygen levels also vary significantly with temperature (see Table 4.08 and Figure 4.28). Cold water holds more oxygen than warm water, and so dissolved oxygen levels fluctuate seasonally and over a 24-hour period. The discharge of hot water into a river from factories or power stations also increases the temperature of the water and lowers its oxygen content. Aquatic animals are most at risk from low dissolved oxygen levels on hot summer days when stream flows are low and water temperatures are high. Each day, they are most vulnerable in the early morning when aquatic plants have not been producing oxygen since the last sunset.

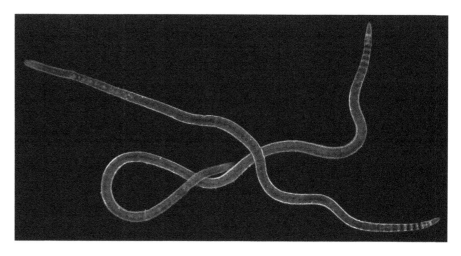

Image 4.14 *Tubifex tubifex* (the sludge worm), can survive in very low levels of oxygen.

4.04 Water pollution

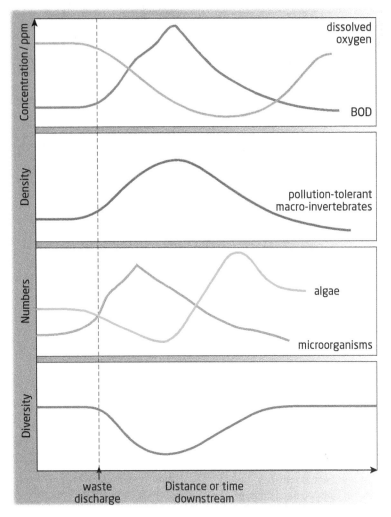

Figure 4.27 Graphs to show the effect of organic pollution.

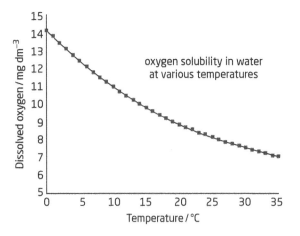

Figure 4.28 Graph to show the maximum amount of water that will dissolve in water at various temperatures.

SELF-ASSESSMENT QUESTIONS 4.04.01

1. Define biochemical oxygen demand.
2. What are the main determinants of biochemical oxygen demand?
3. Suggest why the number of microorganisms decreases downstream from the waste discharge shown in Figure 4.26.
4. **Research idea:** Find out how polluted the nearest river to your school is. What are the possible causes of this pollution?

4.04.05 Indicator species and measuring pollution using a biotic index

Water quality assessments can be based on chemical analysis, but these methods only reflect the conditions when the sample is taken. Most macro-invertebrates have a life cycle of at least a year. They also do not move great distances, and are more or less confined to the area of stream being sampled. The macro-invertebrate community of a stream lives with the stresses and changes that occur in the aquatic environment, both natural and human-induced. They are ideal biotic indicators of water quality.

A **biotic index** is used to assess the quality of an environment by indicating the types of organisms in that environment. Not all aquatic organisms react the same way to poor water quality. Some species are pollution tolerant while others are very pollution sensitive. A healthy water body will have many different organisms, but **indicator species** become absent as the level of organic pollution in a river increases. From such knowledge and understanding, a scale can be developed to determine water quality based on the aquatic organisms found in a water body. A number of different biotic indices are used by environmentalists. The one described here is the Citizen Monitoring Biotic Index, which was compiled by the University of Wisconsin and the US Department of Natural Resources.

> A **biotic index** is an indirect measurement of pollution that assays the impact on species according to their tolerance, diversity and abundance.
>
> **Indicator species** are species that by their presence or absence can be indicative of polluted water (or other systems).

The Citizen Monitoring Biotic Index

The purpose of this index, and others like it, is to enable groups of people to be involved in monitoring the health of aquatic systems. Volunteers are asked to complete the index twice a year, once in spring and once in the autumn. Only about 45 minutes is needed to take the measurements.

The recommended equipment is:

- D-frame kick net
- two white buckets
- white plastic spoons
- white ice-cube trays
- magnifying glass
- tweezers
- plastic cups.

Aquatic macro-invertebrates are collected using frame nets from sample areas of a body of water. They have general characteristics that make them useful in assessing the health of a body of water:

- They are abundant and found in rivers, streams and lakes throughout the world. Because there are lots of them in the water, they are fairly easy to sample.
- They are not very mobile and therefore they are exposed on a continuous basis to water quality in that stream or river.
- They carry out part or all of their life cycle within the water body.

For each sample, tasks A–I on the recording form for the Citizen Monitoring Biotic Index (see Figure 4.29) are carried out and animals identified using the 'life in the river' key (see Figure 4.30). Not all the macro-invertebrates that are listed on the identification key are found on the scoring sheet. This is because some do not rely on oxygen in the water for survival. Many are able to collect air from the atmosphere and are unaffected by water oxygen levels. The calculations provide an index score, divided into categories from 'poor' to 'excellent'. For example, if the index score is 3.2, the health of the water body is said to be good. A single number that characterises the health of a stream is useful to the specialist biologist as well as to those non-specialists charged with managing stream health.

4.04 Water pollution

Recording Form for the Citizen Monitoring Biotic Index

Name:_____   Date:_____

Watershed and Stream Names:_____   Time:_____

Location:_____   Site ID:_____
(County, Township, Range, Section, Road, Intersection, Other)

At this point, you should have collected a wide variety of aquatic macro-invertebrates from your three sites. You will now categorise your sample, using the chart (other side) to help you identify the macroinvertebrates found. The number of animals found is not important; rather, the variety of species and how they are categorised tells us the biotic index score. Before you begin, check off the sites from which you collected your sample (see right).

☐ Riffle Sampling
☐ Snag Areas, Tree Roots, Submerged Logs
☐ Leaf Packs
☐ Undercut Banks

1. Check the basin with the debris to see if any aquatic macroinvertebrates crawled out. Add these animals to your prepared sample.

2. Fill the ice cube tray half-full with water.

3. Using plastic spoons or tweezers, (be careful not to kill the creatures – ideally, you want to put them back in their habitat after you're finished) sort out the macroinvertebrates and place same species together in their own ice cube tray compartments. Sorting and placing same species together will help ensure that you find all varieties of species in the sample.

4. Refer to the 'Life in the River Key' and the Citizen Monitoring Biotic Index to identify the macroinvertebrates:

   A. On the back of this page, circle the animals on the index that match those found in your sample.

   B. Count the number of circled animals in each category and write that number in the box provided.

   C. Enter each boxed number in work area below.

   D. Multiply the entered number from each category by the category value.

   E. Do this for all categories.

   F. Total the number of animals circled.

   G. Total the values for each category.

   H. Divide the total values by the total number of animals: total values (b) / total animals (a).

   I. Record this number.

   SHOW ALL MATHS (Use space below to do your maths computations)

No. of animals from group 1 _____ x 4 = _____

No. of animals from group 2 _____ x 3 = _____

No. of animals from group 3 _____ x 2 = _____

No. of animals from group 4 _____ x 1 = _____

   TOTAL ANIMALS(a)_____ TOTAL VALUE(b)_____

Divide total value (b)_____ by total no. of animals (a) _____ for index score:

Call your local Monitoring Coordinator if you have questions about sampling or determining the biotic index score.

Return form to:

Index score:

How healthy is the stream?
Excellent _____ 3.6+
Good _____ 2.6–3.5
Fair _____ 2.1–2.5
Poor _____ 1.0–2.0

Figure 4.29 Part of a recording form for the Citizen Monitoring Biotic Index.

Indicator species and measuring pollution using a biotic index

Group 1: These are sensitive to pollutants. Circle each animal found.

   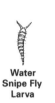 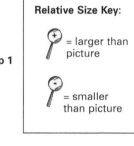

Stonefly Nymph | Dobsonfly Larva | Alderfly Larva | Water Snipe Fly Larva

No. of group 1 animals circled:

Relative Size Key:
⊕ = larger than picture
⊖ = smaller than picture

Group 2: These are semi-sensitive to pollutants. Circle each animal found.

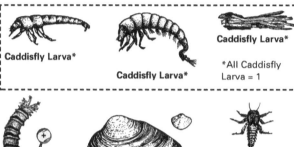  

Caddisfly Larva* | Caddisfly Larva* | Caddisfly Larva* *All Caddisfly Larva = 1 | Dragonfly Nymph | Crawfish

Cranefly Larva | Freshwater Mussels or Fingernail Clams | Mayfly Nymph | Damselfly Nymph | Water Penny | Riffle Beetle

No. of group 2 animals circled:

Group 3: These are semi-tolerant of pollutants. Circle each animal found.

    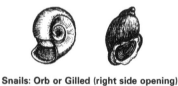

Blackfly Larva | Non-Red Midge Larva | Snails: Orb or Gilled (right side opening) | Amphipod or Scud

No. of group 3 animals circled:

Group 4: These are tolerant of pollutants. Circle each animal found.

Pouch Snail (left side opening) | Isopod or Aquatic Sowbug | Bloodworm Midge Larva (red) | Leech | Tubifex Worm

No. of group 4 animals circled:

© Spring 2001 University of Wisconsin. This publication is part of a six-series set, "Water Action Volunteers- Volunteer Monitoring Factsheet Series." All record forms are free and available from the WAV coordinator. WAV is a cooperative program between the University of Wisconsin-Cooperative Extension & the Department of Natural Resources. For more information, call (608) 265-3887 or (608) 264-8948.

Figure 4.30 'Life in the river' key.

4.04 Water pollution

> ### SELF-ASSESSMENT QUESTIONS 4.04.02
>
> 1. What is a biotic index?
> 2. Why are macro-invertebrates used to assess the health of a body of water?
> 3. On the Citizen Monitoring Biotic Index, what is the health of water bodies with the following scores?
>   - a 1.4
>   - b 3.9
>   - c 2.3
> 4. **Discussion point:** To what extent do you agree that the Citizen Monitoring Biotic Index is a straightforward exercise that could be carried out by a significant number of people?

4.04.06 Eutrophication

How eutrophication occurs

Eutrophication refers to the natural or artificial addition of nutrients, particularly nitrates and phosphates, to a body of water, resulting in depletion of the oxygen content. Human activity can accelerate the process by the addition of sewage, detergents and agricultural fertilisers.

Eutrophication is a natural process that occurs in aquatic systems as plants grow and die and their organic remains are recycled by decomposers. In an unpolluted system, negative feedback ensures that the system remains in equilibrium. If the addition of nutrients is too rapid or on too large a scale, positive feedback may cause an imbalance in the system and a deviation from equilibrium. These effects are summarised in Table 4.09.

Negative feedback and eutrophication	Positive feedback and eutrophication
Rising levels of nutrients lead to an increase in biomass as nutrients are taken up by plants. This causes a reduction in nutrients in water so equilibrium is restored.	More nutrients lead to more algal growth. Death and decomposition of algae leads to increased nutrients and further deviation from equilibrium.
Rising levels of algae lead to a rise in other species which feed on them, so numbers of algae decrease and equilibrium is restored.	Rising growth of algae blocks light to other plants and leads to their death and therefore more nutrients. More nutrients cause more growth of algae and therefore a further deviation from equilibrium.
Rising levels of dead organisms lead to an increase in numbers of decomposers which feed on them. More decomposers lead to a decrease in organic matter, so that equilibrium is restored.	Increase in bacteria feeding on dead remains leads to a rise in biochemical oxygen demand and decrease in organisms that need oxygen. This leads to a greater biochemical oxygen demand and more deviation from equilibrium.

Table 4.09 The role of negative and positive feedback in the process of eutrophication.

The stages in eutrophication

Eutrophication is the one of the most serious impacts that human activity exerts on the quality of water. When its effects are severe, eutrophication can be considered a form of pollution. The most common causes of eutrophication are the addition of nitrates and phosphates to water systems.

The stages that occur during eutrophication are shown in Figure 4.31

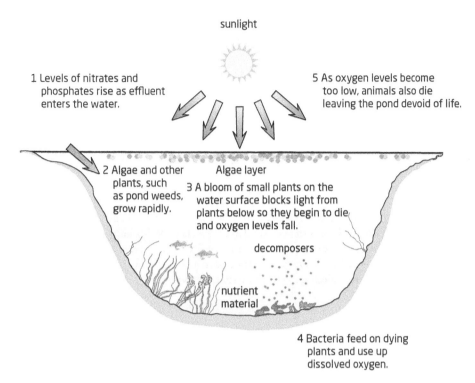

Figure 4.31 Stages in eutrophication in a pond ecosystem.

> **CONSIDER THIS**
>
> In the UK, an Environment Agency report has estimated that the main sources of phosphorus entering surface waters are agriculture (45 per cent), human and domestic wastes in sewage (24 per cent) and detergents (19 per cent). The report estimated that 70 per cent of total nitrogen input is from diffuse sources – agriculture, precipitation and urban-runoff in order of decreasing importance.

Phosphates are a common constituent of agricultural fertilisers, manure, and organic wastes in sewage and industrial effluent, and are also found in household detergents, which become part of domestic sewage. Soil erosion can also add phosphates, as the soil contains residues of both manure and fertiliser. Algae thrive in runoff water from these sources and often appear as a green scum or blanket on the surface of a body of water. A large accumulation of algae on surface water is known as an **algal bloom**.

Nitrates are also an important component in fertilisers that can run off into water. Animal excreta are rich in both phosphorus and nitrogen and, in communities where livestock concentrations are high, large quantities of slurry are often spread on the fields as a source of fertiliser. This contributes significantly to concentrations in rivers and lakes.

> **Algal bloom** is a large accumulation of algae on the surface of a body of water.

> **CONSIDER THIS**
>
> Raw sewage produced by human activity contains many pathogens. If it is released into rivers and streams that are used for drinking water or bathing, diseases such as cholera and typhoid can easily be spread. Countries with limited access to clean water often have higher incidences of these illnesses.

4.04.07 Impacts of eutrophication

Eutrophication can cause a range of biological changes (see Image 4.15). Human activities worldwide have caused the nitrate and phosphate content of many rivers to double. In some cases, increases of up to 50 times have been recorded. Eutrophication causes a change in both plant and animal species present in water, and often observation of plant species can indicate an excess of nitrates and phosphates in water.

The typical impacts of eutrophication are as follows:

- Increase in turbidity
- Increase in primary productivity: often as a dense green sludge, but an algal bloom can be other colours such as yellow-brown or red depending on the species of algae present
- Decrease in dissolved oxygen

4.04 Water pollution

- Accumulation of organic compounds to concentrations that are toxic to mammals and sometimes to fish
- Loss in the diversity of fish species
- Reduction in the length of food chains and loss of species diversity.

Eutrophication not only affects inland waters, it can also impact significantly on coastal marine ecosystems with similar effects. In the Gulf of Mexico there is now a large hypoxic **dead zone**, where the low oxygen content of the water has led to the death of ocean species. The cause has been traced to nutrient runoff from the agricultural states of the mid-western USA, carried to the Gulf by the Mississippi River. Agriculture in this region of the USA is characterised by high fertiliser use. Coastal eutrophication is often characterised by the appearance of so-called 'red tides' (see Image 4.16). These are caused by the algal blooms of red algal species present in the ocean.

Higher frequency of algal blooms in fresh water increases the costs of filtration for the domestic water supply. Algae blooms may cause detectable tastes and odours in water supplies and the production of large quantities of fine detritus, which clogs filters and can support communities of aquatic organisms such as nematode worms and various insects. Some algal blooms cause health problems ranging from skin irritation to pneumonia.

> **Dead zones** are areas in oceans or fresh water where there is not enough oxygen to support life.

Extension

Challenge yourself to manage a river catchment using this game:
http://catchmentdetox.net.au/play-game

Image 4.15 Dead fish floating on a eutrophic lake.

Image 4.16 Red tide caused by a bloom of red algae.

SELF-ASSESSMENT QUESTIONS 4.04.03

1. State one way in which negative feedback maintains equilibrium and prevents harmful eutrophication in an aquatic system.
2. State two sources of phosphate that enters aquatic systems.
3. Define a dead zone.
4. **Research idea:** What is the evidence of eutrophication in the region in which you live?

4.04.08 Pollution-management strategies

Reducing human activities that cause pollution

The following are some examples of changes in human activities that can reduce pollution:

- Use alternative methods to increase farm yields, such as growing vegetative buffer strips on farms to prevent organic and chemical fertiliser from running off into the water.
- Change to organic farming, which does not use artificial fertilisers, and reduces the amount of phosphorus and nitrates entering adjacent water bodies.
- Replace laundry and dishwashing detergents with phosphate-free equivalents. Many dishwashing detergents still contain between 9 and 30 per cent phosphate because of its cleansing properties. Phosphate-free dishwasher detergents are more readily available in some countries than others.

Reducing the release of pollution into the environment

The following are some examples of ways in which the release of pollution into the environment can be reduced:

- Use advanced sewage-treatment processes that remove nitrates and phosphates from the waste before water is returned to waterways.
- Reduce the use of manure on farmland. For example, Dutch farmers are subject to manure quotas and may only spread a certain amount of manure per hectare. For any manure used beyond that amount, the farmer must pay the costs of removal and processing.
- Reduce the amount of pollutants that reach water, using the protection of forest cover which reduces soil erosion significantly. Trees planted as a buffer between farmland and water courses or lakes can soak up the nutrients that cause eutrophication.

Clean-up and restoration

Clean-up and restoration of heavily polluted water bodies can take a number of forms including:

- pumping mud from eutrophic lakes
- reintroducing plant and fish species that have been lost.

Theory of knowledge
4.04.01

The impact of eutrophication on water bodies is clear to see on the water surface. Many different parameters are used to test water quality. From these measurements, judgements are made about causes and effects of water quality.

1. How can we identify cause–effect relationships?
2. How can we be sure our judgements are correct if we only observe correlations?

4.04 Water pollution

CASE STUDY 4.04.01

Lago Paranoá, Brazil

The 4000 ha Lago Paranoá is an artificial lake constructed in 1959 by Brasilia, the capital city of Brazil (see Figure 4.31). By the early 1970s, the lake had become highly eutrophic, with significantly deteriorating water quality, due to the inflow of inadequately treated domestic sewage from two sewage plants. In the late 1970s, severe algal bloom caused a high rate of mortality among the fish population as oxygen levels fell.

The Lago Paranoá restoration programme began in 1975 through a United Nations Development Programme cooperation programme. The main strands of the project were:

- training Brasilia Water and Sewerage Corporation staff
- constructing a laboratory to permanently monitor water quality
- collecting sewage, since domestic sewage was identified as the main source of eutrophication.

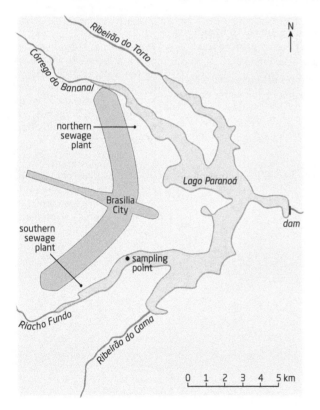

Figure 4.32 Map showing location of sampling point in Lago Paranoá.

One significant step in the restoration of Lago Paranoá occurred between 1993 and 1994, when two new sewage plants capable of processing, $2400\,ls^{-1}$ of sewage from the city were built. A complementary sewage-collecting system composed of 46 km of sewers, 15 pump stations, 13 km of pressure sewers and 417 m of underwater pipelines was also constructed. The overall cost of this new sewage treatment facility reached US$200 million.

Other management strategies, including biomanipulation and hydrological monitoring, have added to the success of the programme.

Biomanipulation has involved manipulating a key component of the food chain to improve water quality. The first step in biomanipulation was an echo-sounding survey, carried out in 1998. It estimated the fish stock in the lake and identified specific areas with large, unbalanced fish populations. A pilot experiment to remove large fish was carried out in May 1999 in the isolated mouth of the Riacho Fundo tributary, the area of the lake where the fish numbers, mainly tilapia, were particularly high. A total of 20 000 adult fish weighing 4000 kg (85 per cent of which were tilapias) were removed by professional fishermen in six days. Controlling the tilapia population will prevent an excessive increase in the phytoplankton biomass, because herbivorous invertebrates (zooplankton; the large fish's prey) can now increase in numbers. The herbivorous invertebrates graze on the phytoplankton and so reduce its biomass, which in turn leads to clearer water. This is an example of negative feedback maintaining equilibrium in the aquatic system.

The hydrological monitoring programme has shown a gradual reverse in eutrophication in Lago Paranoá.

The case study of Lago Paranoá shows that a careful assessment and identification of main sources of eutrophication combined with an efficient long-term monitoring programme are key factors in implementing a successful restoration programme that directly benefits 2 million inhabitants in the capital of Brazil.

Questions

1. Suggest why the construction of an efficient sewage system was important in restoring the lake ecosystem.
2. Describe how biomanipulation has been used as a management technique in Lago Paranoá.
3. Outline two other potential benefits of this restoration programme.

End-of-topic questions

1. **a** What is the difference between stores and transfers in the hydrological cycle? [2]

   b How does the process of urbanisation impact on surface runoff and infiltration? [4]

2. **a** What is water security? [2]

   b Discuss the physical and human factors influencing access to safe drinking water. [2]

3. **a** Outline what is meant by the Gulf of Mexico hypoxia? [1]

   b What role do land-based nutrients play in the hypoxia of the Gulf of Mexico? [3]

4. The two maps in Figure 4.33 show Toxics Release Inventory (TRI) releases per year (the TRI records releases of toxic chemicals), and median incomes for Santa Clara County in the USA.

   a Identify the relationship between toxic releases and income. [2]

   b Suggest *two* reasons for the relationships you have identified. [2]

   c List *three* measures that could be taken to reduce the release of toxic substances in an area. [3]

5. **a** Describe the process of eutrophication.

   b Outline the role of human activity in making the problem worse?

   c Suggest *three* ways in which eutrophication affects natural aquatic systems.

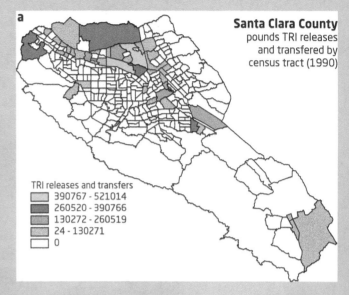

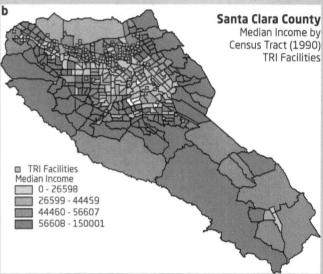

Figure 4.33 (a) TRI releases and (b) median incomes for Santa Clara County, USA, in 1990.

Topic 5
SOIL SYSTEMS AND TERRESTRIAL FOOD PRODUCTION SYSTEMS AND SOCIETIES

Introduction to soil systems

5.01

LEARNING OBJECTIVES

After reading this chapter you should be able to:

- understand that the soil system is a dynamic ecosystem that has inputs, outputs, storages and flows
- appreciate that the quality of soil influences the primary productivity of an area.

KEY QUESTIONS

How does a soil profile illustrate the soil system?
What are the processes involved in the soil stystem?
How do the structures and properties of sand, clay and loam soils differ?

5.01.01 Soil systems integrate aspects of living systems

Soil systems

Soil can be defined as the naturally occurring unconsolidated material on the surface of the Earth that has been influenced by parent material, climate, macroorgainisms and microorganisms, and relief, all acting over a period of time. It is thus a mixture of inorganic mineral particles and organic material from decomposed flora and fauna that covers the underlying bedrock, and in which a wide variety of terrestrial plants grow. Soils are complex systems which carry out a wide range of functions that are critical to the functioning of the Earth as a whole system. The soil system may be illustrated by a **soil profile** (see Figure 5.03).

> **Extension**
>
> Great introductory gallery presentation from national geographic
> http://ngm.nationalgeographic.com/2008/09/soil/richardson-photography
>
> Related quiz
> http://ngm.nationalgeographic.com/2008/09/soil/quiz-interactive

Soils form due to the interaction between the lithosphere, biosphere, atmosphere and hydrosphere (see Figure 5.01). The lithosphere is the solid, rocky crust covering the Earth. It is inorganic and is composed of minerals. Weathering of the lithosphere provides the inorganic component of a fertile soil, providing elements such as phosphorous and potassium. When a soil develops from the **parent rock** beneath, the minerals in the rock are susceptible to different processes and rates of weathering. Figure 5.02 illustrates the weathering of granite and the different soils this can produce. Parent rock may be the main factor determining the soil type in a region. Parent rock contributes to a soil's:

- depth
- texture
- drainage
- quality
- colour.

Soil is a mixture of inorganic mineral particles and organic material from decomposed flora and fauna that covers the underlying bedrock, and in which a wide variety of terrestrial plants grow.

The **soil profile** is the vertical succession down through a soil which reveals distinct layers or horizons in the soil.

Parent rock is the upper layer of rock on which soil forms under the influence of biological and biochemical processes and human activity. The properties of the parent rock are changed in the process of soil formation through the effect of other soil-formation factors, but to a large extent the properties of the parent rock still determine the properties of the soils.

5.01 Introduction to soil systems

Soil formation

Topography (the shape of the landscape) can have a considerable effect on the amount of soil present in an area. For example, steep slopes struggle to hold soil because of their susceptibility to soil erosion.

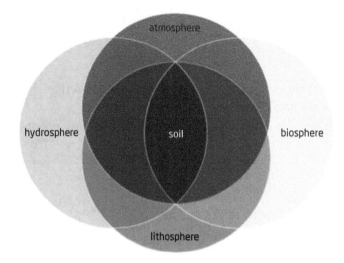

Figure 5.01 Soil formation and the interaction between lithosphere, biosphere, atmosphere and hydrosphere.

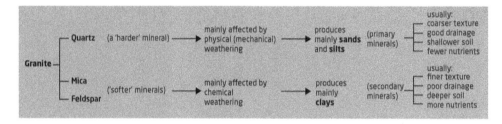

Figure 5.02 The influence of a parent rock, granite, on soil formation.

Living organisms provide the organic component of a **fertile soil** (see Image 5.01) through their death and decay. Soils that have a very low organic component have limited fertility (see Image 5.02). The breakdown of soil organic matter provides nutrient elements such as nitrogen and sulfur. Soil contains a wide range of animal life. In the top 30 cm of 1 hectare of soil there are on average 25 tonnes of soil organisms, comprising:

- 10 tonnes of bacteria
- 10 tonnes of fungi
- 4 tonnes of earthworms
- 1 tonne of soil organisms such as springtails, mites, isopods, spiders, snails, mice and other creatures.

Earthworms alone can make up between 50 and 70 per cent of the total weight of fauna in arable soils. Microorganisms can exist in even greater numbers. Just 25 g of soil can contain over 100 million microorganisms. Larger living organisms help to mix the various soil elements through their movement within the soil and through their bodily functions. Such mixing is an important process in the development of soil fertility. A typical cultivated soil will be composed of 50–60 per cent mineral particles, 1–5 per cent organic matter, and 40 per cent pore spaces between the particles, which will have varying amounts of air and water in them.

Fertile soil is soil that is rich in the nutrients necessary for basic plant nutrition, including nitrogen, phosphorus and potassium.

Zonal classification of soils is the subdivision of the world into broad soil regions based primarily on differences in climate.

Soil systems integrate aspects of living systems

Image 5.01 Bales of hay on flat, fertile farmland, County Wexford, Ireland.

Image 5.02 Goat herders on the margins of the Gobi desert, where soils are very infertile and grazing extremely limited.

The atmosphere is a vital part of the soil formation process. The **zonal classification of soil** states that on a global scale soils are determined by climate. Soil is more prevalent in environments of high moisture and high temperatures than in cold, dry regions. This is because the increased moisture contributes to erosion, and higher temperature contributes to a more rapid breakdown of organic material. When organic material breaks down, it forms **humus** – a dark, crumbly substance that is very fertile for plant growth. An average soil is approximately half minerals and half water and air.

Precipitation effectiveness is the balance between precipitation and potential evapotranspiration. It determines the direction of water movement within a soil. When precipitation is greater than potential evapotranspiration, the process of **leaching** occurs. Soils that are characterised by a net downward movement of water are called pedalfers.

A soil water deficit occurs if precipitation is less than potential evapotranspiration. This results in water being drawn to the surface, bringing with it calcium carbonate. Soils with a soil water deficit are called pedocals and are common in arid and semi-arid regions.

A generalised soil profile

The vertical succession down through a soil is known as a soil profile. A soil profile (see Image 5.03) consists of various layers (horizons) which result from the balance between inputs and outputs into the soil system, and from the redistribution of and chemical changes in the soil constituents.

Soil formation begins first with the breakdown of rock into **regolith**. Regolith can be defined as the irregular cover of loose rock debris that covers the Earth. As weathering continues, soil horizons form leading to the development of a *soil profile*. A **soil horizon** is a specific layer that is parallel to the surface and possesses physical characteristics that differ from those of the layers above and beneath.

Figure 5.03 illustrates a typical soil profile, which is in fact a living system.
Soil horizons vary in terms of texture, structure, colour, pH and mineral content.

- *O horizon* The uppermost layer of soil is known as the organic (O) horizon and consists of leaf litter in various stages of decomposition. The litter helps to prevent erosion, holds moisture, and decays to form humus, a vital element in soil fertility.
- *A horizon* Immediately below is the A horizon, a mixed mineral–organic layer where seeds germinate and plant roots grow. The A horizon has a layer of dark decomposed organic materials, which is called humus. The A horizon is commonly known as topsoil. It is the zone in which most biological activity occurs, as soil organisms are concentrated here, often in close association with plant roots.

> **CONSIDER THIS**
>
> An average soil sample is 45 per cent minerals, 25 per cent water, 25 per cent air, and 5 per cent organic matter.

Humus is a dark brown or black organic substance made up of decayed plant or animal matter. It provides nutrients for plants and increases the ability of soil to retain water.

Effective precipitation is the amount of precipitation that is actually added and stored in the soil.

Leaching is a natural process by which water-soluble substances such as calcium are washed out from soil. This reduces the fertility of a soil.

Regolith is the irregular cover of loose rock debris that covers the Earth.

The **soil horizon** is a specific layer of soil that is parallel to the surface and possesses physical characteristics that differ from those of the layers above and beneath.

5.01 Introduction to soil systems

Image 5.03 Soil profile above chalk bedrock, Dorset, UK.

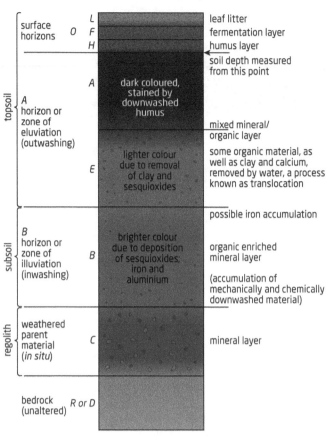

Figure 5.03 A typical soil profile.

- *E horizon* This is the eluvial or leached horizon. Leaching removes material from the E horizon, making it brighter in colour. It is made up of sand and silt, having lost much of its minerals and clay.
- *B horizon* This is the deposition or illuvial horizon, which receives material removed from the overlying E horizon, such as iron, humus and clay. The B horizon is often referred to as the subsoil. Plant roots penetrate through this layer, but there is very little humus in this horizon.
- *C horizon* Here the bedrock has begun to break up into blocks, forming a transition zone between the R horizon and the layers of soil above. This horizon shows little or no sign of soil formation.
- *R horizon* The R horizon comprises the parent material (hard bedrock).

SELF-ASSESSMENT QUESTIONS 5.01.01

1. Define the terms 'soil profile' and 'soil horizon'.
2. What is leaching?
3. **Research idea:** Take a soil sample from the grounds of your school. What is the evidence of organic and inorganic material in the soil?

The processes of soil formation

Soils should be viewed as open systems in a state of dynamic equilibrium. They vary as the factors and processes that influence them change. All soils have processes in common but they differ in terms of the rates and types of processes. These processes cause transfers of material (including deposition), which results in reorganisation of the soil. As with any system, inputs, transformation (processes) and outputs can be clearly recognised (see Figure 5.04). It takes several thousand years to build a thin layer of fertile topsoil.

The inputs into a soil include:

- organic material from decaying flora and fauna
- precipitation, gases and solid particles from the atmosphere
- gases from the respiration of soil fauna
- excretions from plant roots
- minerals from the breakdown of parent material.

The outputs from a soil include:

- nutrients taken up by plants growing in the soil
- nutrient losses through leaching
- losses of soil through soil erosion and mass movement (e.g. soil creep)
- evaporation.

Transformations include:

- decomposition
- weathering
- nutrient cycling – for example, some of the plant nutrients lost as an output will return as leaf litter in time.

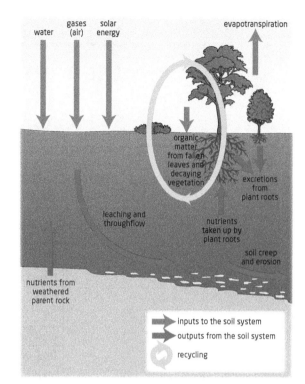

Figure 5.04 The open soil system.

SELF-ASSESSMENT QUESTIONS 5.01.02

1. Why should soils be viewed as open systems in a state of dynamic equilibrium?
2. State three inputs into a soil.
3. Give three outputs from a soil.
4. **Discussion point:** Discuss the role of transformations (processes) in a soil.

Theory of knowledge 5.01.01

Soil knowledge has developed since the earliest known practice of agriculture around 11 000 BC. By the 4th century AD, peoples around the world showed various levels of soil understanding. Early knowledge of soils was based largely on observations of nature. Many famous scientists, including Francis Bacon and Charles Darwin, wrote about soil issues. However, the knowledge and understanding of soils did not become a true science until the 19th century with the development of genetic soil science. The maturing of soil science as a distinct academic discipline first required scientific advances in related fields including biology, chemistry, physics, geography and geology.

1. When do you think the development of a body of knowledge such as the knowledge and understanding of soils can be classed as a distinct academic discipline?
2. Should we attempt to make infertile soil productive when formation factors have limited the soil's productivity?

5.01 Introduction to soil systems

5.01.02 Sand, clay and loam soils, and their influence on primary productivity

Soil texture

<aside>Soil texture is the look and feel of a soil, and is determined by the size and type of particles that make up the soil. While soil texture is affected by organic material, it mostly refers to the inorganic material in the soil.</aside>

Soil texture refers to the size of the solid particles in a soil. The three main mineral components of soil are: clay, silt and sand. Fertile soils generally have a good balance between these three components. The relative proportion of each component gives the soil its texture. Referred to as 'soil separates', the mineral components have specific ranges of soil particle size. Table 5.01 shows the diameter limits for clay, silt and sand according to the United States Department of Agriculture (USDA). Sand particles are subdivided into five divisions from very fine to very course. Particles larger than sand are described as stones.

Name of soil separate	Diameter limits / mm (USDA classification)
clay	< 0.002
silt	0.002–0.05
very fine sand	0.05–0.10
fine sand	0.10–0.25
medium sand	0.25–0.50
coarse sand	0.50–1.00
very coarse sand	1.00–2.00

Table 5.01 Soil separates.

The way a soil is described is derived from the size of the primary constituent particle (e.g. clay, silt) or a combination of the sizes of the main particles (e.g. silty clay, sandy clay). An additional term, 'loam', is used to describe a roughly equal mixture of the soil separates (e.g. sandy loam, clay loam). Figure 5.05 shows the differences between the soil classes according to the USDA.

Soil texture is very important because it affects:

- drainage and moisture content
- aeration (the circulation of air)
- retention of minerals and nutrients
- biota and potential to hold organic matter
- ease of cultivation
- root penetration of crops and other vegetation.

Sandy soils

How a soil feels when it is touched and gently rubbed between the fingers gives good evidence about its texture. Sandy soils have poor structure. They are gritty in nature and lacking in cohesion. They feel dry in comparison with clays and loams. The composition of sand is highly variable, depending on the local rock sources and conditions. Sandy soils are free draining (see Table 5.02) and thus they dry out rapidly. These soils frequently lack important nutrients, because they have been washed out by the downward movement of water. On the positive side, they are easy to cultivate because they are so light in nature and they permit crop roots, such as carrots, to penetrate. Another advantage is that they warm up quickly in spring, which helps to provide a longer growing season. Sandy soils need significant amounts of fertiliser to make them productive.

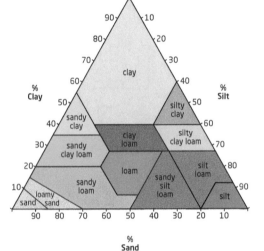

Figure 5.05 A soil textural triangle – clay, silt and sand.

Historically, regions dominated by sandy soils have often been ignored in terms of agriculture if better options have existed in the vicinity (see Image 5.04). For example, in southern England, many areas of sandy soil have been left as heathland, while the clay and loam soils in the region have long been developed for agricultural purposes. Often farming was attempted in these sandy areas, but low yields and declining fertility usually led to its abandonment after a relatively short time.

Loam soils

Loams are generally considered the ideal soils for farming. Loams have greater cohesion than sandy soils and hold together better when a handful is picked up. They are soft and rich to the touch. Loam is composed of sand, silt and clay in about 40–40–20 per cent concentration respectively, but this can vary. Different proportions of sand, silt and clay give rise to types of loam soil: sandy loam, silty loam, clay loam, sandy clay loam, silty clay loam, and loam.

Image 5.04 Sandy heathland with poor soil fertility in southern England.

Generally containing more nutrients and humus than sandy soils, loams have better infiltration and drainage than silty soils. While drainage is good, this type of soil retains sufficient amounts of water. Loams are easier to cultivate than clay soils and thus a very popular option with farmers and gardeners. Loamy soils may be wet in winter as water tables rise, but they are usually well drained in summer. Loam soils are generally the least susceptible to erosion.

Clay soils

Clay is the heaviest of the three soil types. The fine particles and the resulting few air spaces give this soil a high level of cohesion. Because of this, clay drains poorly and feels lumpy and sticky when it is wet. In the field or garden, it tends to stick to footwear and tools. Clay soils feel smooth (not gritty) when a piece is rubbed between finger and thumb.

Clay is generally heavy to cultivate, because it forms big clods that are difficult to separate, but if drainage is improved crops can grow very well because clay holds more nutrients than many other soils. Soil that consists of over 50 per cent clay particles is referred to as heavy clay. Clay tends to be relatively rich in nutrients, because the particles that make up clay soil are negatively charged. They attract and pick up positively charged particles such as calcium, potassium and magnesium. The fact that clay holds moisture well is another important advantage for this soil type. However, its slow draining can lead to waterlogging in periods of sustained precipitation.

Clay compacts easily, which can make it difficult for plant roots to grow well. Under very dry conditions, clay can become extremely hard. It can shrink into a series of blocks separated by fissures. Under such conditions, cultivation is extremely difficult. In comparison with sandy soils, clay warms slowly in spring, which impacts on the length of the growing season. Clay also has a tendency to be alkaline and to heave in winter (i.e. to swell upwards if frozen), and both of these characteristics can make life more difficult for farmers.

Characteristic	Sandy soil	Loam soil	Clay soil
mineral content	high	high	intermediate
drainage	very good	good	poor
water-holding capacity	low	intermediate	high
air spaces	large	intermediate	small
biota and potential to hold organic matter	low	high	intermediate
primary productivity	low	high	intermediate

Table 5.02 Summary and comparison of the characteristics of sandy, clay and loam soils.

5.01 Introduction to soil systems

Studying your local soil

To see what the soil in your local area is composed of, put a couple of handfuls in a jar of water. Shake the jar and let the soil settle. The particles of sand, which are heavier and larger than the silt and clay, will fall to the bottom first forming a clear layer (see Figure 5.06). After a couple of hours the silt layer should settle, forming a relatively clear boundary with the sand particles below. If the jar is left undisturbed for 24 hours, the clay layer should have formed with clear water above it. Mark the levels of each layer on the outside of the jar and work out the percentage of sand, silt and clay. Use the soil texture triangle to classify your soil.

> **Soil conditioners** are materials added to soil to improve soil fertility.

5.01.03 Soil type, primary productivity and development

Soil type can have a major impact on the viability of agricultural communities. All other factors being equal, the better the soil, the higher the rate of primary productivity. Thus in most circumstances, primary productivity would be greatest in loam soils. Clay soils would be in second place, with sandy soils trailing a very clear third. However, **soil conditioners** can improve soils significantly, and soils that appear to have limited fertility at first may provide reasonable yields with the application of appropriate soil science.

The wealth and level of technology in a country have a major influence on agricultural practice and primary productivity. For example, farmers in more economically developed countries (MEDCs) are much more likely to be able to afford the mechanical equipment to work heavy clay soils. In contrast, such work in less economically developed countries (LEDCs) may have to be done by hand, thereby making it very time-consuming and thus resulting in very low productivity. Terracing is a technique employed in both richer and poorer economies to make farming possible and to improve soil fertility (see Image 5.05).

> **CONSIDER THIS**
>
> The Netherlands, which is a relatively small country, is the world's second-largest exporter of agricultural products after the USA. Together with the USA and France, the Netherlands is one of the world's three leading producers of vegetables and fruit. It supplies a quarter of the vegetables that are exported from Europe.

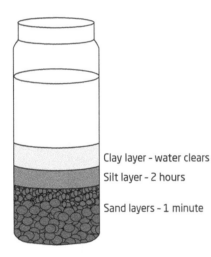

Figure 5.06 Classifying your local soil.

Image 5.05 Terracing is the answer to cultivating steep slopes in Nepal.

CASE STUDY 5.01.01

The world's arable land

Arable land (see Figure 5.07) is generally thought of as land that is capable of being cultivated and used to grow crops. The Food and Agriculture Organization (FAO) of the United Nations goes further and classes arable land as cropland that is actually being farmed. Arable land accounts for only about 11 per cent of the total land area of the world. The proportion of arable land in a country or region is constrained by a range of physical and human factors, which include:

- soil constraints
- mountainous landscapes
- inhospitable climates
- land degradation
- urban encroachment
- unequal land distribution.

For example, Russia is a much larger country than the USA, and Canada is slightly larger than the USA. However, the northerly latitude (and thus cold climate) of much of Russia and Canada compared with the USA means that the latter has a much larger area of arable land. With a much larger area of temperate farmland, the USA can produce much more arable produce in general, but it can also grow a much wider range of crops.

The countries with the largest amounts of arable land in 2011 were:

1. USA – 1 602 000 km²
2. India – 1 574 000 km²
3. Russia – 1 215 000 km²
4. China – 1 116 000 km²

Figure 5.07 World map showing the percentage of land that is arable in each country.

5. Brazil – 719 000 km²
6. Australia – 477 000 km²
7. Canada – 430 000 km²

Arable production is not just affected by the amount of arable land available, but also by its productivity. The actual fertility of the soil is the prime influence on productivity, but human influences are also important.

The differences in arable soil availability around the world have sociopolitical, economic and ecological influences. For example, where arable land is scarce, countries have to import much of the arable produce they need, if they are able to afford it. China has about 20 per cent of the world's population but only 7 per cent of its arable farmland, and it is becoming an increasing net importer of food. The current per-capita cultivated farmland in China is about 0.092 hectares, which is only about 40 per cent of the global average. Food security is becoming an increasing concern in China.

A lack of arable land also reduces the possibilities for employment in agriculture and the contribution of the agricultural sector to the national economy.

Questions

1. Give four constraints on the proportion of arable land in a country.
2. Which four countries have the largest amounts of arable land in the world?
3. Why is China becoming increasingly worried about food security?

5.01 Introduction to soil systems

> **SELF-ASSESSMENT QUESTIONS 5.01.03**
>
> 1. Define soil texture.
> 2. Why is soil texture so important in terms of cultivation?
> 3. Briefly discuss the link between soil type and primary productivity.
> 4. **Research idea:** What is the main soil type in the area of farmland nearest to you? What agricultural activities are conducted on this land?

Terrestrial food production systems and food choices

5.02

LEARNING OBJECTIVES

After reading this chapter you should be able to:

- understand that the sustainability of food production systems is influenced by sociopolitical, economic and ecological factors
- appreciate that consumers have a role to play through their support of different food production systems
- be aware that the supply of food is inequitably available and land suitable for food production is unevenly distributed among societies, and that this can lead to conflict and concerns.

KEY QUESTIONS

What are the factors that influence the sustainability of terrestrial food production systems?

What are the different reasons for food waste in LEDCs and MEDCs?

Why is the per-capita availability of land for food production decreasing?

5.02.01 Inequalities in food production and distribution

'Malnutrition is the number one cause of disease in the world. If hunger were a contagious disease, we would have already cured it.'

Jose Graziano da Silva, FAO Director-General, November 2014

The total amount of food produced around the world today is enough to provide everyone with a healthy diet. The problem is that, while some countries produce a food surplus or have enough money to buy food elsewhere, other countries are in food deficit and lack the financial resources to buy enough food abroad. Figure 5.08 shows the number of people affected by undernourishment by world region in 1990–1992 and 2012–2014. Although the global figure declined by about 20 per cent between these two periods, about one in nine of the world's population remain chronically undernourished. The global distribution of undernourishment has changed significantly in recent decades.

Forecasts of famine tend to appear every few decades or so. In 1974, a world food summit held in Rome met against a background of rapidly rising food prices and a high rate of global population growth. The major concern was that the surge in population would overwhelm humankind's ability to produce food in the early 21st century. The next world food summit, again hosted by Rome, was held in 1996. It too met against a background of rising prices and falling stocks. But new concerns, unknown in 1974, had appeared. Global warming threatened to reduce the productivity of substantial areas of land, and many scientists were worried about the long-term consequences of genetic engineering.

5.02 Terrestrial food production systems and food choices

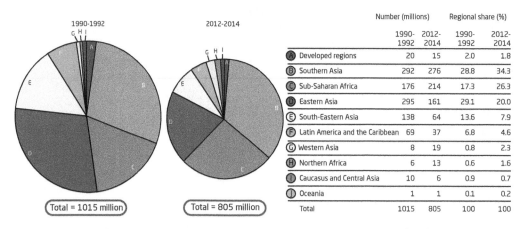

		Number (millions)		Regional share (%)	
		1990-1992	2012-2014	1990-1992	2012-2014
A	Developed regions	20	15	2.0	1.8
B	Southern Asia	292	276	28.8	34.3
C	Sub-Saharan Africa	176	214	17.3	26.3
D	Eastern Asia	295	161	29.1	20.0
E	South-Eastern Asia	138	64	13.6	7.9
F	Latin America and the Caribbean	69	37	6.8	4.6
G	Western Asia	8	19	0.8	2.3
H	Northern Africa	6	13	0.6	1.6
I	Caucasus and Central Asia	10	6	0.9	0.7
J	Oceania	1	1	0.1	0.2
	Total	1015	805	100	100

Figure 5.08 Pie chart showing the number of people affected by undernourishment in 1990–1992 and 2012–2014.

Rapidly rising food prices (see Image 5.06) in recent years and a range of other problems associated with food production have resulted in the frequent use of the term 'global food crisis'. At the World Summit on Food Security held in Rome in November 2009, the FAO Director-General Jacques Diouf referred to the over 1 billion hungry people in the world as 'our tragic achievement in these modern days'. He stressed the need to produce food where the poor and hungry live and to boost agricultural investment in these regions.

There is a huge geographical imbalance between food production and food consumption, resulting in a lack of **food security** in many countries. The three main strands of food security are:

- food availability: sufficient quantities of food available on a consistent basis
- food access: having sufficient resources to obtain appropriate foods for a nutritious diet
- food use: appropriate use based on knowledge of basic nutrition and care, as well as adequate water and sanitation.

> **Food security** is when all people at all times have access to sufficient, safe, nutritious food to maintain a healthy and active life.

In terms of agricultural production, the nations of the world can be placed into three groups:

- *The haves* – Europe, North America, Australia and New Zealand (see Image 5.07). These have sufficient cropland to meet most of their food needs, and efficient farm production systems enabling the production of more food from the same amount of land.
- *The rich have-nots* – a mixed grouping of countries that includes land-short Japan and Singapore, along with rapidly economically developing countries such as Indonesia and

Image 5.06 Selling local produce in a Vietnamese food market.

Image 5.07 Dairy cattle in New Zealand grazing on good-quality grass growing on fertile loam soil.

China, Chile, Peru, Saudi Arabia and the other Gulf States. These countries are unable to grow enough food for their populations but can afford to purchase imports to make up the deficit.
- *The poor have-nots*, consisting of the majority of the developing world. These countries with over three billion people are unable to produce enough food for their populations and cannot afford the imports to make up the deficit.

The current food crisis presents three fundamental threats, which are:
- pushing more people into poverty
- eroding the development gains that have been achieved in many countries in recent decades
- endangering political stability in some countries (a strategic threat); a significant number of countries have experienced food-related riots and unrest in recent years.

Table 5.03 summarises some of the current adverse influences on food supply and distribution. The LEDCs have long complained about the subsidies that the EU and other MEDCs give to their farmers and the import tariffs they impose on food products coming from elsewhere. This denies valuable markets to many LEDCs. On the other side of the coin, production for local markets has declined in some LEDCs because of increasing production for export markets. Agricultural transnational corporations are often the driving force behind this trend, but governments in LEDCs are also anxious to increase exports to obtain foreign currency.

Nature of adverse influence	Effect of adverse influence
Economic	• Demand for cereal grains has outstripped supply in recent years. • Rising energy prices and agricultural production and transport costs have pushed up costs all along the farm-to-market chain. • Serious underinvestment in agricultural production and technology in LEDCs has resulted in poor productivity and underdeveloped rural infrastructure. • The production of food for local markets has declined in many LEDCs as more food has been produced for export.
Ecological	• Significant periods of poor weather and a number of severe weather events have had a major impact on harvests in key food-exporting countries. • Problems of soil degradation have increased in both MEDCs and LEDCs. • Declining biodiversity may impact on food production in the future.
Sociopolitical	• The global agricultural production and trading system, built on import tariffs and subsidies, creates great distortions, favouring production in MEDCs and disadvantaging producers in LEDCs. • The international system of monitoring and deploying food relief is inadequate. • Disagreements have arisen over the use of transboundary resources such as river systems and aquifers.

Table 5.03 Adverse influences on global food production and distribution.

Short- and long-term effects of food shortages

The effects of food shortages are both short and longer term. **Malnutrition** can affect a considerable number of people, particularly children, within a relatively short period when food supplies are significantly reduced. With malnutrition, people are prone to a range of deficiency diseases and more likely to fall ill. Such diseases include beriberi (vitamin B1 deficiency), rickets (vitamin D deficiency) and kwashiorkor (protein deficiency). People who are continually starved of nutrients never fulfil their physical or intellectual potential. Malnutrition reduces people's capacity to work, so land may not be properly tended and other forms of income may not be successfully pursued. This threatens to lock at least some LEDCs into an endless cycle of ill-health, low productivity and underdevelopment.

Malnutrition is insufficiency in one or more of the nutritional elements necessary for health and well-being.

5.02 Terrestrial food production systems and food choices

> **SELF-ASSESSMENT QUESTIONS 5.02.01**
>
> 1. Explain the term 'food security'.
> 2. Define malnutrition.
> 3. How does malnutrition hinder development?
> 4. **Research idea:** Find out about the extent of malnutrition in one least developed country.

5.02.02 The sustainability of terrestrial food systems

'The world's agricultural system faces a great balancing act. To meet different human needs, by 2050 it must simultaneously produce far more food for a population expected to reach about 9.6 billion, provide economic opportunities for the hundreds of millions of rural poor who depend on agriculture for their livelihoods, and reduce environmental impacts, including ecosystem degradation and high greenhouse gas emissions.'

World Resources Institute, December 2013

The sustainability of terrestrial food production systems is influenced by a wide range of factors. These include:

- *Industrialisation, mechanisation and fossil fuel use*: Industrial farming that is highly mechanised with a high fossil fuel input can be extremely productive but can also have significant adverse impacts on the environment.
- *Scale*: Industrialised agriculture requires economies of scale to operate at a high level of efficiency. For example, in many farming regions, hedgerows have been destroyed to increase the size of fields, with the subsequent loss of wildlife habitats.
- *Seed, crop and livestock choices*: The choice of seeds is an important part of agricultural decision-making. High-yielding varieties of seed give the greatest returns, but often certain disadvantages accompany this. Farming systems that produce crops (see Image 5.08) are much more energy efficient than those that produce livestock. Within each general type of system (crop or livestock), levels of sustainability vary in the balance between productivity and environmental impact.
- *Water use:* The right choices in terms of the use of the available water in farming regions is crucial to productivity and sustainability. Where irrigation is required, there is a massive variation in the efficiency of available systems. However, the most efficient systems are much more costly (at least in the beginning) than less efficient systems. Large-scale irrigation systems can have a considerable environmental impact.
- *Fertilisers:* In the last half a century there has been a huge increase in the global use of artificial fertilisers. Use of fertilisers has risen in all world regions, but to varying degrees. However, research has shown that the heavy and sustained use of artificial fertiliser can result in serious soil degradation. Pollution from fertilisers occurs when more is applied than crops can absorb or when they are washed or blown off the soil surface before they can be incorporated. Excess nitrogen and phosphates can leach into groundwater or run off into waterways. This nutrient overload causes eutrophication of water bodies.

Image 5.08 Woman drying grain in a Nepalese village.

- *Pest control*: Insecticides, herbicides and fungicides are also applied heavily in many developed and developing countries, polluting fresh water with carcinogens and other poisons that affect humans and many forms of wildlife. Pesticides also reduce biodiversity by destroying weeds and insects and hence the food species of birds and other animals.
- *Pollinators*: Pollination is an important ecosystem service. About 70 per cent of the world's most produced crop species rely, to some extent, on insect pollination. A decline in pollinating insect numbers has been linked to factors including climate change, habitat loss, pollution, pesticides, harmful parasites, disease and insufficient food.
- *Antibiotics*: A high proportion of the antibiotics added to animal feed is excreted in urine or manure. Once excreted, these antibiotics can enter surface water and/or groundwater from manure-applied lands. Most of the antibiotics are strongly adsorbed in soils and are not readily degraded. A significant environmental concern is the presence of antibiotics in sources of potable water.
- *Legislation*: Governmental and inter-governmental legislation with regard to agriculture can have a major impact on the types and levels of production, and on subsequent impacts on the environment. (For example, Canada's Agricultural Growth Act C-18 enshrined the farmers' right to save, store and clean their own seeds.)
- *Levels of commercial versus subsistence food production:* Subsistence food production systems are usually small in scale and have developed over thousands of years. While farming techniques are simple and productivity is low, this system of farming is much more ecologically friendly than commercial farming.

The Sustainable Food Systems Programme

The Sustainable Food Systems Programme (SFSP), established in 2011, has been jointly developed by the United Nations Environment Programme (UNEP) and the FAO. Its objective is to accelerate the shift to **sustainable agriculture** (production) and consumption in both MEDCs and LEDCs. The programme will focus on the key issues of food security and nutrition, climate change, food losses and waste, biodiversity and habitat loss, and water scarcity and efficiency.

Food waste

The FAO estimates that one-third of food produced annually for human consumption worldwide is lost or wasted along the chain that stretches from farms to food-processing factories, market places, retailers, restaurants and household kitchens. The FAO states: 'At 2.8 trillion pounds (weight), that's enough sustenance to feed three billion people.'

Other pertinent facts about food waste include:

- Twenty-eight per cent of the world's agricultural area is used to produce food that is lost or wasted.
- Most food waste ends up in landfill, representing a large part of municipal solid waste. Methane emissions from landfill are a significant source of greenhouse gas emissions.
- The carbon footprint from food waste is estimated at 3.3 billion tonnes of CO_2 equivalent released into the atmosphere each year.
- The total volume of water used each year to produce food that is lost or wasted is equivalent to three times the volume of Lake Geneva.

Food waste occurs towards the end of the food chain, at the retail and consumer level. In general, the richer the nation, the higher its per-capita rate of food waste. Food waste is much more prevalent in MEDCs than in LEDCs. According to the FAO, MEDCs waste an amount of food almost equal to the entire net food production of sub-Saharan Africa.

In contrast, **food loss** occurs mainly at the front of the food chain, during production, post-harvest and processing. Food loss is less prevalent in MEDCs than in LEDCs.

> **Sustainable agriculture** refers to agricultural systems emphasising biological relationships and natural processes that maintain soil fertility, thus allowing current levels of farm production to continue indefinitely.
>
> **Food waste** is when perfectly edible foodstuffs are thrown away at the retail and consumer level.
>
> **Food loss** refers to food that spills, spoils, incurs an abnormal reduction in quality, such as bruising or wilting, or otherwise gets lost before it reaches the retailer and consumer. Food loss typically takes place at the production, storage, processing and distribution stages in the food value chain.

5.02 Terrestrial food production systems and food choices

Figure 5.09 Reducing food waste.

The latter tend to lack the infrastructure to deliver all of their food in good condition to consumers. In Africa, with limited storage facilities, refrigeration and transportation, 10–20 per cent of the continent's sub-Saharan grain is lost to hazards such as mould, insects and rodents. Facing similar challenges, India loses an estimated 35–40 per cent of its fruit and vegetables.

The UNEP, together with the Waste and Resources Action Programme (WRAP), FAO and the SAVE FOOD initiative, have recently provided guidance to reduce food waste and conserve natural resources. The focus is on prevention of food waste in households and in the food supply chain. Figure 5.09 shows how individual households can reduce food waste. In recent years, consumers in MEDCs have become more aware of the issue of food waste, and as a result consumer behaviour is beginning to change. For example, there have been campaigns to get supermarkets to stop offers of 'buy one, get one free' and instead reduce the price on individual items.

SELF-ASSESSMENT QUESTIONS 5.02.02

1. Define sustainable agriculture.
2. What is the difference between food waste and food loss?
3. What are the disadvantages associated with the high rate of food waste and loss currently experienced by the world?
4. **Research idea:** Find out the extent of food waste and loss in the country in which you live.

5.02.03 Choices of food production systems

'Every really successful system of agriculture must be based on the long view, otherwise the day of reckoning is certain.'
The Waste Products of Agriculture by Sir Albert Howard and Yashwant Wad, 1931

A wide range of factors combine to influence agricultural land use and practices on farms.

Physical factors

Physical factors set broad limits as to what can be produced. The farmers' decisions are then influenced by economic, social/cultural and political factors.

Key physical factors influencing possible types of farming are:

- temperature
- precipitation
- soil type and fertility
- locally – aspect, angle of slope and wind intensity may also be important factors in deciding how to use the land.

Economic factors

Economic factors include transport, markets, capital and technology. The cost of growing different crops or keeping different livestock varies. The market prices for agricultural products will vary also and can change from year to year. The necessary investment in buildings and machinery can mean that some changes in farming activities are very expensive. In most countries there has been a trend to fewer but larger farms. Large farms allow economies of scale to operate. Distance from markets has always been an important influence on farming. In general, the closer a farm is to its market(s), the less the cost of transporting its produce and the greater the returns from the sale of the produce. Thus land closest to the market tends to be the most intensively farmed.

Extension
Read more about unfair trade using the following link:
www.oxfam.org.hk/EN/unfairtrade.aspx

The development and application of **agricultural technology** requires investment. The status of a country's agricultural technology is vital for its food security and other aspects of its quality of life. An important form of aid is the transfer of agricultural technology from more advanced to less advanced nations.

> **Agricultural technology** is the application of techniques to control the growth and harvesting of animal and vegetable products.

Political factors

The influence of government on farming has steadily increased in many countries. For example, in the USA the main parts of federal farm policy over the past half century have been:

- price support loans
- production controls
- income supplements.

Thus the decisions made by individual farmers are heavily influenced by government policies. However, in centrally planned economies the state has far more control. This was the case for many years in the former Soviet Union and China. An agricultural policy can cover more than one country, as evidenced by the EU's Common Agricultural Policy.

Social/cultural factors

There is a tendency for farmers to stay with what they know best and often a sense of transgenerational responsibility to maintain family farming tradition. Legal rights and **land tenure** are heavily influenced by culture. In the past, inheritance laws have had a huge impact on the average size of farms. In some countries, it has been the custom on the death of a farmer to divide the land equally between all his sons (but rarely between daughters). The reduction in the size of farms by these processes often reduced them to operating at only a subsistence level. In most societies, women have very unequal access to, and control over, rural land and associated resources.

> **Land tenure** refers to the ways in which land is or can be owned.

Extension
Find out more about farming in the third world using the following link:
http://3rdworldfarmer.com/

Greater consumer power

Consumers are having a greater influence on agricultural systems than at any time in the past. In recent years, more and more consumers have given greater consideration to the food they buy. Political pressure on governments has resulted in more detailed food labelling on many food products. For example, more consumers are purchasing:

- organic produce
- free-range eggs
- grass-fed beef
- local produce.

5.02 Terrestrial food production systems and food choices

5.02.04 Links between social systems and food production systems

The culture of a social system is a major factor in how that society makes decisions. Culture can be defined as the total of the inherited ideas, beliefs, values and knowledge that constitute the shared basis of social action. How a society feeds itself is arguably the most important decision or set of decisions that it can make. Some societies have a very strong relationship with the physical environment; it may, in fact, be an important element of a society's religion. Societies that hold the physical environment in high value are much more likely to care for it and appreciate its limits.

Shifting cultivation

Shifting cultivation (see Image 5.09) is a traditional farming system that developed a long time ago in tropical rainforests. An area of forest is cleared to create a small plot of land which is cultivated until the soil becomes exhausted. The plot is then abandoned and a new area cleared. Frequently the cultivators work in a circular pattern, returning to previously used land once the natural fertility of the soil had been renewed. Shifting cultivation is also known as 'slash and burn'.

Shifting cultivation is necessary in the Amazon rainforest because of the limited fertility of the fragile soils. About a hectare of forest is cleared, although sometimes a few larger trees are left standing to protect crops from the intense heat and heavy rain of the region. In addition, trees providing food such as kola nut and bananas are left in place. Once the felled vegetation is dry it is burnt. This helps to clear the land, remove weeds and provide ash for fertiliser. A drawback is that burning also destroys useful organic material and bacteria. Crops such as yams, manioc, beans, tobacco, coca and pumpkins are then planted.

The very low population density of the Amazon basin and the belief systems of the cultivators are major influences on the way this farming system operates. Tradition and ritual are important in choosing a site for cultivation and how the site is cleared. When a previously cultivated site is returned to because it has regained its fertility, it is often regarded as an important historical event when ancestors are remembered. Many traditional shifting cultivators believe in the spiritual nature of the forest, which instils a strong respect for this environment. Low population density allows such groups to move on from a plot of land when its fertility declines, allowing that land to recuperate over time.

Shifting cultivation can be regarded as very energy-efficient and a sustainable form of forest management. However, very few people are involved in shifting cultivation, as with other traditional forms of farming, today compared with a century ago. The exploitation of the rainforests for a variety of commercial reasons, including logging, mining and clearing for ranches, has forced cultivators deeper into the untouched forest areas or onto designated reservations.

Image 5.09 Shifting cultivators in the Democratic Republic of the Congo.

5.02.05 Decrease in available land per capita

In 1960, when the global population was 3 billion, about 0.45 hectares of cropland per capita was available. This was considered the minimum area necessary for the production of a diverse, healthy, nutritious diet of plant and animal products like that enjoyed widely in MEDCs. With a global population of 7.2 billion today, significant soil degradation over the past half century, and other factors such as urbanisation reducing available agricultural land, available per-capita cropland is now about 0.23 hectares. This trend is forecast to continue (see Figure 5.10).

5.02.06 The efficiency of terrestrial food production systems

Food production systems can be compared with regards to their trophic levels and efficiency of energy conversion. The amount of available energy declines at every step in a food chain. A certain amount of energy is used in the growth of biomass and the production of offspring, but more energy is used up through respiration, waste production and other non-productive uses.

In terrestrial systems, most food is harvested from relatively low trophic levels (producers and herbivores). This means that terrestrial systems make fairly efficient use of solar energy. However, two relatively distinct terrestrial systems can be recognised. Farming systems that produce crops are much more energy efficient than those that produce livestock, as Figure 5.11 shows. Clearly, many more people can be fed on 7.5 tonnes of wheat than on 0.3 tonnes of beef. Where **diet** is dominated by the consumption of arable products, energy has only passed through one step. In contrast, the energy in meat passes through two steps before human consumption. Thus the yield of food per unit area from lower trophic levels compared with higher trophic levels is greater in quantity, lower in cost and generally requires fewer resources.

Diet is the kinds of food that a person, animal or community habitually eats.

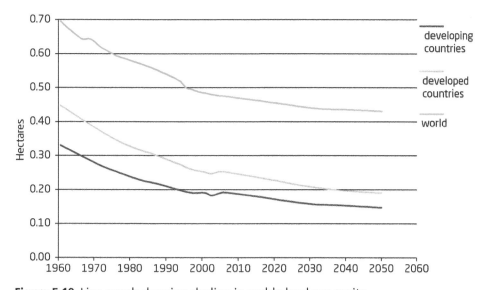

Figure 5.10 Line graph showing decline in arable land per capita.

5.02 Terrestrial food production systems and food choices

Newly industrialised country (NIC) A country that has undergone rapid and successful industrialisation since the 1960s.

High levels of meat consumption have been the norm in MEDCs for some time. However, as incomes have increased in **newly industrialised countries (NIC)**, the demand for meat has risen rapidly. This is placing much greater demands on global food production, resulting in soil degradation and other environmental problems in various parts of the world.

It should, of course, be remembered that livestock can produce more than just meat. Milk production is an important aspect of farming in many countries, and animals may also provide other valuable products such as wool and hides. In LEDCs, livestock may also provide an important source of labour.

Research from Cambridge and Aberdeen universities published in 2014 stated that:

- Greenhouse gases from food production will increase 80 per cent if meat and dairy consumption continue to rise at their current rates.
- More and more forest land or arable fields will be changed for use by livestock (see Figure 5.12).
- The average efficiency of livestock at converting plant feed to meat is less than 3 per cent.
- Agricultural practice is not necessarily at fault, but our choice of food is.

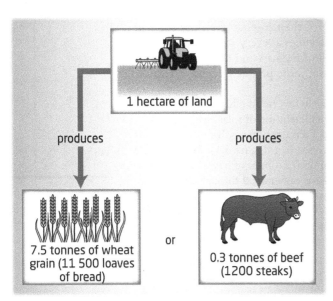

Figure 5.11 Comparing the efficiency of energy conversion in arable and pastoral farming.

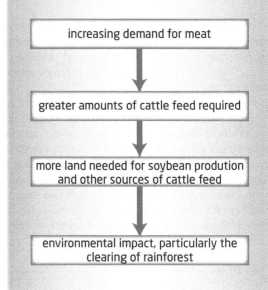

Figure 5.12 Flowchart showing the environmental impact of the increasing demand for meat.

SELF-ASSESSMENT QUESTIONS 5.02.03

1. What groups of factors combine to influence agricultural land use and practices on farms?
2. To what extent has the amount of cropland per capita declined globally since 1960?
3. Why are farming systems that produce crops more energy efficient than those that produce livestock?
4. **Discussion point:** Do you think there is a limit on the amount of meat that the world can produce?

5.02.07 Agro-industrialisation

Agro-industrialisation, or industrial agriculture, is the form of modern farming that involves the industrialised production of livestock, poultry, fish and crops. It is strongly related to modern urban society and to Western corporate capitalism. This type of large-scale, capital-intensive farming originally developed in Europe and North America and then spread to other parts of the developed world. Agro-industrialisation has been spreading rapidly in many developing countries since the beginning of the **green revolution**.

Industrial agriculture is heavily dependent on oil for every stage of its operation. The most obvious examples of oil uses are in fuelling farm machinery, transporting produce and producing fertilisers and other farm inputs.

The general characteristics of agro-industrialisation include:

- very large farms
- concentration on one farm product (monoculture) or a small number of farm products
- a high level of mechanisation (see Image 5.10)
- low labour input per unit of production
- heavy usage of fertilisers, pesticides and herbicides
- sophisticated ICT management systems
- highly qualified managers
- frequent ownership by large agribusiness companies
- frequent vertical integration with food processing and retailing.

Regions where agro-industrialisation is clearly evident on a large scale include the Canadian prairies, the corn and wheat belts in the USA, the Paris basin, East Anglia in the UK, the Russian steppes, and the pampas in Argentina.

Agro-industrialisation is a consequence of the globalisation of agriculture, the profit ambitions of large agribusiness companies, and the drive for cheaper food production. Over the last half a century, every stage in the food industry has changed in an attempt to make the industry more efficient (in an economic sense). Vertical integration has become an increasingly important process, with increasing linkages between the different stages of the food industry (see Figure 5.13). Farming and food production around the world are becoming more dominated by large biotechnology companies, food brokers and huge industrial farms. The result is a complex movement of food products around the world.

In countries where agro-industrialisation dominates, very few people are actually employed on the land. This lack of a close connection means that many people simply don't think about how their food is produced. However, there has been an increasing reaction to high-input farming as more and more people have become concerned about the use of fertilisers, pesticides, herbicides and other high-investment farming practices that are having a significant impact on the environment. The main evidence of this concern is the growth of the organic food market and the increasing sale of food resulting from more 'gentle' farming practices.

> **Agro-industrialisation** is industrialised farming that is typically large scale and capital intensive.
>
> **Green revolution** refers to the introduction of high-yielding seeds and modern agricultural techniques in LEDCs.
>
> An **agro-ecosystem** is a form of modern farming which involves industrialised production of livestock, poultry, crops and fish.

5.02.08 Organic farming

Organic farming does not use manufactured chemicals and thus occurs without chemical fertilisers, pesticides, insecticides and herbicides. Instead, animal and green manures are used along with mineral fertilisers such as fish and bonemeal. Thus, organic farming requires a higher input of labour than mainstream

Image 5.10 An element of an **agro-ecosystem** in Spain.

5.02 Terrestrial food production systems and food choices

Theory of knowledge 5.02.01

Many advances in agriculture and other scientific areas have not been quite as beneficial as they first seemed. The green revolution is a case in point, as some unexpected consequences have occurred. For example in the early 1990s, nutritionists noticed that, even in countries where average food intake had risen, incapacitating diseases associated with mineral and vitamin deficiencies remained commonplace and in some instances had actually increased. The problem is that the high-yielding varieties introduced during the green revolution are usually low in minerals and vitamins. Because the new crops have displaced the local fruits, vegetables and legumes that traditionally supplied important vitamins and minerals, the diet of many people in less economically developed countries is now extremely low in zinc, iron, vitamin A and other micronutrients.

1. Is there a general lesson here, that we should be trying to work with nature rather than ruthlessly exploiting the planet for the short-term benefit of our own species?
2. What specific disadvantages have critics of the green revolution highlighted?

Food miles are the distance food travels from the farm where it is produced to the plate of the final consumer.

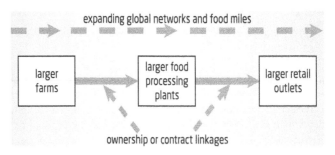

Figure 5.13 Agro-industrialisation – increasing vertical integration.
Source: Geography for the IB Diploma: Global Interactions

farming. Weeding is a major task with this type of farming. Organic farming is less likely to result in soil erosion and is less harmful to the environment in general. For example, there will be no nitrate runoff into streams and much less harm to wildlife. Organic farming tends not to produce the 'perfect' potato, tomato or carrot. However, because of the increasing popularity of organic produce, it commands a substantially higher price than mainstream farm produce.

SELF-ASSESSMENT QUESTIONS 5.02.04

1. What is agro-ecosystems?
2. State five characteristics of agro-industrialisation.
3. **Discussion point:** What do you feel about the food production systems closest to where you live, and why?

5.02.09 Food miles and increasing air freight

The term 'food miles' was first used in the 1990s by Dr Tim Lang, Professor of Food Policy at London's City University, as part of the debate on sustainable agriculture. **Food miles** can be defined as the distance food travels from the farm where it is produced to the plate of the final consumer. This measure is an indication of the environmental impact of food consumption.

In the UK, 95 per cent of fruit and half of vegetables are imported. Increasingly, these food imports have been transported by plane. The growing volume of air freight has resulted in increasing emissions of various pollutants, in particular carbon dioxide and nitrogen oxides. Fruit and vegetables are the main food products transported by air, but other food products such as animal feed are increasing in volume. Due to rising public concern, some large retailers now label fresh food transported by air with an airplane sticker.

Sometimes, food may be produced very close to the point where it is consumed, but because it may have to pass through a supermarket's national distribution system the food miles incurred may be far in excess of the straight-line distance between the points of production and consumption. Supermarket distribution now frequently operates on the 'just-in-time' principle, allowing the maximum possible floor space to be used for sales and keeping storage areas to the minimum. The result is a high-volume and complex network of vehicle deliveries. Such a system has pushed up the food miles total considerably.

CASE STUDY 5.02.01

Commercial farming in the Canadian prairies

Commercial farming is farming for profit, where food is produced for sale in the market.

'Depending upon one's perspective, the Canadian prairies is either a breadbasket for the world or one of the world's most "disturbed" ecological systems because of the high proportion of land converted to agriculture.'

Prairie Soils and Crops Journal, 2010

The Canadian prairies provides an example of extensive commercial cereal production (see Case study 5.03.01). This major agricultural region straddles the three Canadian provinces of Manitoba, Saskatchewan and Alberta. This region is often referred to as the breadbasket of Canadian agriculture, containing nearly half of Canada's farms and much larger shares of its cropland and grassland. The hot, dry summers and generally fertile soils of the prairies favour wheat production. However, feed grains (oats, rye, barley and corn) are also grown. There are also dairy and beef farms scattered throughout the region. Gradual and distinct changes in climate and natural vegetation have resulted in distinct soil zones.

Wheat was first cultivated on the prairies in the early 1600s. Since then many changes have occurred. The family farm has grown from a few hand-worked hectares to a large, highly mechanised unit. Average farm size in Saskatchewan is 675 ha, or over two times the national average. In Alberta and Manitoba, farms average 472 ha and 459 ha respectively.

New strains of wheat have been developed that can withstand lower temperatures and a shorter growing season. Today's wheat is resistant to many diseases (e.g. wheat streak mosaic disease) that once destroyed huge areas of crops. Inputs of fertiliser, herbicides and pesticides have steadily risen as demand for the region's food products has increased. Farming has become more and more capital intensive as machinery has replaced human labour wherever possible. A modern combine harvester allows one person to harvest more than 2000 bushels a day. The Canadian Department of Agriculture spends a substantial amount of money on research and development for wheat and Canada's other main agricultural products to ensure the industry remains productive and competitive in world markets. Production of wheat in the prairies, based initially on the improved variety Marquis, trebled from 1908 to 2008.

Much of the grain produced in the prairies is sold in other parts of Canada or abroad. Large grain elevators (see Image 5.11) are located at regular intervals along railway lines so that trains can take the grain to major transport terminals at Vancouver, Prince Rupert, Thunder Bay and Churchill. Here it is either loaded on board ship or placed in huge terminal storage elevators. There are about 5000 elevators sited at 2000 shipping points in the prairies. The amount and quality of grain required at the shipping terminals is decided six weeks before delivery. Efficient transportation is a vital part of this high-technology food production system, although the system has been put under severe strain in years of very high production. Exporting prairie grain involves a complicated relationship between farmers, elevator companies, railways, and port terminals.

Image 5.11 Grain elevator on the Canadian prairies.

Figure 5.14 summarises the inputs, processes and outputs of a typical wheat farm on the prairies. The range of inputs is wide, indicating the large annual capital investment required on such farms. A range of expensive machinery is needed to carry out the various processes quickly and efficiently. Farm managers have to have a detailed knowledge and understanding of a wide range of issues relating to modern agriculture. Computer literacy is just one of the skills they require.

Outputs can be subdivided into main products, by-products and waste products. By-products can provide valuable additional income. Although the yields per hectare in monetary terms are not as high as in commercial intensive farming, the productivity

5.02 Terrestrial food production systems and food choices

CASE STUDY 5.02.01 (continued)

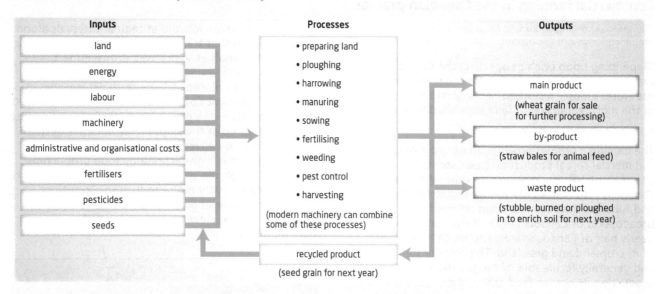

Figure 5.14 Systems diagram for a wheat farm.

per worker is high due to the very high level of mechanisation.

Such large-scale farming invariably has a substantial impact on the environment. Soil degradation poses a considerable challenge. The almost total dominance of commercial farming in the prairies has left little in terms of habitat for wild native species, and some scientists are greatly concerned about the loss of such biodiversity in the long term. Another issue receiving much attention is the high level of fertiliser use and its environmental impact, particularly in terms of greenhouse gas emissions. Canada has one of the highest per-capita usages in the world of nitrogenous fertiliser, an issue that environmental groups have protested about. The first genetically modified (GM) crops were planted in the prairies in the mid-1990s. Environmental groups such as Greenpeace have been constant critics, pointing to a mounting range of evidence about the adverse impact of GM production. This is an issue that will not go away.

Prairie farmers are increasingly adopting environmentally friendly tillage practices, including more no-tillage and less summer fallow, to reduce soil salinisation and erosion.

Questions

1. Where are the Canadian prairies located?
2. Describe the characteristics of farming in this region.
3. What are the main environmental issues associated with prairie farming?

CONSIDER THIS

Rice forms the basis of many Indian dishes. At weddings it is customary to throw rice at the couple and for the bride to cook rice as the first meal, because rice is associated with prosperity and fertility. A popular Indian saying is: 'Two brothers should be like a grain of rice, close but not stuck together.' There is frequent mention of rice and rice cultivation in Indian literature.

CASE STUDY 5.02.02

Subsistence farming in the lower Ganges valley

Subsistence farming is the most basic form of agriculture, where the produce is consumed entirely or mainly by the family who work the land or tend the livestock. If a small surplus is produced, it may be sold or traded.

Intensive subsistence rice cultivation dominates the lower Ganges valley (see Figure 5.15) in India and Bangladesh. Rice cultivation began in the lower Ganges valley thousands of years ago as a small but important part of the culture. References to rice have been found in early Sanskrit texts. The suitability of this crop to the prevailing environmental conditions ensured its expansion as the population grew. Rice is the world's most important grain next to wheat and is crucial to global food security.

The crop is grown on very small plots of land using a very high input of labour. Rice cultivation by small farmers is sometimes referred to as 'pre-modern intensive farming' because of the traditional techniques used, in contrast to intensive farming systems in MEDCs, such as market gardening, which are very capital intensive.

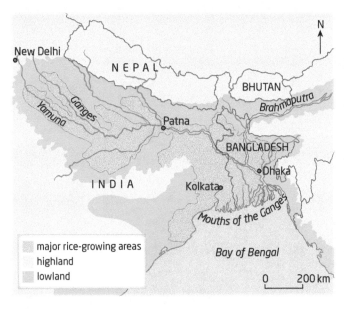

Figure 5.15 Map of the lower Ganges valley.

The Ganges valley is one of the most heavily populated regions in the world. High population density means that the land has to be cultivated as intensively as possible with the available technology, and that plots of land are invariably very small indeed. With such high population pressure on the available natural resources, the spirit of cooperation evident in most rural communities is an important aspect of culture that has evolved over a long period. In most cases, every member of the family will have their individual duties to ensure the best possible yield from their land.

Rice contributes over 75 per cent of the diet in many parts of the region. This is not surprising, considering its high nutritional value. The physical conditions in the lower Ganges valley and delta that make the area very suitable for rice cultivation are:

- temperatures of 21 °C and over throughout the year, allowing two crops to be grown annually; rice needs a growing season of only 100 days
- monsoon rainfall of over 2000 mm, providing sufficient water for the fields to flood, which is necessary for wet rice cultivation
- rich alluvial soils built up through regular flooding over a long period during the monsoon season
- an important dry period for harvesting the rice.

Current rice production systems are extremely water intensive. Ninety per cent of agricultural water in Asia is used for rice production. The International Rice Research Institute estimates it takes 5000 litres of water to produce 1 kg of rice. 'Wet' rice is grown in the fertile silt and flooded areas of the lowlands, while 'dry' rice is cultivated on terraces on the hillsides. Dry rice is easier to grow but provides lower yields than wet rice. Significant areas of wet rice require irrigation to achieve good yields.

The farming system

Padi fields (flooded parcels of land) characterise lowland rice production. Water for irrigation is provided either when the Ganges floods or by means of gravity canals. At first, rice is grown in nurseries. It is then transplanted when the monsoon rains flood the padi fields. The flooded padi fields may be stocked with fish for an additional source of food. The main rice crop is harvested when the drier season begins in late October. The rice crop gives high yields per hectare. A second rice crop can then be planted in November, but water supply can be a problem in some areas for the second crop.

Water buffalo are used for work. This is the only draft animal adapted for life in wetlands. The water buffalo provide an important source of manure in the fields. However, the manure is also used as domestic fuel. The labour-intensive nature of rice cultivation provides work for large numbers of people. This is important in areas of very dense population where there are limited alternative employment opportunities. The low incomes and lack of capital of these subsistence farmers mean that hand labour still dominates in the region. It takes an average of 2000 hours a year to farm one hectare of land.

5.02 Terrestrial food production systems and food choices

CASE STUDY 5.02.02 (continued)

Rice seeds are stored from one year to provide the next year's crop. During the dry season, when there may be insufficient water for rice cultivation, other crops such as cereals and vegetables are grown. Farms are generally small, often no more than one hectare in size. Many farmers are tenants and pay for use of the land by giving a share of their crop to the landlord.

The use of higher-yielding varieties of rice in recent decades has increased production significantly. However, the transfer from traditional varieties to higher-yielding varieties brings some disadvantages as well as the obvious advantages. The average padi rice yield in India increased from 1.5 tonnes per hectare in 1960 to over 3.5 tonnes per hectare in 2013.

The use of traditional farming methods, applying appropriate (intermediate) technology, has helped to avoid serious problems of soil degradation, but rice cultivation is a significant source of atmospheric methane, a greenhouse gas. Methane is 20 times more potent as a greenhouse gas than carbon dioxide. The high water requirement of rice cultivation is another major issue, particularly where irrigation is required.

Questions

1. Why is rice cultivation in the lower Ganges valley considered an intensive form of agriculture?
2. To what extent have rice yields in India increased since 1960?
3. What are the environmental problems associated with rice cultivation in the region?

Soil degradation and conservation

5.03

LEARNING OBJECTIVES

After reading this chapter you should be able to:

- understand that fertile soils require significant time to develop through the process of succession
- appreciate that human activities may reduce soil fertility and increase soil erosion
- understand that soil conservation strategies exist and may be used to preserve soil fertility and reduce soil erosion.

KEY QUESTIONS

What is the relationship between soil ecosystem succession and soil fertility?

What are the consequences of reduced soil fertility?

How can soil management preserve the fertility of soils?

5.03.01 Soil ecosystems develop through succession

Soil ecosystems mature through the process of succession. The initial stage in the formation of a soil is the weathering of bedrock to provide a layer of loose, broken material known as regolith. The second stage, the formation of topsoil, results from the addition of living organisms (biota), decayed organic matter (humus), water and air. Fertile soil contains a community of organisms that work to maintain functioning nutrient cycles and that are resistant to soil erosion. Soils are formed through the interaction of five major factors: climate, parent material, topography, organisms and time (see Figure 5.01). The relative influence of each factor varies from place to place, but the combination of all five factors normally determines the kind of soil developing in any given place. The process of soil development is known as **pedogenesis**.

The Soil Science Society of America states that it takes from 500 to thousands of years to create an inch of topsoil. In order to accumulate enough substances to make a soil fertile, it takes about 3000 years. With such a timespan the soil is considered a non-renewable resource, and that once it has been destroyed it is lost forever.

The time needed to form a soil depends on the latitude. Soil formation is faster in wet, tropical areas than in regions characterised by a mild climate. Soil formation is very slow in cold, dry climates. Young soils are usually easy to recognise because they have little or weak soil horizon development and the horizons commonly are indistinct. Older (mature) soils are strongly developed, generally deeper, and have well-defined soil horizons.

A good example of the development of a soil system and the changes that occur in soil profile can be seen across a section of sand dunes, moving inland from the coast (see Image 5.12).

Pedogenesis is the process of soil development.

CONSIDER THIS

It is thought that roughly 95 per cent of the world's soils have been moved or transported to their present location. Only 5 per cent of the world's soils are 'residual soils', or soils that formed in place from the existing parent material.

5.03 Soil degradation and conservation

Image 5.12 Sand dune transect from the beach to inland.

Image 5.13 Degraded soil in central Asia.

5.03.02 Processes and consequences of soil degradation

Soil degradation, according to the FAO definition, is 'a change in the soil health status resulting in a diminished capacity of the ecosystem to provide goods and services for its beneficiaries'.

Soil degradation (see Image 5.13) is a global process. It involves both the physical loss (erosion) of and reduction in quality of topsoil, associated with nutrient decline and contamination. It impacts significantly on agriculture and also has implications for the urban environment, pollution and flooding. The loss of the upper soil horizons containing organic matter and nutrients and the thinning of soil profiles reduce crop yields on degraded soils.

The extent of soil degradation

Globally it is estimated that 2 billion hectares of soil resources have been degraded. This is equivalent to about 15 per cent of the Earth's land area. Such a scale of soil degradation has resulted in the loss of 15 per cent of world agricultural supply in the last 50 years. For three centuries ending in 2000, topsoil had been lost at the rate of 300 million tonnes a year. Between 1950 and 2000 topsoil was lost at the much higher rate of 760 million tonnes a year. In sub-Saharan Africa, nearly 1 million square miles of cropland have shown a 'consistent significant decline', according to a March 2008 report by a consortium of agricultural institutions. Some scientists consider this a slow-motion disaster.

The loss of the ability of degraded soils to store carbon is receiving significant attention. Over the last 50 years or so global soils have lost about 100 billion tonnes of carbon in the form of carbon dioxide to the atmosphere due to the depletion of soil structure.

The Global Assessment of Human-induced Soil Degradation (GLASOD) is the only global survey of soil degradation that has been undertaken. Figure 5.16 is a generalised map of the findings of this survey. It shows that substantial parts of all continents have been affected by various types of soil degradation. In temperate areas, much soil degradation is a result of market forces and the attitudes adopted by commercial farmers and governments. In contrast, in the tropics much degradation results from high population pressure, land shortages and lack of awareness. The greater climate extremes and poorer soil structures in tropical areas give greater potential for degradation in such areas compared to temperate latitudes. This difference has been a significant factor in development or the lack of it.

Processes and consequences of soil degradation

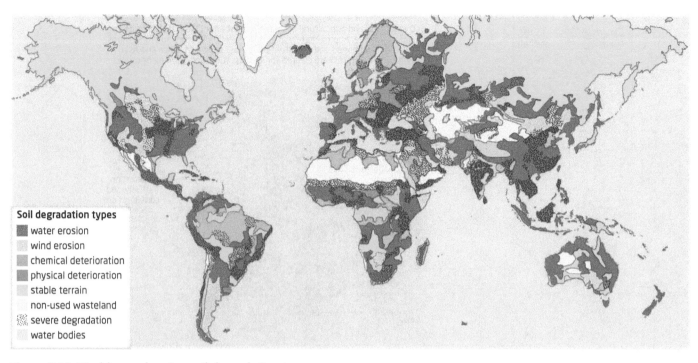

Figure 5.16 World map showing soil degradation types.

SELF-ASSESSMENT QUESTIONS 5.03.01

1. **a** Define soil degradation.
   b What is the evidence in Image 5.13 that the soil is degraded?
2. Briefly state the different reasons for soil degradation in temperate and tropical areas.
3. **Research idea:** Try to find out how degraded the soils are in the areas of farmland nearest to where you live.

The processes of degradation

The main cause of soil degradation is the removal of the natural vegetation cover, leaving the surface exposed to the elements. Figure 5.17 shows the human causes of degradation, with deforestation and overgrazing as the two main problems. The resulting loss of vegetation cover is a leading cause of wind and water erosion.

Deforestation occurs for a number of reasons including the clearing of land for agricultural use, for timber, and for other activities such as mining. Such activities tend to happen quickly, whereas the loss of vegetation for fuelwood, a massive problem in many developing countries, is generally a more gradual process. Deforestation means that rain is no longer intercepted by vegetation, with rain splash loosening the topsoil and leaving it vulnerable to removal by overland flow.

Overgrazing is the grazing of natural pastures at stocking intensities above the livestock-carrying capacity. Population pressure in many areas and poor agricultural practices have resulted in serious overgrazing. This is a major problem in many parts of the world, particularly in marginal ecosystems. The process occurs in this way:

1. Trampling by animals (and humans) damages plant leaves.
2. Some leaves die away, reducing the ability of plants to photosynthesise. Now there are fewer leaves to intercept rainfall, and the ground is more exposed. Plant species sensitive to trampling quickly disappear.

> **Deforestation** is the process of destroying a forest and replacing it with something else, especially by an agricultural system.
>
> **Overgrazing** is the grazing of natural pastures at stocking intensities above the livestock-carrying capacity.

5.03 Soil degradation and conservation

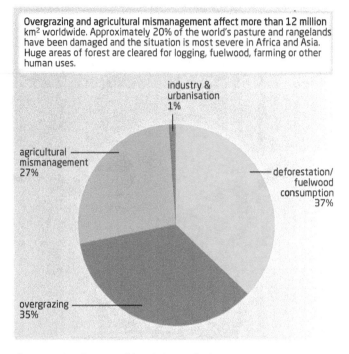

Figure 5.17 Causes of land degradation.

3   Soil begins to erode when bare patches appear. Trampling will have compacted the soil and damaged its structure.
4   Loose surface soil particles are the first to be carried away by either wind or water.
5   The loss of soil structure means that less water can infiltrate to the lower soil horizons. The growth rate of plants is reduced, and it is more difficult for damaged plants to recover.

Agricultural mismanagement is also a major problem, due to a combination of lack of knowledge and the pursuit of short-term gain in favour of consideration of longer-term damage. Such activities include shifting cultivation without adequate fallow periods, absence of soil conservation measures, cultivation of fragile or marginal lands, unbalanced fertiliser use and the use of poor irrigation techniques.

Soil degradation is more directly the result of the following:

- Erosion by wind and water (see Table 5.04). These two agents of erosion account for approximately 80 per cent of the world's degraded landscapes.
- Physical degradation: loss of structure, surface sealing and compaction.
- Chemical degradation through various forms of pollution. Changes in pH, **acidification**, and **salinisation** are examples of chemical degradation.
- Biological degradation through loss of organic matter and biodiversity.
- Climate and land use change, which may accelerate the above factors.

Acidification is the change in the chemical composition of soil, which may trigger the circulation of toxic metals.

Salinisation is the condition in which the salt content of soil accumulates over time to above normal levels. It occurs in some parts of the world where water containing high salt concentration evaporates from fields irrigated with standing water.

Physical degradation

The signs of physical deterioration are soil crusting, sealing and compaction, which can be caused by several factors such as compaction through heavy machines or animals. These problems occur in all continents, in nearly all climates and soil physical conditions. Soil crusting and compaction tend to increase runoff, decrease the infiltration of water into the soil, prevent or inhibit plant growth and leave the surface bare and subject to other forms of degradation. Severe crusting of the soil surface because of breakdown of soil aggregates can inhibit water entry into the soil and prevent seedling emergence.

Water erosion	Wind erosion
Rainfall intensity and runoff: the impact of raindrops on the soil surface can break down soil aggregates and disperse the aggregate material. Lighter aggregate materials, such as very fine sand, silt, clay and organic matter, can be easily removed by raindrop splash and runoff water; greater raindrop energy or runoff amounts might be required to move the larger sand and gravel particles. Runoff can occur whenever there is excess water on a slope that cannot be absorbed into the soil or trapped on the surface. The amount of runoff can be increased if infiltration is reduced due to soil compaction, crusting or freezing. Runoff from agricultural land may be greatest during spring months when the soils are usually saturated, snow is melting and vegetative cover is minimal. Gully and rill erosion are the dominant forms of water erosion. They provide flow paths for subsequent flows, and the gullies or rills are in turn eroded further. This process leads to the self-organised formation of networks of erosional channels.	**Erodibility of soil:** very fine particles can be suspended by the wind and then transported great distances. Fine and medium-sized particles can be lifted and deposited, while coarse particles can be blown along the surface (commonly known as the saltation effect).
Soil erodibility: an estimate of the ability of soils to resist erosion, based on the physical characteristics of each soil. Generally, soils with faster infiltration rates, higher levels of organic matter and improved soil structure have a greater resistance to erosion. Sand, sandy loam and loam-textured soils tend to be less erodible than silt, very fine sand, and certain clay-textured soils.	**Soil surface roughness:** soil surfaces that are not rough or ridged offer little resistance to the wind. Excess tillage can contribute to soil structure breakdown and increased erosion.
Slope gradient and length: naturally, the steeper the slope of a field, the greater the amount of soil loss from erosion by water. Soil erosion by water also increases as the slope length increases due to the greater accumulation of runoff.	**Climate:** the speed and duration of the wind have a direct relationship to the extent of soil erosion. Soil moisture levels can be very low at the surface during periods of drought, thus releasing the particles for transport by wind.
Vegetation: plant and residue cover protects the soil from raindrop impact and splash, tends to slow down the movement of surface runoff and allows excess surface water to infiltrate.	**Unsheltered distance:** the lack of windbreaks (trees, shrubs, residue, etc.) allows the wind to put soil particles into motion for greater distances, thus increasing abrasion and soil erosion. Knolls are usually exposed and suffer the most.
	Vegetative cover: the lack of permanent vegetation cover in certain locations has resulted in extensive erosion by wind. Loose, dry, bare soil is the most susceptible. The most effective vegetative cover for protection should include an adequate network of living windbreaks combined with good tillage, residue management and crop selection.

Table 5.04 The factors influencing the erosion of soil by water and wind.

Chemical degradation

Chemical deterioration involves loss of nutrients or organic matter, salinisation, acidification, soil pollution and fertility decline. The removal of nutrients reduces the capacity of soils to support plant growth and crop production and causes acidification. Acidification results from a change in the chemical composition of soil.

Acid rain can impact significantly on some soils in particular. Some soils are naturally acidic, but this can be considerably increased by acid rain or dry deposition of acid gases and particles. The combustion of fossil fuels is the main source of acidity in the atmosphere.

In arid and semi-arid areas, problems due to accumulation of salts can arise, which impedes the entry of water in plant roots. Salinisation is the concentration of abnormally high levels of salts in soils due to evaporation. It frequently occurs in association with irrigation and leads to the death of plants and loss of soil structure. Salt-affected soils are common in arid areas, coastal zones and in marine-derived sediments, where capillary action brings salts to the upper part of the soil.

5.03 Soil degradation and conservation

Soil toxicity is the extent of the presence of toxic chemicals in a soil.

Soil toxicity can be brought about in a number of ways, but typical causes are municipal or industrial wastes, oil spills, the excessive use of fertiliser, herbicides and insecticides, or the release of radioactive materials and acidification by airborne pollutants.

Biological degradation

Research has shown that the heavy and sustained use of artificial fertiliser can result in serious soil degradation. In Figure 5.18a, the soil profile illustrates the problems that can result. In contrast, the soil profile in Figure 5.18b shows a much healthier soil treated with organic fertiliser.

Many key soil functions are underpinned by biodiversity and organic matter. Organic matter enters soils mainly from plant remains and the addition of organic manure. The loss of organic matter degrades the soil and in particular its ability to produce reasonable crop yields. Loss of organic matter also reduces the stability of soil aggregates which, under the impact of rainfall, may then break up. This may result in the formation of soil crusts, which reduces infiltration of water into the soil. This increases the likelihood of runoff and water erosion happening. Loss of soil structure can also occur because of compaction from agricultural machinery and cultivation in wet weather.

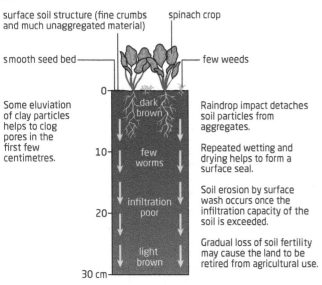
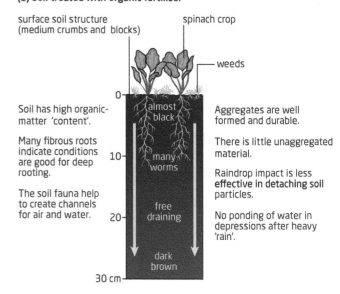

Figure 5.18 Two soil profiles: (a) with artificial fertilisers, (b) with organic fertiliser.

SELF-ASSESSMENT QUESTIONS 5.03.02

1. Define: **a** deforestation, **b** overgrazing.
2. What is salinisation?
3. **Discussion point:** Why does overgrazing take place when the consequences of this process must be obvious?

Climate and land use change

Changes to how land and soil are managed may be more important than changes in the soil due to climate modification. For example, in the UK wetter winters may lead to increased muddy flooding unless land use changes are made. Muddy flooding occurs when bare soil is left exposed at the wettest time of the year. In the UK's South Downs, changes in land management practices occurred in response to the floods in the early 1990s. This included the reversion of some winter cereal fields to permanent grassland under the EU's Set Aside scheme. No muddy floods have occurred locally since these measures were adopted.

Drier summers may result in wind erosion becoming more of a problem as soils dry out. This may cause problems with air quality and visibility and possible adverse health implications.

In low-lying coastal areas, land degradation from seawater inundation is a potential concern. River deltas and low-lying islands are particularly at risk. Salinity can reach levels where farming becomes impossible.

Local soil degradation

Figure 5.19 illustrates how a combination of causes and processes can operate in an area to result in soil degradation. The diagram shows a range of different economic activities that impact on the soil. Can you think of other economic activities that you could reasonably expect to find in such an area? What impact would these activities have on the soil? Notice how the diagram shows an increase in the area characterised by the soil sealing as the urban area expands at the expense of farmland.

Urban soils can be degraded by pollution, removal or burial. Soil can be contaminated by building waste and the whole range of pollutants that are particularly concentrated in urban areas. The role of soils in urban areas has only received limited attention to date.

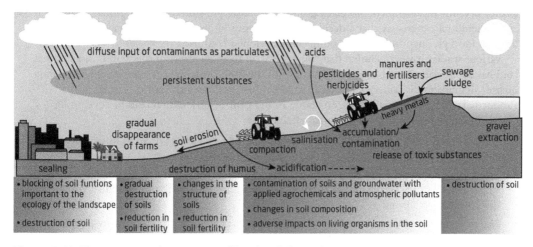

Figure 5.19 The causes and processes of local soil degradation.

5.03.03 The consequences of soil degradation

The environmental and socioeconomic consequences of soil degradation are considerable. Such consequences can occur with little warning, as damage to soil is often not perceived until it is far advanced.

Desertification

Desertification is the gradual transformation of habitable land into desert. It is arguably the most serious environmental consequence of soil degradation. Desertification is usually caused by climate change and/or by destructive use of the land. It is a considerable problem in many parts of the world, for example on the margins of the Sahara Desert in North Africa and the Kalahari Desert in southern Africa.

In semi-arid areas such as the edge of the Kalahari Desert, a combination of low and variable precipitation, nutrient-deficient soils and heavy dependence on subsistence farming makes soil degradation a significant threat. The problem has been exacerbated in recent decades by political

Desertification is the transformation of land that could once support human life into desert.

5.03 Soil degradation and conservation

changes that have disrupted traditional communal land ownership patterns, social networks and agricultural practices.

In this fragile ecosystem, soil disturbance and the removal of vegetation cover through grazing or preparing the ground for planting increases the probability of wind and water erosion. Finer and organic-rich particles are more readily removed, reducing soil nutrient levels and damaging soil structure.

Other degradation processes contributing to desertification in this region are salinisation and acidification. Bush encroachment is also a problem in some areas. This occurs where the heavy grazing of grasslands leads to them becoming overgrown by woody shrubs.

Dust storms, which can seriously damage crops, may also be a problem in such areas. Dust storms (see Image 5.14) occur naturally wherever dry soils and strong winds combine, but human activity can increase their severity significantly. These human activities are removal of vegetation, overgrazing, over-cultivation and surface disturbance by vehicles. All these practices can add to the severity of the problem. In the Sahel, the increase in dust-storm frequency has been shown to coincide with periods of severe drought.

> **A dust storm** is a severe windstorm that sweeps clouds of dust across an extensive area, especially in an arid region.

Soil degradation: a threat to food security?

The increasing world population and the rapidly changing diets of hundreds of millions of people as they become more affluent are placing more and more pressure on land resources. Some soil and agricultural experts say that a decline in long-term soil productivity is already seriously limiting food production in less economically developed countries.

Various studies have concluded that soil quality on three-quarters of the world's agricultural land has been relatively stable since the middle of the 20th century. However, on the remaining quarter, degradation is widespread and its pace has increased over the last half century. Productivity has fallen significantly on about 16 per cent of agricultural land in developing countries. The consequences for the areas affected are:

- reduced food supply, which can have a devastating effect in areas where food security is already a problem
- lower farming income and economic growth – a reduction in farm income can adversely impact on all other aspects of the rural economy; such a reverse-multiplier effect can be very difficult to turn around
- higher food prices, which can put some staple foods beyond the reach of many people
- increased child malnutrition – children and the elderly usually bear the brunt of food shortages and higher food prices
- rural-to-urban migration, which often results in the transfer of population pressure from rural to urban areas.

Image 5.14 Darkhan, Mongolia (a) on a clear day and (b) during one of the regular dust storms that are contributing to soil degradation.

On a global scale, degradation poses a modest rather than a severe threat to world food supply over the next decade or so. This is because of the global capacity for supply substitution and the dominance of the less degraded temperate regions in world food production and trade. However, a greater impact may be felt in terms of higher food prices and malnutrition. Countries that depend on agriculture as their dominant economic sector are likely to suffer the most hardship.

SELF-ASSESSMENT QUESTIONS 5.03.03

1 What is desertification?
2 What are the reasons for desertification?
3 **Research idea:** Find out the extent of desertification in one arid area of your choice.

5.03.04 Soil conservation measures

A range of management strategies can be employed to reduce soil degradation. This section examines some of the most widely used measures.

Soil conditioners

A **soil conditioner**, or a combination of conditioners, corrects the soil's deficiencies in structure and/or nutrients. The type of conditioner added depends on:

- the existing composition of the soil
- the climate of the region
- the type of crop to be cultivated.

Some soil conditioners have been in use for a long time, as they are tried and tested, while advances in soil science have produced new combinations. Lime is commonly used to reduce soil acidity by increasing the pH level. Agricultural lime or ground limestone is the most widely available and cheapest source of lime for farming; this is almost pure calcium carbonate, which is finely ground. Apart from counteracting soil acidity, the addition of lime also:

- provides important plant nutrients – calcium and magnesium
- reduces the solubility and toxicity of certain elements in the soil, such as aluminium, manganese and iron; this toxicity could reduce crop growth under acid conditions
- promotes the availability of major plant nutrients such as zinc, copper and, especially, phosphorus; calcium acts as a regulator and aids in bringing about the desirable range of availability of many plant nutrients
- increases bacterial activity and thus improves soil structure.

Fertilisers can add depleted plant nutrients. Increasing the organic content of the soil by applying animal manure, compost or sewage sludge can enable soil to hold more water, preventing aerial erosion and stabilising soil structure. The addition of gypsum releases nutrients and improves soil structure. However, research has shown that the heavy and sustained use of artificial fertiliser can result in serious soil degradation. In artificially fertilised soil, the ability of soil to infiltrate water can be compromised by the breakdown of soil aggregates to fine particles which seal the soil surface. This can result in ponding in surface depressions followed by soil erosion.

> **Soil conditioners** are materials added to soil to improve soil fertility.

Wind-reduction techniques

Wind-reduction techniques have been employed over a long period to conserve soil. The planting of trees in **shelter belts** (see Image 5.15) and the use of hedgerows can do much to dissipate the impact of strong winds, reducing the wind's ability to disturb topsoil and erode particles.

> **Shelter belts** are hedgerows and trees grown by farmers to protect soil from prevailing winds. The vegetation dissipates the impact of strong winds.

5.03 Soil degradation and conservation

Image 5.15 Shelter belts protecting agricultural land, South Island, New Zealand.

Shelter belts shelter the soil by reducing wind and evaporation and thus increasing soil temperature. They provide roots at the boundaries of the field, supplying valuable organic matter.

Hedgerows provide a habitat for a whole range of animal life, adding to the general fertility of the fields or parcels of land they surround. Research in the Philippines and elsewhere has shown that hedgerows have proved very effective in reducing soil erosion.

Strip cultivation is growing crops in a systematic arrangement of strips across a field. The strips are arranged at right angles to the direction of the prevailing wind for maximum effectiveness. A close-growing crop or strip of grass is alternated with a strip of less protective cover. Effective use of strip-cropping can:

- reduce wind erosion
- protect crops from damage by wind-borne soil particles
- reduce soil erosion from water
- reduce the transport of sediment and other water-borne contaminants
- increase infiltration and available soil moisture.

> **Strip cultivation** is the planting of alternate strips of a close-growing crop or grass with a crop less protective of the soil, arranged at right angles to the direction of the prevailing wind.
>
> **Contour ploughing** is a pattern of ploughing that ensures that the ridges and furrows are at right angles to the slope, preventing moisture from running downhill and thus reducing erosion considerably.
>
> **Terracing** is the conversion of a steep slope into a series of flat steps with raised outer edges so that cultivation can be practised.

Cultivation techniques

Various cropping techniques can be employed to reduce soil degradation (see Figure 5.20).

- **Contour ploughing** – a tried and trusted technique which prevents or diminishes the downslope movement of water and soil. Such ploughing ensures that the ridges and furrows are at right angles to the slope, preventing moisture from running downhill, and reducing erosion considerably.
- **Terracing**. Where slopes are too steep for contour ploughing, then terracing may be practised. Here the steep slope is converted into a series of flat steps with raised outer edges (bunds). The monsoon regions of South East Asia exhibit widespread terracing, particularly for rice cultivation, where water is harvested at the same time. Terracing is common in viticulture (wine). Some studies have concluded that terracing can reduce erosion twenty fold.
- Converting land from arable to pastoral uses. The planting of grass helps to bind soil particles together, reducing the action of wind and rain compared with the effect on bare soil surfaces.
- Including grasses in crop rotations.
- Leaving unploughed grass strips between ploughed fields.
- Keeping a crop cover on the soil for as long as possible, thus minimising the 'bare soil' period.
- Selecting and using farm machinery carefully – in particular, avoiding where possible the use of heavy machinery on wet soils, to prevent damage to the soil structure, and using low ground pressure set-ups on machinery when available.
- Leaving the stubble and root structure in place after harvesting.

All of these techniques have a proven track record in reducing soil degradation, with very few disadvantages. There may well be short-term decreases in food production, but these will be strongly outweighed by long-term protection of the structure of the soil. Also, the introduction of a new soil-protection technique requires appropriate knowledge and may involve the cost of employing specialist advice. New machinery that has a low impact on the soil can also be a significant additional cost.

Efforts to stop ploughing of marginal lands

Marginal lands are by their very nature fragile ecosystems. Such lands will not be cultivated if sufficient higher quality land is available. The cultivation of marginal lands gives farmers lower yields than higher quality land, while at the same time increasing the risks of soil degradation. As countries have become more aware of how rapidly marginal lands can deteriorate when they are not cultivated very carefully, increasing restrictions have been placed by national governments on the farming of such lands, alongside investment in educating farmers about sustainable farming techniques.

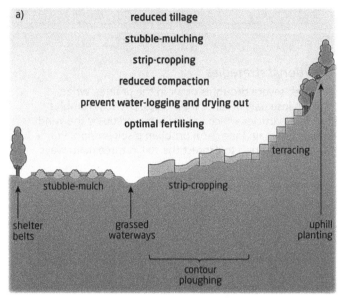
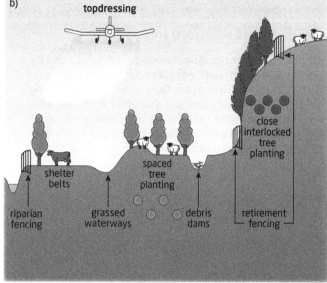

Figure 5.20 Diagram of soil conservation measures: (a) for cropland, (b) for grassland.

In 2008, a $180-million five-year project to improve sub-Saharan Africa's depleted soils was launched in Nairobi. The Alliance for Green Revolution in Africa's Soil Health Program aims to work with 4.1 million farmers to regenerate 6.3 million hectares of land. The programme will give particular attention to the role of women in promoting sustainable farming. The idea is to adapt farming techniques to local conditions to avoid repeating the mistakes of the past. Soil depletion on long-established farmland has pushed farmers to clear forests and savannahs in the search for new arable land, placing new pressures on fragile ecosystems.

SELF-ASSESSMENT QUESTIONS 5.03.04

1 **a** What is soil conditioning? **b** Explain how the application of lime can improve a soil.
2 How can shelter belts help to conserve soil?
3 **Discussion point:** Why is it important to carefully monitor cultivation on marginal lands?

5.03.05 Soil management systems in commercial and subsistence farming systems

Theory of knowledge 5.03.01

It is sometimes difficult to convince farmers to undertake soil conservation measures, even though they appreciate that they are important in the long term. There are a number of reasons for such a reluctance to change. One reason is that there may be financial costs in the short term, which some farmers may be unwilling to bear. Another reason is tradition – they are used to a system of working that has been passed down from previous generations. In addition, some farmers may just hope that the soil-degradation problems they face will right themselves naturally.

1 How can people holding such views be persuaded to change?
2 Should governments pass laws to make it illegal to continue with farming practices that are damaging the environment?

CASE STUDY 5.03.01

A commercial farming system: Canadian prairies

Three-quarters of Canada's farmland lies in the prairie provinces of Alberta, Saskatchewan and Manitoba (see Figure 5.21). This is one of the world's largest areas of extensive commercial cereal farming (see Image 5.16). Here, farms averaging 300 hectares in size stretch for almost 1500 km from east to west. The agricultural

5.03 Soil degradation and conservation

CASE STUDY 5.03.01 (continued)

productivity of the prairies is vital to the Canadian economy. As such, soil degradation has long been taken very seriously indeed. Over time, prairie farming has adapted to overcome soil and other problems as they have presented themselves. In recent years new techniques have been employed to counter escalating soil-degradation problems.

Traditional strategies

At times, severe droughts occur in the prairies, when farmers lose part or all of their crops. Dry soil crumbles into fine particles which are easily picked up by the wind. In fact, the main long-term problem is soil erosion. Prairie farmers have tried to protect the soil in three main ways.

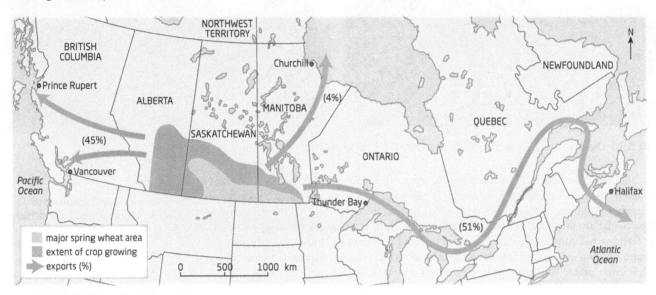

Figure 5.21 Map of prairies showing grain export routes.

Image 5.16 Commercial farming in the prairies.

- Crop-fallow rotation – fields are left fallow every other year to replenish soil moisture. The crop stubble is left to further reduce the effect of the wind.
- Ripping – a caterpillar tractor is used in winter to cut the frozen soil and knurl it into 0.3 m chunks. These chunks break the effect of the wind close to the soil surface.
- Strip farming – the farmer leaves narrow strips of fallow land perpendicular to the prevailing wind, between seeded fields.

Recent management strategies

Other changes have also been in evidence over the last decade or so. Traditional monoculture cereal cropping systems that rely on frequent summer fallowing and use of mechanical tillage for weed control on fallow areas and for seedbed preparation are being replaced by extended and diversified crop rotations together with the use of conservation tillage (minimum- and zero-tillage) practices.

Traditional monoculture cereal cropping has led to considerable soil loss by water and wind erosion, and deterioration in the quantity and quality of soil organic material, and has contributed to soil salinisation.

CASE STUDY 5.03.01 (*continued*)

In turn, these trends have:

- lowered the soil's resistance to further erosion losses
- reduced permeability and water-holding capacity
- contributed to the depletion of plant nutrients
- reduced overall soil productivity.

Because these problems had become so widely recognised, it was not too difficult to persuade farmers to consider new but proven conservation techniques. The key issues for farmers were that these conservation techniques were:

- suited to the soil and climate of the region
- practical to implement
- capable of producing the quality and quantity of grain expected
- able to maintain or enhance the quality of soil, water and air resources.

Including oilseed and pulse crops in rotations that have traditionally been cereal monoculture and reducing the frequency of summer fallow has contributed to higher net farm incomes in most parts of the prairies, even though production costs have increased because of the new practices. However, in some of the drier parts of the prairies where economic risk associated with stubble cropping is high, completely eliminating fallow from the rotation has occurred much more slowly.

Higher net farm incomes have been obtained because of the production of higher-value crops. The mixed cropping system has had a number of other clear benefits, including:

- lower disease and weed pressures
- greater residual soil nutrients and moisture reserves
- reduced soil losses.

The rise of no-till farming

The traditional practice of turning the soil before planting a new crop is a leading cause of soil degradation. An alternative is no-till farming, which minimises soil disruption. Here, famers leave crop residue on the fields after harvest, where it acts as a mulch to protect the soil and provide nutrients. To sow seeds, farmers use seeders that penetrate through the residue to the undisturbed soil below. This important sustainable approach to farming is spreading, but so far it has been mainly confined to major farming nations because of the high equipment costs involved in changing from traditional practices.

The top five countries with the largest areas of no-till are the USA, Brazil, Argentina, Canada and Australia. Approximately 85 per cent of no-till farmland is located in North and South America. Figure 5.22 compares no-till farming with conventional tillage and conservation tillage. The latter is a half-way stage between no-till and conventional tillage.

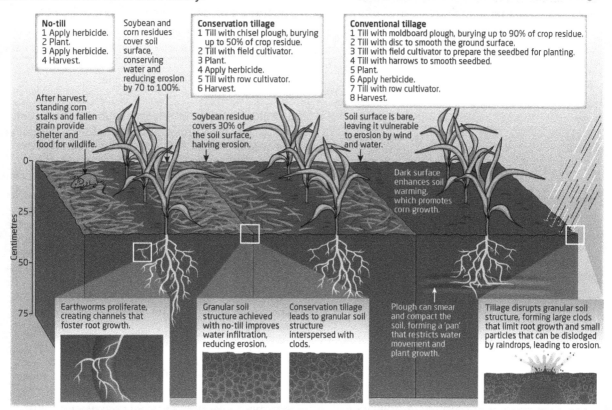

Figure 5.22 Diagram comparing no-till, conservation tillage and conventional tillage.

5.03 Soil degradation and conservation

CASE STUDY 5.03.01 (continued)

Questions

1. Briefly explain how soil conservation methods have changed in the prairies.
2. What is no-till farming?
3. Explain why some of these methods would not be available to less economically developed countries.

Prairie soil loss in Kyrgyzstan
www.youtube.com/watch?v=KDJIIE_vTOg

CASE STUDY 5.03.02

A subsistence farming system: Arumeru, Arusha, Tanzania

The FAO has stated that crop production levels in Tanzania as a whole are generally below potential. Agricultural production in the country is largely smallholder subsistence farming. An example of smallholder subsistence farming is Arumeru, which is one of eight districts in Arusha region in north-eastern Tanzania (see Figure 5.23). The population of the district is almost 520 000.

The farmers of Arumeru attempt to make a living for themselves and their families in a very difficult environment. Among the problems they face are steep slopes, erosion hazards, variable soils and low and unreliable rainfall. Soil fertility and nutrient status is generally precarious after a significant period of soil degradation, much of it associated with population pressure.

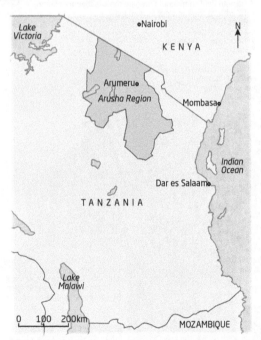

Figure 5.23 Map showing the location of Arumeru in Tanzania.

The People, Land Management and Environmental Change (PLEC) project has attempted to diversify the soil management methods used by farmers in Arumeru to raise soil productivity and enhance soil conservation. PLEC aims to promote sustainable agricultural production by recognising the indigenous knowledge accumulated by small-scale farmers over decades and even centuries, and also by showing how scientifically proven practices from other subsistence regions can work on farmers' small plots of land. A major objective is that successful farmers using sustainable methods will train other farmers, thus steadily increasing the land area covered by sustainable farming.

The three basic soil conservation principles of the project are:

- minimum soil disturbance or, if possible, no tillage at all
- soil cover, permanent if possible
- crop rotation.

By following these principles the objective is to enhance soil fertility by improving water retention, increasing soil organic matter and reducing soil degradation. Conservation agriculture aims to increase farm production, household food security and farm income.

The main land use in the sub-humid Arumeru district is the coffee/banana/maize/beans/trees agroforestry system, which has survived for over 200 years. Coffee and bananas are planted under various trees that are grown for timber, fruit, medicine, animal fodder and shade. Some farmers also keep a few dairy cattle. This historic farming system has been under stress from low fertility, soil erosion, seasonable moisture problems, land pressure and the lack of money to purchase commercial fertilisers.

What the PLEC project has called 'agrodiversity' has helped the farmers of Arumeru cope with the difficult conditions of their environment and make a modest living in circumstances that are far from easy. Table 5.05 shows the soil-related constraints at a selection of PLEC sites in Arumeru, and the

CASE STUDY 5.03.02 (continued)

corresponding soil-management strategies undertaken by farmers. Farmers have responded to their soil problems with a mixture of management practices that reflect the resources they have available, traditional knowledge and the awareness of new techniques provided by PLEC. While conservation agricultural practices in Arumeru are still at a relatively early stage, the PLEC project has concluded that there has been enough evidence of successful implementation to undertake similar work in other areas.

Site	Soil constraints	Land use type	Soil management strategy
Olgilai/ Ngiresi	low soil fertility	coffee/bananas/ maize/beans	manure application; incorporation of crop residues, house refuse, weeds and ashes; planting agro-forestry trees (e.g. *Sesbania sesban, Leucaena leucocephala*); compost application (few); incorporation of decomposed trashline* materials; heaping banana stems round coffee tree trunks; application of mineral fertilisers on coffee; import stover from distant support plots
		maize/beans	farmyard manure application, green manuring; incorporation of leaf litter (e.g. *Grevillea*), trashlines, mineral fertilisers, crop residues; crop rotations
		pastures	planting grass–legume mixtures (N fixation), nutrient recycling trees; tethering animals
		home gardens	intensive manuring, mulching, mineral fertilisers, ashes
		planted forests	controlled harvesting of trees; controlled bushfires; incorporation of crop residues and decomposed forest litter to planted crops
		natural forests	controlled tree/firewood harvesting; controlled bushfires
		water source microcatchments	fallowing
	soil moisture stress	coffee/bananas/ maize/beans	mulching; incorporation of crop residues; protective canopy from agro-forestry system; incorporation of decomposed trashline materials; green cover crops (e.g. *Vigna* and *Mucuna*); biophysical structures
		maize/beans	self-mulching; incorporation of crop residues and decomposed trashline materials; protective intercropped canopy; timely planting; weed control
		pastures	rotational grazing
		home gardens	irrigation during dry periods; mulching; farmyard manure application; construction of sunken beds; application of crop residues and mineral fertilisers
		planted forests	adequate tree spacing; protective crop and tree canopy; decomposition of litter and crop residues
		natural forests	controlled harvesting; maintenance of understorey; litter decomposition
		water source microcatchments	fallowing; area enclosure
	soil erosion	coffee/bananas/ maize/beans	construction of trashlines; mulching; rain interception by tree canopy; planting hedges of flowers and/or fodder plants; planting agro-forestry trees
		maize/beans	trashlines; crop and tree canopy; crop rotation using spreading plants (e.g. sweet potatoes); application of ashes; incorporation of crop residues; fodder grass strips
		home gardens	small scale near homestead; sunken beds; large application of manure and/or compost

(continued)

5.03 Soil degradation and conservation

CASE STUDY 5.03.02 (continued)

Site	Soil constraints	Land use type	Soil management strategy
Olgilai/ Ngiresi	soil erosion	planted forests	controlled harvesting; prevention of trespassing
		natural forests	controlled harvesting; prevention of trespassing; restricted grazing
		water source microcatchments	vegetation regeneration

*Trashlines are formed by placing crop residues in lines across a field. This reduces runoff and improves infiltration.

Table 5.05 Soil constraints and soil management techniques in Arumeru.

Questions

1 Comment on the agricultural problems that face subsistence farmers in Arumeru.
2 How has the PLEC project tried to improve soil conservation?
3 How successful has the programme of soil conservation been in Arumeru?

End-of-topic questions

1 Look at Figure 5.24.

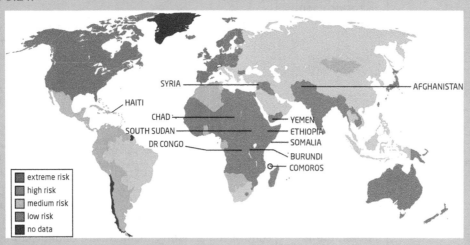

Figure 5.24 World map showing the Food Security Risk Index, 2013.

a What is the food security risk index attempting to show? [2]

b Describe the distribution of countries with high and extreme risk. [3]

c Which parts of the world are at low risk? [3]

2 a What is soil? [4]

b Discuss the characteristics of loam soils. [3]

3 a Explain *two* human activities that have had a serious impact on soil fertility. [4]

b Describe *two* soil conservation measures that have been successful in reducing the rate of soil erosion. [4]

4 a How significant is the problem of food waste? [3]

b What are the different reasons for food waste in MEDCs and LEDCs? [4]

Topic 6
ATMOSPHERIC SYSTEMS AND SOCIETIES

6.01 Introduction to the atmosphere

LEARNING OBJECTIVES

After reading this chapter you should be able to:

- explain that the atmosphere is a dynamic system that is essential to all life on Earth
- understand the behaviour, structure and composition of the atmosphere
- appreciate that these characteristics of the atmosphere influence variations in all ecosystems.

KEY QUESTIONS

What are the main elements of the atmosphere as a system?

How have human activities impacted on the composition of the atmosphere?

Why is the greenhouse effect a natural and necessary phenomenon in maintaining suitable temperatures for living systems?

6.01.01 A dynamic system

The **atmosphere** is an envelope of gas that surrounds the Earth, becoming increasingly thinner with distance from the Earth's surface and held in place by the Earth's gravitational pull.

The **atmosphere** in relation to the Earth itself has been likened to the outer skin of an onion, because it is so thin. The radius of the Earth is 6371 km. In comparison, half of the mass of the atmosphere is contained in a layer within 5.5 km of the Earth's surface. The latter is equivalent to only 0.08 per cent of the radius of the Earth. Although atmospheric gases exist much further out from the Earth's surface, they are at very low concentrations.

The atmosphere is a dynamic system as opposed to a static system. This means that the atmosphere does not stay the same but changes over time. Such changes have occurred over billions of years, but for the past 200 million years the atmosphere has been broadly similar to the way it is today. Figure 6.01 shows the main elements of the atmosphere as a system.

The Earth's original atmosphere consisted mainly of carbon dioxide, methane, ammonia, neon and water vapour; however, it lacked oxygen. Over time, the proportion of carbon dioxide decreased while the proportion of oxygen increased. In terms of very recent change, the level of carbon dioxide in the atmosphere has increased significantly again since the beginning of the Industrial Revolution. This will be examined in more detail in Chapter 7.02.

> **Extension**
> To learn more about the earth's atmosphere visit the following website:
> http://forces.si.edu/atmosphere/interactive/atmosphere.html

The biotic components of the Earth (its living organisms) have transformed the composition of the atmosphere and vice versa throughout geological time. For example, vegetation and soil (and other surfaces on the planet) control how much energy from the Sun is returned to the atmosphere. Some of this energy is reflected from the surface and some is absorbed. Of the energy absorbed:

- some is re-radiated, heating the atmosphere
- some evaporates water, returning it to the atmosphere.

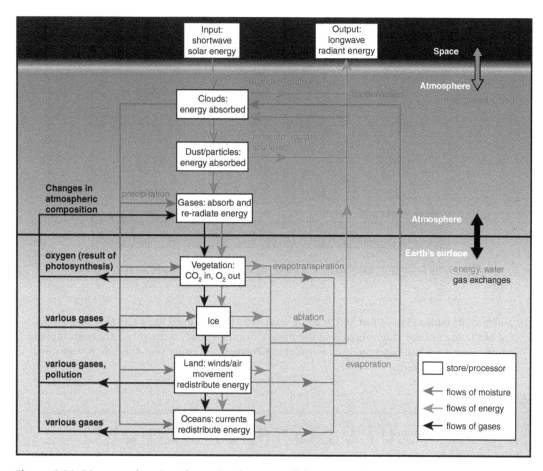

Figure 6.01 Diagram showing the main elements of the atmosphere as a system.

Extension

For another helpful visual on global energy flows visit the website below:
www.skepticalscience.com/pics/Global_Energy_Flows.jpg

The atmospheric system has:
- Inputs – by far the most important input is the energy from the Sun, which produces the movements or currents in the atmosphere.
- Outputs – precipitation is a major output from the atmosphere.
- Flows – the movement of air masses causes variations in weather and climate around the world.
- Storages – for example the concentration of water in clouds.

6.01.02 Atmospheric pressure

Although temperature varies with altitude, pressure falls continually. Atmospheric pressure is the result of the pull of gravity. Pressure is expressed in millibars (mb). The average pressure over the Earth's surface at sea level is 1013.25 mb. This is a useful figure to remember when looking at weather maps, as 1000 mb would be considered low pressure and 1020 mb would be high pressure. Of course, atmospheric pressure at the Earth's surface can be much higher and lower than these figures. The **pressure gradient** is the change in pressure over a horizontal distance. The pressure gradient causes air to flow from areas of low pressure to areas of high pressure (see Figure 6.02).

The **pressure gradient** is the change in atmospheric pressure per unit of horizontal distance.

6.01 Introduction to the atmosphere

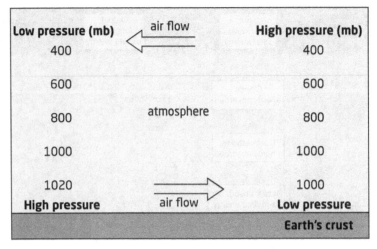

Figure 6.02 Diagram showing air flowing from high pressure to low pressure.

CONSIDER THIS

Atmospheric pressure at sea level is about 1013 mb. At a height of 10 km, roughly the height of Mount Everest, air pressure drops to 265 mb. It is not surprising that most people who attempt to climb Mount Everest need bottled oxygen to breathe!

While atmospheric gases can be traced up to about 1000 km from the Earth's surface, 99 per cent of the mass of the atmosphere is in the lowest 40 km and half of the mass of the atmosphere is in the lowest 5.5 km. Figure 6.03 shows that the change in pressure with altitude is very steep in the lower atmosphere.

6.01.03 Composition of the atmosphere

The atmosphere is composed mainly of a mix of gases, but it also contains liquids and solids. Air is a mechanical mixture of gases, not a chemical compound. It is highly compressible such that its lower layers are much more dense than those above.

The average composition of the dry atmosphere by volume is shown in Table 6.01.

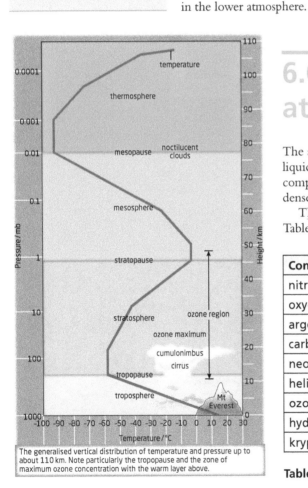

The generalised vertical distribution of temperature and pressure up to about 110 km. Note particularly the tropopause and the zone of maximum ozone concentration with the warm layer above.

Figure 6.03 Changes in temperature and pressure with altitude.

Constituent	Percentage of dry atmosphere
nitrogen	78
oxygen	21
argon	0.93
carbon dioxide	0.039
neon	0.0018
helium	0.0005
ozone	0.00006
hydrogen	0.00005
krypton, xenon and methane	trace

Table 6.01 The average composition of the dry atmosphere by volume.

Thus the atmosphere is mainly a mixture of nitrogen and oxygen, but that is not to say that the minor gases by volume do not play an important role in the atmospheric system. The various gases are relatively constant in the lower atmosphere, but they can vary at higher levels, as already mentioned.

Although the list above does not include water vapour, air can contain as much as 5 per cent water vapour, although the amount usually ranges between 1 and 3 per cent. This makes water vapour the third most common gas (which alters the other percentages accordingly). Most of the atmosphere's water vapour is held in the lowest layer, the **troposphere**. Warm air can hold more water vapour than cold air, and above the troposphere the air is too cold to hold water vapour and there is insufficient mixing or turbulence to transport water to higher levels. Water vapour is the most highly variable gas in the atmosphere.

The solids in the atmosphere include dust, smoke (see Image 6.01), salt and volcanic ash. The movement of air in the atmosphere can keep small particles aloft for a considerable time. Large particles can be transported in strong winds. The solids in the atmosphere play a vital role by acting as hygroscopic nuclei in the condensation process, leading to cloud formation and precipitation. Laboratory experiments clearly illustrate how difficult it is for condensation to occur in 'clean' air – that is, in air without minute solid particles. Shine a torch in a dark room to get some indication of the minute particles that are suspended in the atmosphere. Also, think of the amount of dust that can accumulate on furniture if a room has not been cleaned for a long time.

The structure of the atmosphere

Figure 6.03 shows that the atmosphere can be divided into a number of layers in terms of temperature variation. This is known as **thermal stratification**. A decline in temperature with altitude is known as a temperature lapse, while an increase in temperature with altitude is a **temperature inversion**. An isothermal layer occurs when there is no change in temperature with altitude for a significant distance. In general the Sun heats the Earth, which in turn heats the atmosphere, so a continuous temperature lapse from the surface to the outer atmosphere might be expected. However, the presence of certain gases in the stratosphere and the thermosphere results in the absorption of solar radiation, creating temperature inversions.

Thermal stratification

The troposphere is the lowest layer of the atmosphere, which contains most of the air, water vapour and solid particles. Thus most clouds in the atmosphere form in the troposphere (see Image 6.02). This concentration of clouds in the troposphere plays an essential role in the Earth's **albedo** effect.

The vertical extent of the troposphere varies from 16 km in the tropics to 9 km at the poles. The temperature lapse in the troposphere can fall as low as −80 °C around the equator. The next

> **Thermal stratification** is the division of the atmosphere into a number of layers in terms of temperature variation.
>
> **Temperature inversion** is an increase in the temperature of the atmosphere with altitude.
>
> **Albedo** is the proportion of radiation that is reflected by a surface.

Image 6.01 Vegetation being burnt in Nepal (2014). Fires are a major contributor to solid particles in the atmosphere.

Image 6.02 The top of clouds in the troposphere, seen from an aircraft flying in the lower stratosphere.

6.01 Introduction to the atmosphere

time you travel by air, look at the flight information available on the screen in front of you. This will tell you the altitude of the aircraft, its speed, the air temperate outside the aircraft and the strength of any headwinds the aeroplane may encounter. Wind speeds generally increase with height, because of reduced friction.

The tropopause is the boundary between the troposphere and the stratosphere. Temperature stops decreasing at this point, and for a while a zero lapse rate (an isothermal layer) occurs.

The **stratosphere** is the second major layer of the atmosphere and extends upwards from the tropopause to about 50 km. Temperature increases to about 0 °C at the stratopause, which is the boundary between the stratosphere and the mesosphere. The weather in the stratosphere is comparatively stable, as warmer air overlies colder air in this layer. Jet aircraft often fly at this altitude because of the relative stability. The good visibility means that passengers can look down on the cloud below in the troposphere (see Image 6.02). This rise in temperature is essentially due to the absorption of **ultraviolet (UV) radiation** from the Sun by ozone molecules. The ozone has an important screening function for the health of humans and other life on Earth, by shielding the surface from harmful radiation.

The troposphere and the stratosphere together make up what is generally referred to as the lower or inner atmosphere. Most reactions concerning living systems occur in the inner atmosphere.

The mesosphere is a 30 km layer of the atmosphere where temperature declines to a maximum of about −90 °C at 80 km altitude. The mesopause forms the boundary with the next layer, the thermosphere.

Atmospheric density is extremely low at the altitude of the thermosphere. Temperatures increase with altitude due to the absorption of ultraviolet radiation by atomic oxygen. The mesosphere and the thermosphere together constitute the upper atmosphere.

> **Ultraviolet (UV) radiation** comprises invisible electromagnetic rays with a wavelength shorter than that of visible light but longer than that of X-rays.

SELF-ASSESSMENT QUESTIONS 6.01.01

1. What is thermal stratification?
2. Define: **a** temperature lapse, **b** temperature inversion.
3. Why does temperature increase in the stratosphere and thermosphere?
4. **Discussion point:** Why is the concentration of solid particles in the atmosphere generally greater in urban areas than rural areas?

Atmospheric moisture

Water is a liquid compound that can be changed by an increase in temperature into a gas (water vapour) and by a decrease of temperature into a solid (ice particles). All three states exist in the atmosphere and are often found in close proximity when large cloud systems form. Such cloud systems can cover a considerable vertical extent with a large temperature variation.

The presence of water in the atmosphere is vital in terms of:

- maintaining the Earth's flora and fauna
- absorbing, reflecting and scattering a proportion of incoming **insolation** to maintain a habitable temperature at the Earth's surface
- transferring surplus energy from tropical areas, either horizontally to polar regions or vertically into the atmosphere, to balance the heat budget.

> **Insolation** (incoming solar radiation) is the heat energy from the Sun, consisting of the visible spectrum together with ultraviolet and infrared rays.

The water cycle (see Chapter 4.01) illustrates water storage and the movement of water between the atmosphere and the Earth and the movement of water around the world. Most water vapour is held in the lower 10–15 km of the atmosphere. The US Geological Survey (USGS) has described the atmosphere as 'the superhighway used to move water around the globe'. Clouds are the most obvious evidence of atmospheric water, in the form of very small droplets of water and ice particles. However, water vapour exists throughout the lower atmosphere which is not generally visible to the human eye. If all the water in the atmosphere fell to the Earth at once, it would only cover the world to a depth of 2.5 cm!

CASE STUDY 6.01.01

Cloud formation and cloud types

Clouds are visible masses of very fine water droplets or ice particles suspended in the atmosphere.

Relative humidity is the amount of water vapour present in air, expressed as a percentage of the amount needed for saturation at the same temperature.

Most **clouds** form in the troposphere and play an important role in the albedo effect of the planet, which regulates global average temperature. Clouds reflect and absorb a significant amount of incoming solar radiation. Cloud albedo is a measure of the reflectivity of a cloud. High values mean that a cloud can reflect more solar radiation. Cloud albedo can vary from less than 10 per cent to more than 90 per cent. The ability to forecast weather with a high degree of accuracy depends to a large extent on our understanding of clouds and which cloud types will produce precipitation.

The overall albedo of the Earth is about 30 per cent. This means that 30 per cent of incoming solar radiation is reflected back into space. It has been estimated that, if all clouds were removed from the atmosphere, global albedo would fall to around 15 per cent. Thus, global cloud cover has a clear overall cooling effect on the planet.

Clouds form when air rises in the atmosphere. The molecules in rising air gradually expand because air pressure decreases with altitude (see Figure 6.03). As the air molecules move further apart in the process of expansion, energy is used up and the air cools. As air cools, its **relative humidity** increases. If cooling is sufficient for the relative humidity to reach 100 per cent, the air is said to be saturated and the process of condensation begins. Condensation produces water droplets and/or ice crystals. The temperature at which saturation occurs in a parcel of rising air is known as the dew-point temperature. Thus the base of a cloud is formed at the dew-point temperature. The top of a cloud marks the level in the atmosphere where the rising air is no longer warmer than the air around it, having cooled to the temperature of its surroundings.

Air is forced to rise in three ways, when:

- an air mass is forced to rise over hills or mountains
- warm air is forced to rise over cold air at a weather front
- the Earth's surface becomes very hot and heats the air above it, causing it to rise in convection currents.

Because of these factors, some regions of the world have high levels of cloud cover while other regions have low levels of cloud cover. In general, the greater the level of cloud cover, the higher the precipitation. Cloud cover reduces the level of insolation reaching the Earth's surface and the amount leaving it. Rainforest areas astride the equator, which are characterised by thick cloud cover (see Image 6.03), experience days of about 30 °C and nights of about 20 °C. In contrast, temperatures in desert regions where cloud cover is extremely low might reach 40 °C during the day and drop to around freezing point at night. This huge difference is because humid air absorbs heat by day and retains it at night.

Image 6.03 Cloud layering, including precipitating clouds over land and the surrounding sea, in Indonesia.

Figure 6.04 shows the differences in height and shape of the main cloud types. These cloud types are separated into three broad categories according to the height of their base above the ground.

6.01 Introduction to the atmosphere

CASE STUDY 6.01.01 (continued)

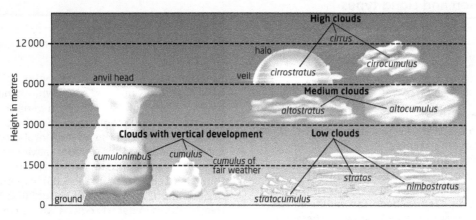

Figure 6.04 Cloud types.

Questions

1. Define the terms: **a** albedo, **b** relative humidity.
2. How do clouds play an important role in the albedo effect of the planet?
3. Briefly explain the process of cloud formation.

6.01.04 Solar radiation

The Sun has a surface temperature of nearly 6000 °C. It constantly emits radiant energy, or insolation. Solar energy comes from nuclear reactions within the Sun's hot core, while visible radiation (light) originates from the Sun's surface. However, only a very small proportion of solar output, about 0.0005 per cent, actually reaches the Earth, as the Sun's energy is emitted in all directions. Solar radiation is transmitted in the form of short waves because the Sun is so hot. About 45 per cent of solar radiation can be perceived as light. The rest is made up of ultraviolet and infrared waves, which cannot be seen. The Sun's insolation is vital in maintaining the Earth's climate and life-support systems. In contrast to the Sun, the much cooler Earth emits longwave infrared, or 'thermal', radiation. Figure 6.05 shows the contrasting wavelengths of solar and terrestrial radiation.

The greenhouse effect of the atmosphere

The greenhouse effect is a normal and necessary condition for life on Earth. Greenhouse gases have always existed and played a vital role in the Earth–atmosphere system. Until relatively recent evidence of human-induced global warming, the atmosphere was not getting any hotter. Thus there has been an **energy balance** (budget) between inputs (insolation) and outputs (re-radiation from the Earth). This relationship is shown in Figure 6.06.

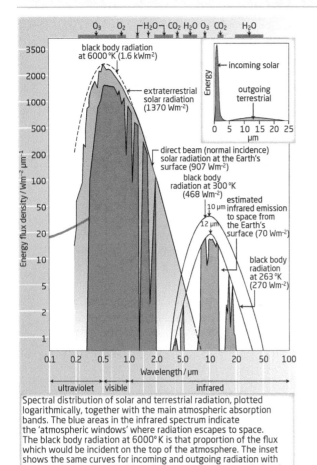

Spectral distribution of solar and terrestrial radiation, plotted logarithmically, together with the main atmospheric absorption bands. The blue areas in the infrared spectrum indicate the 'atmospheric windows' where radiation escapes to space. The black body radiation at 6000° K is that proportion of the flux which would be incident on the top of the atmosphere. The inset shows the same curves for incoming and outgoing radiation with the wavelength plotted arithmetically on an arbitrary vertical scale.

Figure 6.05 Diagram showing wavelengths of solar and terrestrial radiation.

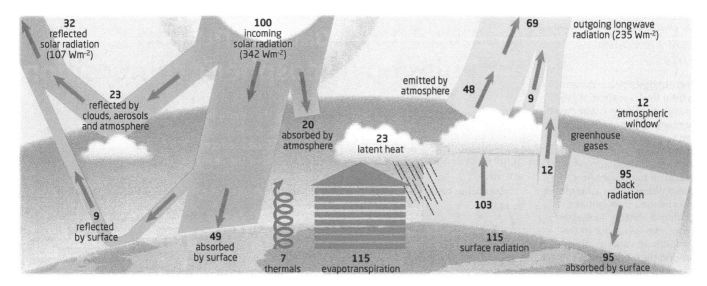

Figure 6.06 Transfers of energy in the Earth–atmosphere system.

The amount of insolation received at the outer edge of the atmosphere is known as the **solar constant**. The Earth receives approximately 342 watts of solar energy for each square metre of its spherical surface. For a stable climate on Earth, the planet must radiate the same amount back into space. However, this would mean an average global surface temperate of −19 °C rather than the 14 °C it actually is. The 33 °C difference is due to the natural greenhouse effect.

The natural greenhouse effect arises because several gases in the atmosphere absorb and emit infrared radiation. The main mechanism by which these gases absorb infrared radiation is through the vibrations of their molecules. The main greenhouse gases are water vapour, carbon dioxide, methane, chlorofluorocarbons (CFCs), nitrous oxides and ozone. These gases allow incoming solar radiation to pass through unaffected, but they trap infrared radiation emitted from the surface of the Earth. As a result, there is reduced radiation back to space and the Earth is warmed until a new balance is achieved. The overall effect is that the atmosphere acts like a blanket, which is the natural greenhouse effect. The contribution that each of the greenhouse gases makes to the total effect depends on two main factors: how efficient the gas is at absorbing outgoing longwave radiation, and its atmospheric concentration.

Figure 6.06 shows the net effect of energy transfers in the Earth–atmosphere system. Of all the radiation received at the edge of the atmosphere, 32 out of every 100 units of solar radiation are reflected back to space. This figure is a combination of reflection from both clouds (23 units) and the Earth's surface (9 units). This process is known as the albedo effect.

Forty-nine units of solar radiation are absorbed at the Earth's surface. In addition, 95 units of 'back' radiation are absorbed by the surface because of the greenhouse effect. So the total surface gain is 144 units. These units are lost as outgoing radiation (115 units), latent heat transfer (23 units) and sensible heat transfer (7 units). This results in a surface energy balance.

Within the atmosphere, Figure 6.06 shows that there are gains of 20 units from solar radiation, 103 units from surface radiation, 23 units from latent heat and 7 units from thermals. This comes to a total gain (after rounding) of 152 units. Losses in the atmosphere are thermal radiation back to the surface (95 units), and thermal radiation from clouds (9 units) and from the atmosphere itself to space (48 units). Again, after rounding there is a balance.

At the edge of the atmosphere, the gain of every 100 units of incoming solar radiation is balanced by (after rounding) the loss of 69 units of outgoing longwave radiation and 32 units that are reflected from clouds and the Earth's surface.

> **Energy balance** is the balance between incoming solar radiation and outgoing terrestrial radiation.
>
> The **solar constant** is the amount of solar energy received per unit area, per unit time, on a surface at right angles to the Sun's beam at the edge of the Earth's atmosphere.

6.01 Introduction to the atmosphere

Theory of knowledge 6.01.01

The atmosphere is a dynamic system that is susceptible to change because it is affected by so many different variables. We can see this in both temporal terms (over time) and spatial terms (over geographical space). In temporal terms, the composition of the atmosphere has changed throughout geological history. In spatial terms, the behaviour of the atmosphere varies significantly around the world and is subject to considerable changes over short periods. The increasing occurrence of extreme weather in recent decades is testimony to this fact.

1 Although knowledge of the behaviour of the atmosphere has increased significantly in recent decades, leading to much greater accuracy in weather forecasting, do you think that it will ever be possible to completely understand the workings of the atmosphere and thus be able to present weather forecasts that are 100 per cent accurate?
2 What would be the benefits of much more accurate weather forecasts?

6.01.05 The impact of human activities on atmospheric composition

Human activities can change the composition of the atmosphere by altering the inputs, outputs, flows and storages of the dynamic system of the atmosphere. Significant changes in the composition of atmospheric gases can have considerable effects on ecosystems all over the world. Some ecosystems are under intense stress as a result of the changes brought about by human activities.

According to the US Environmental Protection Agency, the main causes of air pollution are, in rank order:

1 vehicle emissions
2 fuel combustion
3 dust
4 industrial processes
5 solvent use
6 gasoline terminals, stations and gas cooking
7 fires
8 agriculture
9 waste disposal
10 radioactive waste.

The impact of pollutants resulting from human activities can vary significantly around the Earth–atmosphere system. The most obvious example of such geographical variation is the destruction of the ozone layer over the polar regions.

SELF-ASSESSMENT QUESTIONS 6.01.02

1 In which three states can water exist in the atmosphere?
2 Define the terms: **a** energy balance, **b** solar constant.
3 **Discussion point:** Discuss how the greenhouse effect regulates the Earth's temperature.

Stratospheric ozone

6.02

LEARNING OBJECTIVES

After reading this chapter you should be able to:

- appreciate that stratospheric ozone is a key component of the atmospheric system because it protects living systems from the negative effects of ultraviolet radiation from the Sun
- understand how human activities have disturbed the dynamic equilibrium of the formation of stratospheric ozone
- describe the pollution-management strategies employed to conserve stratospheric ozone.

KEY QUESTIONS

Which are the ozone-depleting substances and what impact have they had on stratospheric ozone?

What impact has the depletion of stratospheric ozone had on human health?

How successful has the Montreal Protocol been in reducing the production and release of ozone-depleting substances?

6.02.01 Ozone and absorption of ultraviolet radiation

During a process evolving over billions of years, oxygen became available in the atmosphere. During the Earth's formation, oxygen was scarce in the atmosphere, and carbon dioxide was probably the dominant gas. Eventually some oxygen was converted into **ozone**, an important gas with the capacity to absorb ultraviolet radiation and convert it into heat energy (see Figure 6.07). Ozone is a faintly blue-tinged odourless gas.

Although the amount of ozone in the atmosphere is very small in terms of total mass, averaging about three molecules of ozone for every 10 million air molecules, it is vital to the health of the planet. Most of the ozone in the atmosphere is in the stratosphere (see Figure 6.08), with only about 10 per cent in the troposphere. The ozone layer is an area of high concentration of the ozone molecule (O_3) in the stratosphere, with a maximum concentration at an altitude of about 20–25 km. The fall-off in ozone concentration either side of this altitudinal range is steep on both sides.

The ozone molecules in the stratosphere and the troposphere are chemically identical because they consist of three oxygen atoms. However, they have different impacts on the planet. Stratospheric ozone, which is sometimes referred to as 'good ozone', absorbs most of the biologically damaging ultraviolet sunlight, allowing only a limited amount to reach the Earth's surface. It can be viewed as a radiation shield. In the lower troposphere, ozone comes into direct contact with life forms, causing a range of environmental problems because high levels of ozone are toxic to living systems. Ozone at this level is sometimes referred to as 'bad ozone', because of the disadvantages it brings. Various studies have illustrated the harmful effects of ozone on forest growth, crop production and human health.

Ozone is a faintly blue-tinged odourless gas with the capacity to absorb ultraviolet radiation and convert it into heat energy.

6.02 Stratospheric ozone

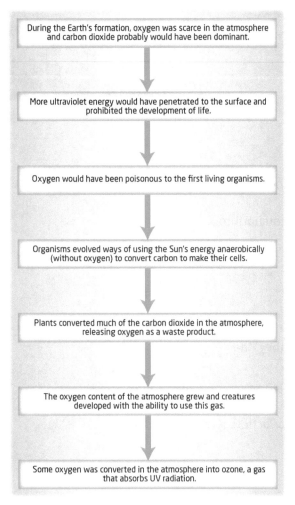

Figure 6.07 The development of oxygen and ozone.

There is considerable concern about increases in ozone in the troposphere. In particular, ozone near the Earth's surface is a major component of photochemical smog, a major issue in many of the world's large cities. Ozone is a very reactive molecule capable of oxidising many substances. It is an irritant to the eyes, nose, throat and lungs in humans. In contrast, the main concern about ozone in the stratosphere is its decline, which reduces the Earth's radiation shield.

Figure 6.09 illustrates how the ozone layer intercepts ultraviolet radiation, which is a form of energy travelling through space. Ultraviolet radiation has a wavelength between 100 and 400 nanometres (nm), whereas the wavelength of visible radiation is between 400 and 780 nm.

Ultraviolet radiation is absorbed during the formation of ozone from oxygen and during its destruction. The oxygen we breathe has two atoms in each molecule of the gas (O_2), while ozone is a form of oxygen with three atoms (O_3). Ozone is created naturally by the combining of atomic oxygen (O) with molecular oxygen (O_2), with the process being activated by sunlight. The process of ozone formation requires the kinetic energy involved in the reaction to be taken up by another molecule, and this is usually molecular nitrogen (N_2). The nitrogen molecules move faster with this additional energy and as a result they become hotter, thus warming the stratosphere.

Conversely, ozone is destroyed naturally by the absorption of ultraviolet radiation:

$$O_3 + \xrightarrow{\text{UV radiation}} O_2 + O$$

Thus, ozone is constantly being created and destroyed in the atmosphere in a balanced way, termed 'dynamic equilibrium'. However, human activities have interfered with the natural processes operating in the stratosphere, resulting in a reduction in atmospheric ozone. Most ozone is formed within the tropics, where solar radiation is at its greatest. However, winds in the stratosphere carry ozone towards the poles, where it tends to concentrate.

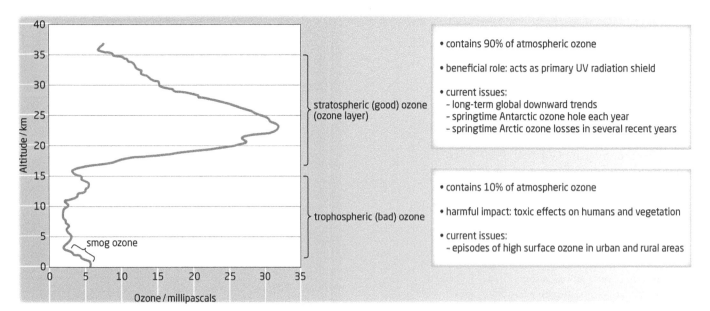

Figure 6.08 The location of ozone in the atmosphere.

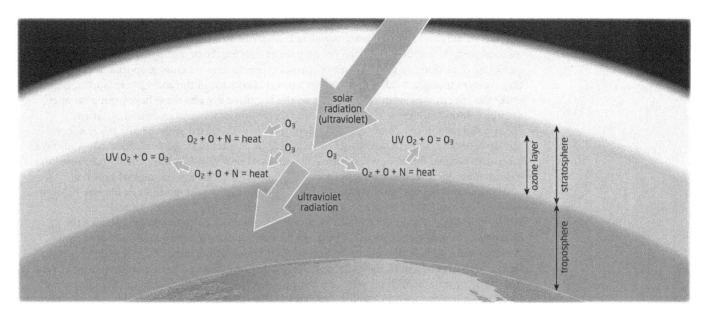

Figure 6.09 How the ozone layer intercepts ultraviolet radiation.

Three types of ultraviolet radiation can be recognised according to wavelength:
- UV-A: The near ultraviolet – the closest to visible light – at 320–400 nm. It causes little damage, with only 5 per cent absorbed by ozone.
- UV-B: The far ultraviolet, between 280–320 nm. Causes sunburn, genetic damage and skin cancer over prolonged exposure. Almost all UV-B (95 per cent) is absorbed by the ozone layer.
- UV-C: The extreme ultraviolet – the closest ultraviolet light to X-rays. All UV-C (100 per cent) is absorbed by the ozone layer.

SELF-ASSESSMENT QUESTIONS 6.02.01

1. How does the wavelength of ultraviolet radiation compare with that of visible radiation?
2. Describe the variations in atmospheric ozone shown in Figure 6.08.
3. **Discussion point:** Briefly discuss the difference between 'good ozone' and 'bad ozone'.

CONSIDER THIS

The National Oceanic and Atmospheric Administration (NOAA) uses satellite, airborne and ground-based systems to continuously monitor stratospheric ozone as well as the chemical compounds and atmospheric conditions that affect it.

6.02.02 Ozone and halogenated organic gases

While ozone is created and destroyed by natural processes, human activity has added significantly to the destruction of ozone. Scientists have been concerned about the destruction of ozone since the early 1970s. For example, in 1974 two American scientists, Mowlina and Rowland, were the first to suggest that CFCs caused ozone depletion. Their ideas were initially treated with scepticism, but as ozone became the subject of an increasing volume of research, they were proved to be correct.

In a 1985 publication, scientists from the British Antarctic Survey concluded that between 1955 and 1985 the amount of ozone over Antarctica decreased by about 50 per cent. Thus, a hole in the ozone layer had been created. In the early 1990s, satellites revealed that the loss of ozone over Antarctica was becoming more severe. However, the loss of ozone was not confined to the southern hemisphere. In the late 1980s, scientists became aware of ozone loss over the Arctic

6.02 Stratospheric ozone

region. Although smaller in scale, the loss of ozone over the Arctic has caused considerable concern because the area is closer to densely populated countries. In March 2011, a record ozone layer loss was observed, with about half of the ozone present over the Arctic having been destroyed.

Ozone depletion can be the result of natural causes such as volcanic eruptions. Volcanic dust removes nitrogen oxides and increases carbon monoxide in the atmosphere. Such changes lower the effectiveness of the ozone layer. In the last three decades there have been a number of large volcanic eruptions. However, the impact of human activity has caused by far the greatest destruction of ozone. Ozone-depleting substances, which are generally the result of human economic activity, have had a considerable impact on stratospheric ozone. Most of these substances contain the halogens chlorine, fluorine and bromine. The halogens are contained in industrial products such as CFCs, hydrochlorofluorocarbons, halons and methyl bromide.

Halogenated organic gases are very stable under normal conditions but can liberate halogen atoms when exposed to ultraviolet radiation in the stratosphere. These atoms react with monatomic oxygen and slow the rate of ozone re-formation. Such pollutants enhance the destruction of ozone, thus disturbing the equilibrium of the ozone production system.

CFCs were initially thought to be a group of harmless substances. CFCs were developed to replace ammonia as a refrigerant in the 1930s. At the time, their properties were found to be safer than those of ammonia. CFCs were also used for aerosol propellants and in the manufacture of foam, including that used in fast-food packaging. However, by the 1980s the destructive power of CFCs had become clear. CFCs can be broken down by ultraviolet light and so release chlorine ions into the atmosphere, and this breakdown occurs faster on the surface of ice crystals. CFCs are immune to destruction in the troposphere and eventually float upwards to the stratosphere. CFCs are very inert, which means that they may last for about 100 years in the stratosphere, continuing to break down ozone. Much of the ozone depletion occurring now is the result of CFCs that were released into the atmosphere decades ago. Although CFC concentrations in the atmosphere are declining, it has been estimated that measurable concentrations of CFCs will persist in the atmosphere until the 24th century. However, the ozone layer should return to its original concentration by around 2050.

> **CONSIDER THIS**
>
> F. Sherwood Rowland, Mario Molina and Dutch researcher Paul Crutzen were awarded the 1995 Nobel Prize for Chemistry for their work on ozone depletion. Rowland, who died in March 2012 at the age of 84, continued to teach chemistry at the University of California Irvine until he was 80.

Extension

CFCs and ozone. For more information on the ozone layer use the links below:
www.bom.gov.au/lam/Students_Teachers/ozanim/ozoanim.shtml
https://www.youtube.com/watch?v=k2kpz_8ntJY
www.metlink.org/secondary/a-level

CASE STUDY 6.02.01

Antarctica – the hole in the ozone layer

Every spring in the stratosphere over Antarctica (September to November), ozone is rapidly depleted by chemical processes (see Figure 6.10). With the arrival of winter, a vortex of winds develops around the pole and isolates the polar stratosphere. The polar stratosphere is separated from the tropical stratosphere by a belt of strong westerly winds known as the polar vortex. When temperatures fall below –78 °C, thin clouds of ice, nitric acid and sulfuric acid mixtures form. Chemical reactions on the surface of ice crystals in the clouds release active forms of CFCs. Ozone depletion begins and the ozone 'hole' appears. (The ozone hole area is defined as the size of the region with total ozone below 220 Dobson units (DU). A Dobson unit is a unit of measurement that refers to the thickness of the ozone layer in a vertical column from the surface to the top of the atmosphere, a quantity called the total column ozone amount.) During a two- to three-month period, about 50 per cent of the total column amount of ozone in the atmosphere disappears. At some levels, the losses approach 90 per cent. This is known as the Antarctic ozone hole. In spring, temperatures begin to rise, the ice evaporates and the ozone layer begins to recover.

In some years the ozone hole has been so large that parts of it have temporarily affected populated areas such as southern Chile and New Zealand. Figure 6.11 shows the extent of the ozone hole in 1998, 2008 and 2010. The 2010 hole was the tenth smallest since 1979. Record ozone holes were recorded in both 2000 and 2006, at 29 million square kilometres.

CASE STUDY 6.02.01 (continued)

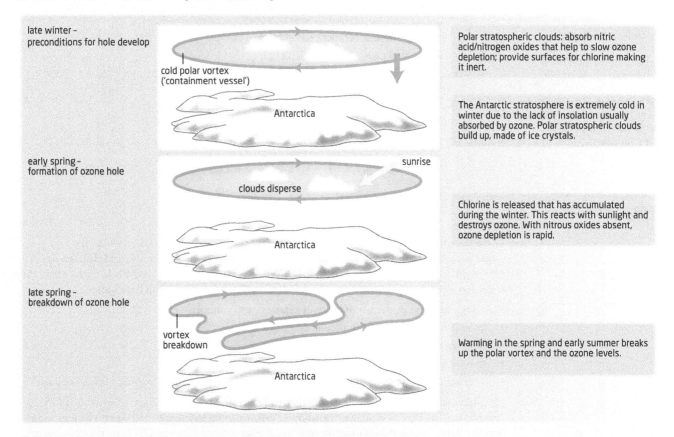

Figure 6.10 Stages in the development of the ozone hole over Antarctica.

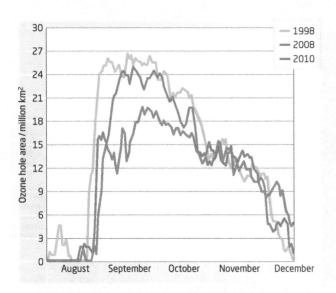

Figure 6.11 Ozone hole area in 1998, 2008 and 2010.

Encouragingly, a United Nations report published in September 2014 found that the ozone layer is starting to heal and should be back to its 1980 state by 2050.

Questions

1 Describe the stages in the development of the ozone hole over Antarctica.

2 Compare the extent of the ozone hole in 1998 and 2010.

3 The populations of which countries have been affected at times by the ozone hole over Antarctica?

6.02 Stratospheric ozone

CONSIDER THIS

Although reference is generally made to the 'hole' in the ozone layer over Antarctica, it is really a 'thinning' of the ozone layer in the stratosphere rather than an actual hole. It was first detected in 1976 and was so unexpected that scientists thought their instruments were malfunctioning. It was not until 1985 that scientists were certain they were seeing a major problem.

SELF-ASSESSMENT QUESTIONS 6.02.02

1. **a** List the major ozone-depleting substances.
   b Where do these substances come from?
2. Name a natural cause of ozone depletion.
3. **Research idea:** With reference to Figure 6.11, find out how the extent of the ozone hole has changed since 2010.

6.02.03 Ultraviolet radiation and living tissues

Ultraviolet radiation has multiple effects on living tissues and biological productivity. These include mutation and subsequent effects on health (see Image 6.04) and damage to photosynthetic organisms (see Figure 6.12). While research is still in progress in many areas, there is general agreement on the main impacts of ultraviolet radiation, which include:

- a large increase in the occurrence of cataracts and sunburn
- a significant increase in the incidence of skin cancer
- suppression of immune systems in organisms
- an adverse impact on crops and animals
- a reduction in the growth of phytoplankton (tiny marine organisms that photosynthesise).

Ultraviolet radiation can cause mutation to DNA, the basic hereditary material of all living organisms. One type of damage caused by UV-B is shown in Figure 6.12. Such changes mean that the living cells cannot correctly read the message carried by the DNA, so the cells become dysfunctional or die.

Image 6.04 Beach in southern Turkey, where summer temperatures can reach very high levels.

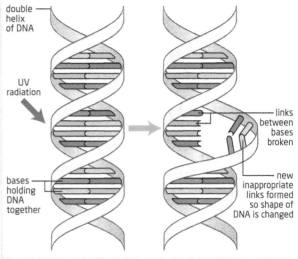

Figure 6.12 The impact of UV-B radiation on DNA.

Impact on humans

The most obvious effect on humans is that the skin burns with a much shorter exposure to sunlight. In some years in Antarctica, the time taken for sunburn to occur without the application of sunblock has been measured in minutes. Small amounts of ultraviolet radiation are beneficial to the human condition and can help prevent diseases such as rickets, psoriasis and jaundice. However, beyond a certain point, ultraviolet radiation poses dangers to human health (see Figure 6.13). Estimates made by the World Health Organization (WHO) about ultraviolet-related mortality and morbidity are that annually about 1.5 million disability-adjusted life years (DALYs) are lost through excessive ultraviolet exposure. The United Nations Environment Programme (UNEP) estimates that a sustained 1 per cent depletion of ozone will ultimately lead to a 2–3 per cent increase in the incidence of non-melanoma skin cancer as well as dramatic increases in cataracts, lethal melanoma cancers and damage to the human immune system.

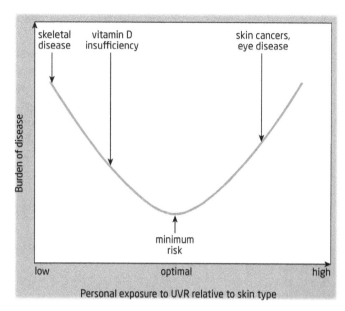

Figure 6.13 Relationship between exposure to ultraviolet radiation and the burden of disease.

Impact on plant and animal life

Although some species respond positively, research has shown that hundreds of species of plants and animals display negative effects from an increase in ultraviolet radiation. Ozone depletion and the resultant increase in ultraviolet radiation received by the surface of the Earth has the potential to decrease the productivity of phytoplankton in marine ecosystems. Phytoplankton productivity is restricted to the upper layer of the water, where sufficient light is available, and thus phytoplankton are susceptible to ultraviolet radiation. This has major implications as phytoplankton are at the bottom of most marine food chains. Declines in their number can significantly decrease productivity rates of zooplankton and higher trophic levels. There are two reasons why this is important:

- More than 30 per cent of the world's animal protein for human consumption comes from seas and oceans.
- Aquatic ecosystems account for roughly half the global production of carbon. A reduction in phytoplankton can lead to a decrease in the ability of these ecosystems to absorb carbon dioxide.

With less organic matter in the upper layers of the water due to a decline in phytoplankton, ultraviolet radiation can penetrate deeper into the water and affect more complex plants and animals living there. Solar ultraviolet radiation directly damages fish, shrimps, crabs, amphibians and other animals during their early development. In the Antarctic, increased exposure to UV-B radiation due to the appearance of the ozone hole commonly results in at least a 6–12 per cent reduction in photosynthesis by phytoplankton in surface waters.

A 2010 report by the Institute of Zoology in London found that whales off the coast of California have shown a distinct increase in sun damage, with scientists fearing that the thinning ozone layer is to blame.

Another significant issue relevant to the human population is the effect of ultraviolet radiation on plants and food crops. UV-B impairs photosynthesis in many species. Greater exposure of plant life to UV-B can result in a decrease in production rates, meaning less food is available worldwide. Plants minimise their exposure to ultraviolet radiation by limiting the surface area of foliage, which in turn impairs growth. Scientists are working to increase the level of ultraviolet radiation resistance in staple crops such as rice, where some species are extremely sensitive to ultraviolet radiation.

Overexposure to ultraviolet radiation can change the flowering times of some kinds of plants and therefore affects the animals that depend on them.

6.02 Stratospheric ozone

> **SELF-ASSESSMENT QUESTIONS 6.02.03**
>
> 1. State three adverse impacts of ultraviolet radiation.
> 2. Explain the relationship illustrated by Figure 6.13.
> 3. How can ultraviolet radiation affect marine life?
> 4. **Research idea:** Briefly summarise one item of recent research into either the impact of ultraviolet radiation on human health or the effects of ultraviolet radiation on plant and animal life.

6.02.04 Reducing manufacture and release of ozone-depleting substances

Ozone-depleting substances are substances that cause the deterioration of the Earth's protective ozone layer.

A **refrigerant** is a substance used in a heat cycle usually including a reversible phase change from a liquid to a gas.

Science has provided a range of methods to reduce the manufacture and release of **ozone-depleting substances**, with research still in progress in a number of areas. Sources of ozone-depleting substances include the production of:

- refrigerants
- gas-blown plastics
- propellants in aerosol cans
- pest control chemicals.

Recycling refrigerants

In earlier years, fluorocarbons, particularly CFCs, were used as **refrigerants**. The main use of refrigerants is in refrigerators/freezers and air-conditioning systems. Work on alternatives for CFCs in refrigerants began in the late 1970s after the first warnings of damage to stratospheric ozone were published. The Montreal Protocol's directives, as well as national and local laws for the disposal of any appliances like refrigerators that were manufactured before CFC phase-outs, have had a major impact on this source of ozone-depleting substances.

A key element of the responsible use and stewardship of refrigerants is the recovery, recycling and reclamation of used refrigerants so that they can be reprocessed for further commercial use or destroyed.

- Refrigerant recovery involves the removal of a refrigerant from a system and the placement of that refrigerant into a container.
- Refrigerant recycling involves processing used refrigerants to reduce contaminants, then reusing the refrigerants.
- Refrigerant reclamation involves purifying used refrigerants to meet industry product specifications.

Most companies now recycle the refrigerants used rather than producing entirely new ones. There are clear environmental and economic benefits from the recycling process, which include:

- minimising atmospheric emissions
- expanding the market opportunity for used refrigerants
- reducing environmental compliance costs
- lowering requirements for new refrigerant
- increasing the lifetime of refrigeration equipment by contaminant removal.

Alternatives to CFCs include ammonia and carbon dioxide refrigerants. Ammonia, for example, is a natural refrigerant that is environmentally benign in the atmosphere. Ammonia is also an efficient and cost-effective refrigerant. It is not an ozone-depleting substance and it does not contribute to global warming. Ammonia refrigeration also has a very good safety record.

Alternatives to gas-blown plastics

A gas-blown plastic is created when a plastic is 'blown' with a gas to create a foam that has a large number of voids incorporating the gas. Foam insulation is a major type of gas-blown plastic. Traditionally many plastics were blown with CFCs, and then later with HCFCs. Replacing CFCs and HCFCs as blowing agents for plastic foam has been relatively easy in comparison to their replacement as a refrigerant. In the production of polymer foams, liquid carbon dioxide can be used as a blowing agent instead of environmentally hazardous substances. Carbon dioxide is a physical blowing agent with properties deemed ideal for the creation of foam in terms of quality, efficiency and ecology. Complete alternatives to plastics are also being increasingly used. For example, in insulation situations, the following can be used:

- glass wool (fibreglass) made from recycled glass
- rock wool, made from basalt, an igneous rock
- slag wool, made from steel-mill slag.

Several natural fibres are being analysed for their potential insulating properties. The most notable of these include cotton, wool, hemp and straw.

Alternatives to methyl bromide

Methyl bromide is an odourless, colourless gas. It is a potent pest-control chemical that was identified as an ozone-depleting substance in 1992. In 1997, governments around the world established a global phase-out schedule for methyl bromide (see Table 6.02). According to the UNEP (2001), 'The phase out of this toxic chemical widely used in agriculture and other sectors by both large and small enterprises, presents a special challenge.'

MEDCs		LEDCs	
1991	base level	1995–98	average base level
1995	freeze		
1999	75% of base level	2002	freeze
2001	50% of base level	2003	review of reduction
2003	30% of base level	2005	80% of base level
2005	phase out*	2015	phase out*

*Limited exceptions may be granted for critical and emergency uses

Source: Adapted from *Sourcebook of Technologies for Protecting the Ozone Layer: Alternatives to Methyl Bromide*

Table 6.02 Montreal Protocol control schedules for methyl bromide phase-out.

Methyl bromide has been used mainly in the production of high-value crops such as tomatoes and strawberries, while lesser amounts have been used for grain cultivation. It gained widespread popularity as a versatile pesticide, effective against a wide range of pests. It is relatively easy to use and penetrates into the soil to reach the more inaccessible pests. Table 6.03 is a UNEP comparison of methyl bromide with a range of alternative pest control techniques, both chemical and non-chemical. The table shows why methyl bromide became used so widely, but also that a reasonable range of alternatives are available.

Non-chemical alternatives to methyl bromide include solarisation and crop rotation. Soil solarisation is a simple, safe and effective alternative to soil pesticides. A clear polyethylene

Stratospheric ozone

cover is used to trap solar heat in the soil. Over a period of several weeks to a few months, soil temperatures become high enough to kill many of the damaging soil pests and weed seed to a depth of nearly 20 cm. Chemical alternatives include chloropicrin and metam sodium.

	Spectrum of soil pests that can be controlled			
	Nematodes	Fungi	Weeds	Insects
Non-chemical techniques				
biological controls	*	*	*	*
crop rotation	**	**	*	*
grafting	*	*		
resistant varieties	*	*		
soil amendments	**	**	*	*
solarisation	***	**	***	**
steam	***	***	***	***
soil substitutes	***	***	***	***
Chemical treatments				
methyl bromide	***	***	***	***
chloropicrin	**	***	**	**
dazomet	**	***	**	**
1,3-dichloropropene	***	*	*	**
metam sodium	**	***	***	**
MITC	**	***	***	**
nematicides	***			
fungicides		***		
herbicides			***	

*Narrow range controlled, ** intermediate range controlled, ***wide range controlled.

Table 6.03 Range of soil-borne pests controlled by methyl bromide and alternative techniques.

Alternatives to CFCs in aerosols

The alternatives to CFCs in aerosols include hydrocarbons such as propane, butane and isobutane. The US Environmental Protection Agency has produced a list of substitutes for CFCs in aerosols, based on:

- ozone depletion potential
- global warming potential
- toxicity
- flammability
- exposure potential.

The list of substitutes is updated on a regular basis. However, replacements for CFCs are sometimes considered pollutants in their own right.

Extension

For more information about replacements, visit the websites below:
www.Epa.gov/ozone/snap
www.theozonehole.com/cfc.htm
www.ozone.unep.org

> **SELF-ASSESSMENT QUESTIONS 6.02.04**
>
> 1 a What is a refrigerant?
>   b What are the main uses of refrigerants?
> 2 Why is ammonia a good alternative to CFCs as a refrigerant?
> 3 What is a gas-blown plastic?
> 4 **Discussion point:** What are the main alternatives to methyl bromide as a pest control agent?

6.02.05 The role of national and international organisations

'Perhaps the single most successful international agreement to date has been the Montreal Protocol.'

Kofi Annan, former Secretary General of the United Nations

The Montreal Protocol

International cooperation has been vital in reducing the emissions of ozone-depleting substances. The role of the UNEP has been pivotal in this process. Various national organisations have also played an important role. The UNEP implemented the Vienna Convention in 1985, which established mechanisms for international cooperation in research into ozone-depleting chemicals. Following on from the Vienna Convention, the Montreal Protocol on Substances that Deplete the Ozone Layer was negotiated and signed in September 1987. The protocol called for the parties to phase down the use of CFCs, halons and other manmade ozone-depleting chemicals. The protocol stated that the production and consumption of ozone-depleting compounds would be phased out according to an agreed schedule (see Table 6.04).

Ozone-depleting substances	Schedule for MEDCs	Schedule for LEDCs
CFCs	phase out end of 1995	phase out 2010
halons	phase out end of 1993	phase out 2010
carbon tetrachloride	phase out end of 1995	phase out 2010
methyl chloroform	phase out end of 1995	phase out 2015
hydrochlorofluorocarbons	freeze beginning 1996 35% reduction by 2004 65% reduction by 2010 90% reduction by 2015 total phase-out 2020	freeze 2016 at 2015 base level total phase-out 2040
hydrobromofluorocarbons	phase out end of 1995	phase out end of 1995
methyl bromide	freeze 1995 at 1991 base level 25% reduction by 1999 50% reduction by 2001 70% reduction by 2005 total phase-out 2010	freeze 2002 at average 1995–98 base level 20% reduction by 2005 total phase-out 2015

Table 6.04 Summary of Montreal Protocol measures.
Source: Adapted from the Montreal Protocol

6.02 Stratospheric ozone

Ratification of:	Total number of ratifying countries
Vienna Convention	191
Montreal Protocol	191
London Amendment	185
Copenhagen Amendment	176
Montreal Amendment	152
Beijing Amendment	124

Table 6.05 Number of countries signing the Montreal Protocol and its amendments.
Source: Adapted from the Montreal Protocol

The Montreal Protocol was one of the first international agreements that included the potential use of trade sanctions to achieve the objectives of the treaty. It also offered significant incentives for non-signatory countries to sign the agreement.

Figure 6.14 shows the extent to which the consumption of ozone-depleting substances declined between 1989 and 2009, illustrating the success of the Montreal Protocol. Overall concentrations of ozone-depleting substances in the atmosphere peaked in 1994. Figure 6.15 presents a detailed log of the milestones in the history of ozone depletion.

The final section of Figure 6.15 notes the amendments to the Montreal Protocol (see also Table 6.05), with meetings in London, Copenhagen, Vienna, Montreal and Beijing between 1990 and 1992. The main objective of the amendments was to speed up the phasing out of ozone-depleting substances. However, other issues were also addressed. For example, at the Copenhagen meeting in 1992, methyl bromide was added to the list of controlled substances. Also, for all substances controlled under the protocol, phase-out schedules were delayed for less economically developed countries (LEDCs) and phase-out in these countries was supported by transfers of expertise, technology and money from more economically developed countries (MEDCs). It is often the case that amendments are required to maintain international agreement when major agreements are signed following meetings.

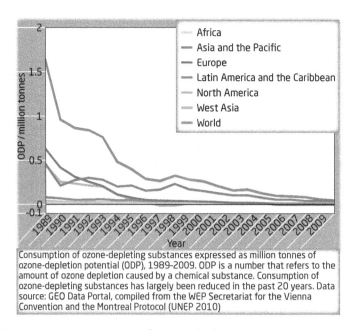

Consumption of ozone-depleting substances expressed as million tonnes of ozone-depletion potential (ODP), 1989–2009. ODP is a number that refers to the amount of ozone depletion caused by a chemical substance. Consumption of ozone-depleting substances has largely been reduced in the past 20 years. Data source: GEO Data Portal, compiled from the WEP Secretariat for the Vienna Convention and the Montreal Protocol (UNEP 2010)

Figure 6.14 Consumption of ozone-depleting substances 1989–2009.

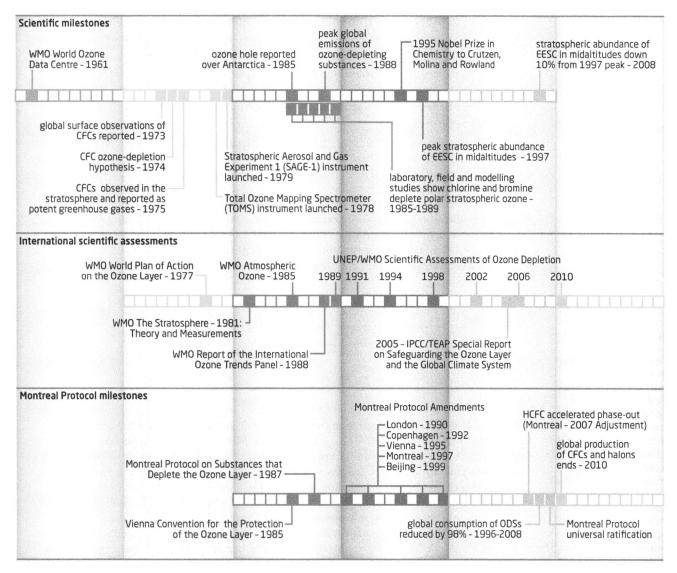

Figure 6.15 Milestones in the history of stratospheric ozone depletion.

The Montreal Protocol has reduced the worldwide stock of CFCs and other ozone-destroying compounds by 98 per cent. CFCs have largely disappeared from the computer manufacturing process, polystyrene packing materials and disposable cups, refrigerators, car and home air conditioners, aerosol sprays, fire extinguishers, degreasing compounds and foam ingredients in furniture. The use of other ozone-depleting substances has also declined significantly.

The role of national governments in implementation

The success of international agreements relies heavily on the proper implementation of such agreements by national governments. Under the Montreal Protocol, individual countries have had to show how they planned to phase out ozone-depleting substances. For example, the detail of India's Country Programme was completed in 1993 with the assistance of the United Nations Development Programme, The Energy and Resources Institute and representatives of various ministries, industries and scientific institutions.

6.02 Stratospheric ozone

The key principles underlying India's Country Programme were:

- to implement phase-outs of ozone-depleting substances without adversely affecting industrial and economic growth
- to meet the demands for substitutes for ozone-depleting substances as far as possible from indigenous sources
- to reflect India's commitment to achieving compliance with the Montreal Protocol's obligations.

As in other developing countries, there was considerable debate in India about the costs and benefits of ratifying the Montreal Protocol. Concerns were raised about equity, state sovereignty, 'green imperialism' and the challenge of regulating the businesses in India that utilise ozone-depleting substances.

Although the Montreal Protocol can be generally regarded as a success, the long life of ozone-depleting substances in the atmosphere means that the problem will be with us for some time.

The Sustainable Scale Project has concluded that, 'Despite the welcomed successes of the Montreal Protocol the level of atmospheric ozone depletion will continue to exceed sustainable scale for several decades. Damage to living things will continue as long as levels remain above sustainable scale.'

6.02.06 The Illegal market in ozone-depleting substances

Illegal trade in CFCs was detected in the USA in 1992, and this trade grew rapidly in the following years. By the late 1990s large amounts of CFCs were being smuggled out of China, where they were readily available at low cost, to a range of destinations. In November 2014, the Environmental Investigation Agency (EIA) stated: 'Far from going away, the threat of ODS [ozone-depleting substance] smuggling and illegal ODS use is increasing.' Such action has decreased the effectiveness of the Montreal Protocol, but only to a limited degree.

Also of concern are apparent loopholes in the Montreal Protocol. An article in *New Scientist* in 2014 noted that a research project had found four types of CFC previously undetected in the atmosphere. All are banned by the Montreal Protocol, but under certain conditions industries can apply for exemptions.

There is a need for constant monitoring of the market in ozone-depleting substances.

Theory of knowledge 6.02.01

The details of international agreements such as the Montreal Protocol are often heavily influenced by the views of the most powerful countries in the world. This may be understandable if they possess much greater scientific and technical knowledge and understanding, and their motives are genuinely altruistic. However, LEDCs sometimes feel they are being coerced into signing international agreements because of the power and influence of the world's major countries. In addition, LEDCs are often faced with significant economic and social costs when implementing agreements such as the Montreal Protocol. An example of this is the cheaper cost of CFCs compared with many alternatives.

1 Is it right that the most powerful countries have the most say in the structuring of international agreements?
2 Should MEDCs have the responsibility to pay more towards the cost of change in LEDCs?

SELF-ASSESSMENT QUESTIONS 6.02.05

1. In which year was the Montreal Protocol signed?
2. By how much has the worldwide stock of CFCs and other ozone-destroying compounds fallen since the protocol was signed?
3. Suggest why there were various amendments to the Montreal Protocol.
4. Briefly explain the role of national governments in implementing the Montreal Protocol.
5. **Discussion point:** To what extent do you agreed with the quote by Kofi Annan at the beginning of this section?

Photochemical smog

6.03

LEARNING OBJECTIVES

After reading this chapter you should be able to:

- understand that the burning of fossil fuels produces primary pollutants that may generate secondary pollutants leading to the formation of photochemical smog
- understand that levels of photochemical smog can vary due to topography, population density and climate
- appreciate that photochemical smog has significant impacts on societies and living systems
- explain how photochemical smog can be reduced by lowering reliance on fossil fuels.

KEY QUESTIONS

What are the primary pollutants that form the basis of photochemical smog?

How is tropospheric ozone formed and what are its impacts on people and the environment?

What pollution-management strategies can reduce the incidence of photochemical smog?

6.03.01 Primary and secondary pollutants

Air **pollution** is one of the major current health risks for the global population, and the problem is getting worse. According to the WHO, urban outdoor air pollution causes 1.3 million deaths per year worldwide. This trend is likely to continue with further urbanisation and industrialisation.

A primary pollutant is an air pollutant emitted directly from a source (see Figure 6.16 and Image 6.05). Primary pollutants are pollutants that are passed into the environment in the form in which they are produced. The combustion of fossil fuels is the main source of primary pollutants, which include carbon monoxide, carbon dioxide, black carbon or soot, unburned hydrocarbons, NO_x and oxides of sulfur.

A secondary pollutant is not directly emitted from a source but forms when other pollutants (primary pollutants) react in the atmosphere in the presence of sunlight (see Figure 6.16). Secondary pollutants are more toxic than primary pollutants.

Pollution is the addition of any substance or form of energy (e.g. heat, sound or radioactivity) to the environment at a rate faster than the environment can accommodate it by dispersion, breakdown, recycling, or storage in some harmless form.

Extension

To read more information about forming photochemical smog visit the following website: http://rhsweb.org/jstewart/assignments/wwwAPES/Semester2/Ch15/AnimationsChapt15/photochemical_anim.swf

6.03 Photochemical smog

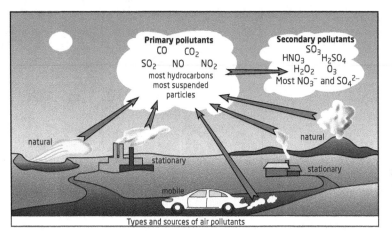

Figure 6.16 Diagram of primary and secondary pollutants.

Image 6.05 Polluting brick kilns on the banks of the Mekong River, Vietnam (2013).

Tropospheric ozone

Tropospheric ozone is an example of a secondary pollutant. You will remember that tropospheric ozone is sometimes referred to as 'bad ozone' because of the damage it can cause.

Some ozone occurs naturally in low concentrations at and close to ground level. The two main sources of this ozone are hydrocarbons released by soil and plants, along with small amounts of stratospheric ozone which can occasionally descend to the surface. The amount of ozone from these sources is so small that it is not considered a threat to human health.

However, ozone created from human activity adds significantly to natural ozone to make it a considerable hazard to human health and the environment. Ozone is a by-product of fossil fuel combustion. Concentrations of ozone are not uniform in the troposphere and are heavily influenced by the level of human activity. In contrast to most other air pollutants, ozone is not directly emitted from any one source. Ground-level ozone results from photochemical reactions between oxides of nitrogen (NO_x) and **volatile organic compounds (VOCs)** in the presence of sunlight. In the northern hemisphere, high levels usually occur from May and September. On a daily basis, ozone levels are highest between noon and early evening after the Sun's rays have had time to react with exhaust fumes from high traffic volumes in the morning rush hour (peak) period. By early evening, the photochemical production process that creates the ozone begins to subside as the Sun's intensity decreases.

Since 1900, the global amount of tropospheric ozone has more than doubled.

> **Volatile organic compounds (VOCs)** are mainly synthetic chemicals that play a major role in the formation of photochemical smog.

Extension

Air pollution in Hong Kong
http://aqicn.org/city/hongkong/central/western/

WHO – the impacts of air pollution on human health
www.who.int/ceh/risks/cehair/en

A good source of articles and videos on air pollution
www.sciencedaily.com

Figure 6.17 shows the sources of volatile organic compounds in Ontario, Canada, in 2006. The transportation sector accounted for 38 per cent of the total, with printing/surface coating and general solvent use in second and third places respectively. These VOCs are not just the result of emissions in Ontario, because significant amounts of ozone and its chemical precursors are carried by wind into Ontario from highly populated, industrial areas in the USA. Airborne pollutants can be transported over extremely long distances.

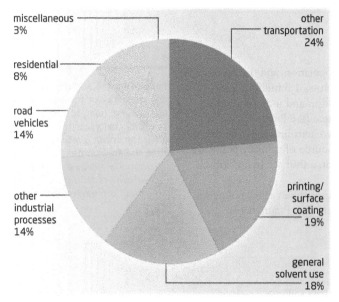

Figure 6.17 Volatile organic compounds emissions by sector in Ontario, Canada (2006).

Effects of tropospheric ozone

Ozone is a toxic gas and an oxidising agent. For humans it is an irritant to eyes and can cause breathing difficulties characterised by chest tightness, coughing and wheezing. It can aggravate asthma and bronchitis, and may increase susceptibility to infection. People with respiratory and heart problems are at a higher risk. Ozone has been linked to increased hospital admissions and premature death. Longer exposure to ozone will increase the negative effects.

In September 2011, Professor Marc Decramer, President of the European Respiratory Society, stated, 'Outdoor air pollution is the biggest environmental threat in Europe. If we do not act to reduce levels of ozone and other pollutants, we will see increased hospital admissions, extra medication and millions of lost working days.'

Ozone is highly reactive and can damage fabrics and other materials. Some elastic materials may become brittle and crack, while paint and fabric dyes can fade more quickly than expected. Ozone damages crops and forests. Noticeable leaf damage occurs in many crops, plants and trees. Thus it can impact considerably on farm incomes. High levels of ozone cause plants to close their stomata, which are the cells on the underside of the plant which allow water and carbon dioxide to diffuse into plant tissue. This results in a slower rate of photosynthesis and slower plant growth.

Air quality indices are numbers used by government agencies to characterise the quality of the air at a given location. Table 6.06 is the index used in Ontario, Canada. It shows the health effects of different air quality index levels due to ground-level ozone. An air quality index of 100 and over can have significant effects on people and the environment. People at risk may change their daily routine to minimise potential adverse health effects.

> The **air quality index (AQI)** is an indicator of air quality, based on air pollutants that have adverse effects on human health and the environment.

Category	AQI level	Pollution concentration breakpoint / ppb*	Health effects
very good	0–15	0–23	• none in healthy people
good	16–31	24–50	• none in healthy people
moderate	32–49	51–80	• respiratory irritation in sensitive people during vigorous exercise • some risk for people with heart/lung disorders • damage to very sensitive plants
poor	50–99	81–149	• respiratory irritation in sensitive people when breathing • possible lung damage in sensitive people when active • damage to some plants
very poor	100+	150+	• serious respiratory effects during light physical exercise • high risk for people with heart/lung disorders • more vegetation damage

* ppb, parts per billion

Table 6.06 Health effects of different air quality index (AQI) levels.
Source: Adapted from airqualityontario.com

SELF-ASSESSMENT QUESTIONS 6.03.01

1. Describe the information illustrated in Figure 6.17.
2. When are ozone levels highest during: **a** the year, **b** the day? In each case, explain why.
3. To what extent has the amount of tropospheric ozone changed since 1900?
4. **Research idea:** Monitor the level of ozone for a week in the nearest large city to where you live and then write a report on your findings.

6.03 Photochemical smog

6.03.02 Photochemical smog

Air quality is determined not only by the level of emissions, but also by current and preceding meteorological conditions. **Photochemical smog** is a particular type of air pollution that is produced when the weather is characterised by stability and strong sunshine. It is associated with **anticyclones** in both winter and summer. Anticyclones can persist for a number of days. When this happens, they are called blocking anticyclones because they block out any approaching depressions.

Visually, photochemical smog appears as a grey-brown haze. Photochemical smog is often accompanied by a characteristic odour (see Image 6.06). It was first described in the 1950s as a product of industrialisation. 'Smog' is a combination of the words 'smoke' and 'fog'. It is a mixture of around 100 primary and secondary pollutants formed under the influence of sunlight. Ozone is the most prominent of this range of pollutants. Photochemical smog is also sometimes known as 'oxidising smog', in that it has a high concentration of oxidising agents such as ozone.

Figure 6.18 illustrates the formation of photochemical smog. In this process, nitrogen is transformed between many different substances in the atmosphere.

Photochemical smog develops when primary pollutants (NO_x and VOCs) interact due to sunlight to produce a multitude of hazardous chemicals known as secondary pollutants. These include peroxyacetyl nitrates and ground-level ozone. For photochemical smog, the starting threshold for ground-level ozone is 82 parts per billion, and ground-level ozone constitutes 90 per cent of all smog found in urban areas. Table 6.07 lists the main toxic constituents of photochemical smog and their environmental effects.

Photochemical smog is air pollution containing ozone and other reactive chemical compounds formed by the action of sunlight on nitrogen oxides and hydrocarbons, especially those in vehicle exhaust. Photochemical smog is a major contributor to air pollution.

An **anticyclone** is a large mass of subsiding air which causes high pressure at the surface. Such weather brings clear skies and little or no wind.

CONSIDER THIS

The term 'smog' was first coined in 1905 in a paper by Dr Henry Antoine Des Voeux to describe the combination of smoke and fog that had badly affected London at that time. Such smog was very different to the **photochemical smog** of today.

Image 6.06 Photochemical smog in Beijing.

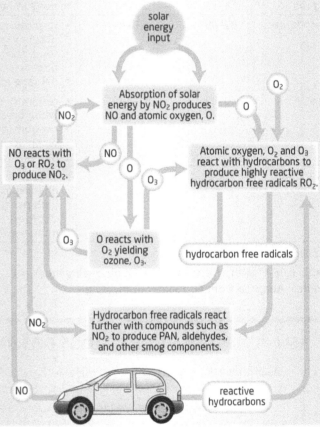

Figure 6.18 Generalised scheme for the formation of photochemical smog.

Toxic chemical	Sources	Environmental effects	Additional notes
nitrogen oxides (NO and NO_2)	• combustion of oil, coal, gas in vehicles and industry • bacteria in soil • forest fires • volcanoes • lightning	• decreases visibility due to yellow colour NO_2 • NO_2 contributes to heart and lung problems • NO_2 can suppress plant growth • decreases resistance to infection • may encourage spread of cancer	• all combustion processes account for only 5% of NO_2 in atmosphere; most is formed from reactions of NO • concentrations likely to rise
volatile organic compounds (VOCs)	• evaporation of solvents • evaporation of fuels • incomplete combustion of fossil fuels • occur naturally in terpenes in trees	• eye irritation • respiratory irritation • some are carcinogenic • decreases visibility due to blue-brown haze	• effects of VOCs depend on the chemical • samples show over 600 different VOCs in the atmosphere • concentrations likely to rise
ozone (O_3)	• formed by photolysis of NO_2 • sometimes results from stratospheric ozone intrusion	• bronchial constriction • coughing, wheezing • respiratory irritation • eye irritation • decreases crop yields • retards plant growth • damages plastics • damages rubber • harsh odour	• concentration of 0.1 ppm* can reduce photosynthesis by 50% • people with asthma and respiratory problems are most affected • only formed in daylight
peroxyacetyl nitrates	• reaction of NO_2 with VOCs • can form naturally in some environments	• eye irritation • respiratory irritation • damaging to proteins • highly toxic to plants	• was not discovered until recognised in smog • more toxic to plants than ozone

*ppm, parts per million

Table 6.07 Major chemical pollutants in photochemical smog.
Source: University of Nebraska

In the modern industrial world, photochemical smog is a common air-quality problem. It is particularly associated with centres of high population density that experience specific climatic conditions. Examples include Mexico City, Beijing, São Paulo and Los Angeles. However, even moderate-sized urban areas can experience photochemical smog if the associated environmental conditions are in place. Thus, apart from the release of large quantities of air pollutants, the other factors that contribute to the formation of smog are as follows:

- Temperature: Most cases of serious photochemical smog are associated with temperature inversions close to the surface.
- Topography: Urban areas located in valleys with surrounding upland areas tend to experience low air circulation and a high level of accumulation of air pollutants compared with cities in more open landscapes.
- Wind patterns: The pattern of wind can affect the frequency of the replacement of local air with fresh air from outside the urban area. Large-scale burning associated with deforestation can contribute significantly to smog in urban areas.

Effect of temperature inversions

When the usual decline in temperature with altitude in the troposphere is reversed, a temperature inversion occurs. Temperature inversions are associated with stable atmospheric conditions. When a temperature inversion occurs, warmer air is found above colder surface air. This happens because the ground is a more effective radiator of heat than the air above. The clear skies associated with anticyclones allow the ground to lose heat rapidly at night. The cold surface air is more dense and is prevented from rising upwards by the warmer air overlying it.

6.03 Photochemical smog

Pollutants contained in the cold air are trapped near the ground, with their levels building up as emissions from vehicles and other sources continue to pollute the environment. Pollutants are not easily dispersed horizontally when there is little air movement. Katabatic winds can add to the temperature inversion problem (see Figure 6.19). These are currents of cold air that blow downslope from surrounding uplands into a valley or other lowland area, thus creating or increasing the mass of cold air in a valley.

Such conditions often occur in winter smogs when cold anticyclonic conditions result in a higher use of energy in homes and businesses, releasing larger than usual amounts of pollutants into the lower atmosphere. However, temperature inversions and associated smogs can also occur at warmer times of the year.

A change in the weather is the usual reason for photochemical smogs to subside. Precipitation cleans the air, and winds disperse the smog. In general, the higher the wind speed, the greater the rate of dispersal. The stability of the atmosphere determines the extent to which vertical motion will mix the pollution with cleaner air above the surface layer.

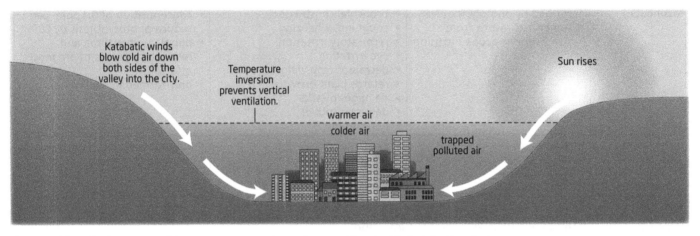

Figure 6.19 Pollution trapped by a temperature inversion combined with katabatic winds.

SELF-ASSESSMENT QUESTIONS 6.03.02

1. Define smog.
2. With the help of Figure 6.18, briefly explain how photochemical smog is formed.
3. **Research idea:** Find out when the last photochemical smog occurred in the closest large urban area to where you live. Produce a brief fact file to describe this event.

6.03.03 Urban air pollution management strategies

Air pollution is a major problem in many urban areas. According to a WHO report published in 2014, the ten most air-polluted cities in the world are:

1. Delhi, India
2. Patna, India
3. Gwalior, India
4. Raipur, India

5 Karachi, Pakistan
6 Peshwar, Pakistan
7 Rawalpindi, Pakistan
8 Khoramabad, Iran
9 Ahmedabad, India
10 Lucknow, India.

The WHO says that just 12 per cent of the world's urban population are breathing clean air, while about 50 per cent are exposed to air pollution at at least two-and-a-half times safe levels, putting them at risk of serious long-term health problems. India has the worst death rate in the world from chronic respiratory disease and also the highest death rate from asthma, which might be expected given its position in the list of the world's most polluted cities.

Air pollutants are usually classified into suspended particulate matter, gaseous pollutants and odours. Pollution particles of less than 10 microns in diameter (PM_{10}) are particularly dangerous to health, penetrating deep into the lungs and bloodstream. WHO standards recommend 20 micrograms per cubic metre of PM_{10} or less.

The main factors influencing trends in air pollution are:

- population growth
- economic activity
- meteorological conditions
- regulatory efforts to control emissions.

Figure 6.20 shows five levels at which action can be taken to prevent or reduce the health effects of environmental hazards including air pollution. Interventions at the top level of the diagram include legal restrictions such as banning the use of lead in petrol, and community-level policies such as promoting greater use of public transport.

A range of strategies have been employed to manage urban air pollution. Arguably the most effective way of reducing the secondary pollutants created in the atmosphere is to cut emissions of the main primary pollutants in photochemical smog – nitrogen oxide and VOCs.

Reducing nitrogen oxide emissions

Catalytic reduction is the main method of reducing the level of nitrogen oxides produced by motor vehicles and industry. Catalytic converters fitted to vehicle exhaust systems convert much of the nitric oxide produced to nitrogen and oxygen. Thus this process converts toxic exhaust emissions into non-toxic substances. Many countries have legislation with regard to the use of catalytic converters. The first widespread introduction of catalytic converters was in the USA, where 1975-model cars were equipped with converters to comply with new US Environmental Protection Agency regulations. Other countries followed this action. For example, in Australia all motor vehicles built after 1985 were required to be fitted with catalytic converters. Although most commonly fitted to motor vehicles, catalytic converters are also used on trains and aeroplanes.

Using less air in combustion can also cut emissions of nitrogen oxides. Lowering the temperature of combustion has a similar effect. Temperatures can be reduced by processes such as two-stage combustion and flue gas recirculation and water injection. Modifying the design of the burner can also reduce the temperature of combustion.

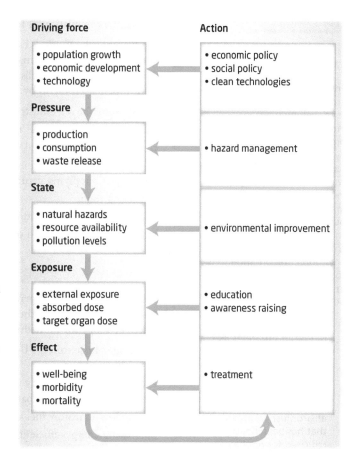

Figure 6.20 Framework for Environmental Health Interventions.

6.03 Photochemical smog

Reducing VOC emissions

VOC emissions from motor vehicles can be cut by the following policies:

- switching from petrol to liquefied petroleum gas (see Image 6.07) or compressed natural gas
- implementing engine and emission controls being developed by manufacturers
- cutting the distances that vehicles travel, by using alternative transport such as bicycles and public transport.

Solvent-based products such as paints also produce VOCs. However, low-VOC paints and paints that are non-VOC are now available on the market. Likewise, buying formaldehyde-free furniture and avoiding synthetic carpets can reduce VOCs in the home.

Image 6.07 Liquefied petroleum gas on sale at a New Zealand petrol station – a cleaner form of energy than petrol.

Other strategies

Other strategies to combat urban air pollution include:

- switching to renewable energy to reduce fossil fuel combustion
- reducing demand for electricity
- using electrostatic precipitators and particulate scrubbers to control the emission of particulates
- land use planning to try to separate industrial and residential areas as far as possible
- government regulation of fuel quality.

For example, the World Resources Institute declared Lanzhou in China the world's most polluted city in 1998. It is a major industrial centre, burning large quantities of coal every day. The city is surrounded by hills, which hinders the dispersal of pollution. The city has addressed its environmental problems by:

- closing some heavy industries
- relocating some industries from inner city to edge-of-city locations
- restricting emissions based on air quality warnings with yellow and red level alerts to reduce factory pollution
- investing in supplying natural gas and cleaner coal
- restricting traffic
- planting trees on surrounding hillsides to reduce dust storms
- building bicycle lanes and developing public transport.

Personal strategies to reduce air pollution

Although most of the management strategies to reduce urban air pollution depend on government legislation and the actions of large companies and other organisations, there remains considerable scope for individuals to play their part. Such individual/household measures include:

- reducing vehicle mileage by walking, using a bicycle (see Image 6.08) and making greater use of public transport
- keeping your car regularly serviced and tyres properly inflated to the manufacturer's specifications to ensure that excessive pollutants are not being emitted
- buying low-emission, fuel-efficient cars
- purchasing energy-efficient household appliances
- not leaving household appliances on standby
- generating your own green energy using solar panels and/or mini wind turbines.

Theory of knowledge 6.03.01

Environmental organisations are sometimes criticised for developing a stance on incomplete knowledge. Such organisations would respond by saying that information can be interpreted in different ways and that waiting for a more complete knowledge base may prevent solutions being put in place in sufficient time to solve a problem. However, critics respond by saying that environmental organisations sometimes distort the truth in order to gain support from politicians and the public.

1. How can we decide who is right in situations like this? Is it just a matter of weighing one kind of risk against another?
2. Can you think of a statement by an environmental organisation that has been heavily criticised for not having enough evidence to back it up?

In some urban areas, air quality is undoubtedly improving, but in many cities significant problems remain. The greatest problems are in the developing world with rapid rates of population growth and economic activity, and with limited resources to tackle air pollution and other environmental issues. Lax regulation is a major problem in many LEDCs. The LEDCs tend to have 'dirtier' cars, factories and power plants. The inhabitants of these cities often burn heavy, polluting fuel for heat and energy, which can produce heavy, thick smoke. In 2009 the Global Atmospheric Pollution Forum stated the need to develop and harmonise high quality air pollution databases across world regions, and to coordinate assessment and control strategies at the hemispheric scale.

Image 6.08 Citibikes in New York City, USA, encouraging people to cycle instead of drive (2014).

SELF-ASSESSMENT QUESTIONS 6.03.03

1. State two ways in which nitrogen oxide emissions can be reduced.
2. How can VOC emissions from motor vehicles be reduced?
3. Comment on the locations of the ten most air-polluted cities in the world.
4. **Discussion point:** What personal/household strategies have you and other people in your class used to reduce air pollution?

CASE STUDY 6.03.01

Air pollution in Delhi

India's capital city, Delhi, suffers from photochemical smog and a range of other types of air pollution. Photochemical smog is common in sunny, warm, dry cities such as Delhi with a large number of motor vehicles and industrial activities burning fossil fuels.

In 2014, the WHO stated that Delhi was the most air-polluted city in the world. The WHO ranking was based on the concentration of $PM_{2.5}$ particles in a study of 1600 cities across 91 countries. These very fine particles, less than 2.5 microns in diameter, are linked to increased rates of heart disease, lung cancer and chronic bronchitis. According to the WHO, Delhi had an average $PM_{2.5}$ level of 153, about ten times that of London, UK. The WHO says that air pollution is the fifth biggest cause of death in India, where respiratory health is deteriorating, especially for babies and infants. However, Indian and Delhi officials have disputed the WHO findings, feeling that the analysis has overestimated the problem in Delhi compared with in polluted cities in other countries, such as Beijing.

The National Capital Region of Delhi has grown rapidly in recent decades. Delhi now covers an area of approximately 900 km². The population of the National Capital Region, which was 1.7 million in 1951, was estimated at 17.8 million in 2014. It is expected to reach at least 22.5 million by 2025. Population growth is the result of natural increase and in-migration. An estimated 200 000–300 000 people arrive in Delhi each year from other parts of India in search of a better life. Such a fast rate of expansion has increased the environmental impacts of transportation (see Image 6.09), industrial activity, power generation, construction, domestic activities and waste generation. Thus Delhi's pollution problems have a wide variety of causes, as Table 6.08 shows:

- Transport is the main source of $PM_{2.5}$, NO_x and VOC emissions.
- Road dust is the main source of PM_{10} emissions.
- Power plants are responsible for half of sulfur dioxide (SO_2) emissions and almost 30 per cent of CO_2 emissions.

6.03 Photochemical smog

CASE STUDY 6.03.01 (continued)

Image 6.09 Advertisement at Dehli airport (2014) – reaching over 35 million passengers a year. The dramatic rise in air traffic is an increasing source of pollution.

Economic growth has created an increasing number of people in the middle-class income bracket who aspire to own their own car. However, not just private cars have significantly increased in number, but also taxis, auto rickshaws, buses and business vehicles of all kinds. There are now an estimated 7.2 million vehicles on Delhi's roads, about twice the number in 2000. Traffic speeds have fallen as the city has struggled to improve its road infrastructure to meet increasing demand. As traffic speeds have fallen, idling time has increased, resulting in higher levels of air pollution.

The inability of the electricity grid to meet spiralling demand has resulted in a proliferation of diesel generators that provide electricity often to entire neighbourhoods. Combustion from diesel is a significant source of ambient air pollution.

Pollution levels are particularly high during winter, due to:

- night-time heating needs: the poor in Delhi use open fires to keep warm in winter
- seasonal weather conditions which trap pollutants very close to the ground. The burning of post-harvest rice stalks from surrounding states is an additional reason for winter smog. Delhi also suffers from dust blown in from the deserts of the western state of Rajasthan.

	Emission / tonnes per year					
	$PM_{2.5}$	PM_{10}	SO_2	NO_X	CO	VOC
Transport	17750 (26%)	23800 (18%)	950 (3%)	329750 (67%)	421450 (28%)	208900 (63%)
Domestic	7300 (11%)	8800 (7%)	2050 (6%)	2350 (1%)	161200 (11%)	18300 (6%)
Diesel	3200 (5%)	4300 (3%)	1050 (3%)	81300 (17%)	85100 (6%)	31600 (9%)
Brick kilns	9250 (13%)	12400 (9%)	4000 (11%)	6750 (1%)	171850 (11%)	24200 (7%)
Industries	9000 (13%)	12650 (9%)	8500 (23%)	41500 (8%)	219600 (14%)	13250 (4%)
Construction	2450 (4%)	8050 (6%)	100 (1%)	2150 (1%)	2700 (1%)	50 (0.01%)
Waste burning	3850 (6%)	5450 (4%)	250 (1%)	1450 (1%)	20050 (1%)	1600 (0.5%)
Road dust	6300 (9%)	41750 (31%)	–	–	–	–
Power plant	10150 (15%)	16850 (13%)	20250 (55%)	27200 (6%)	442150 (29%)	34900 (10%)
Total	69050	133900	37000	492250	1524050	332700

Table 6.08 The sources of emissions in Delhi, India, 2010.
Source: http://urbanemissions.info/emissions-inventory-for-delhi.html

CASE STUDY 6.03.01 (continued)

Tackling the problem

In 2014, the Indian environment ministry launched a national air quality index that will rank 66 Indian cities. It will give real-time information on air quality, to put pressure on local authorities to take concrete steps to reduce pollution. The index will provide details of associated health risks in a colour-coded manner that can be understood by everyone. It is likely that this measure will significantly increase public awareness of the problem.

The relocation of some factories to the outskirts of the city in the early years of the new century temporarily lowered pollution; industry surrounding the city remains a significant source of pollution. Likewise, the benefits of the switch to compressed natural gas-based vehicles about a decade ago has diminished with the huge recent increase in traffic levels.

Investment in public transport has been slow to develop in Delhi compared with other cities of a comparable size. The Delhi Metro Rail only opened in December 2002, with an 8.3 km rail line, and has since been extended to 190 km with the completion of phase 2 in 2011. Construction of phase 3 is now underway, to add a further 103 km. Another significant development has been the Delhi bus rapid transit, which opened along a 5.6 km initial corridor in 2008.

Shutting coal power plants, promoting motor-less transport, and strict penalties for those violating pollution-control norms are among the suggestions that the government is looking at to improve the air quality in the city over the next five years.

Questions

1. Suggest why the ranking of cities according to air pollution can be contentious.
2. To what extent do the sources of air pollution in Delhi vary by pollutant?
3. What is being done to tackle the air pollution problem in Delhi?

6.04 Acid deposition

LEARNING OBJECTIVES

After reading this chapter you should be able to:

- understand that acid deposition can impact living systems and the built environment
- appreciate that the pollution management of acid deposition often involves cross-border issues.

KEY QUESTIONS

How are acidic deposits formed?
What are the effects of acid deposition?
What can be done to reduce acid deposition?

6.04.01 Acidified precipitations

Acid deposition is the mix of air pollutants that together lead to the acidification of freshwater bodies and soils. It can also impact heavily on the built environment.

Acid deposition refers to the mix of air pollutants that together lead to the acidification of freshwater bodies and soils:

- Dry deposition: the direct uptake by the ground of pollutants in the form of particles, aerosols and gases in the absence of precipitation.
- Wet deposition: acid rain, snow, fog and mist.
- Occult deposition: the direct impaction of contaminated cloud water on hills.

Acid deposition (see Image 6.10 and Figure 6.21) began entering the atmosphere in large amounts during the Industrial Revolution. It was first discovered by a Scottish chemist, Robert Angus Smith, in 1852 when he uncovered the relationship between acid rain and atmospheric pollution in Manchester, UK. However, the process did not gain significant public attention until the 1960s.

Extension

For more information on acid rain, visit the website below:
www.epa.gov/acidrain/what/

Acidity is measured on the pH scale, which ranges from 0 to 14 (see Figure 6.22). Pure water is neutral and has a pH of 7. However, natural rainwater is slightly acidic, mainly because of dissolved CO_2 which produces carbonic acid (H_2CO_3). Thus, the pH of unpolluted rainwater ranges from pH 5 to 6. Acid rain is generally viewed as rainwater with a pH of less than 5. In some parts of the northern hemisphere, the pH of the rainwater has been recorded as being as low as 2. In 1982, the pH of a fog on the west coast of the USA was measured at 1.8.

Image 6.10 Damage caused by acid rain to Reims Cathedral, France.

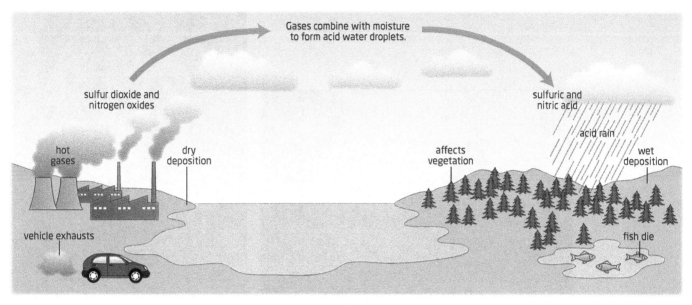

Figure 6.21 Acid deposition.

Rotting vegetation and erupting volcanoes release some chemicals that can cause acid deposition, but the vast majority of these chemicals result from human activity. Acid precipitation is mainly caused by the conversion of sulfur dioxide and NO_x, produced when fossil fuels are burned, into the sulfates and nitrates of dry deposition and the sulfuric acid and nitric acids of wet deposition. Wet deposition is formed when sulfur dioxide and NO_x are released into the atmosphere, where they can react with water, oxygen and other chemicals to form mild solutions of sulfuric acid and nitric acid. The longer that sulfur dioxide and NO_x stay in the atmosphere, the more chance they will be oxidised to sulfuric acid and nitric acid.

Coal-burning power plants are major producers of sulfur dioxide, but all processes that burn coal and oil contribute to emissions of this gas. A range of industries also produce significant amounts of sulfur dioxide. In contrast, NO_x emanates mainly from vehicle emissions. In the USA, roughly two-thirds of all SO_2 and a quarter of all NO_x come from fossil-fuel-burning power plants. Sulfur dioxide and NO_x go through several steps of chemical reactions before they become the acids found in acid rain. There are two phases: the gas phase and the aqueous phase.

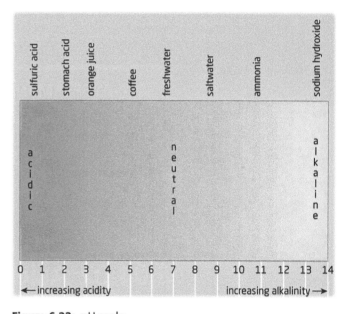

Figure 6.22: pH scale.

Dry deposition is logically more important in areas of low precipitation. When dry deposition falls to the surface, it can stick to the ground, buildings, cars and vegetation. It can be washed from these surfaces by precipitation later. About half of the acidity in the atmosphere falls back to Earth through dry deposition. Dry deposition generally occurs relatively close to emission sources and can cause significant damage to buildings and other structures over time. It can have a devastating effect on historic buildings (see Image 6.11). The rate of deterioration depends to a considerable extent on the type of building material in question. In contrast, wet deposition can be carried great distances, sometimes thousands of miles, and thus it is a significant factor in transboundary pollution.

6.04 Acid deposition

Image 6.11 Acid rain has significantly damaged the ancient monuments at Angkor Wat, Cambodia. Acid rain has darkened the stone, which eventually leads to its decay.

SELF-ASSESSMENT QUESTIONS 6.04.01

1. Distinguish between wet and dry deposition.
2. Which type of activity emits the largest amounts of sulfur dioxide?
3. Compare the acidity of pure water, natural rainwater and acid rain.
4. **Research idea:** For the region in which you live, try to identify a historic building that shows evidence of deterioration due to acid rain. Describe the effect that acid rain has had on this building.

6.04.02 Acid deposition effects on soil, water and living organisms

Acid deposition has a wide range of environmental implications, including:

- increasing the acidity of lakes, streams, wetlands, and other aquatic environments; this causes water bodies to absorb the aluminium that makes its way from soil into lakes and streams; this can make waters toxic to crayfish, clams, fish, and other aquatic animals
- damaging forests, especially those at higher elevations (trees' leaves and needles are harmed by acids)
- robbing the soil of essential nutrients and releasing aluminium in the soil, which reduces water uptake by trees
- reducing crop yields because of increased acidity

- contributing to visibility degradation and harming public health
- accelerating the decay of building materials and paints.

The adverse impact of acid rain can be viewed from a number of perspectives, for example by analysis of:

- direct effects
- toxic effects
- nutrient effects.

The direct effect of acid deposition on forests

Acid deposition does not usually kill trees quickly. Rather, it is more likely to weaken trees by damaging their leaves, limiting the nutrients available to them, or poisoning them with toxic substances slowly released from the soil (see Figure 6.23). Aluminium toxicity is the most widespread problem in acid soil.

Sulfur dioxide interferes with photosynthesis. When acid deposition is frequent, leaves tend to lose their protective waxy coating. Leaves and pine needles turn brown and fall off. Once trees are weak, they can be more easily attacked by diseases or insects that ultimately kill them. Weakened trees may also become injured more easily by cold weather. Coniferous trees are often most at risk from acid rain.

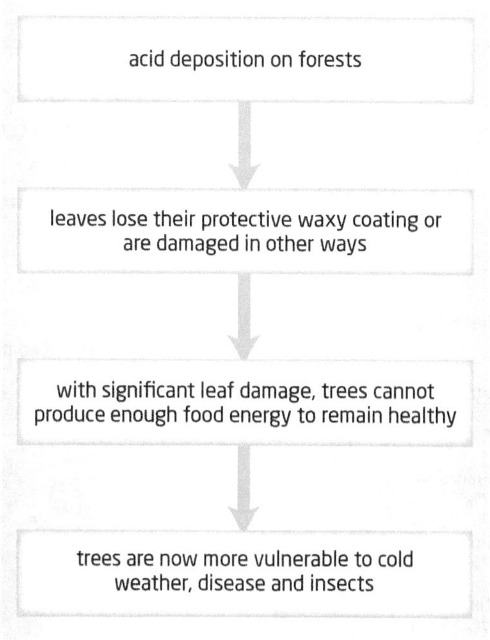

High aluminium concentrations in soil due to acid deposition can prevent the use of nutrients by plants. Acid rain which has seeped into the ground can poison vegetation with toxic substances that are slowly absorbed through the roots.

In the USA, areas that have been significantly affected by acid deposition include the high elevation forests of the Appalachian Mountains from Maine to Georgia. This includes areas such as the Shenandoah and Great Smoky Mountain National Parks. Many lakes and streams in these areas are no longer able to maintain fish populations. Some of the worst effects on forests have been recorded in Europe. It has been estimated that half of the forests in Germany and Poland are damaged, while 30 per cent in Switzerland have been affected.

```
acid deposition on forests
           ↓
leaves lose their protective waxy coating or
       are damaged in other ways
           ↓
with significant leaf damage, trees cannot
produce enough food energy to remain healthy
           ↓
trees are now more vulnerable to cold
   weather, disease and insects
```

Figure 6.23 The impact of acid deposition on forests.

The nutrient effect on soil

With acid deposition, the hydrogen ions in sulfuric acid trade places with the metal ions. The hydrogen ions are retained and neutralised by the soil, but the calcium, potassium and magnesium ions are leached or washed out of the topsoil into the lower, inaccessible subsoil. These ions are then not available as nutrients for vegetation growth. Such leaching occurs naturally, but acid deposition speeds up the process.

The toxic effect on fish

Acid deposition can enter water bodies in a number of different ways. It can interfere with the ability of fish to take in oxygen, salt and nutrients. For freshwater fish, maintaining osmoregulation is vital to stay alive. Osmoregulation is the ability to maintain a state of balance between salt and minerals in the organism's tissues. Acid molecules cause mucus to develop in the gills of fish, hindering the absorption of oxygen. Some fish are unable to maintain their calcium levels when the water they swim in becomes more acidic. This can result in reproduction problems. Spring is a vulnerable time for many species, as this is the time of the year for reproduction.

Species vary in their ability to tolerate acid. For example, while snails are sensitive to changes in pH, frogs may tolerate acidity.

6.04 Acid deposition

> **SELF-ASSESSMENT QUESTIONS 6.04.02**
>
> 1. State one way in which forests can be damaged by acid deposition.
> 2. How can acid deposition adversely affect soils?
> 3. Explain the toxic effect that acid deposition can have on fish.
> 4. **Research idea:** Find out what effects acid deposition has on crops and food.

6.04.03 The regional effect of acid deposition

In the Industrial Revolution, acid deposition generally occurred in urban areas close to the sources of pollution. Since the 1950s, much higher chimneys were constructed on many industrial sites to improve local air quality by dispersing pollution. The downside of this practice was that it polluted regions further away and sometimes sent pollution high into the atmosphere, where it could be transported long distances.

However, the effect of acid deposition is generally considered to be regional rather than global, because most acid pollution emitted in a particular world region is deposited in the same region – although exceptions do occur. Thus most acid deposition in North America can be traced to sources within that continent. This is in contrast to the problems of global warming and ozone depletion, which are considered to be global.

Ecosystems downwind of major industrial regions are most at risk of high levels of acid deposition. Prevailing winds in Europe generally carry pollution from west to east. For example, pollution from power plants and industrial regions in the UK can be carried by winds to Scandinavia and other parts of western Europe. Sweden and Norway attribute much of the sulfur dioxide they receive to other countries, primarily the UK, Germany and Poland. Figure 6.24 shows the situation for two pollutants and two selected countries, with Germany as an example of a net exporter of pollutants and Sweden as an example of a net importer of pollutants. For both countries, the fact is that most of the depositions of sulfur and NO_x come from outside their own territory. Another similarity is that an increasing share of the deposition originates from international shipping.

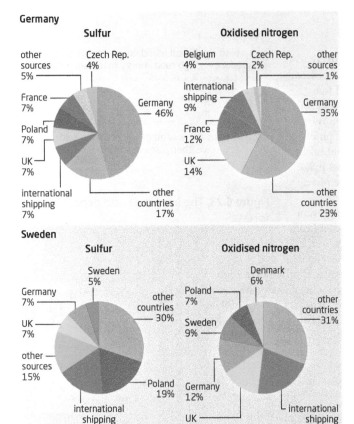

Figure 6.24 The origin of the deposition of sulfur and oxidised nitrogen over Germany and Sweden in 2003.

The effect of geology

Geology can have a considerable impact on the effect that acid deposition has on a region. Lime-rich soils and rocks have a much greater capacity than acidic rocks to absorb and neutralise acid deposition. Some rocks such as granite weather slowly and do not produce much in terms of neutralising chemicals, making them vulnerable to **acidification**.

Acidification is a change in the chemical composition of soil, which may trigger circulation of toxic metals.

Soil has the capacity to neutralise some or all of the acidity of acid rainwater. This ability is known as the **soil's buffering capacity**. Elements such as calcium, potassium and magnesium are base nutrients that help buffer acid inputs. Without such buffering capacity, soil pH can change rapidly. In soils that cannot cope with all the acid deposition they receive, high pH levels help accelerate soil weathering and remove nutrients. High pH also makes some toxic elements such as aluminium more soluble. Shallow soils are often more sensitive than deep soils to acidification.

In Canada, soils in the eastern part of the country tend to have lower buffering capacities than soils in western Canada. However, western regions where granite is the main rock type are at risk from acid rain, because granite cannot effectively neutralise acid. Eastern Canada receives higher levels of acid deposition than the western provinces, due to the concentration of manufacturing, the soil quality and easterly winds.

SELF-ASSESSMENT QUESTIONS 6.04.03

1. Where does most of the acid deposition in Norway and Sweden originate?
2. What is a soil's buffering capacity?
3. Why are areas of granite bedrock at risk from acid deposition?
4. **Research idea:** Use the internet to put together a series of images to illustrate the consequences of acid deposition in a country of your choice.

6.04.04 Pollution management strategies for acid deposition

The replace, regulate and restore model

A number of references to pollution management have already been made in previous sections of this book. These include mention of the 'replace, regulate and restore' model of pollution management, which can be applied to acid deposition.

Replace

Replacement strategies include:

- Reduce the demand for electricity.
- Change the energy mix of a country from fossil fuel dominance to greater use of renewable energy.
- Switch to low-sulfur fossil fuels while renewable energy is not an immediate prospect.
- Encourage people to switch from private cars to public transport.
- Reduce vehicle emissions.

The crux of these measures is to reduce reliance on polluting fossil fuels. This will take time and involve a high level of investment, which some countries are better able to afford than others. To make this even more difficult, the rapidly expanding economies of China and India are based firmly on the large coal reserves in both countries.

Regulate

Examples of regulation include clean-up measures at 'end-of-pipe' locations, which are a very important element of management strategies for acid deposition. Many governments require energy producers to clean smoke stacks by using scrubbers that trap pollutants before they are released into the atmosphere, and catalytic converters in cars to reduce their emissions. Complying with such regulations involves both individuals and organisations paying the costs of

6.04 Acid deposition

Theory of knowledge 6.04.01

Acid deposition is a classic example of transboundary pollution. Its effects on the environment can be extremely damaging and costly to repair.

1. As the scientific knowledge exists to trace the origins of much acid deposition, do you think that the polluting countries should compensate countries that are adversely affected by acid deposition?
2. How might the costs of acid deposition be assessed?

compliance. Opinions vary as to whether it is worth it, but surveys frequently show general public support for such actions.

Restore

The restore element of the model includes examples such as the liming of acid lakes. In the 1950s, it was discovered that fish were disappearing from lakes and waterways in southern Scandinavia. In 1976, the Swedish government initiated liming measures to restore acidified lakes and rivers. Finely ground limestone was added to water bodies to raise the pH and increase resistance to acidification.

Today, some 14 000 Swedish lakes are affected by acidification, with widespread damage to plant and animal life as a consequence. Around 7500 lakes and 11 000 km of waterways are now limed each year, as this has proved to be a cost-effective way of tackling the problem. The ground limestone is sometimes applied by helicopter if other means of access are a problem. This method of application adds to the cost. The application of ground limestone is used not only to restore acidified lakes but also to increase the resistance of lakes that are at risk but not yet affected.

The main problem is that liming has to be repeated on a regular basis to be effective.

CASE STUDY 6.04.01

Acid deposition in China's Pearl River Delta region

'Acid rain hangs over Guangdong province'
    Headline in *China Daily*, 27 August 2012

The Chinese economy has attained such a size and is growing so rapidly that it is now being called 'the new workshop of the world,' a phrase first applied to Britain during the height of its Industrial Revolution in the 19th century. However, in China's main industrial areas the environment is being put under a huge strain, leaving China with some of the worst pollution problems on the planet. One of China's main industrial regions is the Pearl River delta which is in Guangdong province (see Figure 6.25).

Acid deposition is a major problem in the Pearl River delta region. The region's manufacturing industries employ 30 million people, but this number will increase in the future. Major industrial centres include Shunde, Shenzhen, Dongguan, Zhongshan, Zhuhai and Guangzhou.

An environmental report published in late 2012 stated that an increasing part of Guangdong province had been affected by acid deposition in the first six months of the year. Five cities were found to be heavily polluted by acid rain, up from four in the same period the previous year. The high concentration of factories and power stations and the growing number of vehicles in the province is the source of the problem. The rapid increase in the number of motor vehicles has been of particular concern in recent years.

The 2012 report said most acid rain in Guangdong results from high concentrations of nitrogen oxide and sulfur dioxide. Apart from increasing emissions from vehicles, a few cities in Guangdong have been burning more coal to produce power, compensating for a decline in power transmission from western China. The persistence of acid rain in Shaoguan and Qingyuan results mainly from the various polluting industries, such

Figure 6.25 Location map of Pearl River delta.

as cement and ceramics, which have moved into these two cities in recent years.

Guangdong has formulated and issued a series of measures to combat regional acid deposition and other forms of air pollution. These measures include:

Power plants:
- Reducing the sulfur content in fuels
- Reducing reliance on fossil-fuel power plants

CASE STUDY 6.04.01 (continued)

Industrial sector:

- Forbidding new cement plants, ceramics factories and glassworks
- Installing particulate matter control devices for cement plants and industrial boilers
- Upgrading air pollutant emission standards for boilers.

Motor vehicles:

- Upgrading motor vehicle emission standards
- Improving the quality of vehicle fuel and promoting the use of unleaded petrol
- Encouraging green public transport
- Encouraging the use of electric taxis and hybrid buses.

Questions

1. With the aid of an atlas, describe the location of the Pearl River delta.
2. What are the causes of acid rain in the Pearl River delta?
3. Discuss the measures that have been used to tackle the acid rain problem.

6.04.05 International agreements

The *Convention on Long-Range Transboundary Air Pollution* (LRTAP) was signed in 1979 and was the first international agreement to recognise acid rain problems associated with the transboundary flow of air pollution. LRTAP emphasised the need for regional solutions. Cooperative action under LRTAP has contributed significantly to reducing transboundary flows of pollutants. Several binding protocols have been adopted under the convention, including the Gothenburg Protocol, which came into force in 2005.

Introduced across Europe on 1 January 2008, the Large Combustion Plant Directive (LCPD) aims at reducing sulfur emissions by giving coal-fired plants two options: they can either agree to a very limited running programme and close down by 2015, or install the equipment needed to remove sulfur from plant emissions.

As a result of the variety of measures undertaken to reduce acid deposition, levels have fallen significantly in many countries. Figure 6.26 shows the decline in total SO_2 emissions in the USA and Canada between 1980 and 2010. Many European countries show similar trends.

Deindustrialisation The shift of manufacturing from MEDCs to lower-cost NICs and LEDCs.

Deindustrialisation

In some countries, market forces have had a major effect on acid deposition. In the process of **deindustrialisation**, MEDCs have seen the movement of many of their heavy and polluting industries to newly industrialised countries. While MEDCs have lost the advantage of the investment and employment of such industries, they have also lost the disadvantage of the pollution. In contrast, acid rain has emerged as a major problem in LEDCs in recent decades.

Research

Our understanding of the effects of acid deposition on ecosystems has improved in recent decades, but it is still incomplete (see Figure 6.27). Further research should enable better management strategies to be developed.

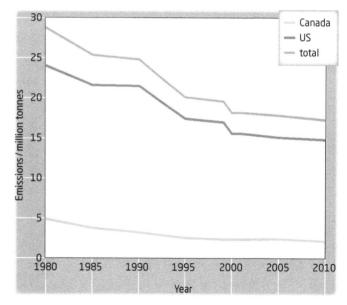

Figure 6.26 Canada and USA – total SO_2 emissions, 1980–2010.

6.04 Acid deposition

> Our understanding of atmospheric pollution and its effects on the environment has increased over the last decade and has provided scientists with fresh perspectives and policy makers with new approaches for dealing with the problem. However, ecologists are still trying to fully understand the long-term effects of acid deposition on ecosystems. This requires long-term studies that include an array of ecosystem types over decades rather than years. For instance, there is still much to learn about the impacts of acid deposition on soils, surface-water chemistry and biology, forest, and ecological response and recovery. The role of nitrogen and its cycling through ecosystems is also a significant area of current research.
>
> Long-term monitoring is a critical component of scientific research. This type of data contributes important information to scientists who study acid deposition. Chemical and biological monitoring enable researchers to evaluate the effectiveness of policies intended to reduce air pollutants as well as to test scientific models of ecosystem responses to acid deposition.
>
> The Ecological Society of America

Figure 6.27 What are scientists doing to better understand acid deposition?

CONSIDER THIS
The Gravestone Project run by the Geological Society of America is using gravestones to better understand how the elements, particularly acid rain, are weathering rocks around the world, and how this has changed over time. The date of death on gravestones generally provides accurate evidence about how long the stone has been open to the elements.

SELF-ASSESSMENT QUESTIONS 6.04.04

1. Name and date the first international agreement to recognise acid rain problems associated with the transboundary flow of air pollution.
2. Describe the elements of the Large Combustion Plant Directive (LCPD).
3. **Discussion point:** Do the advantages of deindustrialisation in terms of reduced pollution outweigh the economic disadvantages?

End-of-topic questions

1. Discuss the role of the albedo effect from clouds in regulating global average temperature. [4]

2. Look at Figure 6.1, which shows changes in ozone concentration in the troposphere and stratosphere.
   a. Describe the changes in ozone concentration shown by the diagram. [2]
   b. Why is the ozone layer in the stratosphere often referred to as 'good ozone'? [1]
   c. How has human activity depleted the ozone layer? [3]
   d. How effective have international agreements been in reducing the emissions of ozone-depleting substances? [4]

3. Apart from the release of large quantities of air pollutants, what other factors contribute to the formation of photochemical smog? [4]

4. a. What is acid deposition? [1]
   b. What impact does acid deposition have on forests? [3]

Topic 7
CLIMATE CHANGE AND ENERGY PRODUCTION

7.01 Energy choices and security

LEARNING OBJECTIVES

After reading this chapter you should be able to:

- understand that there are a range of different energy sources available to societies that vary in their sustainability, cost and socioeconomic implications
- appreciate that the choice of energy sources is controversial and complex, and that energy security is an important factor in making energy choices.

KEY QUESTIONS

Why do fossil fuels still contribute to the majority of humankind's energy supply?

Which sources of renewable energy make the greatest contribution to global energy supply?

How can improvements in energy efficiencies and energy conservation limit growth in energy demand?

7.01.01 Energy resources

Energy security is the uninterrupted availability of energy sources at an affordable price.

Energy poverty is a lack of access to modern energy services – a household's access to electricity and clean cooking facilities (e.g. to fuels and stoves that do not cause air pollution in houses).

Fuel poverty is when a low-income household is living in a home that cannot be kept warm at a reasonable cost.

Non-renewable energy sources are finite, so as they are used up the supply that remains is reduced.

Renewable energy sources can be used over and over again. These resources are mainly forces of nature that are sustainable and usually cause little or no environmental pollution.

Energy mix is the relative contribution of different energy sources to a country's energy production/consumption.

Energy resources are at the core of the global economy. Energy is vital for economic growth and development. However, emissions of greenhouse gases from energy use are the main contributors to human-induced climate change. Energy production and consumption are major issues both within and between countries. Many countries that have to import a significant amount of the energy they use have become more and more concerned about **energy security**. There are growing concerns in some parts of the world that inequitable availability and uneven distribution of energy sources may lead to conflict.

Energy shortages have occurred in different parts of the world on a number of occasions. Such shortages can have major economic and social consequences. But potential shortages of energy are not the only concern. In poor countries, **energy poverty** has a major impact on people's lives and is a major obstacle to development. The concept of **fuel poverty** is becoming an increasingly important issue in many apparently affluent countries as people struggle to pay rising energy bills in harsh winters.

Renewable and non-renewable sources of energy

Non-renewable energy sources are the fossil fuels (coal, oil and natural gas) and nuclear fuel. These energy sources are finite, so as they are used up the supply that remains is reduced. Eventually, these non-renewable resources could become completely exhausted. **Renewable energy sources** can be used over and over again. These resources are mainly forces of nature that are sustainable and which usually cause little or no environmental pollution (see Image 7.01). Renewable energy includes hydroelectric, biomass, wind, solar, geothermal, tidal and wave power. At present, non-renewable resources dominate global energy. The challenge is to transform the global **energy mix** to achieve a better balance between renewables and non-renewables.

The domination of fossil fuels

The demand for energy has increased steadily over time. Figure 7.01 shows a global increase of over 60 per cent between 1987 and 2012. Fossil fuels dominate the global energy situation. Their relative contribution to total world energy consumption in 2012 was: oil – 33 per cent, coal – 30 per cent, natural gas – 24 per cent. In contrast, hydroelectricity accounted for 6.6 per cent, and nuclear energy 4.5 per cent. Renewable energy made up only 1.3 per cent of total global energy supply. Figure 7.01 includes commercially traded energy sources only. It excludes fuels such as wood, peat and animal waste, which, although important in many countries, are unreliably documented in terms of production and consumption statistics.

Image 7.01 Renewable energy: these solar panels are in northern Spain.

Fossil fuels vary widely in the impacts of their production and their emissions. In terms of production, coal has by far the greatest environmental impact in general, but large individual oil spills, such as the Gulf of Mexico/BP oil spill in 2010, have had devastating consequences in different parts of the world. Coal is also the dirtiest fossil fuel in terms of emissions. Natural gas is regarded as the fossil fuel that has least impact on the environment.

Figure 7.02 shows the regional pattern of global energy consumption for 2012. Consumption by type of fuel varies widely by world region.

- *Oil:* Only in Asia Pacific is the contribution of oil less than 30 per cent, and it is the main source of energy in four of the six regions shown in Figure 7.02. In the Middle East it accounts for almost 50 per cent of consumption.
- *Coal:* Only in the Asia Pacific region is coal the main source of energy. In contrast, it accounts for less than 5 per cent of consumption in the Middle East and South & Central America. China was responsible for 50.2 per cent of global coal consumption in 2012.

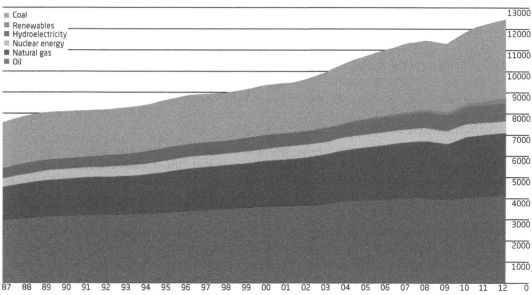

World primary energy consumption grew by a below-average 1.8% in 2012. Growth was below average in all regions except Africa. Oil remains the world's leading fuel, accounting for 33.1% of global energy consumption, but this figure is the lowest share on record and oil has lost market share for 13 years in a row. Hydroelectric output and other renewables in power generation both reached record shares of global primary energy consumption (6.7% and 1.9%, respectively).

Figure 7.01 Changes in world energy consumption by type, 1987–2012.

7.01 Energy choices and security

Regional consumption pattern 2012
Percentage

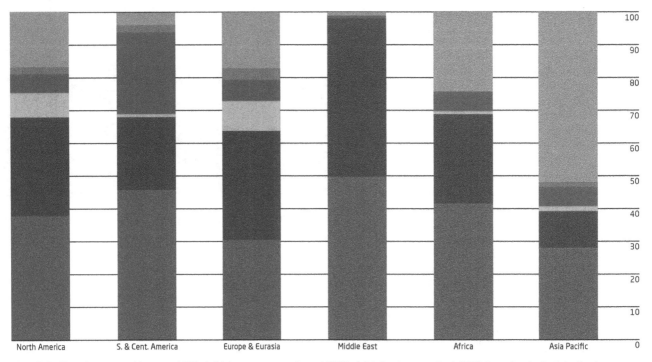

The Asia Pacific region accounted for a record 40% of global energy consumption and 69.9% of global coal consumption in 2012; the region also leads in oil and hydroelectric generation. Europe & Eurasia is the leading region for consumption of natural gas, nuclear power, and renewables. Coal is the dominant fuel in the Asia Pacific region, the only region dependent on a single fuel for more than 50% of total primary energy consumption. Natural gas is dominant in Europe & Eurasia, and oil is dominant in other regions.

Figure 7.02 Divided bar graphs showing regional energy consumption patterns, 2012.

- *Natural gas:* Natural gas is the main source of energy in Europe & Eurasia, and it is a close second to oil in the Middle East. Its lowest share of the energy mix is 11 per cent in Asia Pacific.
- *Hydroelectricity:* The relative importance of hydroelectricity is greatest in South & Central America (25 per cent). Elsewhere its contribution varies from 6 per cent in Africa to less than 1 per cent in the Middle East.
- *Nuclear energy:* Nuclear energy is not presently available in the Middle East, and it makes the smallest contribution of the five traditional energy sources in Asia Pacific, Africa and South & Central America. It is more important in Europe & Eurasia and North America than in other regions.
- *Renewables:* Consumption of renewable energy other than hydroelectricity is rising rapidly, but from a very low base. Renewable energy made the largest relative contribution to energy consumption in Europe & Eurasia.

Figure 7.03 shows per-capita energy consumption around the world. The highest-consumption countries such as the USA and Canada use more than 6 tonnes of oil equivalent per person, while almost all of Africa and much of South America and Asia use less than 1.5 tonnes of oil equivalent per person.

In terms of usage by type of energy, some general points can be made.

- The most developed countries tend to use a wide mix of energy sources, being able to both invest in domestic energy potential and buy energy from abroad.
- The high investment required for nuclear electricity means that only a limited number of countries produce electricity this way. However, many countries that could afford the investment choose not to adopt this strategy.
- Richer nations have been able to invest more money in renewable sources of energy.
- In the poorest countries, **fuelwood** is an important source of energy, particularly where communities have no access to electricity.

Fuelwood is wood and charcoal used to supply energy.

Figure 7.03 World map showing energy consumption per capita in tonnes oil equivalent, 2012.

Fuelwood in developing countries

In developing countries, about 2.5 billion people rely on fuelwood, charcoal and animal dung for cooking. Wood and charcoal used for fuel are collectively called fuelwood and account for just over half of global wood production. Fuelwood provides much of the energy needed for sub-Saharan Africa. Fuel is also the most important use of wood in Asia. In 2010, 1.2 billion people were still living without electricity. Figure 7.04 shows the countries with the greatest number of people lacking access to electricity. Although at least one study claims that the global demand for fuelwood peaked in the mid-1990s, there can be no doubt that there are severe shortages of electricity in many countries. This is a major factor in limiting development.

Income, regional electrification and household size are the main factors impacting on the demand for fuelwood. Thus, forest depletion is initially heavy near urban areas but slows down as cities become wealthier and change to other forms of energy. The more isolated rural areas are most likely to lack a connection to an electricity grid. In such areas, reliance on fuelwood is greatest. Wood is likely to remain the main source of fuel for the global poor in the foreseeable future. The collection of fuelwood does not cause deforestation on the same scale as the clearance of land for agriculture, but it can seriously deplete wooded areas. The use of fuelwood is the main cause of indoor air pollution in developing countries.

In developing countries, the concept of the '**energy ladder**' is important. Here, a transition from fuelwood and animal dung to higher-level sources of energy occurs as part of the process of economic development.

Figure 7.04 Bar graph showing electricity access deficit, 2010.

Millions of people:
- India 306.2
- Nigeria 82.4
- Bangladesh 66.6
- Ethiopia 62.9
- Dem. Rep. Congo 56.9
- Tanzania 38.2
- Kenya 31.2
- Sudan 30.9
- Uganda 28.5
- Myanmar 24.6
- Mozambique 19.9
- Afghanistan 18.5
- South Korea 18.0
- Madagascar 17.8
- Philippines 15.6
- Pakistan 15.0
- Burkina Faso 14.3
- Niger 14.1
- Indonesia 14.0
- Malawi 13.6

> The **energy ladder** is the improvement in energy use due to rising household incomes. As incomes rise, the energy types used by households become cleaner and more efficient, but also more expensive as households move from traditional biomass to electricity.

7.01.02 Renewable energy

Hydroelectricity dominates renewable energy production. The newer sources of renewable energy that make the largest contribution to global energy supply are wind power and biofuels. However, investment in solar energy is now increasing significantly.

- *Hydroelectricity:* The big-four hydroelectric power nations of China (23.4 per cent), Brazil (11.4 per cent), Canada (10.4 per cent) and the USA (7.6 per cent) account for almost 53 per cent of the global total. Most of the best hydroelectric power locations are already in use, so the scope for more large-scale development is limited. However, in many countries there is scope for small-scale hydroelectric power plants to supply local communities.

> **CONSIDER THIS**
>
> The individual countries consuming the most energy in 2012, as a percentage of the world total, were: China (21.9 per cent), the USA (17.7 per cent), Russia (5.6 per cent), India (4.8 per cent) and Japan (3.8 per cent).

7.01 Energy choices and security

- *Wind energy:* The worldwide capacity of wind energy reached almost 240 000 MW in early 2012. Global wind energy is dominated by a relatively small number of countries. The leaders in global wind energy are China, the USA, Germany and Spain. Together, these countries account for over 67 per cent of the world total. In recent years, wind energy has reached the 'take-off' stage (a critical, self-sustaining stage of development), both as a source of energy and a manufacturing industry.
- *Biofuels:* These are fossil fuel substitutes that can be made from a range of agri-crop materials including oilseeds, wheat, corn and sugar. They can be blended with petrol and diesel. The biggest producers of biofuels are the USA, Brazil and China. By increasing biofuel production, these countries have reduced the amount of oil they need to consume, which is the main reason behind biofuel production. In recent years, increasing amounts of cropland have been used to produce biofuels. Initially, environmental groups such as Friends of the Earth and Greenpeace were in favour of biofuels but, as the damaging environmental consequences have become clear, such environmental organisations have been the first to demand a rethink of this energy strategy.
- *Solar power:* From a relatively small base, the installed capacity of solar electricity is growing rapidly. In 2012, global solar power capacity passed 100 000 MW. This amounts to about 0.4 per cent of all global electricity generation. Experts say that solar power has huge potential for technological improvement, which could make it a major source of global electricity in years to come. Germany, Italy, the USA, China and Japan are the leading generators of solar electricity. Solar electricity is currently produced in two ways: photovoltaic systems and concentrated solar power systems.
- *Geothermal energy:* This is the natural heat found in the Earth's crust in the form of steam, hot water and hot rock. Rainwater may percolate several kilometres in permeable rocks, where it is heated due to the Earth's **geothermal gradient** (the rate at which temperature rises as depth below the surface increases). The average rise in temperature is about 30 °C per km, but the gradient can reach 80 °C near plate boundaries. The USA is the world leader in geothermal electricity, but this amounts to only 0.4 per cent of US electricity production. Other countries using geothermal electricity include the Philippines, Italy, Mexico, Indonesia, Japan, New Zealand and Iceland.
- *Tidal power:* Tidal power plants act like underwater windmills, transforming sea currents into electrical current. Tidal power is more predictable than solar or wind, and the infrastructure is less obtrusive, but start-up costs are high. The 240 MW Rance facility in north-western France is the only utility-scale tidal power system in the world. However, the greatest potential is in Canada's Bay of Fundy in Nova Scotia.
- *Wave energy:* This is where generators are placed on the ocean's surface. Energy levels are determined by the strength of the waves. This form of energy, like tidal power, is still in the research stage.

> **Geothermal gradient** is the rate at which temperature rises as depth below the surface increases.

SELF-ASSESSMENT QUESTIONS 7.01.01

1. List the non-renewable sources of energy.
2. What is renewable energy and what are its main sources?
3. How was global energy demand satisfied in 2012?
4. a What is fuelwood?
   b Why is fuelwood such an important source of energy in the developing world?
5. **Research idea:** Look at the latest BP Statistical Review of World Energy (www.bp.com/statisticalreview). Use the energy charting tool to create custom charts for your own analysis of the global energy situation.

7.01.03 Oil: advantages and disadvantages

Oil is the most important of the non-renewable sources of energy. Even though investment in new sources of energy is increasing rapidly, the global economy still relies on oil to a considerable extent. Oil clearly has significant advantages as a source of energy, otherwise it would not have attained the importance it has today. However, its disadvantages have gained increasing recognition in recent decades (see Table 7.01).

Advantages	Disadvantages
A compact, portable source of energy, relatively easy to transport and store.	Non-renewable – takes millions of years to form.
Used for most forms of mechanical transportation.	Burning oil generates CO_2, a greenhouse gas.
Flexible use – can be distilled into different fuel products.	Oil contains sulfur, which, when burnt, forms sulfur dioxide and sulfur trioxide. These combine with atmospheric moisture to form sulfuric acid, leading to acid rain.
Cleaner and easier to burn than coal.	Not as clean or efficient in use as natural gas.
Compared with most other fuel sources it remains one of the most economical sources of energy.	Serious oil spills have occurred from super-tankers and pipelines.
The oil industry has been the source of much advanced technology.	Locating additional reserves requires a high level of investment.
Oil refining produces the world's supply of elemental sulfur as a by-product, used for many industrial applications.	Political instability of some major oil-producing countries and concern about the vulnerability of energy pathways.
Has a well-established global infrastructure.	There are concerns that 'peak oil' (the highest level of oil production) is not far away.
	The price of oil has risen significantly over the past decade.
	Some oil is now being strip-mined in the form of tar sands, which raises serious environmental concerns.

Table 7.01 The advantages and disadvantages of oil.

Rapidly rising demand

Figure 7.05 shows the change in daily oil consumption by world region from 1987 to 2012. From less than 60 million barrels daily in the mid-1980s, global demand rose steeply to 89.8 million barrels a day in 2012. Satisfying such a rapid rate of increase in demand requires a high level of investment and exploration, with environmental and other consequences. The largest increase has been in the Asia Pacific region, which now accounts for 33.6 per cent of consumption (see Image 7.02). This region now uses more oil than North America, which accounts for 24.6 per cent of the world total. In contrast, Africa consumed only 4.0 per cent of global oil.

In 2012, the Middle East accounted for over 48 per cent of global proved reserves. Political instability in the Middle East is a major concern to the countries that import oil from this region. Table 7.02 shows the **reserves-to-production ratio** for the world in 2012. While the reserves-to-production ratio is over 78 years in the Middle East, and even higher in South & Central America, it is only 13.6 years in the Asia Pacific.

The **reserves-to-production ratio** is the reserves remaining at the end of any year divided by the production in that year. The result is the length of time that those remaining reserves would last if production were to continue at that level.

7.01 Energy choices and security

Consumption by region
Million barrels daily

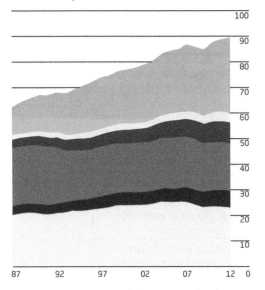

Figure 7.05 Line graph showing oil consumption by world region, 1987–2012.

Image 7.02 Fuel station on the Mekong River, Vietnam.

Region	Reserves-to-production ratio / years
North America	38.7
South & Central America	123.0
Europe & Eurasia	22.4
Middle East	78.1
Africa	37.7
Asia Pacific	13.6
World	52.9

Table 7.02 Oil reserves-to-production ratio at the end of 2012.
Source: Data from *BP Statistical Review of World Energy* 2013

The price of oil increased sharply in the early years of the new century, causing major financial problems in many importing countries. It rose from $10 a barrel in 1998 to more than $130 a barrel in 2008, before falling back sharply in the global recession of 2008–2009. As the global economy has slowly recovered, the price of oil has fluctuated considerably and was at about $50 a barrel in early 2015.

The main reasons for the rapid rise in the price of oil in the mid-2000s were:

- The significant increase in the demand for oil, the largest for almost 30 years. Very high growth in demand in China, India, the USA and some other economies has impacted heavily on the rest of the world. Many emerging economies, such as India and China, subsidise oil, which encourages consumption.
- Insufficient investment in exploration and development over the previous two decades. Low oil prices for most of this period did not provide enough incentive for investment.
- Problems in the Middle East, centred on Iraq. Exports of oil from Iraq are well below potential because of terrorist attacks and the slow pace of reconstruction after the war.
- Major buyers, particularly governments, stocking up on oil to guard against disruptions to supply.
- The impact of hurricanes, particularly on US oil production in the Gulf of Mexico.

- Limited US refining capacity due to inadequate investment in recent decades. US refineries are ageing and thus require more maintenance, which in turn reduces capacity.
- A lack of spare oil-production capacity. In the past, Saudi Arabia has maintained a significant amount of spare capacity (wells it was not pumping oil from) to prevent global supply problems when supplies were interrupted in other producing countries. This spare capacity has declined to a 20-year low.

When will global peak oil production occur?

There has been growing concern about when global oil production will peak and how fast it will decline thereafter. There are concerns that not enough large-scale projects are underway to offset declining production in well-established oil-production areas. The rate of major new oil-field discoveries has fallen sharply in recent years. It takes six years on average from first discovery for a large-scale project to start producing oil. In 2010, the International Energy Agency expected **peak oil production** somewhere between 2013 and 2037, with a fall by 3 per cent a year after the peak. The US Geological Survey has predicted that the peak is 50 years or more away.

However, in total contrast, the Association for the Study of Peak Oil and Gas (ASPO) predicted in 2008 that the peak of global oil production could come as early as 2011, stating, 'Fifty years ago the world was consuming 4 billion barrels of oil per year and the average discovery was around 30 billion. Today we consume 30 billion barrels per year and the discovery rate is now approaching 4 billion barrels of crude oil per year.' ASPO's dire warnings have not (yet) materialised. This is at least partly down to new developments, particularly the rapid growth in the production of shale oil and gas in the USA, which has changed the global energy situation. The current period of slow growth in the global economy has also eased the pressure on energy resources.

Shale oil

The exploitation of **shale oil** is a recent development. It has been concentrated mainly in the USA. The rapid increase in the scale of production in the USA has fundamentally changed global energy markets, as the USA has quickly regained much of its self-sufficiency in energy. Gas can also be obtained from shale, and this has led the exploitation of shale oil by a few years. The exploitation of shale deposits has had a massive impact on US gas and oil production in recent years. Further significant production increases of shale oil and gas are forecast for the USA, and the 'shale revolution' is likely to spread to other parts of the world, albeit with a time lag. The speed of this geographical spread (diffusion) will depend on a number of factors, including the extent of opposition to this process on environmental grounds.

Although the basic technology was originally developed in the USA in the 1940s, the recent 'shale revolution' is the result of technological breakthroughs in horizontal drilling and hydraulic fracturing (fracking), which have made shale deposits economically viable. In the USA, oil production peaked in 1970 at 534 million tonnes, falling to 305 million tonnes in 2008. The subsequent rise to 499 million tonnes in 2013 has been a phenomenal turnaround. Most of this increase has been due to the rapid rise in the production of shale oil. The full extent of recoverable shale oil in the USA is still to be determined, as exploration continues. A recent analysis by the EIA put the total at nearly 8 billion tonnes, a considerable energy resource base indeed!

The geopolitical impact of changes in patterns and trends in oil

Energy security depends on resource availability, domestic and foreign, and security of supply. It can be affected by **geopolitics** and is a key issue for many economies. Because there is little excess capacity to ease pressure on energy resources, energy insecurity is rising, particularly for non-renewable resources.

Peak oil production is the year in which the world or an individual oil-producing country reaches its highest level of production, with production declining thereafter.

Theory of knowledge 7.01.01

In predicting what will happen in the future, reputable organisations often arrive at scenarios that are significantly different. Prediction of 'peak oil and gas' is a case in point. How can you judge which prediction is the most accurate? It is a very difficult task. For some forecasts, there may be general agreement because few factors are involved and good quality information is available. For other forecasts, the information available may be far from perfect and there are many factors influencing what can happen.

1. How can we best deal with uncertainty in areas of such huge importance?
2. Some scientists argue that peak oil will be here soon, while others argue that we have enough oil for a long time to come. Who has the more accurate scientific claim?

Shale oil is oil extracted from reserves, sometimes described as 'tight oil' reserves, held in shales and other rock formations from which it will not naturally flow freely. Shale oil has become more accessible due to advances in technology.

Geopolitics refers to political relations among nations, particularly relating to claims and disputes pertaining to borders, territories and resources.

7.01 Energy choices and security

Strategic petroleum reserves are large reserves of oil held by countries including the USA and China to tide them over for a few months or so if normal oil supplies are disrupted.

Energy pathways are supply routes between energy producers and consumers, which may be pipelines, shipping routes or electricity cables.

The USA, gravely concerned about the political leverage associated with imported oil, began in 1977 the construction of a **strategic petroleum reserve**. The oil was to be stored in a string of salt domes and abandoned salt mines in southern Louisiana and Texas which could be easily linked up to pipelines and shipping routes. The initial aim was to store 1 billion barrels of oil, which could be used in the event of supply discontinuation. The USA's strategic petroleum reserve currently holds 700 million barrels.

The Middle East is the major global focal point of oil exports. The long-running tensions in the Middle East have caused serious concerns about the vulnerability of oilfields, pipelines and oil tanker routes. The destruction of oil wells and pipelines during the Iraq War showed clearly how energy supplies can be disrupted. Middle East oil exports are vital for the functioning of the global economy. Most Middle East oil exports go by tanker through the Strait of Hormuz, a relatively narrow body of water between the Persian Gulf and the Gulf of Oman. The strait at its narrowest is 55 km wide. Roughly 30 per cent of the world's oil supply passes through the strait, making it one of the world's strategically important chokepoints. Iran has indicted that it could block this vital shipping route in the event of serious political tension. This could cause huge supply problems for many importing countries.

Concerns about other key **energy pathways** have arisen from time to time. These include:

- *Nigeria:* At times, local rebel groups have attacked the oil installations in the Niger delta, either out of frustration at the paucity of benefits accruing to the region or in an attempt to gain pay-outs.
- *Venezuela:* The left-wing government has been hostile to the US government. The USA has voiced concern about its reliance on oil supplies from Venezuela. Venezuela is the USA's fourth largest oil supplier, and the USA is the number one buyer of Venezuelan oil.
- *China:* Chinese politicians have expressed concern about the country's energy security situation. Four-fifths of China's oil imports pass through the Strait of Malacca, the busiest shipping lane in the world. China is looking to bypass the Strait of Malacca to a certain extent through the construction of new pipeline systems through neighbouring countries such as Pakistan and Burma.

Oil sands in Canada and Venezuela

Huge oil (tar) sand deposits in Alberta, Canada, and in Venezuela could be critical over the next 50 years as the world's production of conventional oil falls. Such synthetic oil, which can also be made from coal and natural gas, could provide a vital bridge to an era of new technologies.

However, there are serious environmental concerns about the development of oil sands:

- It takes almost 2 tonnes of mined sand to produce one barrel of synthetic crude, leaving lots of waste sand.
- It takes about three times as much energy to produce a barrel of Alberta-oil-sands crude as it does a conventional barrel of oil. Thus oil sands are large sources of greenhouse gas emissions.

Extension

Problems arising from tar sands can be viewed by visiting the website below:
www.foe.org/projects/climate-and-energy/tar-sands

7.01.04 The pros and cons of wind power

Wind power is arguably the most important of the new renewable sources of energy (see Image 7.03). The worldwide capacity of wind energy is approaching 400 000 MW, a significant production mark (see Figure 7.06). The wind industry set a new record for annual installations in 2014. Global wind energy is dominated by a relatively small number of countries. China is currently the world leader, with 31 per cent of global capacity, followed by the USA, Germany,

Spain and India. Together, these five countries account for almost 72 per cent of the global total. In the last five years, for the first time ever, more new wind power capacity was installed in developing and newly industrialised countries than in the developed world.

Wind energy has reached the 'take-off' stage, both as a source of energy and as a manufacturing industry. As the cost of wind energy improves further against conventional energy sources, more and more countries will expand into this sector. However, projections regarding the industry still vary considerably because of the number of variables that will impact on its future.

Costs of generating electricity from wind today are only about 10 per cent of what they were 20 years ago, due mainly to advances in turbine technology. Thus at well-chosen locations, wind power can now compete with conventional sources of energy. Wind energy operators argue that costs should continue to fall due to further technological advances and to increasing economies of scale. One large turbine manufacturer has stated that it expects turbine costs to be reduced by 3.5 per cent a year for the foreseeable future.

Table 7.03 summarises the advantages and disadvantages of wind power.

Public finance continues to play a strong role in the economics of the industry, and this is likely to continue in the foreseeable future, particularly in the light of the current global financial situation and the fragility of commercial banks.

Apart from establishing new wind energy sites, **repowering** is also beginning to play an important role. Repowering means replacing first-generation wind turbines with modern multi-megawatt turbines which give a much better performance. The advantages are:

- more wind power from the same area of land
- fewer wind turbines
- higher efficiency, lower costs
- enhanced appearance, as modern turbines rotate at a lower speed and are usually more visually pleasing due to enhanced design
- better grid integration, as modern turbines use a connection method similar to conventional power plants.

As wind turbines have been erected in more areas of more countries, the opposition to this form of renewable energy has increased, for the reasons listed in Table 7.03.

The recent rapid increase in demand for turbines has resulted in a shortage of supply. New projects now have to make orders for turbines in large blocks up to several years in advance to ensure firm delivery dates from manufacturers. Likewise, the investment from manufacturers is having to rise significantly to keep pace with such buoyant demand.

New developments in wind energy include:

- In 2008, a Dutch company installed the world's first floating wind turbine off the southern coast of Italy in water 110 m deep. The technology is known as the submerged deepwater platform system.
- The Swedish company Nordic has recently brought a two-bladed turbine onto the market.

Image 7.03 Wind farm in Spain.

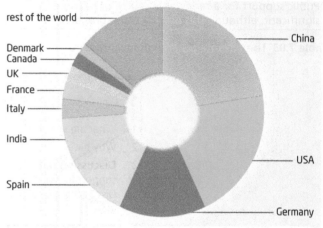

Country	MW	%
China	44 733	22.7
USA	40 180	20.4
Germany	27 214	13.8
Spain	20 676	10.5
India	13 065	6.6
Italy	5 797	2.9
France	5 660	2.9
UK	5 204	2.6
Canada	4 009	2.0
Denmark	3 752	1.9
rest of the world	26 749	13.6
Total TOP 10	**170 290**	**86.4**
World total	**197 039**	**100**

Figure 7.06 Global wind power capacity, end of 2014. (Percentage figures are rounded to the nearest 0.1%.)

Repowering is the replacing of the initial turbines with modern turbines that produce more electricity.

7.01 Energy choices and security

Advantages	Disadvantages
Wind power is a renewable source of energy that can produce reasonable levels of electricity with current technology.	Concerns are growing about the impact on landscapes as the number of turbines and wind farms increases.
Advances in wind turbine technology over the last decade have reduced the cost per unit of energy considerably.	Not in my back yard (NIMBY) protests have been staged by people concerned about the impact of local turbines adversely affecting the value of their properties.
Suitable locations with sufficient wind conditions can be found in most countries.	The hum of turbines can be disturbing for both people and wildlife.
Wind energy has reached the take-off stage both as a source of energy and as a manufacturing industry.	There is debate about the number of birds killed by turbine blades.
There is flexibility of location, with offshore wind farms gaining in popularity.	TV reception can be affected by wind farms.
Repowering can increase the capacity of existing wind farms.	The development of wind energy has required significant government subsidies. Some people argue that this money could have been better spent elsewhere (i.e. there is an opportunity cost).
Public support for a renewable source of power is significant, although this may be waning to an extent.	Many wind farms are sited in coastal locations, where land is often very expensive.

Table 7.03 The advantages and disadvantages of wind power.

SELF-ASSESSMENT QUESTIONS 7.01.02

1. What are energy pathways and why are they important?
2. Why have some countries developed strategic petroleum reserves?
3. **Discussion point:** Until relatively recently there was an extremely high level of public support for wind power. While wind power is still popular, more people are now critical. Discuss the reasons for such criticism.

7.01.05 The nuclear power debate

Until a few years ago the future of nuclear power looked bleak, with a number of countries apparently running down their nuclear power stations. However, renewed fears about energy security have brought this controversial source of power back onto the global energy agenda. Figure 7.07 shows the top ten nuclear-generating countries in 2013.

No other source of energy creates such intense discussion as nuclear power. Concerns include:

- The possibility of power plant accidents, which could release radiation into air, land and sea.
- Radioactive waste storage/disposal – most concern is over the small proportion of 'high-level waste'. High-level waste is so radioactive that it generates heat and corrodes all containers. It would cause death within a few days to anyone directly exposed to it. In the UK, high-level waste amounts to about 0.3% of the total volume of nuclear waste, but about half of the total radioactivity. No country has yet implemented a long-term solution to the nuclear waste problem.
- Rogue state or terrorist use of nuclear fuel for weapons – as the number of countries with access to nuclear technology rises, such concerns are likely to increase.
- High construction and decommissioning costs – recent estimates put an average price of about $6.3 billion on a new nuclear power plant.

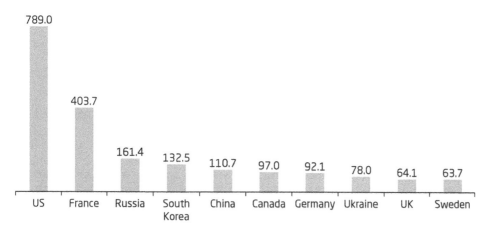

Figure 7.07 Bar graph showing top 10 nuclear-generating countries (billion kWh) in 2013.

- Because of the genuine risks associated with nuclear power and the level of security required, it is seen by some people as less 'democratic' than other sources of power.
- The possible increase in certain types of cancer near nuclear power plants; there has been much debate about this issue, but the evidence appears to be becoming more convincing.

At one time the rise of nuclear power looked unstoppable. However, a serious incident at the Three Mile Island nuclear power plant in Pennsylvania, USA, in 1979, and the much more serious Chernobyl disaster in the Ukraine in 1986, brought any growth in the industry to a virtual halt. No new nuclear power plants have been ordered in the USA since then, although public opinion has become more favourable in recent years as worries about polluting fossil fuels increase.

Most of the recent nuclear power plants constructed have been in Asia. The advantages of nuclear power are:

- It creates zero emissions of greenhouse gases.
- Reliance on imported fossil fuels is reduced.
- Nuclear power is not as vulnerable to fuel price fluctuations as oil and gas – uranium, the fuel for nuclear plants, is relatively plentiful. Most of the main uranium mines are in politically stable countries.
- In recent years, nuclear plants have demonstrated a high level of reliability and efficiency as technology has advanced.
- Nuclear technology has spin-offs in fields such as medicine and agriculture.

This decade will be crucial to the future of nuclear energy, with many countries making final decisions to extend or create nuclear electricity capacity. Nuclear energy is likely to be a major political issue in some countries.

7.01.06 What factors affect the choice of energy sources adopted by different societies?

Global variations in energy supply occur for a number of reasons. These can be broadly subdivided into physical, economic/technical and political/cultural factors.

7.01 Energy choices and security

Type of factor	Specific reason
Physical	• Deposits of fossil fuels are only found in a limited number of locations. • Large-scale hydroelectric development requires high precipitation, major steep-sided valleys and impermeable rock. • Large power stations require flat land and geologically stable foundations. • Solar power needs a large number of days a year with strong sunlight (see Image 7.04). • Wind power needs high average wind speeds throughout the year. • Tidal power stations require a very large tidal range. • The availability of biomass varies widely due to climatic conditions.
Economic/technical	• The most accessible, and lowest cost, deposits of fossil fuels are invariably developed first. • Onshore deposits of oil and gas are usually cheaper to develop than offshore deposits. • Potential hydroelectric sites close to major transport routes and existing electricity transmission corridors are more economical to build than those in very inaccessible locations. • In poor countries, foreign direct investment is often essential for the development of energy resources. • When energy prices rise significantly, companies increase spending on exploration and development. • The level of technological expertise in a country may be a significant factor in decision-making.
Political/cultural	• Countries wanting to develop nuclear electricity require permission from the International Atomic Energy Agency. • International agreements such as the Kyoto Protocol can have a considerable influence on the energy decisions of individual countries. • Potential hydroelectric power schemes on 'international rivers' may require the agreement of other countries that share the river. • Governments may insist on energy companies producing a certain proportion of their energy from renewable sources. • Legislation regarding emissions from power stations favours the use, for example, of low-sulfur coal as opposed to coal with a high sulfur content. • Public perception in some countries may be firmly against certain sources of energy.

Table 7.04 Reasons for global variations in energy supply.

Image 7.04 Solar panel and mini wind turbine powering streetlights.

Table 7.04 shows examples for each of these groupings. Technological change has had a major impact. For example, offshore oil rigs can now drill in much deeper water than 30 years ago. Public perception has become increasingly important. People have never been more aware of environmental and other issues surrounding the development of energy resources.

7.01.07 Variable energy patterns over time

The use of energy in all countries has changed over time, due to a number of factors:

- *Technological development:* For example:
  - Nuclear electricity has only been available since 1954.
  - Oil and gas can now be extracted from much deeper waters than in the past.
  - Renewable energy technology is advancing steadily.
- *Increasing national wealth:* As average incomes increase, living standards improve, which involves the increasing use of energy and the use of a greater variety of energy sources. The term 'energy ladder' is sometimes applied to this process.

- *Changes in demand:* At one time, all of Britain's trains were powered by coal and most people also used coal for heating in their homes. Before natural gas was discovered in the North Sea, Britain's gas was produced from coal (coal gas).
- *Changes in price:* The relative prices of the different types of energy can influence demand. Electricity production in the UK has been switching from coal to gas over the past 20 years, mainly because power stations are cheaper to run on natural gas.
- *Environmental factors / public opinion:* Public opinion can influence decisions made by governments. People are much better informed about the environmental impact of energy sources today compared with in the past.

SELF-ASSESSMENT QUESTIONS 7.01.03

1. State three concerns about nuclear energy and three advantages of this source of power.
2. Identify three physical factors influencing global variations in energy supply.
3. **Discussion point:** Outline some of the ways in which energy patterns have varied over time.

CASE STUDY 7.01.01

The factors affecting energy sources in China

'The fact that China overtook the US as the world's largest energy consumer symbolises the start of a new age in the history of energy.'

Faith Birol, International Energy Agency Chief Economist, 2010

China overtook the USA in total energy usage in 2009 (see Image 7.05). The USA had held the top position in the energy usage league for more than a century. Only a decade before, China's total energy consumption was about half that of the USA. The demand for energy in China continues to increase significantly as the country expands its industrial base. Figure 7.08 shows China's energy consumption breakdown by source in 2011. Coal dominates energy consumption in China, accounting for 69 per cent of total supply, with oil in second place at 18 per cent.

China's energy strategy

China's energy policy has evolved over time. As the economy expanded rapidly in the 1980s and 1990s, much emphasis was placed on China's main energy resource, coal (see Figure 7.08), in terms of both increasing production and building more coal-fired power stations. However, the price was an alarmingly high casualty rate among coal miners.

China was also an exporter of oil until the early 1990s although it is now a very significant importer (see Figure 7.09). This transformation has had a major impact on Chinese energy policy as the country has sought to secure overseas sources of supply. Long-term energy security is viewed as essential if the country is to maintain the pace of its industrial revolution.

Image 7.05 The rapid expansion of air travel in China has been a major source of increasing energy demand.

Coal is the dirtiest of the fossil fuels and thus the environmental consequences of such a heavy reliance on coal were all too predictable. According to Greenpeace, 80 per cent of China's carbon dioxide emissions and 85 per cent of its sulfur dioxide pollution comes from burning coal. China is the world's leading energy-related carbon dioxide emitter.

In recent years, China has tried to take a more balanced approach to energy supply and at the same time reduce its environmental impact. China's 11th five-year plan (2006–2010) focused on two major energy-related objectives: to reduce energy use per unit gross domestic product (GDP) by 20 per cent, and to ensure a more secure supply of energy. Because of the dominant position of coal in China's energy mix, the development of clean coal technology is central to China's energy policy with regard to fossil fuels. China has emerged in the past two years as the world's

7.01 Energy choices and security

CASE STUDY 7.01.01 (continued)

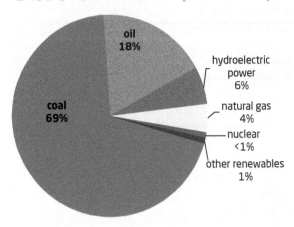

Figure 7.08 Total energy consumption by type in China 2011. (Percentage figures are rounded to the nearest 1%.)

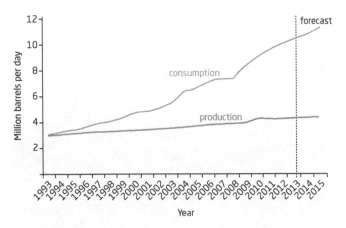

Figure 7.09 Chinese oil consumption and production 1993–2015.

leading builder of more efficient, less polluting, coal power plants. China has begun constructing such clean coal plants at a rate of one a month. The government has begun requiring power companies to retire an older, more polluting power plant for each new one they build.

The further development of nuclear and hydroelectric power is another important strand of Chinese policy. The country also aims to stabilise and increase the production of oil while augmenting that of natural gas and improving the national oil and gas network. Nuclear power reached a capacity of 9.1 GW by the end of 2008, with a target capacity of 40 GW by 2020. By the end of 2009, China had 11 operational nuclear reactors with a further 17 under construction. The World Nuclear Association says that China has a further 124 nuclear reactors on the drawing board. This will lead to a dramatic increase in China's demand for uranium, the raw material of nuclear reactors.

China's strategic petroleum reserve

As part of China's concerns about energy security and its increasing reliance on oil imports the country is developing a strategic petroleum reserve. The plan is for China to build facilities that can hold 500 million barrels of crude oil by 2020 in three phases. This will be equivalent to about 90 days' supply. Phase 1, completed in 2009, consists of four sites with a total storage capacity of 103 million barrels. Phase 2, which will be completed by the end of 2015, will add a further 170 million barrels of storage capacity.

China is following the USA and other countries in building up a petroleum reserve. This will protect China to a certain extent from fluctuations in the global oil price which can arise for a variety of reasons.

Renewable energy policy

China aims to produce at least 15 per cent of overall energy output from renewable energy sources by 2020 as the government seeks to improve environmental conditions. Renewable energy currently contributes more than one-quarter of China's total installed energy capacity, with hydroelectricity by far the largest contributor. China produces more hydroelectricity than any other country in the world, with hydroelectricity accounting for more than 15 per cent of the country's total electricity generation. The world's largest hydroelectricity project, the Three Gorges Dam along the Yangtze River, was completed in 2012. It has 32 generators with a total maximum capacity of 22.5 GW.

China is now the world leader in wind energy, accounting for 31 per cent of global installed wind power capacity. The year 2008 saw the initial development of China's offshore wind farm policy. China's wind-turbine-manufacturing industry is now the largest in the world. Chinese policy is not just to gain the energy advantages of wind energy but also to develop it as a significant industrial sector. China is now also the largest manufacturer of solar photovoltaic cells. The solar hot water market in China has also continued to boom, partly as a result of a new rural energy subsidy programme for home appliances, for which solar hot water qualifies. China aims to increase solar electricity capacity from 3 GW in 2012 to 35 GW by the end of 2015.

Questions

1. Describe China's energy mix.
2. Comment on the trends in China's production and consumption of oil, shown in Figure 7.09.
3. Why is China developing a strategic petroleum reserve?

CASE STUDY 7.01.02

The energy balance in the UK

Table 7.05 shows that the three fossil fuels contributed 87 per cent of UK primary energy consumption in 2011. Natural gas was the single most important source of energy. The energy balance of the UK has changed significantly in the past due to:

- changes in resource availability
- technological progress
- the relative cost of different sources of energy
- consumer behaviour.

The energy balance of the UK will undoubtedly change again in the future for similar reasons alongside increasing desire for cleaner sources of power to radically reduce the environmental impact of energy production.

Fuel	Percentage of total consumption / %
natural gas	38
oil	33
coal	16
nuclear	8
bioenergy and waste	4
wind and hydroelectric power	1

Table 7.05 UK energy consumption by fuel, 2011.
Source: Annual Energy Statement, 2012, Department of Energy and Climate Change

At one time, coal dominated energy supply in the UK, due largely to the country's large domestic resources and the available technology at the time, but today it is only third in order of importance as a source of energy. In contrast the rise of renewable energy, apart from very limited and long-standing hydroelectric capacity, has been a relatively recent phenomenon; however, the trend is likely to continue firmly in the future with strong government backing.

Electricity generation and the decline of coal

Coal remains important to electricity generation in the UK, although its role has decreased as other sources of electricity generation have become more financially and technically viable. In 1970, coal accounted for around two-thirds of all electricity generation, whereas now it accounts for less than half. The void left by the decline in the use of coal has been filled largely by the rapid increase in the use of natural gas. In 2012, UK domestic coal production fell to an all-time low, with most UK coal consumption supplied by imports.

Nuclear power currently accounts for about one-fifth of total electricity generation, but its total contribution has declined as the oldest nuclear power stations have been decommissioned. In the late 2020s, nuclear electricity will generate more power as the UK's new-generation nuclear power stations come online. The government is committed to nuclear power remaining a significant part of the UK's energy mix.

An increasing reliance on energy imports

Until as recently as 2004, the UK was a net exporter of energy because of high levels of production from offshore oilfields and gas fields in UK territorial waters. However, production levels of both oil and gas in the North Sea have declined significantly as major deposits of the most accessible oil and gas have been depleted. The UK became a net importer of natural gas and crude oil in 2004 and 2005 respectively.

In recent years the UK has imported more and more energy, with a dependency level that had risen to over 48 per cent by the end of 2012 (see Figure 7.10). The data showed that around 47 per cent of gas was imported, 87 per cent of coal and 37 per cent of oil (see Image 7.06). This is generally regarded as a very substantial level in terms of energy security. Overall, UK energy production has fallen in each year since 1999. Dependency on imported energy has risen as:

- UK production of oil and gas in the North Sea has fallen rapidly
- domestic coal production has continued its long-term decline
- some nuclear power stations have closed having reached the end of their productive lives, with others due for closure in the next decade.

Government policy has become a major influence on the UK's energy balance. The UK government has stated its intention to develop supplies that are secure, diverse, affordable and low carbon. Successive governments have stated their commitment to reducing carbon dioxide emissions. The Low Carbon Transition Plan launched in July 2009 aimed to generate 30 per cent of electricity from renewable sources and 4 per cent from low carbon content fuels by 2020.

7.01 Energy choices and security

CASE STUDY 7.01.02 (continued)

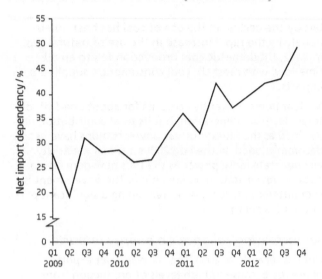

Figure 7.10 Line graph showing the UK's net import dependency, 2009–2012.

Image 7.06 Oil tanker bringing imported oil into Milford Haven, South Wales.

Future developments

Over the past decade, the UK has made a substantial investment in renewable sources of energy, and all the indications are that this trend will continue. The Department of Energy and Climate Change expects that, by 2030, renewables will be by far the biggest source of energy used in electricity generation, making up about 40 per cent of the overall balance. The commitment to include nuclear electricity as a significant element of the UK's energy balance is also clear. This is all part of the overall objective of maintaining a range of sources of energy in the overall energy balance for the future. Old coal-fired power stations will continue to close, and it is clear that the future of coal as a significant element of energy supply lies in the application of clean-coal technology. This new technology has developed forms of coal that burn with greater efficiency and capture coal's pollutants before they are emitted into the atmosphere.

However, a new form of energy may have a major impact on the country's energy balance in the future. In June 2013, the UK shale gas explorer IGas stated that there was enough shale gas in the UK to change the country's energy balance. The UK imports about 1.5 trillion cubic feet of gas a year and consumes an annual total of about 3 trillion cubic feet. IGas says that, assuming 10–15 per cent of the UK's shale gas deposits could be recovered, technically it could move import dependency out for about 10–15 years. Shale gas has already transformed the energy balance in the USA. However, shale gas has yet to be commercially exploited in the UK. There has been considerable opposition to the development of shale gas, due to strong environmental concerns. It remains to be seen whether there is sufficient political will to drive this section of the energy industry forward. Much will depend on the level of opposition and the strength of the arguments put forward by both sides in the ongoing debate.

Questions

1. Describe the UK's energy mix.
2. What factors have caused the UK's energy mix to change in the past?
3. What are the reasons for the UK's increasing reliance on energy imports?

Climate change – causes and impacts

7.02

LEARNING OBJECTIVES

After reading this chapter you should be able to:

- understand that climate change has been a normal feature of the Earth's history but human activity has contributed to recent changes
- appreciate that there has been significant debate about the causes of climate change
- explain that climate change causes widespread and significant impacts on a global scale.

KEY QUESTIONS

What is the enhanced greenhouse effect?
What are the potential consequences of climate change?
What are the feedback mechanisms associated with climate change?

7.02.01 The difference between weather and climate

The word 'climate' comes from the Greek *klima,* meaning 'area'. Climate refers to a region's long-term weather patterns. In this context, 'long-term' generally means 30 years or more. Climate is measured in terms of average precipitation, maximum and minimum temperatures throughout the seasons, sunshine hours, humidity, the frequency of extreme weather, and other meteorological indicators.

In contrast, 'weather' is the way the atmosphere is behaving in the short term (over minutes to weeks, and maybe months), mainly with respect to its effects upon life and human activities. Weather describes the atmospheric conditions at a specific place at a particular time. A way to remember the difference is that climate is what you expect, like a very hot summer, and weather is what you get, like a hot day with sudden thunderstorms (see Table 7.06).

Atmospheric and oceanic circulatory systems

The circulation systems in the atmosphere and the oceans have a major impact on weather and climate. Ocean currents account for about 80 per cent of the global transfer of heat from low to high latitudes, with winds accounting for the remaining 20 per cent.

7.02 Climate change – causes and impacts

	Weather	Climate
Definition	describes the atmospheric conditions at a specific place at a specific point in time	describes the average conditions expected at a specific place at a given time
Time frame	short term: minutes, hours, days or weeks	long term: months, years, decades, or longer
Determined by	real-time measurements of atmospheric pressure, temperature, wind speed and direction, humidity, precipitation, could cover, and other variables	aggregating weather statistics over periods of 30 years ('climate normals')
Study	meteorology	climatology

Table 7.06 Distinguishing between weather and climate.

Source: Modified from www.nasa.gov/mission_pages/noaa-n/climate/climate_weather.html and www.diffen.com/difference/Climate_vs_Weather

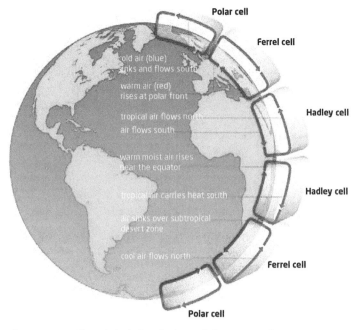

Figure 7.11 The global circulation of the atmosphere.

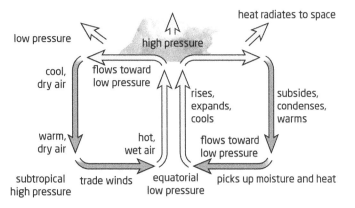

Figure 7.12 The Hadley cells.

The global movement of winds mainly results from the rotation of the Earth and the differences in the amount of solar radiation received in tropical and polar latitudes. Other contributory factors include:

- the position of the continents
- major relief barriers, such as the Himalayas and Rocky Mountains
- the differences between the degree of heating of the atmosphere over land and sea areas.

In each hemisphere there are three cells (see Figure 7.11) in which air circulates through the depth of the troposphere:

- *Hadley cells:* by far the largest cells by volume, extending from the equator to about 30 degrees north and south. Within the Hadley cells the trade winds blow from the high pressure of the subtropics towards the lower pressure of the equator (see Chapter 6.01). These winds then rise near the equator in large convection currents (see Figure 7.12), forming a broken line of thunderstorms. This line forms the Inter-Tropical Convergence Zone, giving the equatorial climate its high level of precipitation. The process of cloud formation (and thus precipitation) is explained in Chapter 6.01. At the summit of these storms in the upper troposphere, the air flows north and south towards higher latitudes. This high-level air gradually subsides to form the subtropical high pressure belt over the world's hot deserts and subtropical oceans.
- *Ferrel cells:* the average motion of air in the mid latitudes, where air flowing near the surface from the high pressure of both the subtropical and polar regions converges at the polar front (50–60 degrees north and south). Warm tropical air rises over cold polar air at the polar front to form clouds and precipitation (see Chapter 6.01).
- *Polar cells:* the smallest and weakest of the three cell systems. Air at high altitudes sinks and flows out towards lower latitudes at the surface.

Ocean currents

The majority of the Sun's radiation reaching the Earth's surface is absorbed by the oceans, which make up about 71 per cent of the Earth's surface and hold 97 per cent of the planet's water. The oceans store radiation and help distribute it around the globe. Ocean water is constantly evaporating, and most rain that falls on land originates in the oceans.

Ocean currents are massive movements of water in a continuous flow. They are created largely by surface winds but also partly by temperature and salinity gradients, the Earth's rotation, and tides. Major current systems flow clockwise in the northern hemisphere and counter-clockwise in the southern hemisphere, in circular patterns that often follow the coastlines.

Warm ocean currents flow from warm tropical to cold polar regions. They increase temperatures in the maritime environments where they flow. Cold currents flow outwards from cold polar regions towards the tropics. They lower the temperature of the maritime regions where they flow. Figure 7.13 shows the movement of the world's major ocean currents.

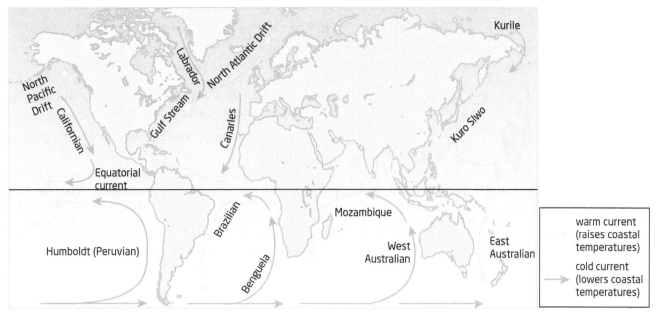

Figure 7.13 The major ocean currents.

The Gulf Stream is a warm ocean current originating in the Gulf of Mexico. It flows north-east along the US coast, and from there to Western Europe. When it crosses the Atlantic Ocean, it is known as the North Atlantic Drift.

Water temperature in the Gulf of Mexico is higher than in Western Europe, therefore it has a considerable warming effect on the maritime countries of Europe – air blowing over the North Atlantic Drift picks up some of its heat and transfers it to adjacent land. However, the air is also quite moist, as it travels over the wide expanse of the Atlantic Ocean. Thus maritime areas in Europe generally have a wetter climate than regions further inland. The North Atlantic Drift keeps the north-west coast of Europe free from ice in the winter.

Natural climate change

There is much evidence to tell us that the global climate changed naturally before humankind was able to influence it with large-scale industrial activity. The ice ages that have occurred throughout geological time are clear evidence of such climate change. The causes of natural changes to the global climate are outlined below.

7.02 Climate change – causes and impacts

Changes in solar radiation

The total amount of solar radiation reaching the Earth can vary due to changes in the Sun's output, such as those associated with sunspots. Research into the amount of radiation emitted by the Sun has found that a solar cycle occurs approximately every 11 years. This is when the Sun undergoes a period of increased magnetic and sunspot activity called the 'solar maximum', which is followed by a quiet period called the 'solar minimum'. Sunspots are seen as small darker spots on the surface of the Sun. They are intense 'bubbles' of magnetic energy which somehow cool down the hot gases within, so they appear dark in relation to the surrounding solar atmosphere.

Slow variations in the Earth's orbit

Glacial periods have occurred when the Earth's orbit has been most circular, and interglacials, or warmer periods, when the orbit has been more elliptical in shape.

Slow changes in the angle of the Earth's axis

The tilt of the Earth's axis has varied in its plane of orbit in geological time by between 21.5 degrees and 24.5 degrees, not by a constant 23.5 degrees as is sometimes believed.

Changes in the albedo of the Earth and atmosphere

The Earth's albedo is the proportion of solar energy reflected from the Earth back into space. Changes in surface albedo occur naturally but climate forcing is also caused by anthropogenic (caused by humans) changes.

Changes in the longwave radiation returned to space

The changing composition of the atmosphere, including its greenhouse gas and aerosol content, is a major internal forcing mechanism of climate change. Such changes have a significant impact on the longwave radiation returned to space. Changes in the greenhouse gas content of the atmosphere can occur as a result of both natural and anthropogenic factors.

Carbon dioxide levels from geological times to the present

Carbon dioxide, temperature and other atmospheric phenomena have changed considerably over geological time. Scientific studies have shown that, in past eras, atmospheric carbon dioxide reached concentrations much higher than the current value (see Figure 7.14).

> **CONSIDER THIS**
> Carbon dioxide is not an environmental polluting agent, because it is not detrimental or poisonous to life. Its relationship with average global temperature is the contentious issue.

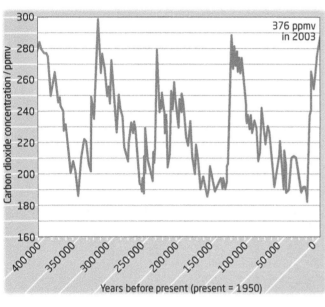

Figure 7.14 Carbon dioxide levels in the recent geological past.

7.02.02 Greenhouse gases and mean global temperature

> **Climate change** is long-term, sustained change in the average global climate.

The Stern Review published in the UK in 2006 (see Figure 7.15) stated that **climate change** is the greatest challenge facing humankind. The Stern Review was a major study into the economics of climate change, commissioned by the UK government in 2005. It was designed to point the way ahead for government policymaking. This major report came to the following conclusions:

- Climate change is the result of the externality associated with greenhouse-gas emissions – it entails costs that are not paid for by those who create the emissions.

Key Messages

An overwhelming body of scientific evidence now **clearly indicates that climate change is a serious and urgent issue. The Earth's climate is rapidly changing, mainly as a result of increases in greenhouse gases caused by human activities.**

Most climate models show that **a doubling of pre-industrial levels of greenhouse gases is very likely to commit the Earth to a rise of between 2 and 5°C in global mean temperatures.** This level of greenhouse gases will probably be reached between 2030 and 2060. A warming of 5°C on a global scale would be far outside the experience of human civilisation and comparable to the difference between temperatures during the last ice age and today. Several new studies suggest up to a 20% chance that warming could be greater than 5°C.

If annual greenhouse gas emissions remained at the current level, concentrations would be more than treble pre-industrial levels by 2100, committing the world to 3-10°C warming, based on the latest climate projections.

Some impacts of climate change itself may amplify warming further by triggering the release of additional greenhouse gases. This creates a real risk of even higher temperature changes.

- Higher temperatures cause plants and soils to soak up less carbon from the atmosphere and cause permafrost to thaw, potentially releasing large quantities of methane.
- Analysis of warming events in the distant past indicates that such feedback could amplify warming by an additional 1-2°C by the end of the century.

Warming is very likely to intensify the water cycle, reinforcing existing patterns of water scarcity and abundance and increasing the risk of droughts and floods.

- Rainfall is likely to increase at high latitudes, while regions with Mediterranean-like climates in both hemispheres will experience significant reductions in rainfall. Preliminary estimates suggest that the fraction of land area in extreme drought at any one time will increase from 1% to 30% by the end of this century. In other regions, warmer air and warmer oceans are likely to drive more intense storms, particularly hurricanes and typhoons.

As the world warms, the risk of abrupt and large-scale changes in the climate system will rise.

- Changes in the distribution of heat around the world are likely to disrupt ocean and atmospheric circulations, leading to large and possibly abrupt shifts in regional weather patterns.
- If the Greenland or West Antarctic Ice Sheets began to melt irreversibly, the rate of sea level rise could more than double, committing the world to an eventual sea level rise of 5-12 m over several centuries.

The body of evidence and the growing quantitative assessment of risks are now sufficient to give clear and strong guidance to economists and polity-makers in shaping a response.

Figure 7.15 Key messages from the Stern Review.

- There is still time to avoid the worst impacts of climate change, if we take strong action now.
- The benefits of strong and early action far outweigh the economic cost of not acting.

The major change in the Earth's climate is **global warming**, which refers to the warming of the Earth's surface outside of the range of normal fluctuations that have occurred throughout the planet's history. Global warming is a scientifically controversial phenomenon that attributes the increase in the average annual surface temperature of the Earth to rising atmospheric concentrations of carbon dioxide and other greenhouse gases.

Although there is not unanimous agreement among scientists, most believe that the **natural greenhouse effect** has been significantly altered by human activity. Now an **enhanced greenhouse effect** is causing temperatures to increase beyond the limits of the natural greenhouse effect. For such scientists, the issue is how far global warming has progressed and whether the **tipping point** has been reached. The tipping point is the level at which the effects of climate change become irreversible to varying degrees. The extent to which temperatures will rise because of a given change in the concentration of greenhouse gases is known as 'climate sensitivity'.

Global warming is the increase in the average temperature of the Earth's near-surface air in the 20th and early 21st centuries and its projected continuation.

The **natural greenhouse effect** is the property of the Earth's atmosphere by which long-wavelength heat rays from the Earth's surface are trapped or reflected back by greenhouse gases in the atmosphere.

An **enhanced greenhouse effect** results from human activities that increase the concentration of naturally occurring greenhouse gases and lead to global warming and climate change.

7.02 Climate change – causes and impacts

Tipping point is the point at which the damage caused to global systems by climate change becomes irreversible.

SELF-ASSESSMENT QUESTIONS 7.02.01

1. Define: **a** climate change, **b** global warming.
2. What do you understand by the term 'tipping point'?
3. **Discussion point:** Suggest why governments such as that in the UK are interested in the economics of climate change.

Image 7.07 The Swiss Alps: the retreat of glaciers in the Alps has reduced the albedo in the region.

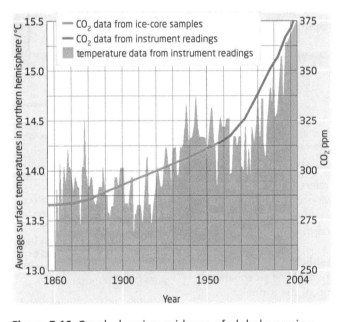

Figure 7.16 Graph showing evidence of global warming.

How human activities add to greenhouse gases

There is no doubt that the Earth is getting warmer (see Image 7.07). Many parts of the world are experiencing changes in their weather that are unexpected. Some of these changes could have disastrous consequences for the populations of the areas affected if they continue to get more severe.

However, the problem is that the present rate of change is greater than anything that has happened in the past. According to a recent NASA research paper, average global temperatures have increased by about 0.8 °C since 1880. Most of this increase took place in the second half of the 20th century and the beginning of the 21st century (see Figure 7.16). January 2000 to December 2009 is the warmest decade on record. The greatest increases have occurred over the middle and high latitudes of the northern hemisphere continents.

Most climate experts believe that this high rate of temperature change is due to human activity leading to an increase in greenhouse gases. The predictions are for a further global average temperature increase of between 1.6 °C and 4.2 °C by 2100.

The argument is that human activity has significantly increased the amount of greenhouse gases in the atmosphere, and this has caused temperature to rise more rapidly than before. The Industrial Revolution, which began in the latter part of the 18th century in the UK, quickly spread to other key countries around the world and seriously increased the levels of greenhouse gas emissions.

Table 7.07 identifies the greenhouse gases and their effects. Together, these gases form only 0.1 per cent of the atmosphere, but they play a major role in the way the atmospheric system works. Carbon dioxide is responsible for four-fifths of the increase in warming by greenhouse gases. Before the Industrial Revolution, the carbon dioxide concentration in the atmosphere was 280 parts per million (ppm). By 2013 it has risen to almost 400 ppm. Most scientists argue that about three-quarters of this increase is due to burning fossil fuels. Some forecasts indicate that, by the end of the current century, carbon dioxide concentrations in the atmosphere could be double those before the Industrial Revolution. Much of this increase will be due to the very high growth rates of the newly industrialised countries (NICs). As carbon dioxide increases, the other greenhouse gases are likely to increase as well. Analysis of ice cores from Greenland and Antarctica shows a very strong correlation between carbon dioxide levels in the atmosphere and temperatures.

Greenhouse gas	Effects
Water vapour	Sometimes not considered in discussions of greenhouse gases, although it actually has the greatest effect on trapping heat energy. The impact of water in the atmosphere varies because of the continuous changes in state between vapour, liquid and ice. Scientific organisations typically state the greenhouse gas contribution of water vapour at about 50%.
Carbon dioxide	Accounts for the largest share of greenhouse gas (apart from water). It is produced by burning fossil fuels in power stations, factories and homes. Vehicle emissions are also a major source. CO_2 is also released into the atmosphere by deforestation and the burning of rainforests. Global CO_2 emissions from the burning of fossil fuels grew at 3.1% a year between 2000 and 2006, more than twice the rate of growth during the 1990s.
Methane	Released from decaying plant and animal remains and from farms (particularly from cattle and rice padi fields). Other sources include swamps, peat bogs and landfill sites. Methane is the second-ranked contributor to global warming with a current annual increase of 1%. Cattle convert up to 10% of the food they eat into methane, emitting 100 million tonnes of methane into the atmosphere annually. It has been estimated that padi fields emit up to 150 million tonnes of methane each year. Of great concern is the fact that bogs trapped in permafrost will melt as temperatures rise, releasing large quantities of methane.
Chloroflourocarbons	These synthetic chemicals destroy ozone as well as absorbing longwave radiation. The main sources are aerosols, refrigerators, foam packaging and air-conditioning systems. CFCs were first used in the 1960s. CFCs have a lifetime of about 20–100 years, and consequently one free chlorine atom from a CFC molecule can do a lot of damage, destroying ozone molecules for a long time. Although emissions of CFCs from the developed world have largely ceased due to international control agreements, the damage to the stratospheric ozone layer will continue for a number of years to come.
Oxides of nitrogen	The main forms of reactive nitrogen in the air are nitrogen monoxide (NO) and nitrogen dioxide (NO_2). Together they are called NO_x. The main sources are power stations, vehicle emissions, fertilisers and burning biomass. Catalytic converters fitted to cars can decrease production of these harmful compounds.
Ozone	Naturally occurring ozone found in the stratosphere has been called 'good' ozone, because it protects the Earth's surface from dangerous ultraviolet light. Ozone can also be found in the troposphere, the lowest layer of the atmosphere. Tropospheric ozone (often termed 'bad' ozone) is human made, a result of air pollution from vehicle and power plant emissions. In Europe, ozone is thought to be increasing by 1–2% a year.

Table 7.07 Greenhouse gases and their effects.

The annual greenhouse gas bulletin from the World Meteorological Organization (WMO) in 2014 included the following statements:

- In 2013, concentrations of carbon dioxide were 142 per cent of what they were before the Industrial Revolution.
- In 2013, concentrations of carbon dioxide increased at their fastest rate for 30 years, reaching 396 ppm. The symbolic 400 ppm is likely to be reached in 2015 or 2016.
- Between 1990 and 2013, the warming effect on the Earth known as 'radiative forcing' due to greenhouse gases increased by 34 per cent.
- The world's seas were becoming more acidic at a rate not seen for at least 300 million years.

Geographical sources of greenhouse gases

Table 7.08 shows the top polluting countries in terms of emissions of carbon dioxide in 2012. In absolute terms, China and the USA are by far the largest polluters. Emissions from China have continued to increase rapidly, rising to a massive 28.6 per cent of the world total in 2012. The USA accounted for 15 per cent of emissions in 2012. In per-capita terms, Australia, the USA,

7.02 Climate change – causes and impacts

Saudi Arabia and Canada were the largest emitters of carbon dioxide. Looking at the period between 1840 and 2004, the USA has been responsible for over 28 per cent of all carbon dioxide emissions. This is way beyond the contribution of any other nation.

Analysis of greenhouse gas emissions in the USA in 2012 showed the following rank order by economic sector:

- electricity 32 per cent
- transportation 28 per cent
- industry 20 per cent
- commercial and residential 10 per cent
- agriculture 10 per cent.

Rank	Country	Percentage of world total carbon dioxide emissions	Annual carbon dioxide emission / thousands of metric tonnes
1	China	23.33	7 031 916
2	USA	18.11	5 461 014
3	India	5.78	1 742 698
4	Russia	5.67	1 708 653
5	Japan	4.01	1 208 163
6	Germany	2.61	786 660
7	Canada	1.80	544 091
8	Iran	1.79	538 404
9	UK	1.73	522 856
10	South Korea	1.69	509 170
11	Mexico	1.58	475 834
12	Italy	1.48	445 119
13	South Africa	1.45	435 878
14	Saudi Arabia	1.44	433 557
15	Indonesia	1.35	406 029
16	Australia	1.32	399 219
17	Brazil	1.30	393 220
18	France	1.25	376 986
19	Spain	1.09	329 286
20	Ukraine	1.07	323 532
	World	100	29 888 121

Table 7.08 Carbon dioxide – top polluting countries in both absolute and relative terms.

CONSIDER THIS

Since 1958, carbon dioxide measurements have been taken continuously at the Mauna Loa laboratory in Hawaii. The Mauna Loa curve, as it is known, shows a steady and relentless rise in carbon dioxide levels. These measurements are also called the Keeling curve in honour of Charles Keeling, a pioneer in monitoring atmospheric carbon dioxide.

SELF-ASSESSMENT QUESTIONS 7.02.02

1. Name the greenhouse gases.
2. Which countries are the largest emitters of carbon dioxide in terms of: **a** total emissions, **b** per-capita emissions?
3. **Discussion point:** Discuss the sources of greenhouse gas emissions in the USA in 2012.

7.02.03 The potential effects of increased mean global temperature

There are many potential consequences of global climate change. Research and reporting have naturally focused on the potential disadvantages of this phenomenon. However, not all the effects might be adverse, and in some parts of the world the benefits may considerably outweigh the costs, although whether such a situation would be long term remains to be seen. There is still much debate about exactly what could happen. The potential changes include:

- biomes shifting
- changes in location of crop growing areas
- changing weather patterns
- coastal inundation
- human health concerns.

Biomes shifting

As climate is the major factor determining the location and character of biomes, changes in climate in different parts of the world will have a significant impact. A study published in 2007 warned that the Earth's tropical belt was expanding north and south. A further 22 million square kilometres are experiencing a tropical climate compared with the climate in 1980. The pole-ward movement of subtropical dry belts could affect agriculture and water supplies over large areas of the Mediterranean, the south-western USA, northern Mexico, southern Australia, southern Africa and parts of South America. The extension of the tropical belt will put more people at risk from tropical diseases.

Tundra ecosystems in Arctic areas are being significantly affected by temperature increase. An area of permafrost spanning 1 million square kilometres has started to melt for the first time since it formed 11 000 years ago at the end of the last ice age. The area, which covers the entire sub-Arctic region of western Siberia, is the world's largest frozen peat bog. Scientists fear that, as it thaws, it will release billions of tonnes of methane, a greenhouse gas 20 times more potent than carbon dioxide, into the atmosphere. Scientists are putting together monitoring networks to measure the release of gases from Arctic soils.

Coral reefs are biologically rich ecosystems, but they are very sensitive to climate change and other forms of stress. An increase in sea temperature along with other factors such as pollution and sedimentation can effectively halt photosynthesis of the zooxanthellae (algae), resulting in the death of the living part of the coral. The zooxanthellae give the coral its colour. The death of the zooxanthellae leaves the coral in an energy deficit and without colour – a process known as **coral bleaching**. Such a change can be caused by a rise in sea temperature of as little as 1–2 °C above the long-term average. The potential for bleaching is greatest in shallow waters.

Many species of wildlife may be wiped out because they will not have a chance to adapt to rapid changes in their environments. The loss of Arctic ice will have a huge effect on polar bears and other species that live and hunt among the ice floes. A 2011 report in the *Nature Climate* journal noted the following:

- Climate change is stunting the growth of polar bears and a wide range of other species. Warmer and drier weather limits plants and animals to smaller sizes. The average size of male polar bears decreased by 11 per cent between 1967 and 2006.
- A study of 1700 plant, insect, bird and amphibian species has shown that 80 per cent are already moving 6.1 km closer to the poles every decade, and 87 per cent are breeding or flowering more than two days earlier each decade.

Coral bleaching is when corals are stressed by changes in their conditions, such as temperature, light or nutrients, and they expel the symbiotic algae living in their tissues, causing them to turn completely white.

Change in location of crop growing areas

As biomes shift towards the poles, significant changes in the location of crop growing areas are likely as the cultivation possibilities increase in previously peripheral areas. For example, vine

7.02 Climate change – causes and impacts

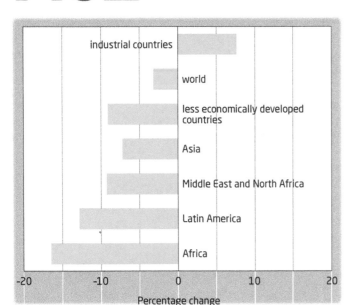

Figure 7.17 Predicted changes in potential cereal output for 2080.

cultivation in Europe will extent northwards, along with arable land in general. This will be seen as an economic advantage in the areas concerned. In Russia, President Putin referred to potential advantages to Russian agriculture in a country that has considerable climatic restrictions. Other countries where the growing season is currently limited are likely to benefit if the growing season is extended.

However, the great concern is that in many parts of the world the disadvantages of climate change will far outweigh any advantages. Plants require certain amounts of moisture and nutrients and can only live in particular temperature ranges. Higher temperatures have already had an impact on global yields of wheat, corn and barley. A recent study revealed that crop yields fall by 3–5 per cent for every 0.5 °C increase in average temperature. Food shortages could begin conflicts between different countries.

In 2008, researchers at the Stanford Center on Food Security and the Environment concluded that many of the world's poorest regions could face severe crop losses in the next two decades because of climate change. Their analysis revealed two hunger hotspots where future climate impacts on agriculture are likely to be particularly severe: southern Africa and south Asia. The research suggests that southern Africa could lose more than 30 per cent of its main crop, maize, in the next two decades. Figure 7.17 shows projected climate change impacts for agriculture around the world, in terms of potential cereal output for 2080.

Changed weather patterns

> A **heatwave** is a prolonged period of excessively hot weather.

In general, higher latitudes and continental regions will experience temperature increases significantly greater than the global average. There will be a rising probability of **heatwaves**, with more days of extreme heat and fewer very cold days. The rise in heatwave frequency will be felt most severely in cities, with temperatures amplified by the heat island effect.

The amount and distribution of rainfall in many parts of the world could change considerably (see Image 7.08). Generally, regions that get plenty of rainfall are likely to receive even more. And regions with low rainfall are likely to get less. The latter will include the poor arid and semi-arid countries of Africa. In 2009, the heaviest rain in 53 years battered Dhaka, the capital of Bangladesh. A total of 33.3 cm of rain fell in 12 hours. Experts predict that storms will hit the UK more frequently. Since the mid-1990s, according to a 2009 report, the number of intense hurricanes has been increasing in the Atlantic Ocean and the size of wildfires has been growing in the western USA.

There is growing concern about El Niño. This phenomenon is a change in the pattern of wind and ocean currents in the Pacific Ocean. This causes short-term changes in weather for countries bordering the Pacific, such as flooding in Peru and drought in Australia. El Niño events tend to occur every two to seven years. There is concern that rising temperatures could increase the frequency and/or intensity of El Niño events.

Image 7.08 Weather station in the Rocky Mountains, USA – part of the global network of recording the weather.

Coastal inundation

Most of the world's major cities were originally built on rivers or coastlines and are thus subject to flooding from rising sea levels and the occurrence of heavier rainfall. A recent article in *Scientific American* stated that 86 per cent of urban residents in rich countries live in low coastal areas that risk flooding from rising sea levels. The respective figures for lower-middle-income countries and low-income countries were 56 per cent and 41 per cent. Sea levels will respond more slowly than temperatures to changing greenhouse gas concentrations.

Sea levels are currently rising at around 3 mm per year, and the rise has been accelerating. Rising sea levels are due to a combination of **thermal expansion** and the melting of **ice sheets** and glaciers (see Image 7.09). Thermal expansion is the increase in water volume due to temperature increase alone. A global average sea level rise of 0.4 m from this cause has been predicted by the end of this century.

Satellite photographs show ice melting at its fastest rate ever. The area of sea ice in the Arctic Ocean has decreased by 15 per cent since 1960, while the thickness of the ice has fallen by 40 per cent. Sea temperature has risen by 3 °C in the Arctic in recent decades. A report in the Science section of *The Guardian* in March 2007 claimed that the Arctic Ocean may lose all its ice by 2040. This would disrupt global weather patterns. For example, it would bring intense winter storms and heavier rainfall to Western Europe. Table 7.09 gives examples of recently observed environmental changes in the Arctic linked to higher temperatures.

The Antarctic continent comprises two main regions separated by the Transantarctic Mountains (see Figure 7.18). Both regions are covered in ice that is, on average, 2000 m thick. East Antarctica is a mountainous plateau as big as Europe. Its huge ice sheet appears, on the whole, to be stable. However, there is much greater concern over the West Antarctic Ice Sheet.

> **Thermal expansion** occurs as sea and ocean temperatures increase, the water molecules near the surface expand, and sea level rises.
>
> The **ice sheet** is a thick layer of ice covering extensive regions of the world, notably Antarctica and Greenland.

Image 7.09 The retreating Athabasca glacier in the Canadian Rockies.

Nature of observed change	Where change observed	Climate link
reduction in extent and thickness of sea ice	all around the Arctic	warmer air
reduction in extent and mass of glaciers	all around the Arctic	warmer air
thawing permafrost	Alaska	warmer air, changes in snow cover
retreating ice sheet and reduction in mass of ice	Greenland	warmer air
increase in flow of rivers, especially in winter	Siberia	reduced permafrost
treeline advancing north	Alaska	warmer air, warmer soil
changed range of birdlife	Alaska	longer growing season for plants
improved calf survival in caribou (but reduced health)	Alaska, northern Canada	warmer air, more insects

Table 7.09 Observed environmental changes in the Arctic.

7.02 Climate change – causes and impacts

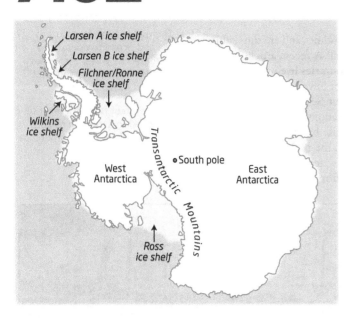

Figure 7.18 Map of Antarctica.

It is inherently less stable, because much of the ice base is under water. Temperatures in western Antarctica have increased sharply in recent years, melting ice shelves and changing plant and animal life on the Antarctic Peninsula.

Two separate studies published in 2014 (NASA, and the University of Washington) stated that the loss of the Western Antarctic Ice Sheet is inevitable, although the collapse of the ice sheet is at least several centuries off. This will cause up to 4 m of additional sea level rise, which will change the coastline in many parts of the world. Scientists have concluded that the causes of ice loss are highly complex, and it is not just due to warmer temperatures causing surface melting of the ice. Contact between the ice and the relatively warmer water at the ocean depths is a major contributory factor.

In 2007, the sea ice around Antarctica had melted back to a record low. At the same time, the movement of glaciers towards the sea has speeded up. A satellite survey between 1996 and 2006 found that the net loss of ice rose by 75 per cent. In 2002, instabilities in the Larsen Ice Shelf led to the collapse of a huge section of the shelf (over 3200 km^2 and 200 m thick) from the Antarctic Peninsula. The Larsen A ice shelf, 1600 km^2 in area, broke off in 1995. In 1998, the 1100 km^2 Wilkins ice shelf broke away. There are now questions about the long-term stability of Antarctica's two biggest ice shelves, the Ross ice shelf and the Filchner ice shelf.

Measurements of the Greenland ice sheet have shown slight inland growth but significant melting and an accumulation of ice flows near the coast. Antarctica and Greenland are the world's two major ice masses. Total melting of these ice masses (which is not predicted at present) could raise global sea levels by 70 m. Ice melting could cause sea levels to rise by a further 5 m (on top of thermal expansion). Hundreds of millions of people live in coastal areas within this range.

Human health concerns

Theory of knowledge 7.01.02

There are elements of climate change that most scientists feel fairly certain about and elements where there is considerable debate. The latter occurs when it is felt that the data available is insufficient to draw concrete conclusions or where various research projects appear to yield contradictory or varied conclusions. Many people feel passionately about climate change for obvious reasons.

1 Do you believe the claim in the Stern Review that climate change is the greatest challenge facing humankind?
2 How could such a claim be disproved?

The Intergovernmental Panel on Climate Change (IPCC) has concluded that climate change would cause:

- increased heat-related mortality and morbidity
- decreased cold-related mortality in temperate countries
- greater frequency of infectious disease epidemics following floods and storms
- substantial health effects following population displacement from sea level rise and increased storm activity.

Many infectious diseases are known to increase or decrease with climatic changes. Insect-borne diseases are especially sensitive to changes in temperature, humidity and rainfall patterns. Insects generally hatch earlier and develop more quickly into adults at higher temperatures, and the adults then move from one host to another. Malaria is one of the diseases that climate change is already affecting. Increasing humidity and rainfall is causing the mosquito that carries malaria to spread over wider areas. It has been estimated that, in South Africa, the area affected by malaria will double, putting an extra 5 million people at risk.

The impact of droughts on health is well documented. Droughts can lead to an increased risk of infectious diseases as well as having other significant impacts on health. Frequent heatwaves already kill many people around the world, and this is likely to increase. The issue of deteriorating air quality on mental health has also received considerable recent attention. Figure 7.19 shows the pathways by which climate change affects human health.

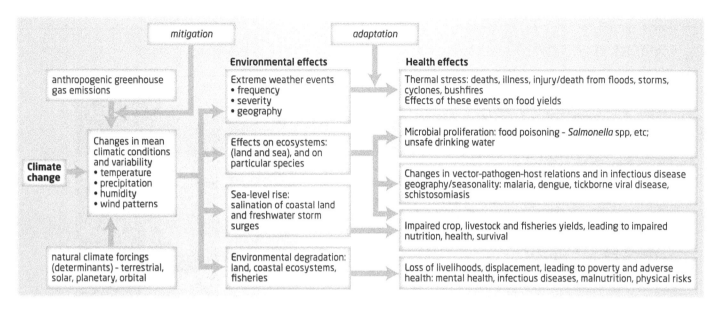

Figure 7.19 The pathways by which climate change affects human health.

SELF-ASSESSMENT QUESTIONS 7.02.03

1. How are tundra ecosystems in Arctic areas being affected by climate change?
2. Why is coral bleaching of such concern to so many people?
3. Why is there growing concern about El Niño?
4. What is happening to the Western Antarctic Ice Sheet?
5. **Research idea:** Go to **wwf.panda.org** to find out more about how climate change is affecting the Arctic region.

CASE STUDY 7.02.01

Climate change in Mongolia

The climate in Mongolia has always presented its people with problems. The country's climate data shows that:

- The mean monthly temperature is below 1 °C over the whole country between November and March.
- Late spring and early autumn frosts leave a short growing season of 80–100 days in the north and 120–140 days in the south.
- The average annual precipitation is only 251 mm, ranging from 400 mm in the north to less than 100 mm in the Gobi Desert. In comparison, London's average annual precipitation is 580 mm.
- Precipitation is very unevenly distributed over the year, with 80–90 per cent of annual rainfall occurring between June and August.
- Strong winds in spring result in high evaporation and soil erosion.
- Low soil moisture and air humidity in spring and early summer impact adversely on agricultural production.

With such extremes of climate and environment, Mongolia is very sensitive to climate change (see Image 7.10). Climate research has shown that the rate of change over the last 40 years is greater than the country has ever experienced before.

Climate change and environmental issues

Mongolian meteorologists have studied annual average air temperatures for Mongolia for the period 1940–2008. This data from 48 weather stations spread over the

7.02 Climate change – causes and impacts

CASE STUDY 7.02.01 (continued)

Image 7.10 Dog sledging on frozen Lake Khovsgol, northern Mongolia. There is concern that the 'frozen period' has declined in recent years.

whole country shows that the average temperature in Mongolia increased by 2.14 °C over this period. However, this figure produced by Mongolian meteorologists is higher than some international estimates that quote a 0.7 °C increase in the last 50 years, which is still substantial. The most extreme changes were recorded in the mountainous areas of the country. During this period, nine of the ten warmest years since 1990 occurred. The latter is a situation that has been recorded in many other countries too.

The study also looked at annual changes in precipitation between 1940 and 2008. During the last ten years of the timeline, above-average precipitation was recorded in only one year. This is of great concern in a country that has large areas of desert and semi-desert (see Image 7.11). In general, winter precipitation is increasing and summer precipitation is decreasing. As most of Mongolia's precipitation falls during the summer, this trend is very disturbing.

The impact of climatic extremes is further exacerbated by significant environmental issues that have resulted from a combination of physical and human processes. These include:

- overgrazing
- deforestation
- soil erosion
- desertification.

Image 7.11 Four-wheel-drive vehicle driving along heavily depleted river bed in southern Mongolia.

The occurrence of droughts is increasing, resulting in depletion of water resources. A 2008 United Nations (UN) report quotes a recent water resources inventory in Mongolia which found that compared with previous knowledge, 22 per cent of rivers and springs and 32 per cent of lakes and ponds had dried up or disappeared. In addition, the area without grass had increased while the forest area had decreased. According to some data sources, 70 per cent of grassland has been affected by desertification. Dust and sand storms are on the increase.

Questions

1. State three facts showing that Mongolia has a challenging climate.
2. What has happened to average temperatures and precipitation during the 1940–2008 study period?
3. What are the major environmental issues in Mongolia that have been caused by a combination of physical and human factors?

7.02.04 Feedback mechanisms associated with an increase in mean global temperature

Feedback mechanisms play a significant role in environmental change. **Feedback** is the return of part of the output from a system as input, so as to affect succeeding outputs. Thus, climate feedback happens when a change in climate causes an impact that changes the climate further. Feedback mechanisms can be both positive and negative; examples of both will be discussed below.

Positive feedback in global warming

Positive feedback is feedback that amplifies or increases change and leads to exponential deviation away from an **equilibrium**. It is a secondary process within a system that feeds back into a primary process and accelerates the rate of change governed by the primary process. An example of positive feedback is increased thawing of permafrost, which results in an increase in methane levels, which in turn increases the mean global temperature.

Another example which is the subject of considerable research is the positive ice albedo effect (see Figure 7.20a). Increasing average temperatures are melting Arctic ice. Until recently, 80 per cent of solar radiation was reflected from the polar ice caps. As the area covered by ice has reduced, the area of open ocean has increased. Because oceans are darker than ice and snow, they absorb more of the Sun's energy and convert it to heat. This increases the warming effect, which melts even more ice. It is, in fact, a vicious circle, known as the positive ice albedo effect.

Negative feedback in global warming

Negative feedback is feedback that tends to damp down, neutralise or counteract any deviation from an equilibrium and promotes stability. It is a secondary process that feeds back into the primary process, limiting further change initiated by the primary process. Figure 7.20b illustrates how negative feedback works. As sea temperatures rise, evaporation increases. As air rises and cools, more cloud is formed, which helps to block out more radiation from the Sun. Thus, in this example, the feedback process is counteracting deviation from the previous equilibrium and promoting stability.

Problems in accounting for feedback in global warming

Feedback mechanisms associated with global warming may involve very long time lags, which makes full understanding of these processes more difficult. In the physical world, time lags can cover decades, centuries or even longer. The thermal inertia of the oceans is sometimes referred to as climate lag.

> **Feedback** occurs when some of the output of a system returns as an input, exerting some control in the process.
>
> **Positive feedback** increases change and leads to exponential deviation from an equilibrium.
>
> **Equilibrium** is a state in which opposing influences are balanced.
>
> **Negative feedback** tends to damp down, neutralise or counteract any deviation from an equilibrium.

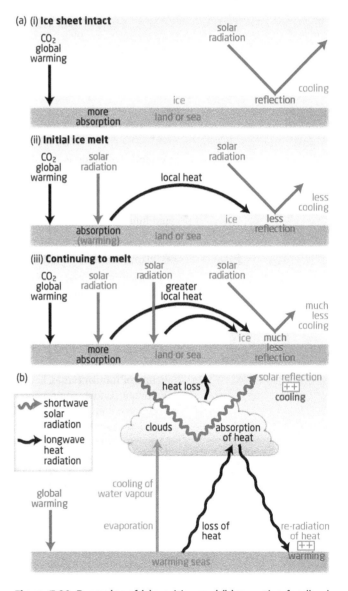

Figure 7.20 Examples of (a) positive and (b) negative feedback.

7.02 Climate change – causes and impacts

The mass of the oceans is about 500 times that of the atmosphere. Thus the time it takes the oceans to warm is measured in decades. The greatest difficulty is in quantifying the rate at which the warm upper layers of the ocean mix with the cooler deeper waters. Because of this (and other factors), there is significant variation in estimates of climate lag. A number of scientific papers have given an estimate of approximately 40 years for climate lag – the time between cause (increased greenhouse gas emissions) and effect (increased temperature).

> ### SELF-ASSESSMENT QUESTIONS 7.02.04
> 1 Define feedback, positive feedback and negative feedback.
> 2 Explain the example of positive feedback illustrated in Figure 7. 20a.
> 3 Why is the thermal inertia of the oceans such an important factor in climate change and the predictions that are made about it?
> 4 **Research idea:** Go to www.metoffice.gov.uk to find out more about climate feedback.

7.02.05 The arguments surrounding global warming

Natural versus anthropogenic causes

As global warming is arguably the greatest issue confronting humankind, it is not surprising that there are conflicting arguments surrounding this controversial topic. While the majority opinion is that human-induced climate change is taking place, opposition opinion is significant and some will argue it is gaining ground. The diversity of opinion is at least partly due to the many natural and human causes of climate change and global warming. Natural causes include:

- variations in the tilt of the Earth's axis
- variations in the Earth's orbit around the Sun
- variations in solar output
- changes in the pattern of ocean currents as a result of continental drift
- changes in the amount of dust in the atmosphere, with volcanic activity a significant component.

Because many of these changes have occurred over a long period of geological time, the evidence that scientists have built up about them is far from complete. Scientists have a clearer picture of human causes, but there is much debate about their degree and significance. Scale is an important issue, encompassing the atmosphere, land masses and oceans. The interactions between these three components of the Earth–atmosphere system are complex and at times difficult to model. Our understanding of feedback mechanisms is improving but remains far from perfect. Many processes are long term, so the impact of changes may not yet have occurred. Global climate models are complex, and there is a degree of uncertainty regarding the accuracy of their predictions.

Majority opinion within countries may be strongly influenced by the structure of the economy and how much it will cost to cut emissions of greenhouse gases. This is arguably the major reason why the USA has been seen by other countries as very slow to act on global warming. However, while the federal government in the USA has been slow to act, many individual states have enacted their own laws to reduce greenhouse gas emissions.

Many scientific and other organisations have clearly stated their views on climate change. Some may appear to have vested interests in projecting one side of the argument or the other. It is important to check for bias as far as you possibly can. If, for example, a group's research is wholly or partially funded by a major energy company, there may be concerns about a conflict of interest. Also, some reports that have been issued in the past have seemed very authoritative at first but under scrutiny were found to be based on flimsy evidence.

The Intergovernmental Panel on Climate Change (IPCC) is the international body established to assess scientific, technical and socioeconomic information on climate change. In 2007, the IPPC concluded:

'Most of the observed increase in global average temperatures since the mid-20th century is very likely due to the observed increase in anthropogenic greenhouse gas concentrations... The observed widespread warming of the atmosphere and ocean, together with ice mass loss, support the conclusion that it is extremely unlikely that global climate change of the past 50 years can be explained without external forcing, and very likely that it is not due to known natural causes alone.'

In contrast, the science group Biology Cabinet sees the current phase of warming as part of the natural cycle and something that we cannot change:

'In the last 20 years, public interest in climate phenomena has grown, especially since the UN-IPCC began its campaign warning of catastrophic climate changes ahead. At Biology Cabinet, we maintain that the changes that we have observed since 1985 have been natural and that human beings cannot delay or stop the advance of these changes, but can only adapt to them. In addition, we have shown that the changes that we observe at present are the result of natural cycles which have occurred many times before.'

Arguments of prominent individuals

Many people are influenced by the opinions of prominent journalists such as George Monbiot (*The Guardian*) and Christopher Booker (*Daily Mail*). The former has for long warned about human-induced global warming and the need to take decisive international action. The latter holds very different opinions, branding the whole issue as a gigantic hoax based on flawed scientific evidence, and also a costly mistake because of the ill-conceived measures employed to try to reduce the 'problem'. Booker argues that constructing wind farms, in particular, is a massive waste of money. Figure 7.21 illustrates the immense gap between the views of the two journalists with regard to global warming.

Much of the debate is to do with the complexity of the problem and the uncertainties contained within global climate models. Many processes have to be considered in climate models, making it inevitable that they become very complex. Large numbers of interacting equations may be required, introducing significant margins of error (see Figure 7.22). Accurately accounting for feedback mechanisms is a major problem in computer modelling. According to Professor Greg O'Hare (*Geography*, spring 2011):

'Some important positive feedbacks which have not been adequately incorporated in the IPCC latest temperature projections, because of the present difficulties of quantifying them, include accelerated ice melt by ice-albedo effects, rapid disintegration of ice sheets by surface and subterranean melt-water and enhanced greenhouse gas emissions from the warming of high latitude permafrost soils, lakes and seas.'

Similarly, negative feedback effects are often inadequately incorporated into models. O'Hare argues that the interaction of positive and negative feedbacks are key to our future climate, stating:

'The importance of such interactions will become more evident in the future as more of them are "discovered" and increasingly better quantified in climate models.'

Global brightening/dimming

The amount of solar radiation reaching the Earth's surface has been recorded since 1923. However, it is only in the last 50 years that a global monitoring network has been taking shape. The data obtained shows that the energy provided by the Sun at the Earth's surface has varied significantly in recent decades, with associated impacts on climate. A reduction in solar radiation causes '**global dimming**', while an intensification of solar radiation causes '**global brightening**'. This is a relatively new field of research.

The amount of solar radiation reaching the Earth's surface fell considerably between the 1950s and the 1980s but has been gradually increasing since 1985. At present it is not clear whether it is clouds or aerosols that trigger global brightening or dimming, or even interactions between the two.

Global dimming is a worldwide decline of the intensity of the sunlight reaching the Earth's surface, caused by particulate air pollution and natural events (e.g. volcanic ash).

Global brightening is an increasing amount of sunlight reaching the Earth's surface.

7.02 Climate change – causes and impacts

6 July 2010

A Bookful of Bookerisms

The climate change deniers are digging themselves an ever deeper hole over 'Amazongate'.

Well this becomes more entertaining by the moment. Those who staked so much on the 'Amazongate' story, only to see it turn round and bite them, are now digging a hole so deep that they will soon be able to witness a possible climate change scenario at first hand, as they emerge, shovels in hand, in the middle of the Great Victoria Desert.

Here's the story so far. In January the rightwing blogger Richard North claimed that the Intergovernmental Panel on Climate Change had 'grossly exaggerated the effects of global warming on the Amazon rain forest'. In 2007 the Panel had claimed that 'up to 40% of the Amazonian forests could react drastically to even a slight reduction in precipitation'. Reduced rainfall could rapidly destroy the forests, which would be replaced with ecosystems 'such as tropical savannahs.'

North asserted that this, 'seems to be a complete fabrication', though he later retracted one of his claims.

His story was picked up by hundreds of other climate change deniers, some of whom went so far as to claim that it destroyed global warming theory. It was also run by the *Sunday Times*, which headlined its report 'UN climate panel shamed by bogus rainforest claim'.

Two weeks ago the *Sunday Times* published a complete retraction and apology, after it found that the story was false.

That, you might think, would be the end of the matter. How wrong you would be. Far from accepting that they had made a mistake, the promoters of this story now seem determined to compound it. On Sunday our old friend Christopher Booker asserted that 'an exhaustive trawl through all the scientific literature on this subject by my colleague Dr Richard North (who was responsible for uncovering 'Amazongate' in the first place), has been unable to find a single study which confirms the specific claim made by the IPCC's 2007 report ... all observed evidence indicates that the forest is much more resilient to climate fluctuations than the alarmists would have us believe.'

There is no doubt that the IPCC made a mistake. Sourcing its information on the Amazon to a report by the green group WWF rather than the abundant peer-reviewed literature on the subject, was a bizarre and silly thing to do. It is also an issue of such mind-numbing triviality, in view of the fact that the IPCC's 2007 reports extend to several thousand pages and contain tens of thousands of references, that I feel I should apologise for taking up more of your time in pursuing it. But the climate change deniers have made such a big deal of it that it cannot be ignored.

It is also true that nowhere in the peer-reviewed literature is there a specific statement that 'up to 40% of the Amazonian forests could react drastically to even a slight reduction in precipitation'. This figure was taken from the WWF report and it shouldn't have been.

But far from 'grossly exaggerating' the state of the science in 2007, as North claimed, the IPCC – because it referenced the WWF report, not the peer-reviewed literature – grossly understated it. The two foremost peer-reviewed papers on the subject at the time of the 2007 report were both published in *Theoretical and Applied Climatology*. They are cited throughout the literature on Amazon dieback.

What do they tell us? That the projection in the IPCC's report falls far, far short of the predicted impacts on the Amazon.

The first paper, by Cox et al, shows a drop in broadleaf tree cover from approximately 80% of the Amazon region in 2000 to around 28% in 2100. That is bad enough, involving far more than 40% of the rainforest. But the forest, it says, will not be largely replaced by savannah:

'When the forest fraction begins to drop (from about 2040 onwards) C4 grasses initially expand to occupy some of the vacant lands. However, the relentless warming and drying make conditions unfavourable even for this plant functional type, and the Amazon box ends as predominantly baresoil (area fraction >0.5) by 2100.'

In other words, the lushest region on earth is projected by this paper to be mostly replaced by desert as a result of global warming (and the consequent reduction in rainfall) this century. I hope I don't have to explain the consequences for biodiversity, the people of the Amazon or climate feedbacks, as the carbon the trees and soil contain is oxidised and released to the atmosphere.

So what does the second paper say? Betts et al go even further. In their model runs: 'By the end of the 21st Century, the mean broadleaf tree coverage of Amazonia has reduced from over 80% to less than 10%.' They are slightly more sanguine about the savannah/desert balance.

'In approximately half of this area, the trees have been replaced by C4 grass leading to a savanna-like landscape. Elsewhere, even grasses cannot be supported and the conditions become essentially desert-like.'

Isn't that reassuring?

Both these papers are referenced elsewhere in the IPCC's 2007 report.

They are not alone. One of the runs in a 1999 paper by White, Cannell and Friend, also published in a peer-reviewed journala, shows almost the entire Amazon basin as desert by the 2080s.

Compare these projections to Booker's claim that 'all observed evidence indicates that the forest is much more resilient to climate fluctuations than the alarmists would have us believe.'

So now the promoters of the Amazongate story have three options. They can persist in claiming that the IPCC was wrong, but this time on the grounds that it underestimated the likely response of the Amazon to climate change. But that would create more problems for them than it solved. They could fall back on their age-old defence and claim that it's all irrelevant, because the scientists' projections for how the Amazon might respond to climate change are based on models. But that would oblige them to suggest a better means of predicting future events. Tealeaves? Entrails? Crystal balls? Or they could quietly slink away before this doomed crusade causes them any more embarrassment, and find something more useful to do.

Booker ends his piece by maintaining that on 'the only occasion' on which I had attempted to expose the misinformation he peddles, I got it wrong and had to apologise to my readers. Yes, I did get one of my claims wrong and I said so as soon as I discovered it. This is where Christopher and I differ: I admit my mistakes, he does not.

Figure 7.21 Booker's work of clanger-dropping fiction (from George Monbiot's blog).

Research continues into these issues and others, including how greatly the effects differ between urban and rural areas, where fewer aerosols are released into the atmosphere. Another challenge to researchers is to incorporate the effects of global dimming/brightening more effectively in climate models in order to understand their impact on climate change better.

7.02.06 Contrasting human perceptions relating to global warming

The computer models used to monitor the carbon cycle have become increasingly complex, thus introducing significant margins of error. One such application involves models that indicate when forest 'carbon sinks' become net carbon generators instead. The results will help pinpoint the effectiveness of trees in offsetting carbon releases that contribute to higher atmospheric temperatures and global climate change. The US Forest Service estimates that the country's forests absorb and store around 750 million tonnes of carbon dioxide each year. Managing forests to optimise carbon sequestration is an important factor in mitigating the effects of climate change.

Figure 7.22 Carbon cycle and computer models.

People hold a variety of perceptions with regard to global warming and other environmental issues. The term 'value system' is sometimes used to describe the package of views that people hold. Personal perception is strongly influenced by the social, economic and political hierarchies that people live in. For example, if you are a coal miner and your family's livelihood and that of most of the people you know depends on coal mining, you are likely to be 'pro-coal' in terms of your views on future energy decisions. However, if you live on a small Pacific island which at its highest point is no more than a few metres above sea level, anything that might increase global warming and thus lead to higher sea levels will be of great concern. Within the social realm, religion may be a significant influence. For example, Buddhism puts a strong emphasis on people living in harmony with all aspects of the world around them.

At the broadest level, environmental viewpoints can be subdivided into:

- an ecocentric worldview, which is life centred
- an antropocentric or technocratic worldview, which is human centred.

In terms of the latter, some people just hope that advances in technology will solve the problem without the need for any substantial changes in the way we lead our lives.

The *Human Development Report 2007/2008* produced a detailed analysis of public perceptions with regard to climate change. It quoted the Pew Global Attitudes Survey, 2006, which found that large majorities of respondents from developed countries had heard of global warming, while awareness remained quite low in several less economically developed countries (LEDCs).

Environmental groups such as Greenpeace and Friends of the Earth actively campaign to change people's perceptions by raising awareness of the changes that are occurring and what individuals can do to help.

SELF-ASSESSMENT QUESTIONS 7.02.05

1. Give three natural causes for climate change.
2. Define: **a** global dimming, **b** global brightening.
3. **Discussion point:** Discuss the very different views of George Monbiot and Christopher Booker.

7.03 Climate change – mitigation and adaptation

LEARNING OBJECTIVES

After reading this chapter you should be able to:

- discuss mitigation attempts to reduce the causes of climate change
- evaluate adaptation attempts to manage the impacts of climate change.

KEY QUESTIONS

How can greenhouse gas emissions be reduced and removed from the atmosphere?

How can societies adapt to reduce the adverse effects of climate change?

What has been the impact of international climate change talks?

7.03.01 Mitigation and adaptation strategies

Management strategies to address the problem of climate change range in scale from the international to the local. **Climate mitigation** measures can be regarded as preventive, while **climate adaptation** measures are clearly reactive. Preventive measures are designed to stop something happening in the first place, while reactive measures tackle an existing problem. Figure 7.23 illustrates the way in which policy responses to climate change have developed.

Mitigation

Mitigation aims to tackle the root causes of climate change by reducing greenhouse gas emissions. The Intergovernmental Panel on Climate Change (IPCC) has defined mitigation as 'an anthropogenic intervention to reduce the sources or enhance the **sinks of greenhouse gases**'.

The more successful mitigation strategies are, the less will be the impacts to which human societies will have to adjust (adaptation). However, both approaches to climate change are necessary, because even if emissions are substantially reduced in the near future, adaptation strategies will be required to address the global changes that have already been set in motion.

Mitigation involves the use of technology and substitution to reduce resource inputs and emissions per unit of output. Some examples of mitigation strategies to reduce greenhouse gases in general are outlined below.

Reducing energy consumption

A report published in 2014 by the American Council for an Energy-Efficient Economy (ACEEE) examined energy usage in 16 of the world's largest economies. Collectively, these countries represent 71 per cent of global electricity consumption. The factors considered included fuel efficiency standards for cars and trucks, efficiency standards for household appliances and the energy consumed to heat a square foot of building.

> **Climate mitigation** refers to strategies that attempt to reduce the causes of climate change.
>
> **Climate adaptation** refers to strategies that attempt to manage the impacts of climate change.
>
> A **greenhouse gas sink** is a process, activity or mechanism that removes a greenhouse gas from the atmosphere.

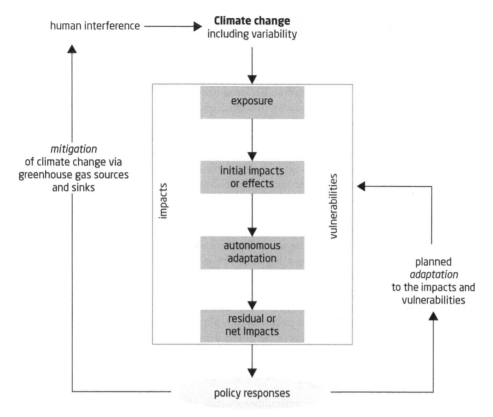

Figure 7.23 Adaptation and mitigation in response to climate change.

The analysis concluded that Germany was the most energy-efficient country, followed by Italy and the EU as a whole (see Figure 7.24). The report stated that: 'Germany is a prime example of a nation that has made energy efficiency a top priority.' In contrast, the report expressed disappointment at the slow rate of progress in the USA, which was ranked 13th out of the 16 countries considered. The report stated: 'The inefficiency in the US economy means a tremendous waste of energy resources and money.'

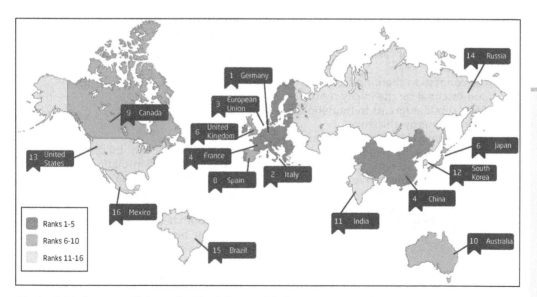

Figure 7.24 Energy efficiency in 16 of the world's largest economies.

> **CONSIDER THIS**
>
> In 2013, emerging economies dominated the global increase in energy consumption, but the increase was below the ten-year average in these countries. China, once again, had the largest growth increment, followed by the USA. Consumption in the EU and Japan fell to the lowest levels since 1995 and 1993 respectively.

7.03 Climate change – mitigation and adaptation

Image 7.12 An example of microgeneration.

A **carbon credit** is a permit that allows an organisation to emit a specified amount of greenhouse gases – also called an emission permit.

Carbon trading is when a company that does not use up the level of emissions it is entitled to can sell the remainder of its entitlement to another company that pollutes above its entitlement.

Community energy is energy produced close to the point of consumption.

Mircogeneration refers to generators producing electricity with an output of less than 50 KW.

A **low-carbon economy** is a country where significant measures have been taken to reduce carbon emissions in all sectors of the economy.

Meeting future energy needs in LEDCs, NICs and more economically developed countries (MEDCs) while avoiding serious environmental degradation will require increased emphasis on radical new approaches. To be effective in the long term, energy solutions must be sustainable. Managing energy supply is often about balancing socioeconomic and environmental needs. We have all become increasingly aware that this requires detailed planning and management. **Carbon credits** and **carbon trading** are an important part of EU environment and energy policies. Under the EU's carbon emissions trading scheme, heavy industrial plants have to buy permits (engage in carbon trading) to emit greenhouse gases over the limit they are allowed, through their carbon credits, by government. However, this could be extended to other organisations such as banks and supermarkets.

Many countries are looking increasingly at the concept of **community energy**, which is invariably renewable energy. Much energy is lost in transmission if the source of supply is far from the point of consumption. Energy produced locally is much more efficient. This invariably involves **microgeneration** (see Image 7.12).

Table 7.10 summarises some of the measures governments and individuals can undertake to reduce the demand for energy and thus move towards a more sustainable situation. Recycling by all sectors of the economy is a vital component along the path to a **low-carbon economy**.

Reducing emissions of oxides of nitrogen and methane from agriculture

Global greenhouse gas emissions from agriculture increased by 13 per cent between 1990 and 2010. Agriculture is the third-largest contributor to greenhouse gas emissions by sector (after energy and transportation). Methane accounts for almost half of agricultural emissions, followed

Government measures	Individual measures
• Improve public transport to encourage higher levels of usage. • Set a high level of tax on petrol, aviation fuel, etc. • Ensure that public utility vehicles are energy efficient. • Set minimum fuel consumption requirements for cars and commercial vehicles. • Use congestion charging to deter non-essential car use in city centres. • Offer subsidies/grants to households to improve energy efficiency. • Encourage business to monitor and reduce its energy usage. • Encourage recycling. • Promote investment in renewable forms of energy. • Pass laws to compel manufacturers to produce higher-efficiency electrical products.	**Transport:** • Walk rather than drive for short local journeys. • Use a bicycle for short to moderate distance journeys. • Buy low fuel consumption/low emission cars. • Reduce car usage by planning more 'multi-purpose' trips. • Use public rather than private transport. • Practise car pooling. **In the home:** • Use low-energy light bulbs. • Install cavity wall insulation. • Improve loft insulation. • Turn boiler and radiator settings down slightly. • Wash clothes at lower temperatures. • Purchase high energy efficiency appliances. • Do not leave appliances on standby.

Table 7.10 Examples of energy conservation measures.

by nitrous oxide (36 per cent) and carbon dioxide (14 per cent). The largest source of methane emissions is the digestion of organic materials by livestock.

Ways to reduce emissions from agriculture include:

- reducing global consumption of meat and dairy products
- applying fertiliser more efficiently
- planting fallow fields with nitrogen-fixing legume crops
- reducing soil tillage.

Increasing the use of renewable energy as an alternative to fossil fuels

The last decade has witnessed a considerable advance in renewable energy technology, along with a reduction in cost per unit of energy. An increasing number of countries have become aware of the disadvantages of burning fossil fuels and have developed strategies to increase the use of renewable energy. However, fossil fuels still dominate the global energy mix (see Chapter 7.01).

Geoengineering

Geoengineering encompasses both solar radiation management, or solar geoengineering, and carbon dioxide removal (CDR) or carbon geoengineering (see Figure 7.25). The effectiveness and cost of such techniques is likely to vary significantly, as Figure 7.25 illustrates. Geoengineering is a science that is very much in its infancy. While some techniques are viable with current technology, most are at the theoretical stage with research still in progress.

> **Geoengineering** is deliberate large-scale intervention in the Earth's natural systems to counteract climate change. Geoengineering techniques can be grouped into two categories: solar radiation management and carbon dioxide removal (CDR).

Solar radiation management techniques

These techniques aim to reflect some of the Sun's energy back into space to try to counteract global warming. According to the Oxford Geoengineering Programme, proposed solar radiation management includes:

- *Stratospheric aerosols:* The idea here is to place small reflective particles into the upper atmosphere to reflect a certain amount of insolation before it reaches the Earth's surface. This would mimic the effect of large volcanic eruptions blocking out solar radiation.
- *Albedo enhancement:* This technique refers to increasing the reflectiveness of clouds and land surfaces (e.g. whitening of roofs) so that more of the Sun's energy is reflected back into space.
- *Space reflectors:* Placed outside the Earth's atmosphere, such reflectors would block a small proportion of the Sun's energy before reaching the Earth.

Carbon dioxide removal techniques

The objective of these techniques is to remove carbon dioxide from the atmosphere to directly reduce the enhanced greenhouse effect and ocean acidification. Widespread application would be required to have a meaningful impact. CRD techniques might include:

- *Afforestation:* Trees are planted on a global scale. This would be the key strategy in protecting and enhancing carbon sinks through land management. An important current programme is the United Nations Collaborative Programme on Reducing Emissions from Deforestation and Forest Degradation in Developing Countries (UN-REDD Programme). It was established in 2008 to help developing countries build capacity to reduce emissions. This programme is based on a partnership among developing countries to embark on low-carbon, climate-resilient development, with developed countries providing funding as a significant incentive. The programme began with nine pilot country programmes.

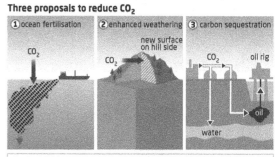

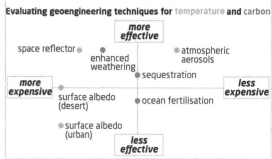

Figure 7.25 Some proposals to reduce temperature and carbon dioxide.

7.03 Climate change – mitigation and adaptation

- *Ambient air capture:* Large machines are constructed that are capable of removing carbon dioxide directly from ambient air and storing it elsewhere.
- *Carbon capture and storage or sequestration (CCS):* This technology can capture up to 90 per cent of carbon dioxide emissions from the use of fossil fuels, thus preventing such carbon dioxide from entering the atmosphere (see Figure 7.26).

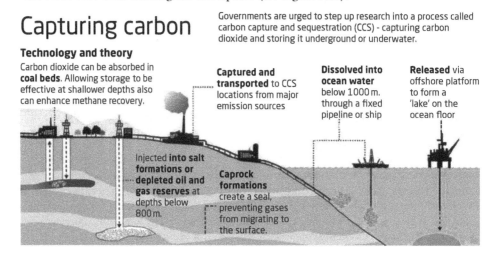

Figure 7.26 Diagram showing carbon capture and storage.

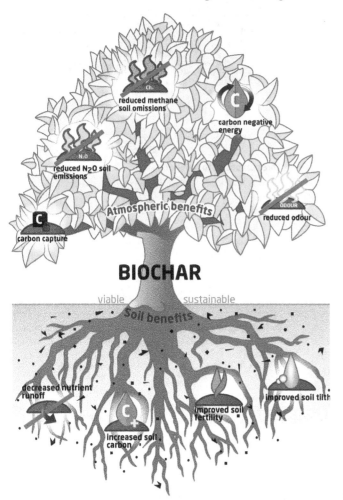

Figure 7.27 Diagram of biochar.

- *Biochar:* Biochar is a solid material obtained from the carbonisation (burning) of biomass. The process has advantages for both the soil and the atmosphere (see Figure 7.27). Biochar converts agricultural waste into a soil enhancer that can:
  - hold carbon
  - boost food security
  - increase soil biodiversity
  - discourage deforestation.

  This practice of burning biomass goes back about 2000 years. The carbon in biochar resists degradation and can hold carbon in soils for many hundreds of years.

- *Bio-energy with carbon capture and sequestration:* Biomass is grown to produce energy and then the carbon dioxide created in the process is captured and sequestered.
- *Enhanced weathering:* Large quantities of minerals that will react with carbon dioxide are exposed. The resulting compound is then stored at sea or below the land surface.
- *Ocean alkalinity enhancement:* Ground-up or dissolved rocks such as limestone are placed in the oceans to increase its ability to store carbon and reduce ocean acidification.
- *Ocean fertilisation:* Nutrients are added to the ocean to increase primary production, which draws down carbon dioxide from the atmosphere. Compounds of nitrogen, phosphorus and iron act to encourage the 'biological pump'. An alternative is to increase upwellings to release nutrients to the surface.

According to the UK's Royal Society, 'There is no single geoengineering "silver bullet" that should be pursued as an all-encompassing solution to climate change. Stratospheric aerosols would appear to offer the best return for the least investment […], but the side-effects they might have on the environment present an unknown risk.' In early 2015, the National Research Council of the

US National Academy of Sciences argued that removing carbon dioxide from the atmosphere is the best option to tackle climate change. The National Research Council thought that the use of geoengineering to reduce the amount of solar energy reaching the Earth was too risky with available knowledge.

Table 7.11, from the Intergovernmental Panel on Climate Change (IPCC), illustrates mitigation technologies, policies and measures by different economic sectors. This detailed summary also considers the constraints and opportunities involved with the different strategies.

There is evidence to show that an effective carbon-price signal could realise considerable mitigation potential in all sectors of an economy. There is also general agreement that mitigation actions can result in near-term co-benefits such as health improvements due to reduced air pollution.

However, even if mitigation strategies drastically cut future emissions of greenhouse gases, past emissions will continue to have an effect for decades to come.

Sector	Key mitigation technologies and practices currently commercially avaliable. *Key mitigation technologies and practices projected to be commercialised before 2030 shown in italics*	Policies, measures and instruments shown to be environmentally effecitve	Key constraints or opportunities (Normal font = constraints; *italics* = *opportunities*)
Energy supply	Improved supply and distribution efficiency; fuel switching from coal to gas; nuclear power; renewable heat and power (hydropower, solar, wind, geothermal and bioenergy); combined heat and power, early applications of carbon dioxide capture and storage (CCS) (e.g. storage of removed CO_2 from natural gas); *CCS for gas, biomass and coal-fired electricity generating facilities; advanced nuclear power, advanced renewable energy, including tidal and wave energy, concentrating solar, and solar photovoltaics*	Reduction of fossil fuel subsidies; taxes or carbon charges on fossll fuels Feed-in tariffs for renewable energy technologies; renewable energy obligations; producer subsidies	Resistance by vested interests may make them difficult to implement *May be appropriate to create markets for low-emissions technologies*
Transport	More fuel-efficient vehicles; hybrid vehicles; cleaner diesel vehicles; blofuels; modal shifts from road transport to rail and public transport systems; non-motorised transport (cycling, walking); land-use and transport planning; *second-generation blofuels; higher efficiency aircraft; advanced electric and hybrid vehicles with more powerful and reliable batteries*	Mandatory fuel economy; biofuel blending and CO_2 standards for road transport Taxes on vehicle purchase, registration, use and motor fuels; road and parking pricing Influence mobility needs through land-use regulations and infrastructure planning; investment in attractive public transport facilities and non-motorised forms of transport	Partial coverage of vehicle fleet may limit effectiveness Effectiveness may drop with high incomes *Particularly appropriate for countries that are building up their transportation systems*
Buildings	Efficient lighting and daylighting; more efficient electrical appliances and heating and cooling devices; improved cooking stoves, improved insulation; passive and active solar design for heating and cooling; alternative refrigeration fluids, recovery and recycling of fluorinated gases, *integrated design of commercial buildings including technologies such as intelligent meters that provide feedback and control; solar photovoltaics integrated in buildings*	Appliance standards and labelling Building codes and certification Demand-side management programmes Public sector leadership programmes, including procurement Incentives for energy service companies (ESCOs)	Periodic revision of standards needed Attractive for new buildings. Enforcement can be difficult Need for regulations so that utilities may profit *Government purchasing can expand demand for energy-efficient products* *Success factor. Access to third party financing*

(*continued*)

7.03 Climate change – mitigation and adaptation

Sector	Key mitigation technologies and practices currently commercially avaliable. *Key mitigation technologies and practices projected to be commercialised before 2030 shown in italics*	Policies, measures and instruments shown to be environmentally effecitve	Key constraints or opportunities (Normal font = constraints; *italics* = *opportunities*)
Industry	More efficient end-use electronic equipment; heat and power recovery; material recycling and substitution; control of non-CO_2 gas emissions; and a wide array of process-specific technologies; *advanced energy efficiency; CCS for cement ammonia, and iron manufacture; inert electrodes for aluminium manufacture*	Provision of benchmark information; performance standards; subsidies; tax credits Tradable permits Voluntary agreements	May be appropriate to stimulate technology uptake. Stability of national policy important in view of international competitiveness Predictable allocation mechanisms and stable price signals important for investments Success factors include clear targets, a baseline scenario, third-party involvement in design and review and formal provisions of monitoring, close cooperation between government and industry
Agriculture	Improved crop and grazing land management to increase soil carbon storage; restoration of cultivated peaty soils and degraded lands, improved rice cultivation techniques and livestock and manure management to reduce CH_4 emissions; improved nitrogen fertiliser application techniques to reduce N_2O emissions; dedicated energy crops to replace fossil fuel use; improved energy efficiency; *improvements of crop yields*	Financial incentives and regulations for improved land management; maintaining soil carbon content; efficient use of fertilisers and irrigation	*May encourage synergy with sustainable development and with reducing vulnerability to climate change, thereby overcoming barriers to implementation*
Forestry/ forests	Afforestation; reforestation; forest management; reduced deforestation; harvested wood product management; use of forestry products for bioenergy to replace fossil fuel use; *tree species improvement to increase biomass productivity and carbon sequestration; improved remote sensing technologies for analysis of vegetation/soil carbon sequestration potential and mapping land-use change*	Financial incentives (national and international) to increase forest area, to reduce deforestation and to maintain and manage forests; land-use regulation and enforcement	Constraints include lack of investment capital and land tenure issues. *Can help poverty alleviation*
Waste	Landfill CH_4 recovery; waste incineration with energy recovery; composting of organic waste; controlled wastewater treatment; recycling and waste minimisation, *biocovers and biofilters to optimise CH_4 oxidation*	Financial incentives for improved waste and wastewater management Renewable energy incentives or obligations Waste management regulations	*May stimulate technology diffusion* Local availability of low-cost fuel Most effectively applied at national level with enforcement strategies

Table 7.11 Examples of key sectoral mitigation technologies, policies and measures.

SELF-ASSESSMENT QUESTIONS 7.03.01

1. Distinguish between climate mitigation and climate adaptation.
2. State three ways in which greenhouse gas emissions from agriculture can be reduced.
3. **Discussion point:** What is the difference between solar radiation management and CDR?

CASE STUDY 7.03.01

The UK's falling energy consumption

The UK's energy balance featured in the case study in Chapter 7.01. You may be surprised to know that energy consumption in the UK has fallen significantly. The UK's primary energy consumption in 2012 was at its lowest since 1985. This has been despite a significant increase in population during this period. Between 1990 and 2012, primary energy consumption in the UK fell by 7 per cent. This is a common trend in developed countries, due largely to considerable improvements in energy efficiency in terms of both producing energy and using it, and the global shift of many energy-hungry manufacturing industries to lower-cost emerging economies. Think how the average home has become more energy efficient in the last 30 years or so with improved insulation, energy-efficient lightbulbs, and more energy-efficient electrical appliances.

Within the UK, the former traditional industrial areas are a shadow of their former selves in terms of industrial production, manufacturing employment and industrial impact on the environment. Such a decline in consumption is in marked contrast to rapid increases in energy demand in newly industrialised countries such as China, India, Brazil and Mexico. The UK economy is now dominated by the service sector, which is considerably less energy intensive than manufacturing. **Energy intensity** is the ratio between energy consumption and the wealth produced by this energy in terms of a country's GDP.

Compared with 1970, consumption of energy by households today has declined by 12 per cent. For industry, the figure is a huge 60 per cent. However, this has been largely offset by a 50 per cent rise in energy use in the transport sector, which has expanded at a substantial rate since 1970. The amount of energy required by transport has increased considerably with the steady rise in the number of vehicles on the UK's roads. In 1970, the number of cars on the country's roads was 10 million, compared with more than 27 million today!

The Department of Energy and Climate Change predicts that energy efficiencies will continue to offset population growth, so the UK is likely to use about the same amount of energy in 2030 as it does today. Over the past decade, the UK has made a substantial investment in renewable sources of energy, and indications are that this trend will continue. The Department of Energy and Climate Change expects that by 2030 renewables will be by far the biggest source of energy used in electricity generation, making up about 40 per cent of the overall balance. The commitment to include nuclear electricity as a significant element of the UK's energy balance is also clear. This is all part of the overall objective of maintaining a range of sources of energy in the overall energy balance for the future. Old coal-fired power stations will continue to close, and it is clear that the future of coal as a significant element of energy supply lies in the application of clean-coal technology. This new technology has developed forms of coal that burn with greater efficiency and captures coal's pollutants before they are emitted into the atmosphere.

In 2014, the UK Department of Energy and Climate Change made an additional £2.5 million available to encourage development of carbon dioxide storage in the North Sea. This is one of two CCS projects under active consideration, the other being in Yorkshire.

Questions

1. By how much did primary energy consumption fall in the UK between 1990 and 2012?
2. Why has energy consumption by manufacturing industry in the UK declined by such a large amount?
3. Comment on the changing energy consumption of the transport sector in the UK.

Adaptation

Adaptation strategies aim to lower the risks posed by the consequences of climate change. The IPCC has defined adaptation as follows: 'Adaptation to climate change refers to adjustment in natural or human systems in response to actual or expected climatic stimuli or their effects, which moderates harm or exploits beneficial opportunities. Various types of adaptation can be

Energy intensity is a measure of the energy efficiency of a nation's economy. It is calculated as units of energy per unit of GDP.

7.03 Climate change – mitigation and adaptation

distinguished, including anticipatory and reactive adaptation, private and public adaptation, and autonomous and planned adaptation.'

Adaptation measures may be planned in advance or put in place immediately in response to a local pressure. They can vary in scale from the relatively small to large-scale infrastructure changes (see Table 7.12). Examples include:

- building coastal defences to protect against rising sea levels
- improving the quality of road surfaces to withstand higher temperatures
- developing drought-resistant crops in regions where precipitation is declining
- vaccination programmes to cope with changing spatial patterns of disease
- planting trees and expanding green spaces in urban areas to moderate increases in temperature
- constructing desalination plants
- behavioural changes such as individual and group decisions to reduce environmental impact.

Adaptation is crucial to reducing vulnerability to climate change. The term **adaptive capacity** is used to describe the potential to adjust in order to minimise negative impacts and maximise any benefits from climate change. Adaptive capacity varies from place to place and can depend on financial and technological resources.

Adaptive capacity is the capacity of a system to adapt if the environment where the system exists is changing.

Sector	Adaptation option/strategy	Underlying policy framework	Key constraints and opportunities to implementation (Normal font = constraints; *italics = opportunities*)
Water	Expanded rainwater harvesting; water storage and conservation techniques; water re-use; desalination; water-use and irrigation efficiency	National water policies and integrated water resources management; water-related hazards management	Financial, human resources and physical barriers; *integrated water resources management; synergies with other sectors*
Agriculture	Adjustment of planting dates and crop variety; crop relocation; improved land management, e.g. erosion control and soil protection through tree planting	R&D policies; institutional reform; land tenure and land reform; training; capacity building; crop insurance; financial incentives, e.g. subsidies and tax credits	Technological and financial constraints; access to new varieties; markets; *longer growing season in higher latitudes; revenues from 'new' products*
Infrastructure/settlement (including coastal zones)	Relocation; seawalls and storm surge barriers; dune reinforcement; land acquisition and creation of marshlands/ wetlands as buffer against sea level rise and flooding; protection of existing natural barriers	Standards and regulations that integrate climate change considerations into design; land-use policies; building codes; insurance	Financial and technological barriers; availability of relocation space; *integrated policies and management; synergies with sustainable development goals*
Human health	Heat-health action plans; emergency medical services; improved climate-sensitive disease surveillance and control; safe water and improved sanitation	Public health policies that recognise climate risk; strengthened health services; regional and international cooperation	Limits to human tolerance (vulnerable groups); knowledge limitations; financial capacity; *upgraded health services; improved quality of life*
Tourism	Diversification of tourism attractions and revenues; shifting ski slopes to higher altitudes and glaciers; artificial snow-making	Integrated planning (e.g. carrying capacity; linkages with other sectors); financial incentives, e.g. subsidies and tax credits	Appeal/marketing of new attractions; financial and logistical challenges; potential adverse impact on other sectors (e.g. artifical snow-making may increase energy use); *revenues from 'new' attractions; involvement of wider group of stakeholders*

(continued)

Sector	Adaptation option/strategy	Underlying policy framework	Key constraints and opportunities to implementation (Normal font = constraints; *italics* = *opportunities*)
Transport	Re-alignment/relocation; design standards and planning for roads, rail and other infrastructure to cope with warming and drainage	Integrating climate change considerations into national transport policy; investment in R&D for special situations, e.g. permafrost areas	Financial and technological barriers; availability of less vulnerable routes; *improved technologies and integration with key sectors (e.g. energy)*
Energy	Strengthening of overhead transmission and distribution infrastructure; underground cabling for utilities; energy efficiency; use of renewable sources; reduced dependence on single sources of energy	National energy policies, regulations, and fiscal and financial incentives to encourage use of alternative sources; incorporating climate change in design standards	Access to viable alternatives; financial and technological barriers; acceptance of new technologies; *stimulation of new technologies; use of local resources*
Note: Other examples from many sectors would include early warning systems.			

Table 7.12 Examples of planned adaptation by economic sector.
Source: http://www.ipcc.ch/publications_and_data/ar4/syr/en/spms4.html

The IPCC views vulnerability to climate change as being determined by three factors:
- exposure to hazards
- sensitivity to those hazards
- the capacity to adapt to those hazards.

Adaptation strategies can reduce vulnerability in varying degrees, for example by lowering sensitivity or improving adaptive capacity.

Urban areas

Much of the effort to reduce greenhouse gas emissions will focus on urban areas. By conservative estimates, the urban areas of the world emit at least 40 per cent of greenhouse gases. However, if their consumption of electricity, food and other commodities is included, the figure is much higher. The planning of new eco cities has received much attention in recent years, but adapting existing metropolises is the more realistic alternative for the majority of the world's urban population. This will involve both high-tech and low-tech changes. Numerous planned eco cities around the world have not gone ahead, mainly because of cost.

International agreements

At the global scale, significant stages in international agreement on climate change include:
- The first global conference on climate change took place in Geneva in 1979.
- The Toronto Conference, in 1988, called for a reduction in carbon dioxide emissions by 20 per cent of 1988 levels by 2005.
- In 1988, the Intergovernmental Panel on Climate Change (IPCC) was established by two UN organisations, the World Meteorological Organization (WMO) and the United Nations Environment Programme (UNEP).
- The United Nations Framework Convention on Climate Change (UNFCCC) was negotiated in 1992 at the Earth Summit held in Rio de Janeiro. The stated objective of the treaty was to stabilise greenhouse gas in the atmosphere at a level that would prevent dangerous anthropogenic interference with the climate system.
- The Kyoto Protocol was established in 1997 and eventually ratified in 2005. It gave all MEDCs legally binding targets for cuts in emissions from the 1990 level by 2008–2012. However, the USA refused to sign the agreement and some other key countries largely disregarded their commitments. These shortcomings allowed emissions to rise at a much faster rate than planned at the time.

7.03 Climate change – mitigation and adaptation

- The UN Climate Change Conference in Bali, Indonesia, in 2007 set out the Bali Action Plan – a two-year process to finalise a binding agreement in Copenhagen in 2009.
- The Copenhagen Climate Change Conference took place in 2009. The weak agreement drew much criticism from many quarters. John Sauven, Executive Director of Greenpeace UK, concluded, 'The city of Copenhagen is a crime scene tonight, with the guilty men and women fleeing to the airport. There are no targets for carbon cuts and no agreement on a legally binding treaty.'
- The Cancun (Mexico) agreements of 2010 stated that global warming should be limited to less than 2.0 °C relative to the preindustrial level. As part of the Cancun agreements, both MEDCs and LEDCs submitted mitigation plans to the UNFCCC.
- In 2011, the Durban Platform for Enhanced Action was adopted.
- In 2012, the Doha (Qatar) Conference agreed to extend the life of the Kyoto Protocol until 2020. Amendments to the Kyoto Protocol were adopted. The wording adopted by the conference incorporated for the first time the concept of 'loss and damage', an agreement in principle that richer nations could be financially responsible to other nations for their failure to reduce carbon emissions.
- In 2014, the Lima climate change talks agreed on a plan to fight global warming that would for the first time commit all countries to cutting their greenhouse gas emissions.

National Adaptation Programmes of Action

A National Adaptation Programme of Action (NAPA) is a request from the least developed countries for financial assistance for adaptation to climate change. A NAPA is a list of ranked priority adaptation activities and projects. According to the UNFCCC, 'NAPAs provide a process for Least Developed Countries to identify priority activities that respond to their urgent and immediate needs to adapt to climate change – those for which further delay would increase vulnerability and/or costs at a later stage.' By the end of 2013, 50 NAPAs had been received by the UNFCCC. Table 7.13 shows the NAPA submitted by Tanzania.

The Least Developed Country Fund (LDCF) was set up to:

- finance the preparation of NAPAs
- implement proposed projects.

The LDCF has financial resources of over $400 million, of which almost $180 million has been approved for 47 projects. More than $550 million of co-financing has been attracted in the process.

Priority project number	Priority project title	Priority project cost/US$
1	Improving food security in drought-prone areas by promoting drought-prone tolerant crops	8 500 000
2	Improving water availability to drought-stricken communities in the central part of the country	800 000
3	Shifting of shallow water wells affected by inundation on the coastal regions of Tanzania mainland and Zanzibar	3 300 000
4	Climate change adaptation through participatory reforestation in Kilimanjaro mountain	3 300 000
5	Community-based mini-hydro for economic diversification as a result of climate change in same district	620 000
6	Combating malaria epidemic in newly mosquito-infested areas	650 000

Table 7.13 NAPA submitted by Tanzania.
Source: UNFCCC

Differences of opinion

It continues to be difficult to achieve meaningful international agreement. It seems that all countries want to reduce emissions, but they want to bear as little as possible of the cost of doing so themselves. In many countries, there is considerable debate about what should be done to reduce greenhouse gas emissions (see Image 7.13). Both MEDCs and LEDCs are concerned about how a rapid reduction in emissions will affect their rates of economic growth. The global financial crisis that began in 2008, the repercussions of which were still being felt in 2015, has made this situation even more difficult. The LEDCs argue that most of the current problem is due to the industrialisation of the MEDCs (see Image 7.13) and as such the MEDCs should pick up most of the bill. However, the counterargument is that increases in current emissions have been driven by rapid economic growth in NICs. Economic growth has been particularly fast in China and India, and the main source of energy in both countries is coal, which is the most polluting of the fossil fuels.

Local and personal initiatives

Look again at Table 7.10. What measures have you undertaken as an individual to change to a greener lifestyle? Is this something you have discussed as a family and, if so, have you acted on it? Has your school/college attempted to reduce its impact on the environment? What about your local community as a whole?

Image 7.13 Pollution from a factory in British Columbia, Canada.

Some institutions (and individuals) have undertaken 'eco audits' to assess their impact on the environment. More and more households are installing smart meters so they can keep a regular check on the energy they are using and how they are using it. The argument here is that, if you can quantify or at least part-quantify environmental impact, this can form the foundation for an action plan. As energy prices continue to rise, the economic incentive to conserve energy also increases.

In recent years the idea of personal carbon allowances has been advanced by some organisations and academics. This system would allow or require each individual person to contribute to reducing emissions. For the average individual in the UK, personal emissions are made up of:

- 45 per cent from household energy use
- 28 per cent from surface transport (car, bus and rail)
- 27 per cent from air travel.

Emissions from international air travel are rising at the fastest rate.

Theory of knowledge 7.03.01

The first global conference on climate change took place in Geneva in 1979. Since then there have been many more international conferences on climate change, each one designed to move forward towards binding agreements to address the most important problem facing global society. After each conference there is a range of opinion about its relative success. The views of national governments and groups of countries often vary considerably. People within individual countries may also differ significantly in how they view climate negotiations.

1. How difficult is it to come to a precise assessment of a particular agreement on climate change?
2. What do you need to do to arrive at an accurate assessment of the value of an agreement?

SELF-ASSESSMENT QUESTIONS 7.03.02

1. Give three examples of climate adaptation strategies.
2. What do you understand by the term 'adaptive capacity'?
3. **Research idea:** What publications are available in your country to help you plan a greener lifestyle?

End-of-topic questions

1. What is energy security and why has it become a matter of increasing concern? [4]

2. Table 7.14 shows US greenhouse gas emissions by economic sector.

Sector	1990	2009
electric power industry	1868.9	2193.0
transportation	1545.2	1812.4
industry	1564.4	1322.7
agriculture	429.0	490.0
commercial	395.5	409.5
residential	345.1	360.1
US territories	33.7	45.5
total emissions	6181.8	6633.2

Table 7.14 US greenhouse gas emissions by economic sector in million metric tonnes (MMT) of carbon dioxide equivalents.
Source: Selected data from table at www.epa.gov/climatechange/emissions

   a. Describe the change in total US emissions of greenhouse gases from 1990 to 2009. [1]

   b. Suggest *two* reasons why emissions from industry declined, while all other sectors increased their emissions. [2]

   c. Emissions from transportation increased by over 17 per cent from 1990 to 2009. Explain why. [2]

   d. Suggest why the US electric power industry creates such a high level of emissions. [2]

3. How can geoengineering mitigate against climate change? [4]

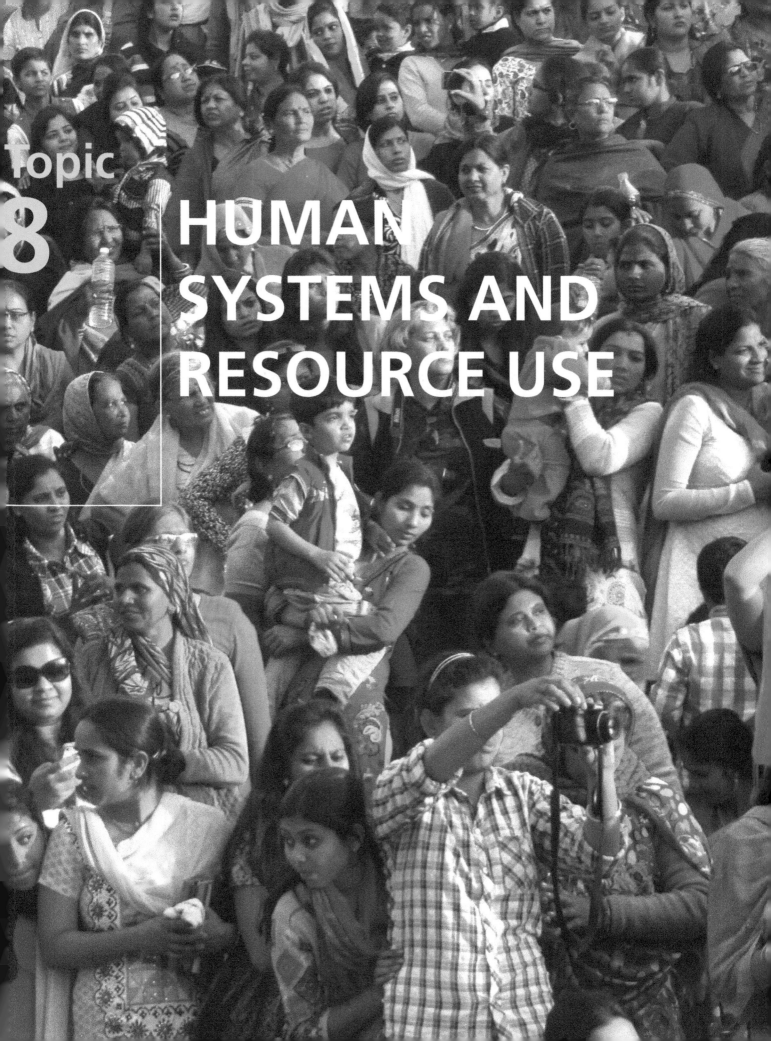

Topic 8
HUMAN SYSTEMS AND RESOURCE USE

8.01 Human population dynamics

LEARNING OBJECTIVES

After reading this chapter you should be able to:

- understand that human population growth rates are impacted by a complex range of changing factors
- appreciate that a variety of models and indicators are employed to quantify human population dynamics.

KEY QUESTIONS

What demographic tools are used to quantify human populations?

How rapidly has the human population increased and what are the forecasts for the future?

How is human population growth placing stress on the Earth's systems?

Carrying capacity is the largest number of individuals in a population that the resources in the environment can support for an extended period of time.

Demographers are professionals who study the characteristics (e.g composition, distribution, trends) of human populations.

8.01.01 The rapid increase in global population

During most of the period since humankind first evolved, global population was very small, reaching perhaps 125 000 people a million years ago, although there is not enough evidence to be precise about population in the distant past. It has been estimated that 10 000 years ago, when people first began to domesticate animals and cultivate crops, world population was no more than 5 million. Known as the Neolithic Revolution, this period of economic change significantly altered the relationship between people and their environments. But even then the average annual growth rate was less than 0.1 per cent per year, extremely low compared with contemporary trends.

However, as a result of technological advance, the **carrying capacity** of the land improved and population increased. By 5500 years ago global population reached 30 million, and by 2000 years ago (see Image 8.01) this had risen to about 250 million.

Demographers estimate that the world population reached 500 million by about 1650. From this time, population grew at an increasing rate. By 1800, global population had doubled to reach one billion (see Figure 8.01).

Extension

For a fun and interesting activity: find your number in the world counter by visiting the website below:
www.bbc.com/news/world-15391515

Figure 8.02 shows the time taken for each subsequent billion to be reached, with the global total reaching 6 billion in 1999. It had taken only 12 years for world population to increase from 5 to 6 billion, a year less than the timespan required for the previous billion to be added. According to the United Nations Population Fund, global population reached 7 billion on Monday 31 October 2011, with another 12-year gap from the

Image 8.01 The Great Wall of China – the history of the Great Wall goes back more than 2000 years when world population was only about 250 million.

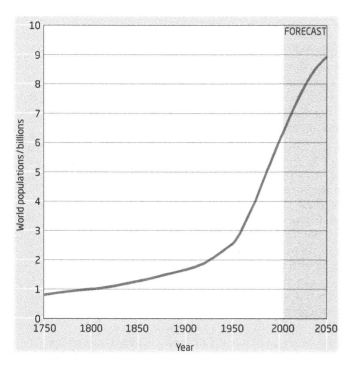

Figure 8.01 Graph: world population growth by each billion.

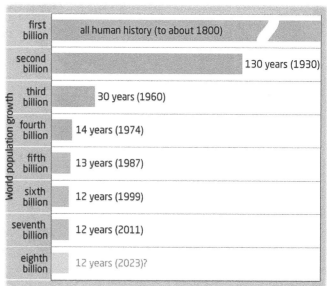

Figure 8.02 Population growth 1750–2050.

previous billion. The population in 2011 was double that in 1967. Table 8.01 shows population change in 2014, with a global population increase of almost 86.6 million in that year. The vast majority of this increase is in the less economically developed countries (LEDCs).

		World	MEDCs	LEDCs
Population		7 238 184 000	1 248 958 000	5 989 225 000
Births per	year	143 341 000	13 794 000	129 547 000
	day	392 714	37 792	354 923
	minute	273	26	246
Deaths per	year	56 759 000	12 328 000	44 432 000
	day	155 505	33 775	121 730
	minute	108	23	85
Natural increase (births – deaths) per	year	86 581 000	1 466 000	85 115 000
	day	237 209	4017	233 193
	minute	165	34	162
Infant deaths per	year	5 507 000	72 000	5 435 000
	day	15 087	197	14 890
	minute	10	0.1	10

Table 8.01 World population clock, 2014.
Source: 2014 Population Reference Bureau, World Population Data Sheet

Extension

For more information on the world's population and its history, visit the world population clock
www.prb.org/Multimedia/Video/2014/world-population-billions.aspx
or
www.prb.org/Publications/Articles/2002/HowManyPeopleHaveEverLivedonEarth.aspx

8.01 Human population dynamics

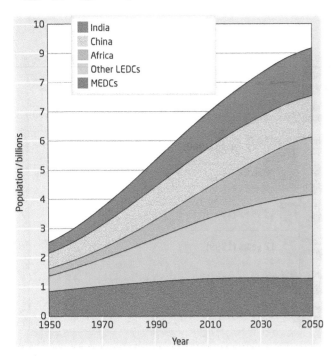

Figure 8.03 Graph – population growth in MEDCs and LEDCs, 1950–2050.

Population momentum is the tendency for population growth to continue beyond the time when replacement-level fertility has been reached, because of a relatively high concentration of people in their childbearing years. This situation is due to past high fertility rates which result in a large number of young people.

Recent demographic change

Figure 8.03 shows that both the total population and the rate of population growth are much higher in LEDCs compared with more economically developed countries (MEDCs). However, only since the Second World War has population growth in LEDCs overtaken that in MEDCs. The MEDCs had their period of high population growth in the 19th and early 20th centuries, while in the LEDCs, high population growth has occurred mainly since 1950.

The highest-ever global population growth rate was reached in the early to mid-1960s, when population growth in LEDCs peaked at 2.4 per cent a year. At this time, the term 'population explosion' was widely used to describe this rapid population growth. But by the late 1990s the rate of population growth was down to 1.8 per cent, and by 2014 it had declined to 1.2 per cent. However, even though the rate of growth had been falling for three decades, **population momentum** meant that the numbers being added each year did not peak until the late 1980s, and they remain at a very high level.

The global population has followed a rapid growth curve that has slowed in recent decades. There is uncertainty about how this may change in the future. The rapid increase in the world's population has placed increasing stress on all the Earth's systems, and there are enormous concerns about how the planet will cope in the future at various levels of population growth.

The demographic transformation, which took a century to complete in MEDCs, has occurred in a generation in some LEDCs. Fertility has dropped further and faster than most demographers foresaw 20–30 years ago. The exception is in Africa, where in over 20 countries families of at least five children are the norm, and population growth is still around 2.5 per cent. Table 8.02 shows the ten largest countries in the world in population size and their population projections for 2050. China and India together currently make up almost 37 per cent of the world's population. Because India's population (Image 8.02) is growing at a faster rate, it should overtake China in population size by about 2030.

2014		2050	
Country	Population / millions	Country	Population / millions
China	1364	India	1657
India	1296	China	1312
USA	318	Nigeria	396
Indonesia	251	USA	395
Brazil	203	Indonesia	365
Pakistan	194	Pakistan	348
Nigeria	177	Brazil	226
Bangladesh	158	Bangladesh	202
Russia	144	Congo, Democratic Republic	194
Japan	127	Ethiopia	165

Table 8.02 The 10 most populous countries in the world in 2014 and 2050.
Source: 2014 Population Reference Bureau, World Population Data Sheet

The concept of exponential growth

Exponential growth occurs when the growth rate of a mathematical function is proportional to the function's current value. This means an increase in number or size at a constantly growing rate. It is also caused geometric growth. The exponential growth model is also known as the Malthusian growth model.

The Reverend Malthus (1766–1834) produced his 'Essay on the Principle of Population' in 1798. He said that the crux of the population problem was 'the existence of a tendency in mankind to increase, if unchecked, beyond the possibility of an adequate supply of food in a limited territory'. Malthus thought that an increased food supply was achieved mainly by bringing more land into arable production. He maintained that, while the supply of food could, at best, only be increased by a constant amount in arithmetical progression (1, 2, 3, 4, 5, 6), the human population tends to increase in geometrical progression (1, 2, 4, 8, 16, 32), multiplying itself by a constant amount each time (see Figure 8.04). In time, population would outstrip food supply until a catastrophe occurred in the form of famine, disease or war. These limiting factors maintained a balance between population and resources in the long term.

Image 8.02 A crowd at a public occasion in Amritsar, India.

Clearly Malthus could not have foreseen the great technological advances that were to unfold in the following two centuries. There have been many advances in agriculture since the time of Malthus which have contributed to huge increases in agricultural production. These advances include: the development of artificial fertilisers and pesticides, new irrigation techniques, high-yielding varieties of crops, cross-breeding of cattle, greenhouse farming and the reclamation of land from the sea.

The rapid increase in the world's population since the mid-18th century is generally thought of as an example of exponential growth. Look at the way the rate of population growth increases in Figure 8.01 and at how the time taken for each billion to be added to world population kept decreasing from the first to the sixth billion. However, it is important to remember that the rate of global population growth has been slowing down since the early to mid-1960s. But, because the absolute increase in population each year remains substantial, the term 'exponential growth' is still often applied, rather loosely in the opinion of some writers. However, the important matter is the pressure placed on the Earth's resources by a high rate of population growth, even though it may no longer be mathematically correct to refer to it as exponential.

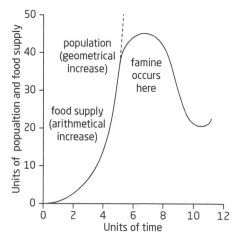

Figure 8.04 Malthus's view on population growth and food supply.

The implications of exponential growth

The most obvious examples of **population pressure** are in the developing world, but the question here is whether these are cases of absolute overpopulation or the result of underdevelopment that can be rectified by adopting remedial strategies over time.

Figure 8.05 summarises the opposing views of the **neo-Malthusians** and the resource optimists such as Esther Boserup (1910-99). Neo-Mathusians argue that an expanding population will lead to unsustainable pressure on food and other resources. In recent years neo-Malthusians have highlighted the following:

- the steady global decline in the area of farmland per person
- the steep rise in the cost of many food products
- the growing scarcity of fish in many parts of the world
- the already apparent impact of climate change on agriculture in some world regions
- the switchover of large areas of land from food production to the production of biofuels, helping to create a food crisis in order to reduce the energy crisis

CONSIDER THIS

Malthus's intellectual heirs include Paul Ehrlich, author of *The Population Bomb*, published in 1968. He warned of the potential for mass starvation in the 1970s and 1980s, advocating immediate action to limit population growth. The book was heavily criticised for its inaccurate predictions, but some people argue that Ehrlich simply got his timing wrong and a population-resources crisis is yet to come.

8.01 Human population dynamics

Growth of a population is exponential when the rate of growth is proportional to the number of individuals.

Population pressure occurs when population per unit area exceeds the carrying capacity.

Neo-Mathusians believe that an expanding population will put unsustainable pressure on food and other resources.

Anti-Malthusians, or resource optimists, believe that humans will continue to overcome problems of limited resources.

The **crude birth rate (CRB)** is the number of live births in a country or region per thousand population in a year.

The **crude death rate (CDR)** is the number of deaths in a country or region per thousand population in a year.

- the continuing increase in the world's population
- the global increase in the level of meat consumption as incomes rise in newly industrialised countries (NICs) in particular.

The **anti-Malthusians**, or resource optimists, believe that human ingenuity will continue to conquer resource problems, pointing to so many examples in human history where, through innovation or intensification, humans have responded to increased numbers. Resource optimists highlight a number of continuing advances, which include:

- the development of new resources
- the replacement of less efficient with more efficient resources
- the rapid development of green technology with increasing research and development in this growing economic sector
- important advances in agricultural research
- stabilising levels of consumption in some MEDCs.

Theory of knowledge 8.01.01

Whenever you are reading about the opinions of a particular writer, it is always important to consider the position from which they are writing and also the time of writing, because these are likely to have a significant influence of their views. The prevailing economic, social and political opinions in a country or region can have a considerable impact on the views of individuals, with the majority likely, broadly, to go along with such views, and a minority going against them to varying degrees.

Also, the breadth and depth of human knowledge increases with time, and, while it is credible to make suggestions about future trends, we cannot expect writers to make accurate predictions. In fact, most predictions of the future have underestimated the rate of human progress rather than overestimated it.

1 What differences between the social positions and life histories of Malthus and Boserup could account for opposing views?

2 Can you think of a recent scenario where the opposing views of two people might be linked to their backgrounds?

Crude birth rate, crude death rate, total fertility rate, doubling time and rate of natural change

The **crude birth rate (CRB)** is determined by taking the number of live births in one year in a country, dividing it by the country's population, and multiplying the number by 1000. Similarly the **crude death rate (CDR)** is determined by taking the number of deaths in a country in one year and dividing it by the country's population and multiplying the number by 1000.

The crude birth rate and the crude death rate are the most basic measures of fertility and mortality. The word 'crude' means that the birth rate applies to the total population, taking no account of gender and age. The crude birth rate and the crude death rate are heavily influenced by the age structure of a population, and in terms of the crude birth rate the male population is considered as well as the female population. However, crude birth and death rates are the most commonly used indicators of fertility and mortality and remain the starting points for most analyses of population.

Imagine a country has a population of 20 million and there are 300 000 births during that year. The calculation for the crude birth rate is:

$$300\,000 / 20\,000\,000 \times 1000/1 = 15 \text{ (15 per thousand)}$$

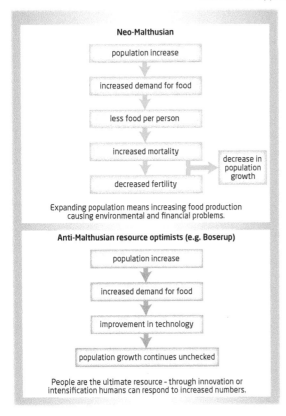

Figure 8.05 The opposing views of the neo-Malthusians and the anti-Malthusians.

If the same country has 200 000 deaths during the year, the calculation for the crude death rate is:

$$200\,000/20\,000\,000 \times 1000/1 = 10/1000$$

If the birth rate of a country is 15/1000, this means that on average for every 1000 people in this country 15 births will occur in a year. If the death rate for the same country is 10/1000, it means that on average for every 1000 people 10 deaths will occur. Think of the 1000 people that live nearest to you, and you will get some feeling about the meaning of these figures. The difference between the birth rate and the death rate is the **rate of natural change**. In this case, it will be 5/1000 (i.e. 15/1000 − 10/1000). Natural change can be positive (natural increase) or negative (natural decrease). Natural decrease occurs when the birth rate is lower than the death rate. The rate of natural change can also be expressed as a percentage. So, in the example given above, a natural change of 5/1000 is equivalent to 0.5 per cent.

Table 8.03 shows how much crude birth and death rates vary by world region. Africa has by far the highest birth rate, at 36/1000, while Europe has the lowest rate, at 11/1000. The rate for Africa is almost three and a half times that of Europe. The global variation in death rates is much less, with a high of 11/1000 in Europe and a low of 6/1000 in Latin America/the Caribbean.

The variation in birth rates between individual countries is significantly greater than that for world regions. According to the 2014 World Population Data Sheet, the crude birth rate varied from a high of 50/1000 in Niger to a low of 6/1000 in Monaco. Sixteen countries in Africa have a birth rate of 40/1000 or more. Europe is in a unique position with the birth rate and death rate being in balance giving a zero rate (0.0 per cent) of natural change.

> The **rate of natural change** is the difference between the birth rate and the death rate. This is usually positive (natural increase) but may also be negative (natural decrease).

Region	Birth rate/ per thousand	Death rate/ per thousand	Rate of natural change/ (a) per thousand and (b) % (to be completed in later activity)
World	20	8	(a) (b)
MEDCs	11	10	(a) (b)
LEDCs	22	7	(a) (b)
Africa	36	10	(a) (b)
Asia	18	7	(a) (b)
Latin America/ the Caribbean	18	6	(a) (b)
North America	12	8	(a) (b)
Oceania	18	7	(a) (b)
Europe	11	11	(a) (b)

Table 8.03 Crude birth and death rates, 2014.
Source: Selected data from 2014 World Population Data Sheet

Fertility rate, total fertility rate and replacement-level fertility

For more accurate measures of fertility than the birth rate, the **fertility rate** and the **total fertility rate** are used. A crucial factor in fertility is the percentage of women of reproductive age. Another important measure is **replacement-level fertility**.

> **Fertility rate** is the number of live births per 1000 women aged 15–44 years in a given year.
>
> **Total fertility rate** is the average number of children that would be born alive to a woman (or group of women) during her lifetime, if she were to pass through her childbearing years conforming to the age-specific fertility rates of a given year.
>
> **Replacement-level fertility** is the level at which the people in each generation have just enough children to replace themselves in the population. Although the level varies for different populations, a total fertility rate of 2.12 children is usually considered as the replacement level.

8.01 Human population dynamics

Image 8.03 Wedding in China – the average age of marriage in a country is an important factor affecting fertility.

The total fertility rate varies from a high of 7.6 in Niger to a low of 1.1 in Taiwan. Table 8.04 shows the variations in total fertility rate by world region alongside data for the percentage of women using contraception for each region. The latter is a major factor influencing fertility, as are other factors such as the average age of marriage (see Image 8.03). Figure 8.06 shows in detail how the total fertility rate varies by country around the world.

Region	Total fertility rate	Women aged 15–49 using contraception / %
World	2.5	63
MEDCs	1.6	70
LEDCs	2.6	61
Africa	4.7	34
Asia	2.2	66
Latin America and Caribbean	2.2	73
North America	1.8	77
Oceania	2.4	62
Europe	1.6	70

Table 8.04 Variations in total fertility rate and percentage women using contraception by world region, 2013.

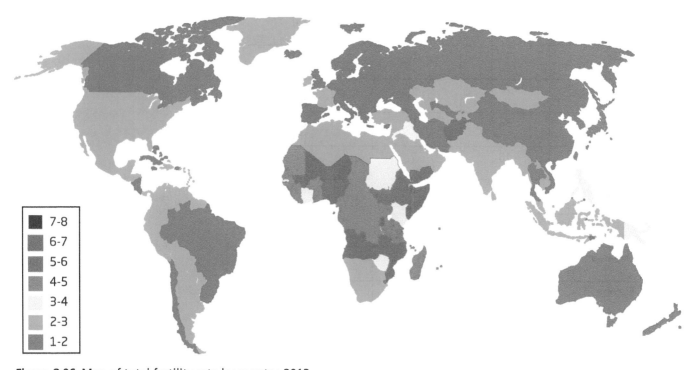

Figure 8.06 Map of total fertility rate by country, 2013.

With global fertility decline (see Figure 8.07) a growing number of countries have reached or fallen below replacement-level fertility. By 2013, almost 90 countries had total fertility rates at or below 2.1. This number is likely to increase in the future. The movement to replacement-level fertility is undoubtedly one of the most dramatic social changes in history, helping to enable many more women to work and children to be educated (see Image 8.04).

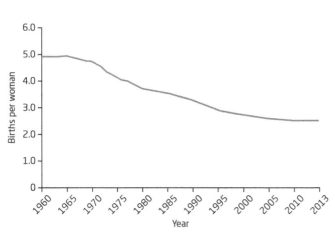

Figure 8.07 Graph showing decline in global fertility rate 1960–2013.

Image 8.04 Children leaving school in rural Cambodia.

Calculating the fertility rate

1 Take the number of live births in a country during a particular year.
2 Divide this figure by the total population of females aged 15–49 at the midpoint of the year.
3 Multiply by 1000.

Calculating the total fertility rate

1 Take the percentage of women of any given age (e.g. 17, 18, 19) who had a child in one particular year.
2 Add all these figures together.
3 Divide by 100.

If, for example, the total percentage was 175 per cent, the total fertility rate would be 1.75.

Population growth rate and doubling time

The **doubling time** of a population is determined by dividing the growth rate into 70. The number 70 comes from the natural log of 2, which is 0.70.

For example, in 2010 a country with a population of 33 million had a population growth rate of 0.9 per cent. Its population doubling time is obtained by dividing 70 by 0.9, giving a value of 77.7 years. So, in 2087, if the current rate of population growth is maintained, the country will have doubled its population from 33 million in 2010 to 66 million.

Knowing the doubling times of populations is useful in making international comparisons and also in examining variations within countries. In terms of the latter, some countries such as India have large regional variations in population growth. The idea of a population doubling itself certainly makes people consider the impact of population growth on resources and the environment in general. However, population growth rates can change significantly over time, and thus figures have to be revised on a regular basis.

> **Doubling time** is the number of years it would take a population to double its size at its current growth rate.

8.01 Human population dynamics

SELF-ASSESSMENT QUESTIONS 8.01.01

1. Using the information in Table 8.05 below, calculate and insert the birth and death rates for Brazil and the UK.

Country	Population/ millions	No. of births/ millions	Birth rate/per 1000 pop.	No. of deaths/ millions	Death rate/ per 1000 pop.
Brazil	196.7	2.95		1.18	
UK	62.7	0.82		0.56	

Table 8.05

2. Complete the final column in Table 8.03 by calculating the rate of natural change for each world region. Show your calculations: **a** per 1000, and **b** as a percentage.

3. Study Table 8.06 below. From the information provided calculate and insert the doubling times for the four countries listed.

Country	Population growth rate/%	Doubling time/years
China	0.5	
India	1.5	
Nigeria	2.5	
USA	0.8	

Table 8.06

3. **Research idea:** Find out how fertility has changed in the last 50 years in the country in which you live. What do you think are the reasons for these changes?

8.01.02 Age–sex pyramids and the model of demographic transition

The model of demographic transition

Although the populations of no two countries have changed in exactly the same way, some broad generalisations can be made about population growth since the middle of the 18th century. These generalisations are illustrated by the model of **demographic transition** (see Figure 8.08), which is based on the experience of north-west Europe, the first part of the world to undergo such changes as a result of the significant industrial and agrarian advances that occurred during the 18th and 19th centuries.

No country as a whole retains the characteristics of stage 1, which applies only to the most remote societies on Earth, such as isolated tribes in New Guinea and the Amazon, which have little or no contact at all with the outside world. All MEDCs are now in stages 4 or 5, most having experienced all of the previous stages at different times. The poorest of the LEDCs (e.g. Niger and Bolivia) are in stage 2 but are joined in this stage by some oil-rich Middle East nations where increasing affluence has not been accompanied by a significant fall in fertility. Most LEDCs that have registered significant social and economic advances are in stage 3 (e.g. Brazil, China and India), while some of the NICs such as South Korea and Taiwan have entered stage 4.

Demographic transition is the historical shift of birth and death rates from high to low levels in a population.

Critics of the model see it as too Euro-centric. They argue that many LEDCs may not follow the sequence set out in the model. It has also been criticised for its failure to take into account changes due to migration.

The basic characteristics of each stage are as follows.

The high stationary stage (stage 1)

The crude birth rate is high and stable, while the crude death rate is high and fluctuating due to the sporadic incidence of famine, disease and war. In this stage, population growth is very slow and there may be periods of considerable decline. Infant mortality is high and life expectancy low. A high proportion of the population are under the age of 15. Society is pre-industrial with most people living in rural areas, dependent on subsistence agriculture.

The early expanding stage (stage 2)

The death rate declines significantly. The birth rate remains at its previous level as the social norms governing fertility take time to change. As the gap between the two vital rates widens, the rate of natural change increases to a peak at the end of this stage. Infant mortality falls and life expectancy increases. The proportion of the population under 15 increases. Although the reasons for the decline in mortality vary somewhat in intensity and sequence from one country to another, the essential causal factors are: better nutrition; improved public health, particularly in terms of clean water supply and efficient sewage systems; and medical advances. Considerable rural to urban migration occurs during this stage. However, for LEDCs in recent decades, urbanisation has often not been accompanied by the industrialisation that was characteristic of MEDCs during the 19th century.

The late expanding stage (stage 3)

After a period of time, social norms adjust to the lower level of mortality and the birth rate begins to decline. Urbanisation generally slows and average age increases. Life expectancy continues to increase and infant mortality to decrease. Countries in this stage usually experience lower death rates than nations in the final stage, because of their relatively young population structures (Image 8.05).

The low stationary stage (stage 4)

Both birth and death rates are low. The former is generally slightly higher, fluctuating somewhat due to changing economic conditions. Population growth is slow. Death rates rise slightly as the average age of the population increases. However, life expectancy still improves as age-specific mortality rates continue to fall.

Image 8.05 Children helping their parents milk goats in central Asia.

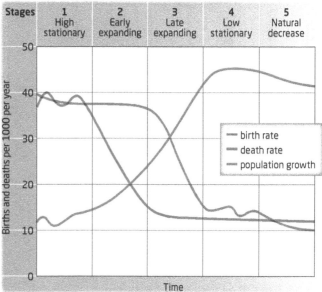

Figure 8.08 Model of demographic transition.

8.01 Human population dynamics

The natural decrease stage (stage 5)

In an increasing number of countries the birth rate has fallen below the death rate, resulting in natural decrease. In the absence of net migration inflows, these populations are declining. Most countries in this stage are in eastern or southern Europe.

Demographic transition in LEDCs

There are a number of important differences in the way that LEDCs have undergone population change compared with the experiences of most MEDCs before them:

- In LEDCs, birth rates in stages 1 and 2 were generally higher. About 12 African countries currently have birth rates of 45/1000 or over. Twenty years ago many more African countries were in this situation.
- The death rate in LEDCs fell much more steeply and for different reasons. For example, the rapid introduction of Western medicine, particularly in the form of inoculation against major diseases, has had a huge impact on lowering mortality. However, AIDS has caused the death rate to rise significantly in some countries, particularly in sub-Saharan Africa.
- Some LEDCs had much larger base populations, and thus the impact of high growth in stage 2 and the early part of stage 3 has been far greater. No countries that are now classed as MEDCs had populations anywhere near the size of India and China when they entered stage 2 of demographic transition.
- For those LEDCs in stage 3, the fall in fertility has also been steeper. This has been due mainly to the relatively widespread availability of modern contraception with high levels of reliability.
- The relationship between population change and economic development has been much more tenuous in LEDCs.

Different models of demographic transition

Although most countries followed the classical, or English, model of demographic transition illustrated in the last section (see Figure 8.08), some countries did not. The Czech demographer Pavlik recognised two alternative types of population change, shown in Figures 8.09b and 8.09c. In France, the birth rate fell at about the same time as the death rate, and there was no intermediate period of high natural increase. In Japan and Mexico, the birth rate actually increased in stage 2 due mainly to the improved health of women in the reproductive age range.

8.01.03 Age and sex pyramids

The structure or composition of a population is the product of the processes of fertility, mortality and migration. The most studied aspects of **population structure** are age and sex. Other aspects of population structure that can also be studied include race, language, religion and social/occupational groups.

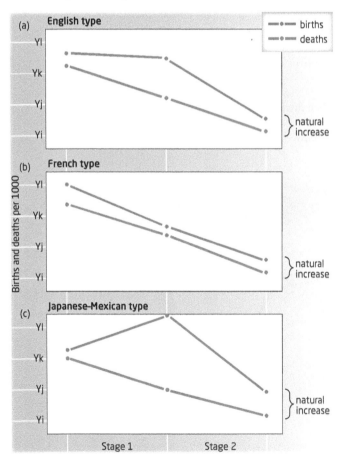

Figure 8.09 Types of demographic transition.

CONSIDER THIS
During stage 2 of demographic transition in England, infant mortality fell from 200 per 1000 in 1770 to 100 per 1000 in 1870. Better nutrition played a substantial role in this decline. Contemporary studies in LEDCs show a strong relationship between infant nutrition and infant mortality.

Population structure is the composition of a population, the most important elements of which are age and sex (gender).

Age and sex structure is conventionally illustrated with the use of **population pyramids**. Pyramids can be used to portray either absolute or relative data. Absolute data shows the figures in thousands or millions, while relative data shows the numbers involved in percentages. The latter is most frequently used, as it allows for easier comparison of countries of different population sizes. Each bar represents a five-year age group. The male population is represented to the left of the vertical axis, with females to the right.

Population pyramids change significantly in shape as a country progresses through demographic transition. Figure 8.10 shows four population pyramids for 2009:

- The wide base in the Niger pyramid reflects extremely high fertility. The birth rate in Niger is 50/1000, the highest in the world. The marked decrease in width of each successive bar indicates relatively high mortality and limited life expectancy. The death rate, at 11/1000, is high, particularly considering how young the population is. Infant mortality has fallen steeply in recent decades to 54/1000. Life expectancy in Niger is 58 years. Fifty per cent of the population are under 15, with only 3 per cent 65 or more. Niger is in stage 2 of demographic transition.

> A **population pyramid** is a bar chart arranged vertically that shows the distribution of population by age and gender.

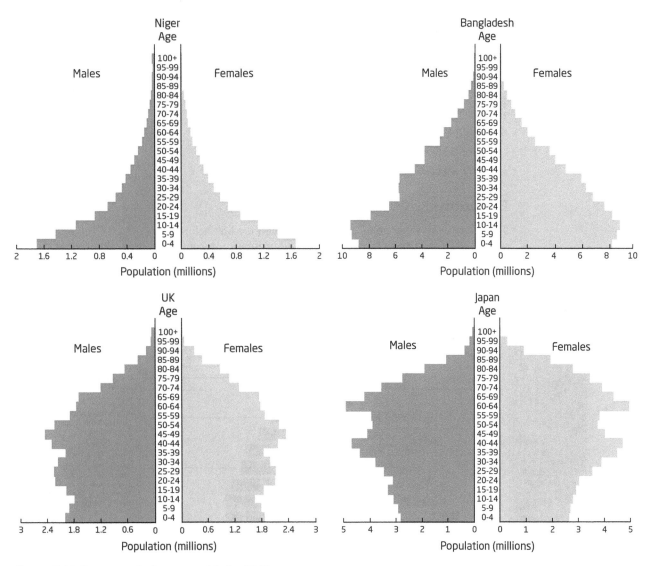

Figure 8.10 Four population pyramids for 2009.

8.01 Human population dynamics

- The base of the second pyramid showing the population structure of Bangladesh is narrower than that of Niger, reflecting a considerable fall in fertility after decades of government-promoted birth-control programmes. The reduced width of the youngest two bars compared with the 10–14 bar is clear evidence of recent falls in fertility. The birth rate is currently 20/1000. Falling mortality and lengthening life expectancy are reflected in the relatively wide bars in the teenage and young adult age groups. The death rate, at 6/1000, is almost half that of Niger. The infant mortality rate is 33/1000. Life expectancy in Bangladesh is 70 years. Twenty-nine per cent of the population are under 15, while 5 per cent are 65 or over. Bangladesh is an example of a country in stage 3 of demographic transition.

- In the pyramid for the UK, much lower fertility than both Niger and Bangladesh is illustrated by very significant narrowing of the base. The birth rate in the UK is only 12/1000. The reduced narrowing of each successive bar indicates a further decline in mortality and greater life expectancy compared to Bangladesh. The death rate in the UK is 9/1000, with an infant mortality rate of 3.9/1000. Life expectancy is 81 years. Eighteen per cent of the population are under 15, while 17 per cent are 65 or over. The UK is in stage 4 of demographic transition.

- The final pyramid (Japan) has a distinctly inverted base, reflecting the lowest fertility of all four countries. The birth rate is 8/1000. The width of the rest of the pyramid is a consequence of the lowest mortality and highest life expectancy of all four countries. The death rate is 10/1000, with infant mortality at 1.9/1000. Life expectancy is 83 years. Japan has only 13 per cent of its population under 15, with 26 per cent aged 65 or over. With the birth rate lower than the death rate, Japan has entered stage 5 of demographic transition.

Figure 8.11 provides some useful tips for understanding population pyramids. A good starting point is to divide the pyramid into three sections:

- Young dependants (i.e. children)
- Economically active population
- Elderly dependants

Extension

To explore more about population pyramids, visit the link below:
www.ined.fr/en/everything_about_population/population-games/tomorrow-population

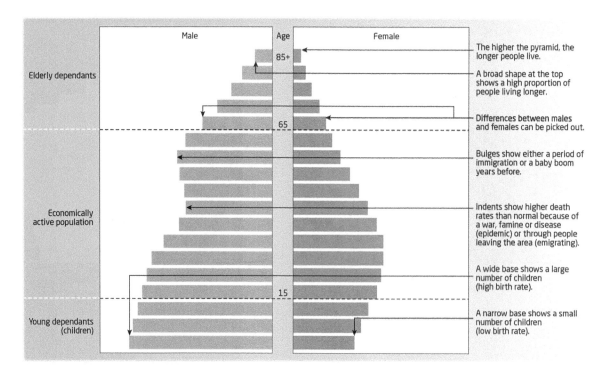

Figure 8.11 Annotated population pyramid.

Population structure: differences within countries

In countries where there is strong rural-to-urban migration, the population structures of the areas affected can be markedly different. These differences show up clearly on population pyramids. Out-migration from rural areas is age-selective, with single young adults and young adults with children dominating this process. Thus the bars for these age groups in rural areas affected by out-migration indicate fewer people than expected in these age groups (see Image 8.06).

In contrast, the population pyramids for urban areas attracting migrants show age-selective in-migration, with substantially more people in these age groups than expected. Such migrations may also be sex-selective. If this is the case, it should be apparent on the population pyramids.

Image 8.06 A small rural school on the coast of northern Vietnam.

Sex structure

The **sex ratio** is the number of males per 100 females in a population. Male births consistently exceed female births for a combination of biological and social reasons. For example, in terms of the latter, more couples decide to complete their family on the birth of a boy than on the birth of a girl. However, after birth the gap generally begins to narrow until eventually females outnumber males, as at every age male mortality is higher than female mortality. For 2014, the global sex ratio was estimated at 107 boys to 100 girls. This process happens most rapidly in the poorest countries, where infant mortality is markedly higher among males than females. Here the gap may be closed in less than a year.

A report published in China in 2002 recorded 116 male births for every 100 female births, because of the significant number of female foetuses aborted by parents intent on having a male child. However, by 2014 the ratio had fallen to 111 male births for every 100 female births, still one of the highest in the world. Even within countries there can be significant differences in the sex ratio.

> The **sex ratio** is the number of males per 100 females in a population.

SELF-ASSESSMENT QUESTIONS 8.01.02

1. Explain the reasons for declining mortality in stage 2 of demographic transition.
2. Why does it take some time before fertility follows the fall in mortality (stage 3)?
3. Suggest why the birth rate is lower than the death rate in some countries (stage 5).
4. Give two limitations of the model of demographic transition.
5. **Research idea:** Find two population pyramids for the same country that show clear changes in age/sex structure over time. Describe and explain these changes.

8.01 Human population dynamics

8.01.04 Effect of national and international development policies and cultural influences on population dynamics and growth

Development policies and falling mortality

A range of policy factors, both direct and indirect, have influenced the growth and size of human populations. Development policies, both international and domestic, have lowered mortality by improving living standards, without often having any significant impact on fertility. Such policies have lowered the death rate through better public health and sanitation, agricultural development, and improved service infrastructure.

The decline in levels of mortality and the increase in life expectancy has been the most important reward of economic and social development. Falling mortality and the corresponding increase in life expectancy have influenced population growth in a significant way. On a global scale, 75 per cent of the total improvement in longevity has been achieved in the 20th century and the early years of the 21st century. In 1900, the world average for life expectancy is estimated to have been about 30 years, but by 1950–1955 it had risen to 46 years. By 1980–1985 it had reached a fraction under 60 years. The current global average is 70 years (males 68, females 72).

The causes of death vary significantly between MEDCs and LEDCs. In LEDCs, infectious and parasitic diseases account for over 40 per cent of all deaths. They are also a major cause of disability and social and economic upheaval. In contrast, in MEDCs these diseases have a relatively low impact. In rich countries, heart disease and cancer are the big killers. Epidemiology is the study of diseases. As countries develop, the major diseases tend to change from infectious to degenerative. This change is known as the **epidemiological transition**.

The Millennium Development Goals established in 2000 by international agreement are probably the most significant major attempt to defeat poverty ever undertaken. The United Nations (UN) set out eight development goals to reduce global poverty substantially by 2015. They are viewed as basic human rights – the rights of every person on Earth to health, education, shelter and security. Measurement of progress is based on 1990 figures. By 2015, all 191 UN member states had pledged to meet these goals.

All of the Millennium Development Goals have an impact on mortality, although some more directly than others. Significant improvements have been made in **child mortality**, with the number of worldwide deaths of children under five falling steadily from 12.6 million in 1990 to 6.3 million in 2013. Sub-Saharan Africa accounts for about half of all under-five deaths around the world.

In some developing regions, there have been major improvements in several key child-survival interventions that should lead to further significant falls in child mortality. These interventions include:

- vitamin A supplementation
- the use of insecticide-treated bed nets
- exclusive breastfeeding
- immunisation.

Maternal mortality is among the health indicators that show the greatest gap between MEDCs and LEDCs. Globally, in 2013 there were 210 maternal deaths per every 100 000, down from 380 in 1990. The lifetime risk of maternal death in industrialised countries is 1 in 4000, versus 1 in 51 in countries classified as least developed. Two regions, sub-Saharan Africa and South Asia, account for 86 per cent of maternal deaths worldwide. Sub-Saharan Africans suffer from the highest maternal mortality ratio – 510 maternal deaths per 100 000 live births. This is nearly two-thirds (62 per cent) of all maternal deaths per year worldwide.

Epidemiological transition The change from mainly infectious diseases, still common in LEDCs, to the degenerative diseases, which have become the main cause of death in MEDCs.

Child mortality The number of deaths of children aged under 5 per 1000 live births in a given year.

Maternal mortality rate The annual number of deaths of women from pregnancy-related causes per 1000 live births.

Obstetric complications, including postpartum haemorrhage, infections, eclampsia, prolonged or obstructed labour, and complications of unsafe abortion, account for the majority of maternal deaths. Fewer than half of pregnant women in developing countries have the benefit of adequate prenatal care.

Fertility change: economic progress and direct policy

Policies with regard to the birth rate may be explicit – that is, a distinct **population policy** – or implicit. It has been noted for some time that birth rates generally decline with economic development. Thus policies that stimulate economic growth may achieve results that are as good as those aimed directly at lowering fertility. Economic growth allows greater spending on health, housing, nutrition and education, which is important in lowering mortality and in turn reducing fertility.

Education, especially improvements in female literacy, is the key to lower fertility. With education comes a knowledge of birth control, greater social awareness and more opportunity for employment; women give birth later in life and have a wider choice of action generally.

Most countries that have population policies have been trying to reduce their fertility by investing in birth control programmes. Within developing countries, the poorest neighbourhoods usually have the highest fertility, due mainly to a combination of high infant mortality and low educational opportunities for women.

In 1952, India became the first developing country to introduce a policy designed to reduce fertility and aid development, with a government-backed family planning programme. Rural and urban birth control clinics rapidly increased in number. Financial and other incentives were offered in some states for those participating in programmes, especially sterilisation. In the mid-1970s, the sterilisation campaign became increasingly coercive, reaching a peak of 8.3 million operations in 1977. Abortion was legalised in 1972, and in 1978 the minimum age of marriage was increased to 18 years for females and 21 years for males. The birth rate fell from 45/1000 in 1951–1961 to 41/1000 in 1961–1971. By 1987 it was down to 33/1000, falling further to 29/1000 in 1995. By 2014 it had dropped to 20/1000.

It was not long before many other developing nations followed India's policy of government investment to reduce fertility. The most severe anti-natalist policy ever introduced has been in operation in China since 1979.

> **Population policy** encompasses all the measures taken by a government aimed at influencing population size, growth, distribution or composition.

CASE STUDY 8.01.01

Anti-natalist policy in China

China, with a population in excess of 1.3 billion, operates the world's most severe family planning programme (see Image 8.07). Although it is the third largest country in the world in land area, 25 per cent of China is infertile desert or mountain, and only 10 per cent of the total area can be used for arable farming. Most of the best land is in the east and south, reflected in the extremely high population densities in these regions. Thus the balance between population and resources has been a major cause of concern for much of the latter part of the 20th century, although debate about this issue can be traced as far back in Chinese history as Confucius (Chinese philosopher and teacher of ethics, 551–479 BC).

Image 8.07 Beijing – crowds at the Forbidden City.

CASE STUDY 8.01.01 (continued)

For people in the West it is often difficult to understand the all-pervading influence over society that a government can have in a centrally planned economy. In the aftermath of the communist revolution in 1949, population growth was encouraged for economic, military and strategic reasons. Sterilisation and abortion were banned and families received a benefit payment for every child. However, by 1954 China's population had reached 600 million and the government was worried about the pressure on food supplies and other resources. Consequently, the country's first birth control programme was introduced in 1956. This was to prove short-lived, for in 1958 the Great Leap Forward began. The objective was rapid industrialisation and modernisation. The government was concerned that progress might be hindered by labour shortages, and so births were again encouraged. By 1962 the government had changed its mind, heavily influenced by a catastrophic famine due in large part to the relative neglect of agriculture during the pursuit of industrialisation. An estimated 20 million died during the famine. Thus a new phase of birth control ensued in 1964. Just as the new programme was beginning to have some effect, a new social upheaval, the Cultural Revolution, got underway. This period, during which the birth rate peaked at 45/1000, lasted from 1966 to 1971.

With order restored, a third family planning campaign was launched in the early 1970s with the slogan 'Late, sparse, few'. However, towards the end of the decade the government felt that the campaign's impact might falter, and in 1979 the controversial one-child policy was imposed. The Chinese demographer Liu Zeng calculated that China's optimum population was 700 million, and he looked for this figure to be achieved by 2080.

Figure 8.12 shows changes in the birth and death rates from 1950. The impact of the one-child policy is clear to see. Some organisations, including the United Nations Population Fund (UNFPA), have praised China's policy on birth control. Many others see it as a fundamental violation of civil liberties. China's total fertility rate has fallen from 4.77 births per woman in the 1970s, just before the one-child policy was introduced, to 1.64 in 2011.

The policy has had a considerable impact on the sex ratio, which at birth in China is currently 119 boys to 100 girls. This compares with the natural rate of 106:100. This is already causing social problems, which are likely to multiply in the future. Selective abortion after prenatal screening is a major cause of the wide gap between the actual rate and the natural rate. But even if a female child is born, her lifespan may be sharply curtailed by infanticide or deliberate neglect.

China's policy is based on a reward and penalty approach. Rural households that obey family planning rules get priority for loans, materials, technical assistance and social welfare. The slogan in China is, 'shao sheng kuai fu' – 'fewer births, quickly richer'.

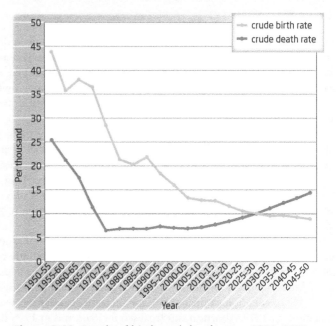

Figure 8.12 Graph of birth and death rates 1950–2050.

The one-child policy has been most effective in urban areas, where the traditional bias of couples wanting a son has been significantly eroded. However, the story is different in rural areas, where the strong desire for a male heir remains the norm. In most provincial rural areas, government policy has now relaxed so that couples can have two children without penalties.

In July 2009, newspapers in the UK and elsewhere reported that dozens of babies had been taken from parents who had breached China's one-child policy and sold for adoption abroad.

A paper published in 2008 estimated that China had 32 million more men aged under 20 than women. The imbalance is greatest in rural areas, because women are 'marrying out' into cities. In recent years, reference has been made to the 'four–two–one' problem whereby one adult child is left having to support two parents and four grandparents.

Questions

1. When was the one-child policy introduced?
2. By how much has China's total fertility rate changed since the introduction of the one-child policy?
3. What impact has the one-child policy had on the sex ratio? Why has this occurred?

Concerns about low fertility

What is perhaps surprising is the number of countries that now see their fertility as too low. Such countries are concerned about:

- the socioeconomic implications of population ageing
- the decrease in the supply of labour
- the long-term prospect of population decline.

Russia has seen its population drop considerably since 1991. Alcoholism, AIDS, pollution and poverty are among the factors reducing life expectancy and discouraging births. In 2008, Russia began honouring families with four or more children with a Paternal Glory medal. The government has urged Russians to have more children, sometimes suggesting it is a matter of public duty.

CASE STUDY 8.01.02

Pro-natalist policy in France

France's relatively high fertility level can be partly explained by its long-term active family policy, adopted in the 1980s to accommodate the entry of women into the labour force. The policy seems to have created especially positive attitudes towards two- and three-child families in France. France has taken steps to encourage fertility on a number of occasions over the last 70 years. In 1939 the government passed the *code de la famille*, which:

- offered financial incentives to mothers who stayed at home to look after children
- subsidised holidays
- banned the sale of contraceptives; this ban stopped in 1967.

More recent measures to encourage couples to have more children include:

- longer maternity and paternity leave – maternity leave, on near full pay, ranges from 20 weeks for the first child to 40 or more for the third child
- higher child benefits
- improved tax allowances for larger families until the youngest child reaches 18
- a pension scheme for mothers/housewives
- a 30 per cent reduction on all public transport for three-child families
- child-oriented policies, for example provision of crèches and day nurseries; state-supported daycare centres and nursery schools are available for infants starting at the age of three months, with parents paying on a sliding scale according to income
- preferential treatment in the allocation of government housing.

Overall, France is trying to reduce the economic cost to parents of having children. In 2006, France overtook Ireland to become the highest-fertility nation in the EU. The 830 900 babies born in 2006 was the highest number since 1981. The 2014 Population Data Sheet put France's total fertility rate at 2.0; this compares with 1.4 in Germany, 1.9 in the UK, 1.4 in Italy and 1.3 in Spain. France is close to the replacement level of 2.1 children per woman. Moderate positive net migration adds to fertility. This net migration is virtually equivalent to a surplus of 75 000 births.

Although the average age of French mothers at childbirth is still rising, it is still lower than in many other European countries. Within France, the highest level of fertility is among the immigrant population. But even for those born in France, the average is 1.8 babies. French economists argue that, although higher fertility means more expenditure on childcare facilities and education, in the longer term it gives the country a more sustainable age structure.

French politicians have talked about demography as a 'source of vitality' for the country. Some French commentators also argue that there is a better work–life balance in France than in many other European countries.

Demography The scientific study of human populations

The central population forecast, based on the stability of fertility and migration at current levels, predicts stability in the population aged 60 or less, while the population aged 60 and over will increase as a consequence of the post-Second World War baby boom.

Questions

1. When did the French government pass the *code de la famille*?
2. State three recent measures introduced in France to encourage couples to have more children.
3. How does France's total fertility rate compare with that in the UK, Germany, Italy and Spain?

8.01 Human population dynamics

Culture, education and urbanisation

In some societies, particularly in Africa, cultural tradition demands high rates of reproduction. Here the opinion of women in the reproductive years may have little influence weighed against intense cultural expectations. In some countries, religion is an important factor. For example, religions generally oppose artificial birth control. However, where these religions are important, more people tend to adhere to religious doctrine in poorer countries compared with richer countries.

In some societies, infant mortality is high and it is usual for many children to die before reaching adulthood. In such societies, parents often have many children to compensate for these expected deaths. In many LEDCs, children are seen as an economic asset because of the work they do and because of the support they are expected to give their parents in old age. In MEDCs, the general perception is reversed and the cost of the child dependency years is a major factor in the decision to begin or extend a family.

Education, especially female literacy, is the key to lower fertility. With education comes:

- a knowledge of birth control
- greater social awareness
- more opportunity for employment
- and a wider choice of action generally.

Overall, education enables women to have greater personal and economic independence, and thus many experts working in the field of development believe that improving educational opportunities for females is the most effective measure for reducing population pressure. An educated mother will be a strong influence on the social development of her children. In this way the benefits of investment in education are passed from one generation to the next.

Urbanisation is as much a social process as it is an economic and geographical process. It is defined as the process by which an increasing proportion of a country's population live in urban areas. Urbanisation changes the role of the family, demographic structures, the nature of work, and other aspects of society. In most countries there is a strong correlation between the levels of urbanisation and fertility.

Children are less of an asset in terms of labour in urban areas compared with rural settlements. They are also more expensive to house and feed. The requirement to pay for goods and services in money (as opposed to through bartering) is much greater in cities than in rural areas, putting greater pressure on both parents to work as many hours as possible. The extended family is less evident in urban as opposed to rural areas, and thus relying on others to help with childcare may be less of an option.

Education about family planning is usually greater in urban than rural areas, and contraception is more easily available. In addition, the social pressures to have a large family that sometimes exist in rural societies may reduce considerably when migration to towns and cities occurs. There is often a considerable difference between the social norms prevalent in urban and rural areas.

SELF-ASSESSMENT QUESTIONS 8.01.03

1. How can development policies lower mortality?
2. Briefly explain the relationship between education and fertility.
3. Why have some countries taken measures to encourage fertility?
4. **Discussion point:** What is the total fertility rate for the country in which you live? Do you think your government should encourage couples to have more or fewer children?

8.01.05 Models and predicting the growth of human populations

Models play a central role in predicting the growth of human populations. Such models might include:

- computer simulations
- statistical and demographic tables
- age–sex pyramids
- graphical extrapolation of population curves.

Population projections

Population projections are the prediction of future populations based on the present age–sex structure, and with present rates of fertility, mortality and migration. The simplest projections are based on extrapolations of current and past trends, but a set of very different projections can be calculated based on a series of different assumptions.

Earlier projections were generally based on detailed analysis of statistical and demographic tables. From such data, population pyramids, line graphs and other cartographic methods were used to plot data and project into the future. The development of computing and the massive increase in computing capacity has allowed increasingly complex permutations to be included in projections. Such technological developments have also allowed frequent updating of projections to reflect changes in trends in fertility, mortality and other relevant factors.

The earliest systematic global population projection dates back to 1945. Population projections differ widely in their geographic coverage, timescale, content and use. Scale can vary from the sub-national (i.e. a city) to the whole world. Local-area projections tend to focus on shorter timescales, frequently ten or 20 years. Global projections may extend for 25, 50 or 100 years. Longer-term projections usually focus on a limited number of variables – frequently total population, age and sex. In contrast, projections for smaller regions often include characteristics such as labour-force characteristics and household type.

Population momentum is a significant factor influencing population projections. Population momentum occurs towards the end of stage 3 of demographic transition. Although the annual rate of population growth may be falling in a country, the natural increase in terms of total number may be rising due to population momentum. On a global scale, the highest annual rates of population increase (as a percentage) were in the early 1960s. However, the highest annual increases in population (millions) did not occur until the late 1980s, a time lag of about a quarter of a century. Thus population momentum is a major factor that must be taken into account when producing population projections.

One reason that demographers are reasonably confident about near- and medium-term population predictions is because, through population momentum, much of the future is built into the current structure of populations. For example, in very high fertility countries, the current population structure guarantees considerable future growth. This would occur even if the total fertility rate fell sharply almost overnight. Such populations would grow for another 50 years or so. In contrast, such low fertility countries as Germany, Italy and Spain do not have this 'positive momentum'. The positive momentum of these populations has been dissipated by decades of fertility below the replacement level.

Projections result from the assumptions made when they are prepared. Assumptions must be made about:

- declines in the future birth rate
- improvements in life expectancy at birth and infant mortality
- migration into and out of an area.

A **population projection** predicts the future population, based usually on current and past trends.

8.01 Human population dynamics

Any population projection is hypothetical. It is always dangerous to assume that various demographic factors will stay the same in the future or change only marginally. The significant demographic changes that have occurred since 1950 in both LEDCs and MEDCs should make us very aware of this point. Many previous population projections have proved to be inaccurate to a considerable degree. Failure to foresee the baby boom after the end of the Second World War, and the end of this boom, is an often-quoted example.

Although total accuracy is almost impossible to achieve, population projections need to be accurate enough to serve as the basis for policies. As D. A. Coleman states in 'The shape of things to come: world population to 2050' (a contribution to the Engelsberg Seminar 2005), 'Population matters because of its associated effects on power, environment and security arising from global and regional, and particularly differential, population growth and composition.' Coleman sees this differential population growth as radically transforming the international political and economic order.

Figure 8.13 shows three global projections to 2050:

- The central forecast expects world population to stabilise at about 9 billion.
- The upper line is the UN's high variant, which assumes slow declines in fertility and faster increase in life expectancy.
- The low variant line shows that the UN now takes seriously the possibility that world population could actually decline before the end of the century.

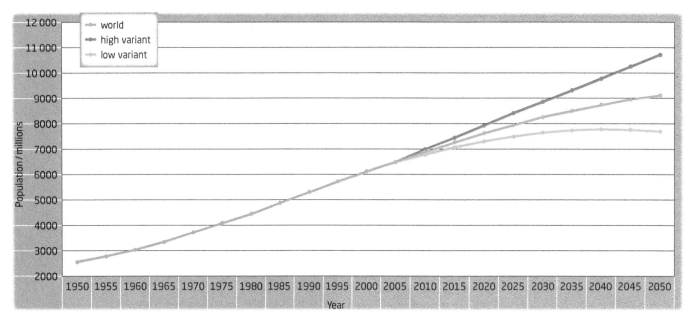

Figure 8.13 World population 1950–2050 – three variant projections.

The UN's 2010 revision

According to the UN's 2010 revision of future population, the world population is expected to hit 10.1 billion by 2100, after reaching 9.3 billion by the middle of the 21st century. Virtually all of the growth will take place in developing countries and will be predominately among the poorest populations in urban areas. Figure 8.14 shows how the population pyramid for the world as a whole is projected to change by 2050 and 2100. By 2100, 22.3 per cent of people will be aged 65 or over, a significant increase from just 7.6 per cent in 2010.

Between 2011 and 2100, the population of high-fertility countries, which includes most of sub-Saharan Africa, is projected to triple from 1.2 billion to 4.2 billion. During the same period, the population of intermediate-fertility countries such as India, the USA and Mexico will increase by just 26 per cent. In contrast, the population of low-fertility countries, which includes most of Europe, China and Australia, will decline by about 20 per cent.

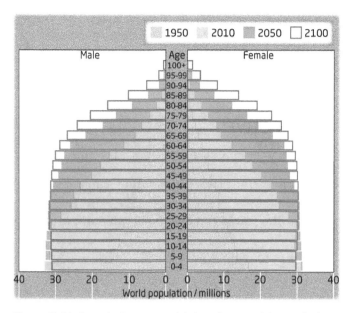

Figure 8.14 Population pyramid showing world population projections.

Eventually, the populations of high-fertility countries are forecast to decline until reaching replacement fertility by the end of the century. However, these projections assume wider availability of family planning services, and small variations in fertility can produce considerable differences in the size of populations over the long run.

SELF-ASSESSMENT QUESTIONS 8.01.04

1. What are population projections?
2. Why do individual countries and international organisations construct population projections?
3. Why are modern population projections likely to be more accurate than projections made 30 years ago?
4. **Research idea:** Find out about the population projections for the country in which you live. Try to obtain a graph showing variant projections, and briefly describe what the graph shows.

8.02 Resource use in society

LEARNING OBJECTIVES

After reading this chapter you should be able to:

- understand that the renewability of natural capital has implications for its sustainable use
- appreciate that the status and economic value of natural capital is dynamic.

KEY QUESTIONS

What is the difference between renewable and non-renewable natural capital?

What are the goods and services provided by natural capital?

How has natural capital been mismanaged?

8.02.01 The concept of resources and natural income

Defining resources and capital

A resource can be defined as any aspect of the environment that can be used to meet human needs (see Image 8.08). Traditionally resources have been defined and thought of in terms of human wealth. However, in recent times the attitudes of many people have changed so that resources are valued in other ways as well. This has gone hand in hand with increasing concerns about what resource exploitation has done to the environment. Sustainability and sustainable development are terms that have moved from the margins to the mainstream of political debate in a matter of decades as public awareness about the relationship between people and environments has intensified.

Jonathon Porritt, in his book *Capitalism as if the World Matters* (2005), argues that more and more people in senior posts in both government and business around the world have grasped the importance of sustainable development. Such change is incremental rather than transformational, but it is beginning to make a difference.

Extension

For an inspiring video on putting a value on nature, listen to the TED talk by visiting the link below:
http://on.ted.com/Sukhdev

Porritt states, 'There need be no fundamental contradiction between sustainable development and capitalism.' This is so, providing less attention is paid to economic growth in itself and more attention to human well-being and the environment. Porritt is an optimist about making progress towards

Image 8.08 The Amazon rainforest.

a green society. Developing this discussion, Porritt argues for an expansion of the traditional view of capital. The latter is viewed in terms of 'stocks' of capital (land, machines and money). Porritt's five-capital framework broadens this view considerably to include:

- *natural capital*, which is required to maintain a functioning biosphere, supply resources and dispose of wastes
- *human capital*, which provides the knowledge and skills to create manufactured capital
- *social capital*, which establishes the institutions that provide the stable human environment within which economic activity can occur
- *manufactured capital*, which is the products and infrastructure that provide people with economic wealth
- *financial capital*, which provides the lubricant to keep the whole system operating.

Developing more sustainable approaches in all of these areas can bring important improvements in well-being and the environment. This involves a change in thinking from preoccupation with short-term economic gain to a longer-term approach that re-engages with the natural world. This can be achieved through a reform agenda rather than revolution. Porritt's views are in contrast to the prevailing views of many radical academics and non-governmental organisations (NGOs) that see the only way forward as a completely different world order.

Natural capital and natural income

The idea of natural capital was introduced in Chapter 1.04. Ecologically minded economists have described resources as **natural capital** for some time. The term 'natural capital' is derived from the economic concept of capital and equates the natural capital of ecosystems with the highly valued capital of economic systems. The concept of natural capital refers to the source of supply of resources and services that are derived from nature. Forests, mineral deposits, fertile soil and fishing grounds are examples of natural capital. Natural capital describes goods and services that have not been processed or manufactured but have value to human populations. Natural capital has financial value, as the use of natural capital drives many businesses.

Four general services are provided by natural capital (see Table 8.07):

- provisioning services (i.e. products from ecosystems)
- regulating services (i.e. benefits from ecosystem processes)
- cultural services (i.e. non-material benefits from ecosystems)
- supporting services (i.e. services that allow other services to operate).

Natural capital refers to the source of supply of resources and services that are derived from nature.

Provisioning services	Regulating services	Cultural services	Supporting services
• food • fresh water • fuelwood • fibre • biochemicals • genetic resources	• climate regulation • disease regulation • water regulation • water purification • pollination	• spiritual/religious • recreation • ecotourism • inspiration • education • cultural heritage	• soil formation • nutrient recycling • primary production

Table 8.07 Services from natural capital.

Resources are forms of wealth that can produce **natural income** indefinitely in the form of valuable goods and services (see Figure 8.15). The stress is very much on management, and the strong implication is that this needs to be sustainable (see Chapter 1.04).

Natural income may consist of:

- marketable commodities or goods such as oil, copper, grain and timber (see Image 8.09)
- **ecological services** such as the erosion protection and flood protection provided by forests. (Ecological services often involve essential processes such as the water cycle and photosynthesis.)

Natural income is the annual yield from sources of natural capital.

Ecological services Attributes of the natural environment that often essential processes such as the water cycle and photosynthesis

8.02 Resource use in society

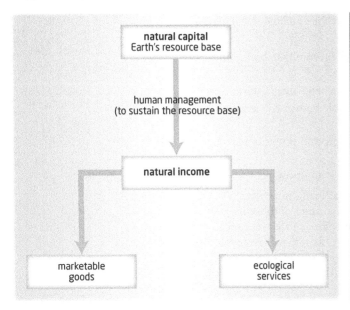

Figure 8.15 Natural capital and natural income diagram.

Image 8.09 Natural income – huge haystacks in northern Spain.

CONSIDER THIS

The term 'natural capital' was first used in 1973 by E. F. Schumacher in his book *Small is Beautiful: A Study of Economics as if People Matter*. The book is basically a critique of Western economies and the impact they have on the natural environment.

When the amount of natural income used up reduces the capacity of natural capital to continue providing the same amount of natural income in the future, this is the point at which sustainable scale has been exceeded.

The relationship between natural capital and economic wealth varies considerably around the world. Some countries with limited natural capital such as Japan and the UK have become wealthy largely through the value of their human resources and their ability to purchase natural capital from other countries. In contrast, many LEDCs, such as Nigeria and Bolivia, that have a good resource base in terms of natural capital have very limited wealth in economic terms. Such natural capital may have been mainly exploited by MEDCs in the past, with very little of the benefit accruing to the LEDCs concerned, or the LEDCs may be currently unable to attract the investment required to develop their resources.

The RNC Alliance – restoring natural capital

What is the restoration of natural capital (RNC)?
The restoration of natural capital (RNC) is any activity that invests in the replenishment of natural capital and thereby improves the flows of ecosystem (natural) goods and services, while enhancing the wellbeing of people. RNC activities restore natural ecosystems and rehabilitate arable land or other production systems in an ecologically sound manner. These activities are intended to enhance the physical, socio-economic, psychological, and cultural aspects of people's quality of life. RNC initiatives include programs to raise awareness of the value of natural capital in people's daily lives.

What is the RNC Alliance?
The RNC Alliance is an international network of individuals and not-for-profit, non-governmental organisations that offer locally appropriate solutions to resolve environmental and economic development problems simultaneously in under developed countries and industrial countries alike.
Research, demonstration, education and outreach are the basic activities of the Alliance.

Figure 8.16 The RNC Alliance – restoring natural capital.

Restoring natural capital

'The restoration of natural capital is an activity that integrates investment in and replenishment of natural capital stocks to improve the flows of ecosystem goods and services, while enhancing all aspects of human well-being. In common with ecological restoration, natural capital restoration is intended to improve the health, integrity, and self-sustainability of ecosystems for all living things.' *Restoring Natural Capital: Science, Business, and Practice* (J. Aronson, S. J. Milton and J. N. Blignaut; 2007).

As we come to appreciate the value of natural capital more than probably at any time in the past, the need to conserve it and, if possible, restore it has moved up the political agenda. Improving our understanding of the relationships and trade-offs among forests, soil, biodiversity, water and food production among other key ecosystem components is driving new developments in applied scientific research. The RNC Alliance is an NGO that is active in the restoration of natural capital (see Figure 8.16). The concept of sustainable scale helps to focus thought on the appropriate balance between the human use of natural capital and the rest of nature's use of the limited amount available.

Identifying critical natural capital

Three characteristics make an ecosystem function or service critical:

- *non-substitutability* – where there is no natural or human-made substitute; for example, the role of the atmosphere in protecting the Earth's surface from solar radiation
- *irrevocable loss* – when degradation reaches such a level that the recovery of an ecosystem is not possible within a human timescale
- *high risk* – where the loss of a function could result in a significant risk to human well-being.

A number of major research projects are examining the specific aspects of natural capital that are critical at the regional and global scales.

> **Overfishing** is a level of fishing resulting in the depletion of the fish stock.
>
> **Tragedy of the commons** is the idea that common ownership of a resource leads to overexploitation because some nations always want to take more than other nations.

SELF-ASSESSMENT QUESTIONS 8.02.01

1. What is a resource?
2. Explain the difference between natural capital and natural income.
3. **Discussion point:** What are the main forms of natural capital in the country in which you live?

CASE STUDY 8.02.01

Mismanagement of natural capital: global fishing

'Valuable fish stocks, as well as a whole host of other marine life, are severely threatened by overfishing.'
WWF Global, 2015

A critical situation

Many of the world's fishing grounds are in crisis because of **overfishing**. In the worst-affected areas it is feared that fish stocks may not recover for a long time, if at all. According to WWF:

- the global fishing fleet is two to three times larger than the oceans can sustainably support
- fifty-three per cent of the world's fisheries are fully exploited
- thirty-two per cent of the world's fisheries are overexploited, depleted or recovering from depletion
- several important fish populations have declined to such an extent that their survival is threatened
- every year, billions of unwanted fish and other animals (dolphins, marine turtles, seabirds, sharks, etc.) die because of inefficient, illegal and destructive fishing practices.

WWF has stated, 'Unless the current situation improves, stocks of all species currently fished for food are predicted to collapse by 2048.' In many parts of the world, fish are vitally important to food security and general living standards. About 1 million people around the world rely on marine fish as an important source of protein.

Image 8.10 Modern fishing trawler.

'The **tragedy of the commons**' is a term used to explain what has happened in many fishing grounds. Because the seas and oceans have historically been viewed as common areas, open to everyone, the capacity of fishing vessels operating in many areas (see Image 8.10) has exceeded the amount of fish available. The result has been resource depletion.

Resource management

The EU's Common Fisheries Policy (CFP) is a significant example of resource management, although perceptions vary widely as to its effectiveness.

8.02 Resource use in society

CASE STUDY 8.02.01 (continued)

The CFP has evolved over a number of stages. Current measures to conserve fish stocks include:

- taking a long-term approach by fixing total allowable catches on the basis of fish stocks
- introducing accompanying conservation measures
- setting recovery plans for stocks below the safe biological limits
- managing the introduction of new vessels and the scrapping of old vessels in such a way as to reduce the overall capacity of the EU fleet
- EU aid for the modernisaton of vessels finished at the end of 2004
- measures to neutralise the socioeconomic consequences of fleet reduction
- measures to encourage the development of sustainable aquaculture.

Fish stocks need to renew themselves due to losses from both fishing and natural causes. Thus it is important that small fish are allowed to reach maturity so that they can reproduce. With this in mind, the CFP sets the maximum quantity of fish or total allowable catch that can be caught each year. This is divided among the member states to give the national quota for each country. The number of small fish caught is limited by:

- specifying minimum mesh sizes
- closing certain areas to protect fish stocks
- banning catches of certain types of fishing
- recording catches and landings in special log books.

Supporters of the CFP argue that it makes a strong contribution to the quest for sustainable fishing. However, environmentalists believe that short-term economic and political concerns override the objective of sustainability.

Questions

1. Why has the rate of fishing in many of the world's fishing grounds exceeded the sustainable yield?
2. What have been the consequences of overfishing?
3. Comment on the ways in which the EU's CFP has tried to rectify this situation.

8.02.02 Renewable, replenishable and non-renewable natural capital

Renewable natural capital

Renewable natural capital (see Image 8.11) is the composition and structure of natural, self-organising ecological systems that, through their functioning, yield a flow of goods and services. These flows are essential to life in general and are very valuable to the human population and all other species. Renewable natural capital can be generated and/or replaced as fast as it is being used. It includes living species and ecosystems that use solar energy and photosynthesis.

Concerns about climate change in particular have highlighted the value of renewable natural capital. The world's forests arguably represent our most important renewable natural capital.

Forests provide the world with many benefits; some are easily quantifiable, for example as a source of construction timber, cooking fuel and medicinal herbs. Forests provide habitat for a wide diversity of plant and animal species. Other, wider ranging, benefits are harder to assess. Ecosystem functions that have particular value for human populations are known as ecosystem services. Such ecosystem services include water cycling, climate regulation and prevention of soil erosion. Forests provide ecosystem services that underpin agriculture and contribute to the formation of topsoil, which serves as an important sink for carbon. Each year, forests lock up 335–365 gigatonnes of carbon, making them a viable solution to climate change mitigation.

Renewable natural capital comprises living species and ecosystems. It is self-producing and self-maintaining and uses solar energy and photosynthesis to produce food and chemical energy.

Image 8.11 Moraine Lake, Alberta, Canada.

Replenishable natural capital

Replenishable natural capital refers to non-living natural resources, such as groundwater, that depend on the Sun's energy for their replenishment. They are continually recycled through their interaction with living resources over long periods. An example is the interaction between surface mineral components and living organisms that produces fertile soil. The condition of renewable natural capital stocks affects the quality, quantity and renewal rate of replenishable natural capital, and the relationship works the other way as well.

Replenishable resources are sometimes viewed as intermediate between renewable and non-renewable resources, because they can be replaced, but often over a long period in human terms. The depletion of groundwater resources is a major issue in many parts of the world. The steadily increasing demand for groundwater means that it is often extracted at a much faster rate than it is recharged.

Non-renewable natural capital

Non-renewable natural capital is either irreplaceable or can only be replaced over geological timescales. Non-renewable resources such as coal and oil take millions of years to form. Thus, in human terms, these resources are fixed in the supply available. **Resource depletion** can occur relatively quickly and supplies eventually become exhausted with no viable further production possible.

Image 8.12 Winding house at the Hetty Pit, Wales. This coal mine closed in 1983.

In the EU, many coalfields that were mining large quantities of coal in the 19th and early 20th centuries no longer produce any coal (see Image 8.12). In many oil-producing countries, large oil fields are declining in annual production, as the most accessible oil has already been pumped out. Some experts are worried that the world is very close to 'peak oil' production, although most of the large oil companies say that we will probably not reach this situation for decades. We do not know how much oil will be discovered in the future, and this goes for other non-renewable resources as well. What we do know is that the price of many non-renewable resources is currently very high and this affects the standard of living of people all over the world.

In recent years, a growing body of research has shown that economic growth in an increasing number of countries has been adversely affected by rapid decline of one or more aspect of natural capital. It is becoming increasingly crucial to understand the relationship between economic growth and natural capital so that intelligent decisions can be made in terms of sustainable development policies.

Recyclable resources can extend the useful lives of renewable, replenishable and non-renewable resources, although most recycling concentrates on reusing non-renewable resources, as these are generally considered to be the most scarce of the world's resource base and hence the most valuable.

> **Replenishable natural capital** consists of stocks of non-living resources. Examples are the atmosphere, fertile soils and groundwater. Such resources are dependent on energy from the Sun for renewal.
>
> **Non-renewable natural capital** consists of subsoil assets such as coal, oil copper and diamonds. Such resources are depleted as they are consumed.
>
> **Resource depletion** is the consumption of non-renewable resources, which will eventually lead to their exhaustion.
>
> A **recyclable resource** is one that can be used over and over, but must first go through a process to prepare it for reuse.

SELF-ASSESSMENT QUESTIONS 8.02.02

1. Define: **a** renewable natural capital, **b** replenishable natural capital, **c** non-renewable natural capital.
2. What are recyclable resources?
3. **Discussion point:** Are there any examples of resource depletion in the country in which you live? If so, what are the reasons for such resource depletion?

8.02 Resource use in society

Image 8.13 Eurostar train in France powered by electricity.

8.02.03 The concept of natural capital is dynamic

Economic, technological, cultural and other factors influence the status of a resource over time and space. For example, once all trains were powered by steam produced from burning coal. Today many are powered by diesel or electricity (see Image 8.13). Technological advance has been the key to:

- developing new resources
- replacing less efficient with more efficient resources.

Changes over time

The global use of resources has changed dramatically over time. Such changes can be illustrated with reference to the UK. The combination of the many advances in technology during the Industrial Revolution, which began in Britain in the latter part of the 18th century, brought about a huge increase in resource use and considerable changes in the demand for different resources. Such changes included:

- Replacement of water power by steam power: This resulted in the rapid development of the UK's coalfields from the mid-1700s. Steam power was the fundamental invention of the Industrial Revolution. A wide range of industries developed on and around the UK's coalfields in a relatively short period.
- Invention of the Gilchrist–Thomas process in the iron and steel industry in 1878: This made it possible to smelt iron from phosphoric iron ores for the first time, leading to the development of the Jurassic iron ore fields of Lincolnshire and the East Midlands. Before 1878, these ores had no economic use. The mining of the Jurassic iron ores led to the construction of steel works in the region and the expansion of urban settlements.

In more recent times, the following changes are notable:

- Intense pressure on food supplies during the Second World War resulted in the ploughing up of large areas of chalk downland for the first time. Advances in agricultural science made it possible to obtain reasonable crop yields with the application of the correct fertiliser mix. Prior to this understanding, the economic use of the chalk downlands was almost totally for sheep farming.
- The development of the nuclear power industry in the UK and other countries found a new use for uranium, which significantly increased its value as a resource.
- The railway system, which was once totally steam driven, was electrified. This major change significantly reduced the demand for coal.
- The location of oil and gas in the deeper parts of the North Sea was established some time before production from these areas commenced. What was required was higher oil prices to justify the costs of deep water production, and advances in deep-sea oil production technology, as oil production had never occurred in such deep waters before. Experience gained in the North Sea has been invaluable in drilling for oil in other deep-water locations around the world.
- Renewable energy technology, particularly the construction of offshore wind farms, is now beginning to utilise flow resources in a significant way.
- Recycling has increased considerably in importance over the last decade, involving a much wider range of materials and products.

As society has changed, attitudes to certain resources and their use have also changed. Attitudes to whaling have changed radically over the last half a century, leading to the closure of many former whaling stations (see Image 8.14). The demand for organic food is much higher today than even ten years ago, and this upward trend is expected to continue with some fluctuations. Some power companies provide green energy options, which are attracting an increasing number of customers. Attitudes to plastic bags and the packaging of goods in general are changing in terms of the resources, like oil, which are used up and the amount of waste created. More and more people are questioning the environmental credentials of the companies from which they purchase goods and services.

Resources, reserves and supply: dynamic interchange

The resources that a country has of any mineral comprise the total amount of that mineral contained in the Earth's crust within its borders. The reserve is that fraction of the total resources of the mineral that can be calculated to exist at commercially exploitable values under known technology. The supply of the mineral is the amount that can be produced and delivered to customers. The level of resources, reserves and supply can change very rapidly, each being partly dependent on and partly independent of the other levels.

Resources can be divided into those that have been identified and those that are as yet undiscovered (see Figure 8.17). Identified resources that are not classified as reserves are classed as being paramarginal or submarginal, depending on the estimated cost of extraction. Paramarginal resources are those recoverable at costs as much as one and a half times those that can be borne now. Submarginal resources are more costly still to extract.

> **CONSIDER THIS**
>
> The human population today extracts and uses around 50 per cent more natural resources than 30 years ago, at about 60 billion tonnes of raw materials a year. People in rich countries consume up to ten times more natural resources than those in the poorest countries.

Image 8.14 Abandoned whaling station, South Georgia, South Atlantic Ocean.

Figure 8.17 US Geological Survey classification of mineral resources.

SELF-ASSESSMENT QUESTIONS 8.02.03

1. Describe two ways in which technological development has changed resource use in the UK.
2. Give an example of the way in which changing public perception can impact on resource use.
3. **Discussion point:** How has your family's use of resources changed in recent years? Which resources do you use more of and less of?

8.02 Resource use in society

8.02.04 Environment and its intrinsic value

> 'One impulse from a vernal wood
> May teach you more of man,
> Of moral evil and of good,
> Than all the sages can.'
>
> *The Tables Turned,* William Wordsworth

The intrinsic value of the environment (see Image 8.15) or any other entity is the value that the entity has in itself. The 'in' (of intrinsic) indicates the inner state of things. Intrinsic value is usually contrasted with extrinsic value, or use value. In her book *The Intrinsic Value of Nature*, Leena Vilkka argues that 'Intrinsic value is a notoriously difficult concept.' She sees the search for intrinsic or natural value as the ultimate ground of nature-conservation philosophy.

Aspects of the environment that are valued on aesthetic or intrinsic grounds may not provide commodities identifiable as either goods or services, and thus remain unpriced or undervalued from an economic perspective. Natural beauty is a typical example of intrinsic valuation, valued for its own sake. Organisms or ecosystems that are appreciated by people for their intrinsic value may be viewed as such from an ethical, spiritual or philosophical point of view. They are valued regardless of their potential use in an economic sense.

The English nature poet William Wordsworth, quoted above, wrote that solitary experiences in nature can engender a sense of fear and beauty that, properly remembered and analysed, can lead to a fuller awareness of the harmony and beauty of nature and ultimately to a reinforced moral sense. Wordsworth was a dedicated walker and often composed poetry while walking.

Most people feel connected to nature in some degree. Some feel a strong spiritual bond that may be rooted in our common biological ancestry. Human cultures around the world profoundly reflect our instinctive attachment to the natural world. Thus cultural diversity is inextricably linked to the Earth's biodiversity. Thousands of cultural groups around the world each have distinct traditions and knowledge in the context of relating to the natural world. For example in rainforests, indigenous peoples who have lived in these environments for thousands of years usually place very different values on the rainforest environment compared with outside groups or relative newcomers to this biome. However, many more people from other environments now visit or want to visit areas of rainforest because they have become aware of their intrinsic characteristics.

Figure 8.18 examines the cultural value of forests. In many parts of the world, natural features such as specific forests, mountains or caves have a religious meaning for various communities. Nature is a common element for many religions and plays a significant role in creating a sense of belonging.

National parks and other protected areas

While most environmental legislation is of relatively recent origin, concern about environmental destruction dates from the mid-18th century, with European nature-romanticism. Such ideas spread to North America, with a growing number of strong-willed individuals advocating environmental protection.

Image 8.15 The spectacular Milford Sound, South Island, New Zealand – a location of great intrinsic value.

By Lara Barbier, Apr 14, 2011

Cultural Values of Forests

There is a particular magic in a walk through woodlands and forests. Perhaps it is the sense of being in the presence of one of the longest living organisms on earth, or maybe it is the abundance and diversity of life that makes its home in these ecosystems or even the quality of light and sound. Whether enjoying woods as a local recreation, visiting an exotic forest as a tourist or valuing them as part of your livelihood, trees have a special affinity to mankind. One need only look at some of the greatest writers and philosophers of our times to see that woods and forests hold a sacred place in cultures across the globe.

TEEB[1] aims to draw attention to the economic value of nature in our daily lives. This is important because decision makers often look at the cost-benefit when making decisions. Too often we are not factoring in the economic value that the many services of nature provide for local and national economies and thus we are only seeing the negative consequences of losing nature's services when the damage is severe or when it is too late. However, applying economic valuation does not seek to ascertain the 'right numbers' that define the total value of nature. The economic value is only one of the many values of nature including also cultural, spiritual and existence values that cannot be expressed in money. Therefore, it is also necessary in decision making to take into account people's cultural and spiritual values concerning nature.

Sacred groves can be found in many parts of the world and have been protected by indigenous peoples and local communities for centuries due to their spiritual and cultural value. There are also indigenous people's protected areas, indigenous people's conserved territories or community conserved areas (ICCA's). Local concern due to cultural significance can be a highly effective way of conserving natural areas that provide important values to local and regional communities. Often the sacred forests, mountains or springs provide multiple ecosystem services that are important for the local community such as recreation, a place for contemplation or drinking water.

[1]TEEB: The Economics of Ecosystems and Biodiversity

Figure 8.18 Cultural value of forests.

Notable among the environmentalists of the 19th century were Ralph Waldo Emerson, Henry David Thoreau, George Perkins Marsh and John Muir. Emerson (1803–1882) in his first book, *Nature*, published in 1836, expressed the view that wild nature was the ultimate source of life and culture. Thoreau's writings stirred many to a new appreciation of nature, and his 1851 proclamation 'In wildness is the preservation of the world' is the motto of the Wilderness Society. Thoreau insisted many times on the ethical and moral value of solitude in nature. Marsh (1801–1882), perhaps more than any other 19th-century writer, identified the extent to which economic activity was ravaging the landscape. Marsh believed that the destruction of the natural world would eventually damn civilisation unless there was a major change in attitudes to nature. John Muir (1834–1914), a renowned explorer and naturalist, founded the Sierra Club in 1892 and was instrumental in the establishment of the national park system in the USA.

The world's first national park was Yellowstone in the western USA (see Image 8.16). Yellowstone was designated in 1872 when members of a scientific expedition brought back photographs and persuaded the federal government to preserve this part of the west in its unspoilt natural condition. This was a major departure from the pervading attitudes of the time, when nature was seen as something to be conquered and exploited for economic gain. The concept of national parks spread around the world. Most countries now have national parks and other protected areas such as national forests and areas of outstanding natural beauty within their borders.

Wilderness areas with the greatest restrictions on access have the highest form of protection. Over time, more parts of the world have benefited from various degrees of protection. Without the designation of national parks and other protected areas, the world's flora, fauna and other resources would be in a much worse state today. Only about 12 per cent of the world's land area is covered by national protection schemes. In four of the largest countries with high biodiversity – Brazil, China, India and Russia – the areas covered by national protection schemes are 7 per cent, 8 per cent, 5 per cent and 8 per cent respectively.

Image 8.16 Bison at Yellowstone National Park.

CONSIDER THIS

The largest national park in the world is the Northeast Greenland National Park, created in 1974 by the Danish government, which is responsible for Greenland. The Northeast Greenland National Park provides a sanctuary for musk, oxen, polar bears and walruses.

8.02 Resource use in society

Theory of knowledge 8.02.01

Nineteenth-century US writers such as Henry David Thoreau, George Perkins Marsh, John Muir and Ralph Waldo Emerson did much to raise environmental awareness in the USA. They were an important influence on the establishment of the early national parks, such as Yellowstone in 1872. The power of their words reached a wide audience as they constantly reminded people of the value of the natural world. Writers and artists in other countries have also done much to raise appreciation of the natural world.

1. How important do you think such contributions are compared with the impact of scientific evidence?
2. Has a work of art or literature raised your appreciation of the natural environment?

However, some protected areas are under severe pressure. The Maasai Mara National Reserve in Kenya is losing animal species at a significant rate, according to a recent scientific study. This is due to increased human settlement in and around the reserve. The study, which was funded by WWF, monitored hoofed species in the Maasai Mara on a monthly basis for 15 years. According to this study, six species – giraffes, impala, warthogs, hartebeest, topis and waterbucks – have declined significantly at an alarming rate in the reserve. The study says that losses were as high as 95 per cent for giraffes, 80 per cent for warthogs, 76 per cent for hartebeest and 67 per cent for impala.

While the main concerns about environmental sustainability relate to areas beyond national parks and other protected areas, the latter are a continuous reminder of the importance of the natural environment and what can be done when there is sufficient political will to act. Much of our knowledge about the environment has been gained from protected areas over a significant period of time.

SELF-ASSESSMENT QUESTIONS 8.02.04

1. What do you understand by the term 'intrinsic value'?
2. When and where was the world's first national park established?
3. **Discussion point: a** Name a natural environment that has particular meaning for you. **b** Give reasons for your choice.

8.02.05 The impacts of extraction, transport and processing of renewable natural capital

The extraction, transport and processing of renewable natural capital can have a huge negative impact, making this natural capital unsustainable. A region of great concern in this respect is the Amazon rainforest in Brazil.

In recent decades, significant areas of rainforest have been cleared. The main reasons have been to:

- provide newly settled smallholders with land and crops
- create huge cattle ranches
- build roads
- exploit mineral deposits
- use wood for fuel
- use wood for furniture manufacture and for pulp and paper
- provide land for urban and industrial uses.

The impact of logging

Amazonia is the world's last great reserve of tropical timber. For a long time, much of the forest was protected by its inaccessibility. However, since the Belem–Brasilia highway was built in 1965, timber companies have flocked to the state of Para in particular. The frontier town of Paragominas has one of the highest concentrations of saw mills in the world. The typical sequence of environmental destruction is as follows:

1. An area of forest is searched for suitable trees to be cut.
2. The entangled nature of the forest means that, when the selected trees crash down, they bring many others with them.

3   More damage is caused when the retrieval team move in with bulldozers to haul out the selected trees.
4   Although a few trees will repopulate the bulldozer trails, nothing grows where the tyre tracks ran. Trees continue to fall along the trails, brought down by winds channelled along the newly created corridors.
5   The area becomes susceptible to fire for the first time. Normally the closed canopy prevents the moist leaf litter on the forest floor from drying enough to allow any chance sparks to spread. But once the canopy has been broken open, fire becomes a real hazard. Throughout Para, fires have wiped out vast areas of logged forest.

However, with careful management, the impact of logging can be much reduced. Vine cutting means that 30 per cent fewer trees may be damaged when felling takes place. Careful route planning can reduce by a quarter the area affected by bulldozers. Smaller gaps in the canopy and less fuel left behind on the forest floor considerably cuts the risk of fire.

Around 42 per cent of Brazil's tropical rainforest is theoretically protected by many different conservation units under varying sets of rules. Although the area is monitored by satellite, there are not enough police and officials on the ground to combat bandit loggers. Corruption among officials and violence towards them are also problems.

The impact of deforestation and the loss of biodiversity in both Brazil and the planet as a whole has been well documented, focusing on the major problems of:

- depletion of the region's genetic bank
- increased levels of carbon dioxide
- reduced levels of oxygen
- decreased precipitation due to reduced vegetation cover
- reduced interception capacity leading to greater likelihood of flooding
- the ultimate alteration of the Amazon's discharge regime and sediment transport pattern
- severe soil exhaustion and erosion
- the adverse impact on the indigenous population.

8.02.06 The concept of sustainability

Societies that deplete their natural capital cannot do so for ever, because their development will be unsustainable (see Chapter 1.04). Much human wealth and general well-being is dependent on the goods and services provided by certain forms of natural capital. However, the rate at which it is used should not exceed the rate at which it is renewed, otherwise natural capital will be depleted. Careful management (see Image 8.17) is required to ensure that this does not happen. Sustainability means living within the means of nature.

Natural income is monetary income derived from natural capital. However, economic theory has largely neglected the protection and appropriate pricing of environmental resources. This is largely because firms did not have to pay for any adverse environmental impact they created. If indiscriminate economic development is allowed to grow unchecked, stocks of natural capital will continue to decline. This will result in problems for natural-life support systems, increased market prices and a decrease in the quality of human life.

Problems related to the protection of natural capital include the following:

- The previous inability of economics to appropriately model and price both market and non-market environmental resources: Modern economists are paying much more attention to such issues today.

Image 8.17 Cattle in West Africa.

8.02 Resource use in society

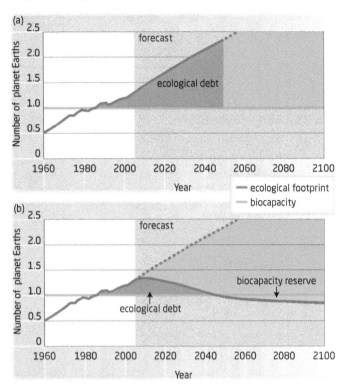

Figure 8.19 (a) Business-as-usual scenario and (b) return to sustainability.

- Lack of willingness to pay: It is perhaps human nature that firms or individuals do not want to pay directly for something that has always been free.
- Lack of knowledge about minimum levels or time spans required for resources to replenish or renew. As the level of research into such vital issues increases, the evidence base will hopefully develop to such an extent that most people will be convinced about the need for sustainable policies.
- Lack of knowledge regarding the interaction and dependences between resources and their true value, usefulness or necessity.
- Poor management of transboundary resources: Many natural capital resources cross political boundaries. While coordinated management is generally improving, in some cases it is still woefully inadequate.
- Inequalities between MEDCs and LEDCs: Developed countries have long been able to purchase natural capital from the developing would at relatively low prices compared with the goods and services that move in the other direction. This has widened the development gap and often made it difficult for developing counties to manage their natural capital in a sustainable way.

8.02.07 Sustainable development

The concept of sustainable development was introduced in Chapter 1.04, where some of the changes in attitudes and values towards consumption of resources by human populations were introduced.

Figure 8.19a, from the WWF's *Living Planet Report 2008*, shows what is likely to happen under the **'business-as-usual'** scenario. Here, only very limited efforts are made in terms of **environmental sustainability** and the problem gets steadily worse. Human populations in the most marginal areas will be affected first, but gradually environmental problems will encompass more and more regions and their populations.

In contrast, Figure 8.19b shows what could happen with environmental sustainability at the forefront of policymaking. The diagram shows ecological debt being gradually reduced until once again the planet has a biocapacity reserve and is living within its means.

Environmental sustainability requires political will by governments on both a national and an international basis and action by all sectors of society. It demands limits on the actions of individuals and organisations whose behaviour damages the environment for personal and organisational gain.

The **business-as-usual** scenario is the scenario for future patterns of production and consumption which assumes that there will be no major changes in attitudes and priorities.

Environmental sustainability refers to meeting the needs of the present without compromising the ability of future generations to meet their needs.

Towards sustainable development in tourism

'It is increasingly apparent that tourism is falling victim, but also contributing, to climate change and the reduction of biodiversity. The path ahead is therefore marked by a different type of growth: more moderate, more solid and more responsible.'

Francesco Frangialli,
Secretary-General of the UN World Tourism Organization (UNWTO)

As the level of global tourism increases rapidly, it is becoming more and more important for the industry to be responsibly planned, managed and monitored. Tourism operates in a world of finite resources (see Image 8.18) where its impact is becoming of increasing concern to a growing number of people. At present, only 5 per cent of the world's population have ever travelled by plane. However, this is undoubtedly going to increase substantially.

Following the 1992 Earth Summit in Rio de Janeiro, the World Travel & Tourism Council (WTTC) and the Earth Council drew up an environmental checklist for tourist development which included waste minimisation, reuse and recycling, energy efficiency and water management. The WTTC has since established a more detailed programme called Green Globe, designed to act as an environmental blueprint for its members.

However, Leo Hickman in his book *The Final Call* claims that the industry is still in a poor state with regard to environmental sustainability. He states:

Image 8.18 Firewood outside tourism ger in Mongolia.

'The net result of a widespread lack of government recognition is that tourism is currently one of the most unregulated industries in the world, largely controlled by a relatively small number of Western corporations such as hotel groups and tour operators. Are they really the best guardians of this evidently important but supremely fragile global industry?'

Environmental groups are keen to make travellers aware of their **destination footprint**. They are urging people to:

- fly less and stay longer
- carbon offset their flights
- consider 'slow travel'.

In slow travel, tourists consider the impact of their activities both for individual holidays but also in the longer term as well. For example, they may decide that every second holiday will be in their own country (not using air transport). It could also involve using locally run guesthouses and small hotels as opposed to hotels run by international chains. This enables more money to remain in local communities.

Virtually every aspect of the industry now recognises that tourism must become more sustainable. **Ecotourism** is at the leading edge of this movement. Ecotourism has helped to bring needed income to some of the poorest parts of the country. It has provided local people with a new, alternative way of making a living. As such, it has reduced human pressure on ecologically sensitive areas.

The **destination footprint** is the environmental impact caused by an individual tourist on holiday in a particular destination.

Ecotourism is a specialised form of tourism where people experience relatively untouched natural environments such as coral reefs, tropical forests and remote mountain areas and ensure that their presence does no further damage to these environments.

The Environmental Sustainability Index

The Environmental Sustainability Index (ESI) benchmarks the ability of nations to protect the environment over the next several decades. The ESI integrates 76 data sets – tracking natural resource endowments, past and present pollution levels, environmental management efforts, and the capacity of a society to improve its environmental performance – into 21 indicators of environmental sustainability. These indicators allow comparison across a range of issues that are grouped into five broad categories:

- environmental systems
- reducing environmental stresses
- reducing human vulnerability to environmental stresses
- societal and institutional capacity to respond to environmental challenges
- global stewardship.

The higher the ESI score, the better the situation with regard to environmental sustainability. Virtually no country scores very high or very low on all 21 indicators.

8.02 Resource use in society

The ESI is compiled by environmental scholars from Yale and Columbia universities in the USA. Figure 8.20 illustrates the global situation by quintile (five equal portions). The most sustainable countries, according to the ESI method, are mainly affluent, sparsely populated nations with significant natural resources. The lowest-ranking countries are North Korea, Iraq, Taiwan, Turkmenistan and Uzbekistan. All these countries face numerous environmental issues, natural and human-induced, and have a poor environmental policy record.

The ESI shows that wealth contributes to the potential for good environmental stewardship although it does not guarantee it, as a number of affluent nations occupy low-ranking positions on the list. The relationship between economic development and environmental sustainability is complex, with countries facing environmental challenges at every level of development.

Development and environmental sustainability

It has become the norm in most countries to consider the environmental impact of a development scheme before planning permission is granted (see Chapter 1.04).

Figure 1.19 illustrates the four key strands of an environmental impact assessment (EIA). These are further described as follows:

- *Eco-friendly:* There has been increasing recognition that long-term development has to be environmentally sustainable. The ability to design sustainable development schemes has improved considerably in recent years as more data has become available from earlier schemes whose merits and limitations have been studied. Environmental science has come a long way in recent decades.

- *Public participation:* Too often in the past this has been a question of going through the motions and then ignoring public opinion if the public's views were contrary to the desires of business people and politicians. Environmentalists advocate bottom-up strategies that frequently involve the use of intermediate technology. If people feel that their opinion matters, they are more likely to actively participate in sustainable projects.

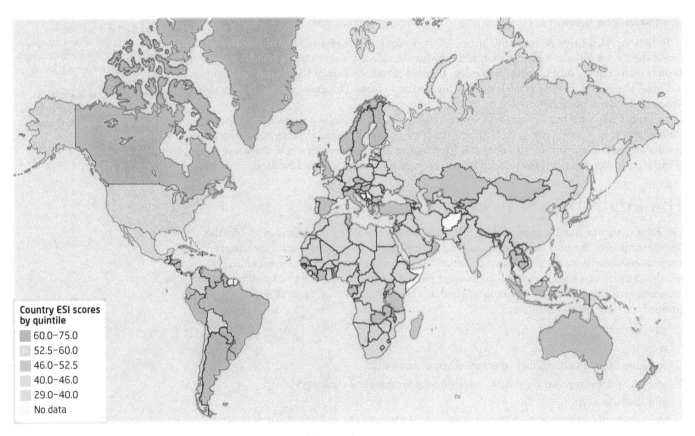

Figure 8.20 World map – 2005 Environmental Sustainability Indices.

- *Social justice and equity:* Development that is just and fair means that poor people become stakeholders in development schemes. Direct involvement with a reasonable share of the benefits available gives a strong incentive to ensure that development is sustainable. Local people want income improvement to be long term and to be available to their children and grandchildren. The concept of **pro-poor strategies** has expanded considerably in recent decades, gaining recognition for its many successes. The development of micro-credit and social business are examples of such strategies.
- *Futurity:* Transgenerational responsibility is central to the concept of environmental sustainability. The development of new technologies to conserve resources has attracted increasing investment in recent years. In some countries, the 'green industrial sector' is now a significant part of the economy, employing an increasing number of people.

> **Pro-poor strategies**
> Development strategies that result in

Changing attitudes to sustainability and economic growth

Global summits have done much to influence attitudes to sustainability around the world. The Rio Earth Summit in 1992 was a significant event in this process. Coordinated by the UN, it produced Agenda 21 and the Rio Declaration. Agenda 21 is a comprehensive plan of action to be taken globally, nationally and locally by UN organisations, governments and major groups in all areas in which there are human impacts on the environment.

The Johannesburg World Summit on Sustainable Development (2002) followed ten years after the Rio Earth Summit. Here the focus was mainly on social issues, with targets set to reduce poverty in its various forms.

The next Earth Summit took place in June 2012 in Rio de Janeiro. Referred to as the Rio+20 or the Earth Summit 2012, the objectives of the summit were to:

- secure renewed political commitment to sustainable development
- assess progress towards internationally agreed goals on sustainable development
- address new and emerging challenges.

The Earth Summit 2012 also focused on two specific themes: a green economy in the context of poverty eradication and sustainable development, and an institutional framework for sustainable development.

Although progress after summits has often been disappointing, significant changes have taken place due to the gradual raising of awareness of the importance of sustainable development at all scales from the individual to the international.

Sustainable yield

Sustainable yield (SY) is the rate of increase in natural capital – that is, the capital that can be exploited without depleting the original stock or affecting its potential replenishment.

Examples of SY include the following:

- *Forestry:* SY is the largest amount of harvest activity that can occur without degrading the productivity of the stock.
- *Fishing:* SY is the amount of fish that can be caught on a regular basis without compromising the ability of the species to reproduce and maintain its population (see Chapter 4.03).
- *Groundwater:* SY is the annual yield of water extraction beyond which an aquifer would risk depletion.

SY can vary over time with the needs of the ecosystem to maintain itself. For example, a forest that has recently suffered a fire will require more of its own ecological yield to sustain and re-establish a mature forest. While the forest is doing so, the sustainable yield may be much less than in previous years.

The related concept of maximum sustainable yield (MSY) is frequently quoted in terms of commercial production (see Chapter 4.03). This is the largest amount of a raw material that can be extracted without permanently depleting the stock. MSY has been exceeded in many of the world's fishing grounds, resulting in various attempts to return to a situation of sustainability.

> **Sustainable yield (SY)** is the rate at which capital can be exploited without depleting the original stock or affecting its potential replenishment.

8.02 Resource use in society

The carrying capacity for a species is dependent on:
- its reproductive capability
- its longevity
- the indigenous resources of the ecosystem or habitat.

Every year or breeding season the population will increase due to new offspring and maybe also due to immigrants. Population loss will occur due to death and emigration. A net increase in population will occur if the number added to the population is larger than the number leaving. If the increase in population (of fish, for example) is harvested, the population will remain the same. This gives the MSY for the population.

SELF-ASSESSMENT QUESTIONS 8.02.05

1. Define environmental sustainability.
2. What is the purpose of the ESI?
3. **Discussion point:** How does the average person have an adverse impact on the environment when they go on holiday?

Solid domestic waste

8.03

LEARNING OBJECTIVES

After reading this chapter you should be able to:

- understand that solid domestic waste (SDW) is increasing as a result of growing human populations and consumption
- appreciate that the production and management of SDW can have a significant influence on sustainability.

KEY QUESTIONS

What is SDW and how have its volume and composition changed?

What are the advantages and disadvantages of the different methods of waste disposal?

How can the management of SDW be improved?

8.03.01 Types of solid domestic waste

There are different types of **solid domestic waste (SDW)** (see Image 8.19), the volume and composition of which changes over time. SDW is produced by households (people living in their homes) and is distinct from waste from other sources such as agriculture, industry and the service sector. It is effectively what families put in their bins. In the USA, the terms 'trash' and 'garbage' are used to describe SDW. In the UK, the terms 'refuse' and 'rubbish' are used. Some different terms are used in other countries. The solid waste from households includes the following classes of material:

- *Paper:* newspapers, magazines, wrapping paper, unwanted posted material, advertising material put through letterboxes, household office waste.
- *Cardboard packaging:* from food products and a variety of other household products.
- *Glass:* bottles and jars from food, drink and cosmetic products mainly.
- *Metal:* cans used to contain drinks and various types of food; foil for use in cooking and keeping food fresh.
- *Plastics:* food and other types of household packaging and containers.
- *Organic waste:* from kitchens (waste from preparing food, and food waste after consumption) and garden waste.
- *E-waste:* used printer cartridges, batteries and electronic equipment.

Domestic waste consists of both organic and inorganic materials. The former comprises food, kitchen waste and garden waste (e.g. lawn clippings and waste from weeding). Inorganic waste is effectively what remains, such as paper, glass and metal. Inorganic waste has been recycled for some time in many countries, and the rate of recycling is on the increase. The rate of household organic waste recycling is considerably less, but the rate is nevertheless rising.

Solid domestic waste (SDW) is waste produced by households as opposed to that produced by other sectors of an economy.

Image 8.19 Types of SDW.

8.03 Solid domestic waste

The amount of SDW produced has increased with population growth and rising living standards. MEDCs produce far more waste in most categories than LEDCs, because of their considerably higher spending power. As living standards develop, the nature of SDW also changes. For example, e-waste is a significant contributor to household waste in MEDCs but only makes a minor contribution in very poor nations. The highest rates of increase in SDW are in the NICs such as China, India and Brazil. SDW is a major environmental problem that is having a significant influence on sustainability. The rate of increase in SDW is creating intense pressures in many parts of the world.

Figure 8.21 shows how the type of waste produced is determined by the ways in which household needs are fulfilled, which can provide a useful framework for collecting data about household waste. Figure 8.22 is a recent estimate of the composition of household waste in the UK, which shows that 50 per cent of the total comprises food waste and paper and board. Accurate household waste data can be difficult to collect, and thus estimates can vary considerably. SDW in the UK amounts to an average of about 300 kg per person. Each year, the average British family throws away six trees worth of paper. The UK uses about 12 billion cans each year. The UK disposes of over 2 million television sets a year; they are now classed as hazardous waste and can no longer go to landfill.

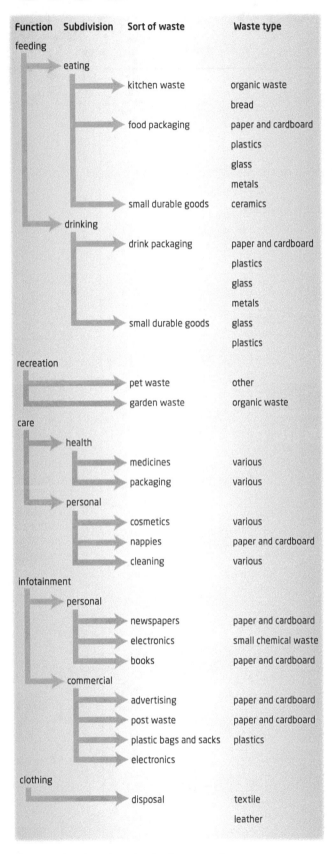

Figure 8.21 Functions in the home and waste types generated.

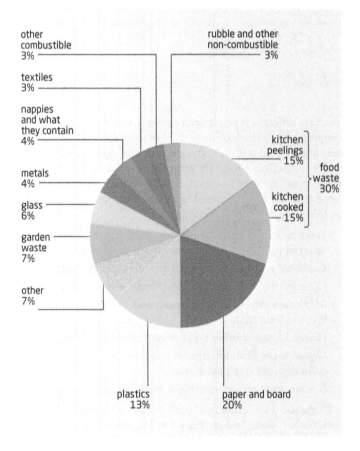

Figure 8.22 Composition of UK household waste.

Table 8.08 shows the composition of domestic waste from five developing countries. While there are some notable differences, there is a clear general pattern. There is no more than a 1 per cent range in the proportion of biodegradable material. Figure 8.23 shows the composition of solid waste for low-income, middle-income and high-income countries, according to United Nations Environment Programme (UNEP). This shows that as income level rises, the proportion of:

- organic/biomass waste decreases
- paper and cardboard increases
- plastic waste increases
- metal and glass waste increases.

	Brazil	Nepal	Nigeria	India	China
Biodegradable material / %	72	71	72	71	72
Plastics / %	11	12	11	9	11
Paper and cardboard / %	7	8	3	9	7
Glass / %	3		4	5	5
Metals / %	3		2	4	1
Textiles / %	2		5	2	2
Miscellaneous / %	2	9	3		2

Table 8.08 Unsorted domestic waste composition from five developing countries.
Source: Adapted from *The Open Waste Management Journal*, 2011, Vol. 4

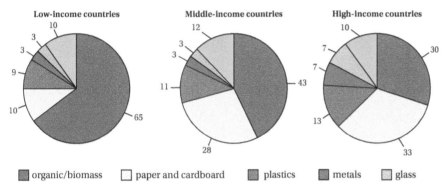

Figure 8.23 Waste composition in relation to the relative wealth of countries.

How much waste does your household generate?

How much waste do you think your household produces in a year (see Image 8.20)? Discuss this topic with other people in your set to see how perceptions vary. Most people probably think or want to believe they produce less waste than they actually do. Keep a waste diary for your household for a week using Figure 8.24. Try to calculate the proportion of your household waste that you recycle. Make lists of the categories you recycle and those that you do not. Do certain members of your family create more waste than others? Cutting down on waste is not only environmentally friendly but can also save households a significant amount of money.

Image 8.20 SDW for both landfill (the large green bin) and recycling (the other containers) outside a house in the UK.

8.03 Solid domestic waste

	MONDAY	TUESDAY	WEDNESDAY	THURSDAY	FRIDAY	SATURDAY	SUNDAY
Paper and cardboard							
Glass							
Metals							
Plastic							
Food							
Books							
Electrical items							
Clothes							

Figure 8.24 A waste diary.

CONSIDER THIS
In the past century, global waste production has risen tenfold. The average person in the USA throws away their body weight in rubbish every month.

SELF-ASSESSMENT QUESTIONS 8.03.01

1 What is SDW?
2 Why does the composition of SDW change as countries undergo economic development?
3 **Research idea:** Contact the organisation responsible for collecting domestic waste in your community. Try to obtain as much information as you can about the amount and composition of domestic waste produced in your community. To what extent has the amount and composition of waste changed over time? What proportion of such waste is recycled and to what extent has this changed over time.

E-waste

'E-waste often ends up dumped in countries with little or no regulation of its recycling or disposal. Historically this has taken place in Asia, but recently the trade has spread to other regions, particularly West Africa.'

Greenpeace, December 2008

E-waste, or electronic waste, is discarded electrical or electronic devices and their parts.

Information and communications technology (ICT) has been an important development tool in many countries, but there is a downside – **e-waste**. Modern electronics contain up to 60 different elements – many valuable, some hazardous, and some both. E-waste is a growing problem in all countries because of the rapid spread of ICT. E-waste includes items such as printers, mobile phones, pagers, laptop computers, toys and televisions. Technological advance is so rapid that there is a high and ever-increasing turnover of products with an escalating amount of equipment becoming obsolete. The lifespan of computing equipment decreased from 4–6 years in 1997 to 2–4 years in 2005.

A report released by the UNEP in February 2010 revealed that developing nations such as China, India, South Africa and Morocco are experiencing a growing pile of e-waste issues with the rise in popularity of electronics. The consequences could be damaging to public health and the local environments if not addressed correctly. Currently, over 40 million tonnes of e-waste are being generated globally per year. This increasing figure has caused UN experts to warn countries that action must be taken to prevent disastrous consequences. The UNEP predicts that, in India, e-waste from old computers will have risen by 500 per cent by 2010, compared with 2007 levels. In South Africa and China, this increase is predicted to be between 200 and 400 per cent.

The report was revealed at the Basel Convention. The report said that China produced about 2.3 million tonnes of e-waste domestically a year, second only to the USA with 3 million tonnes. More e-waste is expected to pile up in developing countries in the wake of rapidly rising sales and aggressive marketing of mobile phones and other electronic appliances. For example, Basel Action Network says that around 50 to 100 containers of e-waste arrive at Chinese ports each year, many from the USA. The network had tracked nine containers of hazardous e-waste from an e-waste collection facility in the USA and found that the containers were shipped to Indonesia, but it managed to foil the attempt by calling Indonesian authorities. The containers were then sent back to the USA.

In August 2008, the environmental organisation Greenpeace highlighted Ghana as a major recipient of foreign e-waste in an article entitled 'Poisoning the poor – Electronic Waste in Ghana'. Previously, Greenpeace had brought the attention of the world to similar practices in China, India and Nigeria. The Ghana analysis was based on samples and observations taken at two e-waste scrapyards. Containers filled with e-waste arrived from Germany, Korea, Switzerland and the Netherlands under the false label of 'second-hand goods'. Exporting e-waste from Europe is illegal, but exporting old electronics for 'reuse' allows unscrupulous traders to profit from dumping old electronics in Ghana. The majority of the containers' contents ended up in Ghana's scrapyards to be crushed and burned by unprotected workers. This waste is often laden with toxic chemicals such as lead, mercury and brominated flame retardants. Most workers, many of them children, wore no protective clothing or equipment. Boys were observed burning electronic cables and other components to melt the plastic and reclaim the copper wiring, releasing toxic chemicals in the process. Plastic and other materials perceived to be of no value were either burnt or dumped. Some samples contained toxic metals as much as 100 times above background levels. A significant number of samples contained chemicals known to interfere with sexual reproduction. Other samples contained chemicals that can affect brain development and the nervous system.

The e-waste problem is becoming particularly serious in developing countries due to:

- lack of legislation and enforcement
- lack of controlled take-back systems
- recycling being done in the informal sector, with limited activity organised by government at various levels
- lack of awareness by government, institutions and the general public
- illegal importation of e-waste from developing countries, often using false documentation
- the increasing outsourcing of ICT from MEDCs to LEDCs.

The Great Pacific garbage patch

The Great Pacific garbage patch is, in fact, two huge accumulations of waste in large areas of the Pacific Ocean, brought together by the North Pacific Subtropical Gyre (see Figure 8.25). Gyres are regions of the oceans where water rotates in a large circular pattern. The eastern garbage patch

8.03 Solid domestic waste

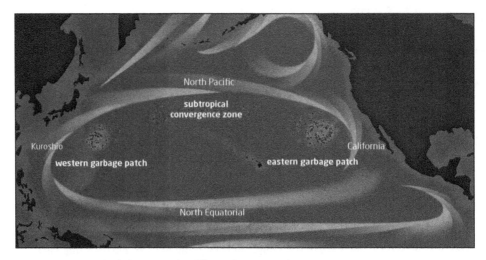

Figure 8.25 Map of the Great Pacific garbage patch.

floats between Hawaii and California; scientists estimate its size as two times bigger than Texas. The western garbage patch forms east of Japan and west of Hawaii. These considerable areas of marine waste went largely unnoticed until the early 1990s. The Great Pacific garbage patch has been described as the largest landfill in the world.

Waste from land, rather than ships, makes up 80 per cent of marine debris. About 60 per cent of this is consumer-used plastics that have not been disposed of properly. The amount of plastic in the North Pacific Subtropical Gyre has increased by 100 times in the last 40 years. A recent study found that 9 per cent of fish in the Great Pacific garbage patch had plastic in their stomachs. Plastic makes up the majority of marine debris seen on shorelines and floating in oceans around the world. This creates a difficult problem, because most plastics are not biodegradable. The UNEP estimated in 2006 that every square mile of ocean hosts 46 000 pieces of floating plastic. About 10 per cent of the plastic produced globally each year ends up in the oceans. It has been estimated that around 70 per cent of this eventually sinks, damaging life on the ocean floor. The rest floats, with a proportion ending up on shorelines around the world.

Garbage patches have also been found in the North Atlantic and the Indian Oceans.

SELF-ASSESSMENT QUESTIONS 8.03.02

1. What is e-waste?
2. Why is the e-waste problem becoming a particularly serious issue in developing countries?
3. What is the Great Pacific garbage patch?
4. **Discussion point:** How much e-waste does your class create during the course of a year?

8.03.02 Pollution management strategies and SDW

The management of SDW has become a critical issue in many parts of the world. Solid waste management is almost always the responsibility of local government, and it is often an expensive budget item. Solid waste can cause considerable pollution problems. Uncollected waste can cause local flooding and air and water pollution.

The traditional method of domestic waste management has been to use landfill, but for many cities, regions and countries the number of remaining available sites has been falling and other strategies have become necessary. Management of solid waste is highly dependent on the composition of rubbish in the waste stream. Reduction of wastes at source is the most efficient management option, because there is no need to transport and handle the material. However, social customs and attitudes are a considerable limiting factor for this option.

Figure 8.26 compares the methods of waste management in different parts of the world. There are large differences between the five world regions:

- Sanitary landfill is highest in North America and Latin America.
- Open dumping is significant in all regions apart from North America.
- Recycling is lowest in Latin America and Africa.
- Incineration is most important in Europe.

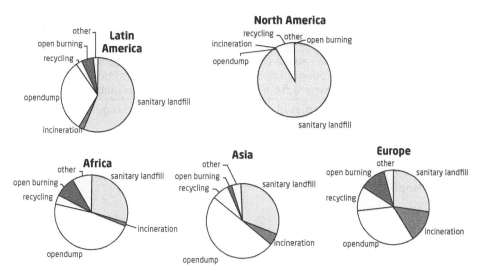

Figure 8.26 Waste management practices in different parts of the world.

Landfill

Landfill is a disposal site where solid waste is buried between layers of dirt in such a way as to reduce contamination of the surrounding land. It involves using a natural depression in the landscape or, more usually, digging a large, deep pit. Modern landfills are often lined with layers of absorbent material and sheets of plastic to keep pollutants from leaking into the soil and groundwater; the economic status of a country is a major determinant of whether this occurs.

Landfill has been the traditional method of waste disposal in most parts of the world because it is relatively cheap and technologically simple. When a landfill is complete, it can be reclaimed for its previous use or another use such as farmland or a park. The quality of landfill sites and their resulting impact on the environment can vary hugely both within and between countries, depending on the level of initial preparation and management and technological inputs. In many LEDCs 'open dumping', without any prior preparation of the surface, rather than landfill is the main method of household waste disposal, as no initial costs are involved. Where investment in modern waste disposal has occurred, such systems have often been confined to the capital city and other large urban areas, with open dumping dominating waste disposal in rural areas.

Table 8.09 summarises the advantages and disadvantages of landfill. In most countries the general perception is that the disadvantages of landfill increasingly outweigh the advantages, and that landfill is an unsustainable method of waste disposal. More and more countries are adopting landfill diversion strategies in order to reduce the amount of landfill as much and as quickly as possible.

Landfill is a waste disposal site where solid waste is buried.

8.03 Solid domestic waste

Advantages	Disadvantages
• Landfill is a relatively cheap method of dealing with waste because environmental costs are generally not taken into account.	• There is a limit to the number of appropriate available sites in many regions. • As sites become scarce, the cost of land increases.
• Landfill is a low technology method of waste disposal that countries at any level of development can use.	• There is increasing public opposition to opening new landfill sites and expanding existing ones because of environmental and health concerns. • Land sites can generate considerable heavy vehicle traffic.
• Landfill occurs in specific locations that can be carefully chosen and monitored.	• Poorly managed landfill sites can pollute water, soil and air. • Leachates can contaminate aquifers and waterways.
• Waste going to well-designed landfills can be processed to remove recyclable materials before tipping.	• The waste in poorly managed landfill can attract animals including insects which may spread disease in the surrounding area.
• Properly managed landfills can capture the methane produced by decomposition.	• The release of methane through the decomposition of biodegradable matter in anaerobic conditions (without oxygen) from landfill sites is a factor in global warming.
	• Completed landfill areas can settle and require maintenance.
	• Landfill has become increasingly recognised as an unsustainable method of waste disposal.

Table 8.09 The advantages and disadvantages of landfill.

The **waste hierarchy** presents the options for waste management in order of desirability, in terms of environmental impact. Waste prevention is at the top, with safe disposal at the bottom.

In the UK, Landfill Tax was introduced in the Finance Act 1996. It was established as the UK's first tax with an explicit environmental purpose. Since then the tax has been raised twice to provide a greater incentive to local authorities to reduce the amount of waste they set to landfill. At the same time, advances in waste treatment, such as mechanical biological treatment, have emerged as competitive alternatives to landfill. The overall objective is to drive waste streams higher up the **waste hierarchy** (see Figure 8.27). The starting point is to reduce disposal, much of which occurs in landfill sites.

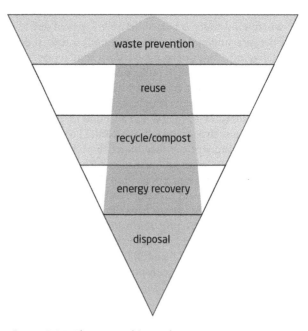

Figure 8.27 The waste hierarchy.

The Waste & Resources Action Programme (WRAP) is the lead agency in the UK in terms of the waste hierarchy. WRAP was originally set up to focus on recycling, but its role has evolved to encompass the higher rungs of the waste hierarchy also. An important principal here is that waste that is not created does not need to be dealt with. When the Welsh Assembly was looking for a means to reduce the country's ecological footprint from a current 5.16 global hectares (gha) per person, waste became a prime consideration, with an action plan based on the waste hierarchy.

The UK-government-funded Envirowise claims that bad waste practices cost UK industry at least £15 billion per year. Commercial waste dominates the total amount of waste the country produces. Approximately 70 per cent of office waste is recyclable, but on average only 7.5 per cent is actually recycled. Envirowise has urged companies to look at their waste hierarchy, beginning with what can be reduced and reused.

Incineration

Incineration (see Image 8.21) is the process of burning waste material in a furnace at high temperatures, so that only ashes, gas and heat remain. The heat may be used to generate electricity. Incineration requires a limited land area compared with landfill and can be operated in virtually all weather conditions. The process reduces the volume of refuse significantly. Incineration takes up much less land than landfill. It also has a good resource recovery rate. Over 400 kilowatt hours of energy can be produced by incinerating 1 tonne of waste.

Image 8.21 An incinerator.

Incineration is the burning of waste material in a furnace at high temperatures, creating only ashes, gas and heat.

On the debit side, incinerators are expensive to build and operate. Incineration requires a high input of energy, and the tall chimney stacks are viewed by most people as a blot on the landscape. The movement of heavy goods vehicles to and from incinerators is also considerable. The ash produced has to be disposed of in landfill. The environmental and health concerns over incineration are long-standing, with sulfur dioxide, nitrogen dioxide, nitrous oxide, carbon dioxide, chlorine, dioxin and particulates being emitted by the process of incineration. However, considerable technological advances have significantly lowered the emission of pollutants into the atmosphere. Thus the environmental impact of incineration varies considerably around the world depending on the age of incinerators.

Incineration is often not feasible in low-income countries, because of the high initial construction costs. Also, because wastes are not high enough in caloric content to sustain the incineration process, more costly fuel must be added.

Composting

Compost is organic material that has been decomposed and then recycled as a fertiliser because of its high nutrient value. It is an important input in organic farming. Local authorities in many countries encourage households to compost as much as possible in their own gardens (if they have a garden), alongside municipal compost operations (see Image 8.22). At the household level composting simply requires making a heap of wetted organic matter on the ground or in a compost bin and waiting for the materials to decompose over a period of months. In contrast, modern municipal composting is a multi-step process with measured inputs of water, air and carbon- and nitrogen-rich materials. The decomposition process is aided by shredding the plant matter, adding water and ensuring proper aeration by regularly turning the mixture. Worms and fungi further break up the material. The success of municipal composting is dependent on careful sorting by households before adding their contribution to the municipal compost heap.

Compost is organic material that has been decomposed and then recycled as high-nutrient-value fertiliser.

Image 8.22 Household compost bins reduce the amount of waste taken to landfill.

Some local authorities supply compost bins to households at a reduced price or even free of charge because they recognise the extent to which this can lower municipal handling costs.

8.03 Solid domestic waste

Image 8.23 Barge carrying SDW containers on the River Thames through London.

Recycling

Recycling is now generally regarded as a key process in providing a liveable environment for the future. However, the cost of recycling is significant. In addition, some wastes cannot be recycled and the separation of useful material from waste can sometimes be difficult.

In LEDCs, formal opportunities for recycling are very limited. In such countries, informal or illegal reuse and recycling is more prevalent than legal recycling. For some people at the bottom of the income and social scale, scavenging on rubbish dumps may be their main or only source of income. For the informal sector's waste pickers, their work on open dumps is labour-intensive and unsafe.

The transport of waste

In and around large cities in particular, the transport of waste involves regular, large-scale movements by road, rail and water. These movements have a significant environmental impact. High volumes of traffic in the vicinity of collecting depots, incineration plants and landfill sites impact on local residents. Image 8.23 shows SDW being taken in containers on a barge along the River Thames in London from a collecting depot in west London to a landfill site by the Thames estuary to the east of London.

SELF-ASSESSMENT QUESTIONS 8.03.03

1. What is landfill?
2. What are the perceived disadvantages of incineration?
3. What is composting?
4. **Discussion point:** Do the advantages of incineration outweigh the disadvantages?

Theory of knowledge 8.03.01

Some local authorities (councils) in some countries impose fines on households that fail to recycle according to local regulations. While many people agree with such policies, others feel that recycling should be a voluntary activity so that individual households can opt in or opt out. At the extreme, some individuals view stringent local authority regulations regarding recycling as an infringement of civil liberties and an example of heavy-handed government.

1. What are the arguments for and against strict local rules with regard to recycling?
2. Are there any other local authority (council) rules that you feel strongly about, for or against?

8.03.03 Management practices and income level

Low-income countries tend to spend most of their SDW management budgets on waste collection, with only a limited investment in waste disposal. This is the opposite of the situation in high-income countries, where the main expenditure is on disposal. Table 8.10 is from a recent World Bank report that examines solid waste management around the world. The table compares management practices in low-income, middle-income and high-income countries.

Energy from waste

Energy from waste techniques have advanced considerably in recent years. New energy from waste techniques can treat non-recyclable waste and produce electricity, heat and fuel, thus saving on the use of fossil fuels. Figure 8.28 describes the four main energy from waste techniques in use. The modern energy from waste industry argues that it should not be damned for the poor environmental performance of older plants that have been closed for some time. The industry also argues that the health concerns of people living close to older technology plants should not be considered in the context of modern facilities.

Activity	Low income	Middle income	High income
Source reduction	No organized programs, but reuse and low per capita waste generation rates common.	Some discussion of source reduction, but rarely incorporated into an organized program.	Organized education programs emphasize the three 'R's' — reduce, reuse, and recycle. More producer responsibility & focus on product design.
Collection	Sporadic and inefficient. Service is limited to high visibility areas, the wealthy, and businesses willing to pay. High fraction of inerts and compostables impact collection— overall collection below 50%.	Improved service and increased collection from residential areas. Larger vehicle fleet and more mechanization. Collection rate varies between 50 to 80%. Transfer stations are slowly incorporated into the SWM system.	Collection rate greater than 90%. Compactor trucks and highly mechanized vehicles and transfer stations are common. Waste volume a key consideration. Ageing collection workers often a consideration in system design.
Recycling	Although most recycling is through the informal sector and waste picking, recycling rates tend to be high both for local markets and for international markets and imports of materials for recycling, including hazardous goods such as e-waste and ship-breaking. Recycling markets are unregulated and include a number of 'middlemen'. Large price fluctuations.	Informal sector still involved; some high technology sorting and processing facilities. Recycling rates are still relatively high. Materials are often imported for recycling. Recycling markets are somewhat more regulated. Material prices fluctuate considerably.	Recyclable material collection services and high technology sorting and processing facilities are common and regulated. Increasing attention towards long-term markets. Overall recycling rates higher than low and middle income. Informal recycling still exists (e.g. aluminium can collection). Extended product responsibility common.
Composting	Rarely undertaken formally even though the waste stream has a high percentage of organic material. Markets for, and awareness of, compost lacking.	Large composting plants are often unsuccessful due to contamination and operating costs (little waste separation); some small-scale composting projects at the community/ neighborhood level are more sustainable. Composting eligible for CDM projects but is not widespread. Increasing use of anaerobic digestion.	Becoming more popular at both backyard and large-scale facilities. Waste stream has a smaller portion of compostables than low- and middle-income countries. More source segregation makes composting easier. Anaerobic digestion increasing in popularity. Odor control critical.
Incineration	Not common, and generally not successful because of high capital, technical, and operation costs, high moisture content in the waste, and high percentage of inerts.	Some incinerators are used, but experiencing financial and operational difficulties. Air pollution control equipment is not advanced and often by-passed. Little or no stack emissions monitoring. Governments include incineration as a possible waste disposal option but costs prohibitive. Facilities often driven by subsidies from OECD countries on behalf of equipment suppliers.	Prevalent in areas with high land costs and low availability of land (e.g. islands). Most incinerators have some form of environmental controls and some type of energy recovery system. Governments regulate and monitor emissions. About three (or more) times the cost of landfilling per tonne.

(continued)

8.03 Solid domestic waste

Activity	Low income	Middle income	High income
Landfilling/ Dumping	Low-technology sites usually open dumping of wastes. High polluting to nearby aquifers, water bodies, settlements. Often receive medical waste. Waste regularly burned. Significant health impacts on local residents and workers.	Some controlled and sanitary landfills with some environmental controls. Open dumping is still common. COM projects for landfill gas are more common.	Sanitary landfills with a combination of liners, leak detection, leachate collection systems, and gas collection and treatment systems. Often problematic to open new landfills due to concerns of neighboring residents. Post closure use of sites increasingly important, e.g. golf courses and parks.
Costs (see Annex E)	Collection costs represent 80% to 90% of the municipal solid waste management budget. Waste fees are regulated by some local governments, but the fee collection system is inefficient. Only a small proportion of budget is allocated toward disposal.	Collection costs represent 50% to 80% of the municipal solid waste management budget. Waste fees are regulated by some local and national governments, more innovation in fee collection, e.g. included in electricity or water bills. Expenditures on more mechanized collection fleets and disposal are higher than in low-income countries.	Collection costs can represent less than 10% of the budget. Large budget allocations to intermediate waste treatment facilities. Upfront community participation reduces costs and increases options available to waste planners (e.g. recycling and composting).

Table 8.10 Comparison of solid waste management practices by income level.
Source: 'What a waste', The World Bank, 2013

Anaerobic digestion (AD)
AD is one of the most talked about technologies following a government backing in the English Waste Strategy 2007. The biological process recycles feedstock with a high moisture content – such as food waste, animal manure and sewage sludge – into a biogas, which can be upgraded to pipeline-quality gas and biofertiliser.

Gasification and pyrolysis
Gasification and pyrolysis are both thermal processes and are identified as advanced conversion technologies (renewables obligation). Gasification, sometimes described as partial combustion, occurs when a hydrocarbon-carbon based substance is heated in a restricted amout of air and produces a synthetic gas (syngas); pyrolysis occurs when air is completely excluded and typically produces liquids.

Mechanical biological treatment (MBT) and biological mechanical treatment (BMT)
During MBT, materials such as metals, glass and plastics are mechanically removed. The leftover organic fraction can then be sent for aerobic composting or AD (see above).

BMT differs in that the materials are separated after the biological treatment. Waste is dried to reduce its mass and water content. Materials such as metals and glass are then removed and the remaining residue can be processed into a refuse derived fuel (RDF) or solid recovered fuel (SRF) and used for energy production.

Incineration
Waste-to-energy plants, or incinerators, with energy recovery, treat municipal waste (household and similar wastes), which remains after waste prevention, reuse and recycling activities. They can produce both electricity and heat for industrial and household users.

Figure 8.28 Energy from waste techniques.

A number of companies are working on advanced systems of gasification which will both get rid of rubbish and produce energy. Although standard techniques for gasification have been around for some time, the current cost of waste disposal is making more viable the possibility of disposing of household waste using higher-energy methods once reserved for hazardous materials such as medical waste and asbestos.

CASE STUDY 8.03.01

Beijing

Beijing is a megacity of over 22 million people. In July 2009, the Beijing municipal government launched a strategy for the improved disposal of the 18 410 tonnes of domestic refuse generated by the city every day. These measures include:

- building more environmentally friendly disposal sites
- improving incineration technologies
- enforcing household waste separation
- recycling.

Beijing aims to increase its processing capacity to 23 100 tonnes by the end of 2015. By 2014, Beijing had 37 waste disposal facilities with a daily processing capacity of 22 000 tonnes. This included:

- nine waste transfer stations
- four incineration plants
- sixteen sanitary landfill sites
- six composting plants
- two food waste treatment plants.

Landfill

Before 2009, more than 90 per cent of SDW went into landfill, using 500 acres of land a year. By 2014, the amount of SDW going to landfill had reduced to 80 per cent. However, Beijing will soon run out of space for landfill.

More than a third of Chinese cities are facing a similar crisis. At present, many landfill sites are without emission or leachate controls. A significant number of landfill sites operate illegally. Some of these illegal sites have been cleaned up after the problem was highlighted in a recent local documentary entitled 'Beijing besieged by waste'. Before the release of the documentary in 2011, Beijing had around 500 illegal dump sites that formed a ring around the city.

Incineration

Incineration requires substantial investment and there is always considerable opposition from people living close to the sites selected for new incinerators. This has been the case in Beijing, where the city government's plans to develop incineration have moved more slowly than forecast because of substantial opposition. The main concern is the significant amounts of heavy metals and dioxins released into the atmosphere by incinerators. However, the city government has promised to improve incineration standards to reach the level required in the EU.

The world's largest incinerator (the Lujiashan incinerator) began operating at full capacity in Beijing in late 2014 (see Figure 8.24). Located in a western suburb of the city, it can process 3000 tonnes of household waste a day. This is about one-sixth of the daily domestic waste generated in the city. The plant will also generate 420 million kilowatt hours of power a year, which is equal to the power generated by 140 000 tonnes of coal.

Image 8.24 The Lujiashan incinerator, Beijing.

Bio-treatment

Through bio-treatment, refuse is turned into fertiliser, which is another method Beijing is keen to develop. The Nangong garbage composting plant, which opened in 1998, was Beijing's first garbage composting plant.

Recycling

Less than 4 per cent of Beijing's SDW is recycled. To help improve this situation, the last Saturday of every month has been designated as a recyclable resources collecting day. However, people in China are not yet as recycling aware as populations in richer countries. Much needs to be done in terms of environmental education to improve this situation.

The household SDW disposal system

Each person in Beijing produces 0.8–1.0 kg of waste every day. The city wants to reduce this amount at source. In 2010, the city began a garbage classification scheme in 3000 neighbourhoods in an attempt to increase reuse and recycling.

Questions

1. How much domestic waste is generated in Beijing each day?
2. Describe the extent of Beijing's waste disposal facilities in 2014.
3. How important is the Lujiashan incinerator to waste disposal in Beijing?

8.04 Human population carrying capacity

LEARNING OBJECTIVES

After reading this chapter you should be able to:

- understand that human carrying capacity is difficult to quantify
- appreciate that the ecological footprint is a model that makes it possible to determine whether human populations are living within carrying capacity.

KEY QUESTIONS

Why is it difficult to estimate human carrying capacity?
How is the ecological footprint calculated?
Why may ecological footprints vary significantly by country and individual?

8.04.01 Carrying capacity and human populations

The world's resources are finite (see Image 8.25), but the continuing growth of population has been placing more and more pressure on these resources. How can such a situation be resolved? This was a question a group of scientists, commissioned by the 'Club of Rome' in 1970, attempted to answer in a book entitled *Limits to Growth* (1972). The model they presented predicted growing resource scarcity, increasing pollution and eventual population decline, all prior to 2100. The five basic factors that *Limits to Growth* examined were:

- population
- natural resources
- agricultural production
- industrial production
- pollution.

The team of experts used system dynamics theory and a computer model called World3. This enabled them to set up and analyse 12 scenarios that showed different possible patterns of global development with associated environmental outcomes. In 1972, global resource use was still within the Earth's **carrying capacity**, but the gap was being closed rapidly. In 1992, *Limits to Growth* was updated in a new book entitled *Beyond the Limits*, which argued that in many areas the Earth had overshot its limits. (To overshoot means to go so far that limits are exceeded.) The question was how the planet could move back into sustainable territory. In a further analysis published in 2002, *Limits to Growth: The 30-Year Update*, the overall conclusion was that the Earth was in a dangerous state of overshoot with many opportunities to mitigate the situation neglected.

All three publications were major analyses at their time, involving the collection and analysis of vast amounts of data. It is certainly a very demanding task to attempt to assess the Earth's carrying capacity. It is arguably even more difficult to apply the concept of carrying capacity to

The **carrying capacity** is the largest number of individuals in a population that the resources in the environment can support for an extended period of time.

local human populations. Various attempts have been made to do this with animal populations, but the range of resources used by humans is much greater than for any other species. The following factors have complicated the situation even more:

- Human ingenuity has enabled resource substitution to overcome problems when a particular resource has become depleted, by coming up with a replacement. Some companies spend considerable sums of money on research and development to find substitutes for resources that are in short supply and thus very expensive. There is particular concern about rare-earth metals, which are vital components of smartphones and computers. China produces about 90 per cent of the world's rare-earth metals.
- The resource requirements of local human populations vary significantly because of different lifestyles and levels of development. For example, the image that many people still have of Chinese cities is of swarms of bicycles, but China is now the fastest-growing car market in the world. This massive change in lifestyle in a short period has resulted in a huge change in the range of resources used by China's population.
- Lifestyle value systems are an important influence on resource use. Some societies are more materialistic than others, even when there is not much difference in the degree of affluence.
- Technological developments can impact considerably on resource requirements and availability. For example, very few railway systems in the world now use coal, when at one time they all did. Modern railways are powered by electricity or diesel.
- Affluent local populations can import resources from other geographical areas. The oil-rich Middle East nations are an obvious example. While importing resources increases the carrying capacity of the local population, it has no impact on global carrying capacity. This is because the total global availability of resources has not changed.
- Degradation of the environment, together with the consumption of finite resources, is expected to limit human population growth.

Image 8.25 Forested landscape.

All these factors make it extremely difficult to provide reliable estimates of carrying capacities for human populations. However, various signs of stress will be evident before carrying capacity is reached, and as such signs of stress increase it will be clear that the local environment is reaching its limits.

Extension

To understand more about population growth visit the website below:
www.ted.com/talks/hans_rosling_on_global_population_growth

> **Optimum population** is the population that achieves a given aim in the most satisfactory manner. This is generally viewed in economic terms as the population that would produce the highest average standard of living.

The carrying capacity is the largest population that the resources of a given environment can support. An important related concept is that of optimum population. The idea of **optimum population** has been mainly understood in an economic sense (see Figure 8.29). At first, an increasing population allows for a fuller exploitation of a country's resource base, causing living standards to rise. However, beyond a certain level, rising numbers place increasing pressure on resources and living standards begin to decline. The highest average living standard marks the optimum population. Before that population is reached, the country or region can be said to be underpopulated. As the population rises beyond the optimum, the country or region can be said to be overpopulated.

There is no historical example of a stationary population having achieved appreciable economic progress, although this may not be so in the future. In the past it is not coincidental that periods of rapid population growth have paralleled eras of

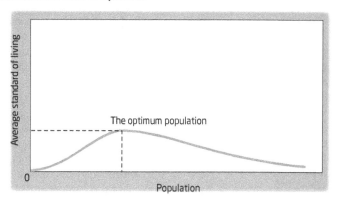

Figure 8.29 The optimum population.

8.04 Human population carrying capacity

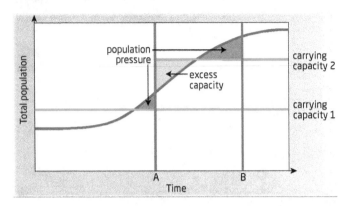

Figure 8.30 Optimum rhythm of growth.

technological advance that have increased the carrying capacity of countries and regions. Thus we are led from the idea of optimum population as a static concept to the dynamic concept of **optimum rhythm of growth** (see Figure 8.30), whereby population growth responds to substantial technological advances. For example, Abbe Raynal (*Revolution de l'Amerique*, 1781) said of the USA, 'If ten million men ever manage to support themselves in these provinces it will be a great deal.' Yet today the population of the USA is over 300 million, and hardly anyone would consider the country to be overpopulated.

The most obvious examples of population pressure are in LEDCs, but the question here is: are these cases of absolute overpopulation or the results of underdevelopment that can be rectified by adopting remedial strategies over time?

Optimum rhythm of growth is the level of population growth that best utilises the resources and technology available. Improvements in the resource situation or/and technology are paralleled by more rapid population growth.

SELF-ASSESSMENT QUESTIONS 8.04.01

1. Define carrying capacity.
2. State three reasons why it is difficult to apply the concept of carrying capacity to local human populations.
3. **Discussion point:** How close do you think the local human population is to carrying capacity in the region in which you live? Give reasons for your answer.

8.04.02 Human carrying capacity and reductions in energy and material use

The carrying capacity in terms of human population is determined by the:
- rate of resource consumption
- level of pollution created by human activity
- extent of human interference in global life-support systems.

Reductions in demand and the reuse and recycling of products and materials can reduce all these impacts, as well as increasing human carrying capacity. Reducing demand for energy can mean that new power stations need not be built and less pollution is created. The same is true for a range of other resources that are used on a regular and large-scale basis.

The literature on sustainable consumption highlights recycling, but some attention is also given to promoting slower consumption through focusing on product lifespan, design for reuse and reuse through various second-hand markets. A number of studies have focused on participation rates in recycling. Such studies have highlighted '**recycling deserts**' alongside areas of high participation.

Product stewardship is an approach to environmental protection in which manufacturers, retailers and consumers are encouraged or required to assume responsibility for reducing a product's impact on the environment. Also called extended producer responsibility, it is a growing aspect of recycling. In many cases this is a system of environmental responsibility whereby producers take back a product, recycling it as far as possible, after the customer has finished with it. For manufacturers, this includes planning for, and if necessary paying for, the recycling or disposal of the product at the end of its useful life. This may be achieved, in part, by redesigning products to use fewer harmful substances and be more durable, reusable and recyclable, and to make products from recycled materials.

Recycling deserts Areas where rates of recycling are well below the national or regional average.

Product stewardship A system of environmental responsibility whereby producers take back a product, recycling it as far as possible, after the customer has finished with it.

Many states in the USA require retailers who sell lead acid batteries to accept used batteries for recycling when consumers purchase new batteries. Germany was the first nation to institute a comprehensive product stewardship programme, passing a law in 1991 requiring manufacturers to assume the costs of collecting and recycling used packaging. Schemes for product stewardship for beverage containers operate in a number of countries.

Substitution is the use of common and thus less valuable resources in place of rare, more expensive resources. An example is the replacement of copper by aluminium in the manufacture of a variety of products. Historically, when non-renewable resources have been depleted, new technologies have been developed that effectively substitute for the depleted resources. New technologies have often reduced pressure on these resources even before they are fully depleted. For example, fibre optics have substituted for copper in many electrical applications.

SELF-ASSESSMENT QUESTIONS 8.04.02

1. What is product stewardship?
2. How can substitution help to conserve important resources?
3. **Discussion point:** What products, if any, have you reused? If you have not reused any, can you think of any possibilities for reuse in the future?

Ecological footprints

'For more than 40 years, humanity's demand on nature has exceeded what our planet can replenish. We would need the regenerative capacity of 1.5 Earths to provide the ecological services we currently use. "Overshoot" is possible because we cut trees faster than they mature, harvest more fish than oceans replenish, or emit more carbon into the atmosphere than forests and oceans can absorb.'

WWF Global, 2014

The enormous growth of the global economy in recent decades has had a phenomenal impact on the planet's resources and natural environment. Many resources are running out and waste sinks are becoming full. The remaining natural world can no longer support the existing global economy, much less one that continues to expand. The main responsibility lies with the rich countries of the world. The richest 20 per cent of the world's people earn 75 per cent of the income, consume 80 per cent of the world's resources and create 83 per cent of the world's waste. The poorest 50 per cent of the world's population have 1 per cent of the world's household wealth.

In Chapter 1.04, the concept of ecological footprint and its importance to sustainable development was introduced. The **ecological footprint** has arguably become the world's foremost measure of humanity's demands on the natural environment. It is a model that makes it possible to determine whether human populations are living within carrying capacity. The ecological footprint for a country has been defined as 'the sum of all the cropland, grazing land, forest and fishing grounds required to produce the food, fibre and timber it consumes, to absorb the wastes emitted when it uses energy, and to provide space for its infrastructure' (*Living Planet Report 2008*). Thus the ecological footprint, calculated for each country and the world as a whole, has six components (see Figure 8.31):

- built-up land
- fishing ground
- forest
- grazing land
- cropland
- carbon footprint.

Theory of knowledge 8.04.01

Energy conservation is a sensitive topic. Individuals are being encouraged to make energy savings while at the same time major international events, which admittedly give pleasure to many, go ahead unchallenged. In addition, many large office blocks, both public buildings and those owned by private companies, are sometimes fully lit at night. Again, the incentive to save energy often seems to be placed more on the individual than on large organisations.

1. Are there any events that you think should be discontinued or changed significantly in format to save energy? Justify your views.
2. Do you think that people take national campaigns less seriously if they feel we are not all in it together?

Substitution is the use of common and thus less valuable resources in place of rare, more expensive ones (e.g. replacement of copper by aluminium in a variety of products).

Ecological footprint is a sustainability indicator that expresses the relationship between population and the natural environment. It sums the use of natural resources by a country's population.

8.04 Human population carrying capacity

Biocapacity is the capacity of an area or ecosystem to generate an ongoing supply of resources and to absorb its wastes.

Global hectare (gha) is the equivalent of 1 hectare of biologically productive space with world average productivity.

Carbon footprint is the total set of greenhouse gas emissions caused directly and indirectly by an individual, organisation, event or product.

In previous years, an additional component reflecting the electricity generated by nuclear power plants was included in ecological footprint accounts. This component is no longer used, because the risks and demands of nuclear power are not easily expressed in terms of **biocapacity**. Biocapacity is the capacity of an area or ecosystem to generate an ongoing supply of resources and to absorb its wastes (see Image 8.26).

Carbon from burning fossil fuels has been the main component of the global ecological footprint for more than half a century and remains on an upward trend. In 1961, carbon was 36 per cent of the total global total footprint; by 2010, it had risen to 53 per cent.

Technological advances, agricultural inputs and irrigation have increased the average yields per hectare of productive area, especially for cropland. These advances have raised the planet's total biocapacity from 9.9 to 12 billion gha between 1961 and 2010.

The ecological footprint is measured in **global hectares (gha)**. A global hectare is a hectare with world-average ability to produce resources and absorb wastes. In 2010, the global ecological footprint was 18.1 billion global hectares (gha) or 2.6 gha per person. This can be viewed as the demand side of the equation. On the supply side, the total productive area, or biocapacity, of the planet was 12.0 billion gha, or 1.7 gha per person. With demand greater than supply, the Earth is living beyond its environmental means.

The ecological footprint can be quoted in absolute (total) and relative (per capita) terms. Figure 8.31 shows the ecological footprint of countries with the highest per-capita figures and how the footprint of each country is made up. Kuwait, Qatar, United Arab Emirates, Denmark and Belgium have the highest ecological footprints per person in the world. Thirteen countries have figures above 6 gha per person. Nations at different income levels show considerable disparities in the extent of their ecological footprint. The lowest per-capita figures were attributed to Bangladesh, Pakistan, Afghanistan, Haiti, Eritrea, Occupied Palestinian Territory and Timor-Leste. All these countries have an ecological footprint well below 1.0 gha per person. Footprint and biocapacity figures for individual countries are calculated annually by Global Footprint Network.

In many of the countries illustrated in Figure 8.31, the **carbon footprint** is the dominant element of the six components that comprise the ecological footprint, but in others such as Denmark, Uruguay and Mongolia, other aspects

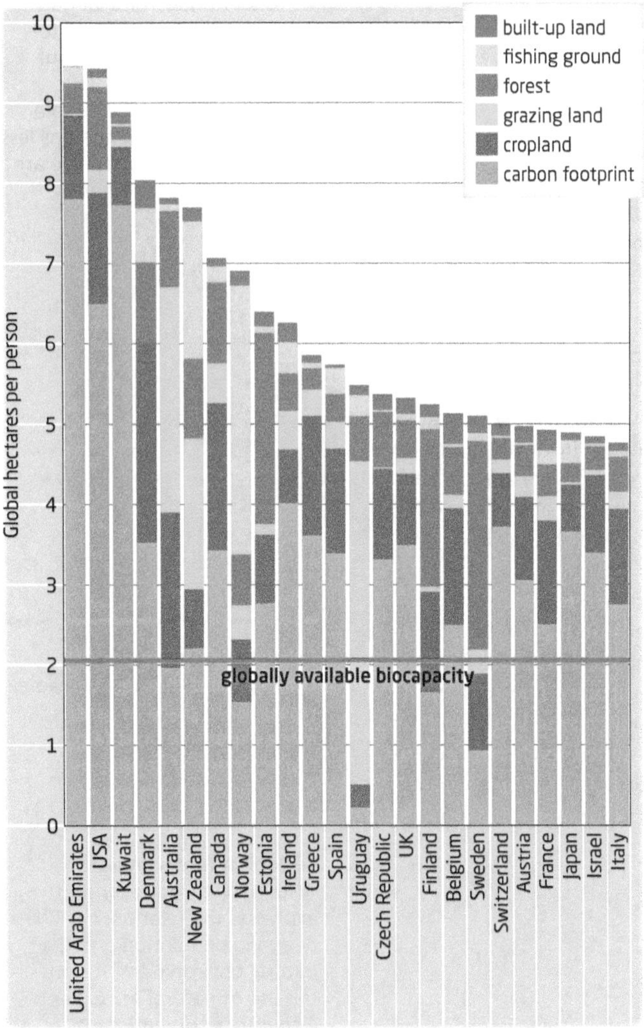

Figure 8.31 Ecological footprint per person for the highest-footprint countries, 2010.

Image 8.26 Uninhabited island in Indonesia producing biocapacity for the heavily populated parts of the country.

of the ecological footprint are more important. In Uruguay, the demand on grazing land is by far the dominant component of the ecological footprint. In Denmark, the demands on its cropland are the country's major impact on the natural environment. In general, the relative importance of the carbon footprint declines as the total ecological footprint of countries falls. In many sub-Saharan African countries, the contribution of carbon to the total ecological footprint is extremely low.

The ecological footprint in absolute terms is strongly influenced by the size of a country's population. The other main influences are the level of demand for goods and services in a country (the standard of living), and how this demand is met in terms of environmental impact. International trade is taken into account in the calculation of a country's ecological footprint. For each country its imports are added to its production, while its exports are subtracted from its total.

The expansion of world trade has been an important factor in the growth of humanity's total ecological footprint. In 1961, the first year for which full datasets are available, global trade accounted for 8 per cent of the world's ecological footprint. By 2010, this had risen to more than 40 per cent.

The ecological footprint includes only those aspects of resource consumption and waste production for which the Earth has regenerative capacity and where data exists that allows this demand to be expressed in terms of productive area. For example, toxic releases do not figure in ecological footprint accounts. Ecological footprint calculations do not:

- attempt to predict the future
- indicate the intensity with which a biologically productive area is being used
- evaluate the social and economic dimensions of sustainability.

Assessing human pressure on the planet is a vital starting point. The ecological footprint can be calculated at the full range of scales from the individual to the total global population. Knowing the extent of human pressure on the natural environment helps us to manage ecological assets more wisely on both an individual and a collective basis. It is an important tool in the advancement of sustainable development.

Figure 8.32 shows how humanity's ecological footprint increased from 1961 to 2013. According to the *Living Planet Report*, the global ecological footprint now exceeds the planet's regenerative capacity by about 30 per cent. This global excess is increasing, and as a result

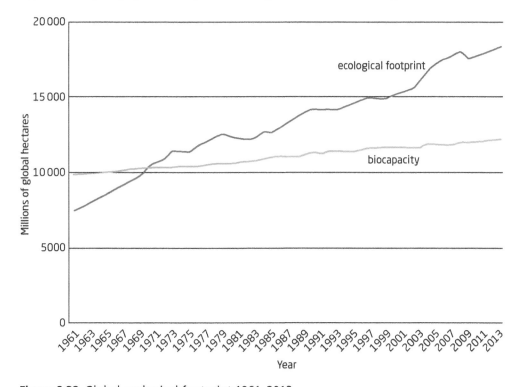

Figure 8.32 Global ecological footprint 1961–2013.

8.04 Human population carrying capacity

CONSIDER THIS
The Global Footprint Network's data shows that Africa's resource demand is likely to outstrip domestic availability as early as 2015. The continent as a whole would then be in a state of ecological overshoot. Africa's ecological footprint is expected to double by 2040.

ecosystems are being run down and waste is accumulating in the air, land and water. The resulting deforestation, water shortages, declining biodiversity and climate change are putting the future development of all countries at risk.

Human demand on the Earth has more than doubled over the past 45 years due to a combination of population growth and rising living standards which have involved greater individual consumption. In 1961, most countries in the world had more than enough biocapacity to meet their own demand. But by the mid-1980s, humankind's ecological footprint had reached the Earth's biocapacity. Since then, humanity has been in ecological overshoot, with annual demand on resources each year exceeding the Earth's regenerative capacity. The WWF calculates that it now takes the Earth one year and four months to regenerate what the global population uses in a year. This is a significant threat to both the well-being of the human population and the planet as a whole.

The world can be divided into eco-debt and eco-credit countries. The former are living beyond their ecological means, while the latter still have spare biocapacity. The eco-debt nations are only able to meet their needs by importing resources from other countries and by using the atmosphere as a dumping ground. The WWF estimates that, if present trends continue (the so-called 'business as usual' scenario), the global population will require two planets to satisfy the demand for goods and services by the early 2030s. However, there are many effective ways to change this situation. For example, the *Living Planet Report* argues that:

- technology transfer and support for local innovation can help emerging economies maximise their well-being while leap-frogging resource-intensive phases of industrialisation
- cities can be designed to support good lifestyles while at the same time minimising demand on both local and global ecosystems
- moving to clean energy generation and efficiency based on current technologies could allow the world to meet the projected 2050 demand for energy services with large reductions in associated carbon emissions
- empowerment of women through education and access to voluntary family planning can slow or even reverse population growth.

The *Living Planet Report* argues that countries with ecological reserves can view their biological wealth as an asset. In 2010, China and the USA had the largest footprints, using 19 per cent and 13.7 per cent of the world's biocapacity respectively (see Figure 8.33). However, China had a much smaller per-capita footprint than the USA, although its impact on the global natural environment is growing at a faster pace. India is in third place, using 7.1 per cent of the planet's total biocapacity.

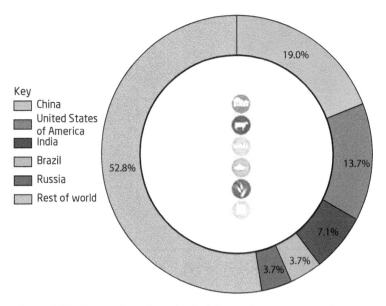

Figure 8.33 Share of total ecological footprint 2010 – top five countries.

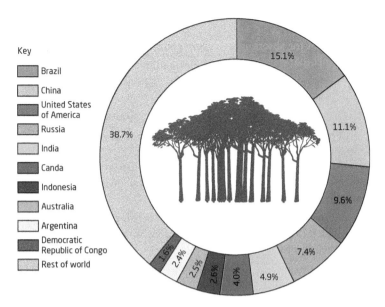

Figure 8.34 Top 10 national biocapacities in 2010. (Percentages are rounded to the nearest 0.1%.)

Ten countries alone accounted for more than 60 per cent of the Earth's total biocapacity in 2010 (see Figure 8.34). For most countries with a high biocapacity per capita, the forest land element makes up the largest proportion of total biocapacity.

SELF-ASSESSMENT QUESTIONS 8.04.03

1 Define: **a** ecological footprint, **b** global hectares (gha).
2 Give the six components that make up the ecological footprint.
3 **Discussion point:** What do you think will happen as humanity's ecological footprint continues to rise?

How can the ecological footprint be calculated?

The basis of ecological footprint calculations is the conversion of human consumption and waste production into the land area required for these purposes. In an ideal situation, every single human impact on the environment would be calculated as part of the ecological footprint. This is, however, wishful thinking, as comprehensive data covering the whole world does not exist for many human actions. Data that exists for one or a number of countries may not be available for other countries. Also, the more factors taken into account to achieve a conclusive figure, the more complex the task becomes.

An approximation of the ecological footprint can be reached through the steps below. This is a simplified version of the ecological footprint introduced in the previous section and explained in more detail below. This simplified calculation takes account of only:

- food consumption and production
- carbon dioxide emissions and carbon fixation.

Equation 1

$$\frac{\text{per capita land requirement}}{\text{for food production (ha)}} = \frac{\text{per capita food consumption (kg yr}^{-1}\text{)}}{\text{mean food production per hectare of local arable land (kg ha}^{-1}\text{ yr}^{-1}\text{)}}$$

8.04 Human population carrying capacity

Equation 2

$$\text{per capita land requirement for absorbing waste } CO_2 \text{ from fossil fuel (ha)} = \frac{\text{per capita } CO_2 \text{ emission (kg C yr}^{-1}\text{)}}{\text{net carbon fixation per hectare of local natural vegetation (kg C ha}^{-1}\text{ yr}^{-1}\text{)}}$$

ecological footprint = (Equation 1 + Equation 2) × total population

The six components used to calculate the full ecological footprint of a country were introduced in the last section.

A number of websites allow you to calculate your own ecological footprint, such as www.bestfootforward.com. Here you are asked to respond to a range of simple questions relating to:

- Your main mode of travel: travelling by car creates the highest footprint, which can be reduced by using public transport and cut to a minimum by walking or using a bicycle.
- Your usual type of holiday: going to far-away destinations and using air transport contribute to a high footprint.
- The size of house you live in and how many people you share the house with: large detached houses are usually the most energy-inefficient, particularly when the density of occupation is low.
- Whether your energy comes from renewable or non-renewable sources: some energy companies in various countries offer customers a choice concerning the source of the energy they use.
- To what extent you conserve energy: there are many ways of reducing energy use, such as cavity wall and loft insulation, and using low-energy lightbulbs.
- How much meat you eat; meat production uses far more energy than arable production (see Figure 8.35).
- How much of the food you eat is produced locally: the term 'food miles' was first used in the 1990s by Dr Tim Lang, Professor of Food Policy at London's City University, as part of the debate on sustainable agriculture. Food miles can be defined as the distance food travels from the farm where it is produced to the plate of the final consumer. It is an indication of the environmental impact of food consumption.
- How much domestic waste you produce and how much is recycled: a high level of recycling can significantly reduce the ecological footprint.

Your answers to most of these questions will be approximations, but by comparing the figure produced for your ecological footprint with those of other people in your group you will be able to see how your ecological footprint could be reduced.

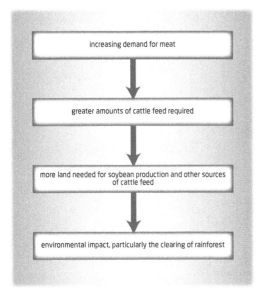

Figure 8.35 Environmental impact of the increasing demand for meat.

CASE STUDY 8.04.01

Ecological footprints of an LEDC, an NIC and an MEDC

There are very large differences between the ecological footprints of MEDCs, NICs and LEDCs, the result mainly of the considerable differences in wealth between these three broad groups of countries. However, ecological footprints are dynamic rather than static – that is, they are subject to change for a variety of reasons. Just think of the six general factors that are used to calculate the ecological footprint and you will see that it is not hard to think how changes can occur even from one year to the next. By far the greatest changes are occurring in a group of countries described as newly industrialised countries (NICs). These are countries where rapid rates of economic growth are moving countries out of LEDC status and towards MEDC status. The NIC group includes China, India, South Korea, Brazil and Mexico.

Figure 8.36 compares the ecological footprint of the USA (an MEDC), China (an NIC) and Bangladesh (an LEDC). China is about 25 per cent above the per-capita globally available biocapacity, whereas the USA's ecological footprint is about four times this level. In comparison,

CASE STUDY 8.04.01 (continued)

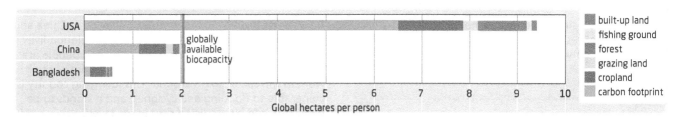

Figure 8.36 Ecological footprint by sector for the USA, China and Bangladesh.

the footprint for Bangladesh is extremely low, at less than half of global available biocapacity.

Not only are the ecological footprints of these three countries very different in terms of size, they also vary significantly in contribution by sector (see Figure 8.36). For example, the USA's carbon footprint accounts for over two-thirds of its overall ecological footprint. The fraction of the overall ecological footprint made up by the carbon footprint is considerably less for China and lower still for Bangladesh. While the carbon footprint is the largest component for both the USA and China, in Bangladesh it is second to the footprint from cropland.

The extremely high per-capita usage of energy in all sectors of the economy largely explains the USA's carbon footprint. Its farming footprint (cropland and grazing land) is typical of MEDCs in general, where about twice as much energy in the diet is provided by animal products than in LEDCs. The very high levels of inputs required by the agro-ecosystems of MEDCs clearly result in a large farming footprint. The rising incomes of a growing middle class in China have resulted in an increasing number of people moving up the food chain, eating more meat, eggs, milk and farmed fish.

Figure 8.37 shows how the ecological footprints for all three countries changed between 1961 and 2010. In the USA, after a period of relative stability, the country's ecological footprint declined modesty to narrow the gap with biocapacity. For China it has been a very different picture, with a rapidly rising ecological footprint over the last decade. The ecological footprint in Bangladesh has been rising for the past two decades at a significant rate, although not as fast as for China.

Table 8.11 shows actual figures for a number of individual resources for the three countries. Per-capita energy use is a good indicator of contrasts in economic development.

Per-capita energy consumption in the USA is almost three and a half times higher than in China. In turn, per-capita energy consumption in China is almost ten times that of Bangladesh. The contrast in meat and water consumption is also considerable, although, as with energy, consumption in China of all these resources is rising rapidly.

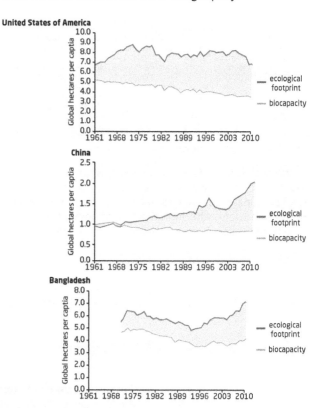

Figure 8.37 Ecological footprints and biocapacity for the USA, China and Bangladesh, 1961–2010.

	Energy use/kg of oil equivalent per capita, 2011	Meat consumption/kg per person per year, 2009	Average water use/dm^3 per person per day, 2006
USA	7032	120.2	576
China	2029	58.2	86
Bangladesh	205	4	46

Table 8.11 Individual resource use for the USA (an MEDC), China (a NIC) and Bangladesh (an LEDC).

8.04 Human population carrying capacity

CASE STUDY 8.04.01 (continued)

Figure 8.38 is a simplified model showing the contrast in the level of resource use and in the rate of change in resource use between LEDCs, NICs and MEDCs.

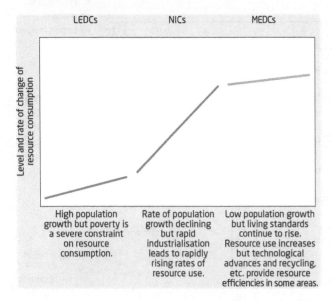

Figure 8.38 Model of the relationship between resources and the level of economic development.

Bangladesh is one of the most densely populated countries in the world, with a population density of 1046 people per square kilometre according to the 2011 Population Sheet. Although the rate of population growth has fallen, it was still estimated at 1.5 per cent a year in 2011. This compares with 0.5 per cent in China and in the USA. Almost half of Bangladesh's population live on less than one dollar a day. The majority of people are employed in agriculture, but there is simply not enough work in this sector to go around. This low-lying country is vulnerable to flooding and cyclones, and it stands to be badly affected by predicted rises in sea levels.

The government is trying to attract foreign investment into the manufacturing and energy sectors, and has achieved some degree of success. A large pool of low-cost labour is certainly an attraction to foreign transnational corporations, particularly when labour costs in other Asian countries are rising. If Bangladesh can transform itself into one of the next generation of NICs, its demand for a wide range of resources will increase rapidly and, along with this, its ecological footprint.

Questions

1. Compare the ecological footprints of the USA, China and Bangladesh to the average available global biocapacity.
2. How do the ecological footprints of the three countries vary in composition?
3. How have the ecological footprints of the three countries changed since 1961?
4. Research idea: Visit an appropriate website to calculate your own ecological footprint. How does your ecological footprint differ from those of other people in your class?

8.04.03 Population–technology–resources and human carrying capacity

> 'For the first time in history, global economic prosperity, brought on by continuing scientific and technological progress and the self-reinforcing accumulation of wealth, has placed the world within reach of eliminating extreme poverty altogether.'
>
> J. D. Sachs, UN Millennium Project

Technological progress has had a phenomenal impact on the development of humankind, playing a greater and greater role in human life as time has evolved. A range of technological developments have allowed higher resource use per unit area of land. In many cases, this has increased carrying capacity and material economic growth at the same time. Carrying capacity should be understood as the maximum load an environment can permanently support. Load refers not just to the number of people but to the total demands they make upon the environment. For human societies, as for populations of other species, the relation of load to carrying capacity is crucial in shaping our future.

Some economists argue that human carrying capacity can be expanded continuously through technological innovation. For example, if resources in general can be used twice as efficiently, the carrying capacity could be doubled or the use of energy of the current population could be doubled. However, this is a great simplification because the majority of people in the developing world currently endure living standards well below those in the affluent nations. The development process alone, with poorer countries raising their living standards considerably, will have a major impact on carrying capacity. This has led some experts to suggest that efficiency would have to be raised by a factor of four to ten to remain within global carrying capacity. Even the lower end of this range would prove to be a formidable challenge!

Figure 8.39 summarises the population–technology–resources relationship. The level of technology available strongly influences the extent to which resources can be utilised to support a population. For example, undersea oil and gas can be drilled at much greater depths today than 20 years ago. If underwater technology advances further, will humankind be able to access resources from deep ocean waters? As oceans cover 70 per cent of the Earth's surface, this could revolutionise the population–resource relationship. Can agricultural science deliver at least one more agricultural revolution?

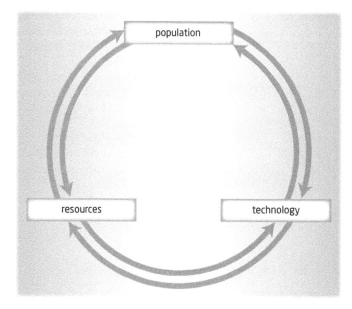

Figure 8.39 The population–technology–resources relationship.

The level of technology also influences the size of the resource base and the size of the population that is supported by the resource base. Advances in birth control techniques have had a major influence on fertility. Will the development of a male birth pill achieve widespread usage around the world? Medical breakthroughs have been vital in lowering mortality rates, and more are likely in the future. Technology is a major influence on population numbers.

The population–technology–resources relationship is a dynamic relationship that affects changes in carrying capacity over time. Figure 8.40 shows the relationship between these three factors in more detail. It shows that cultural and political factors can exert a strong influence. For example, attitudes to family size and birth control affect the rate of population growth. Political decisions about how to allocate a country's financial resources can significantly affect the rate of economic development. The expected standard of living in a society is another factor to be

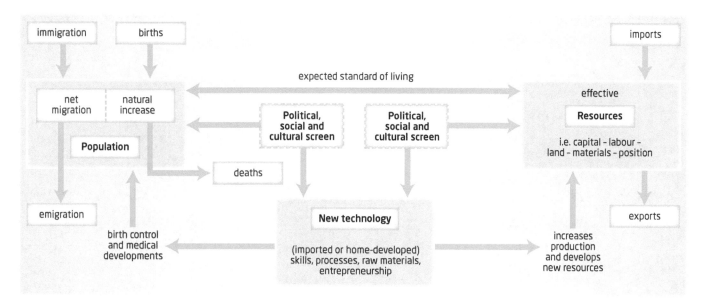

Figure 8.40 Population–resource–technology relationships.

8.04 Human population carrying capacity

taken into account. If a society is reasonably content, there will probably be less pressure on the resource base than if the opposite is true. For individual countries, external influences may impact significantly on each of the three factors in the relationship.

When discussing the population–resource–technology relationship, it is almost inevitable that the focus is on cutting-edge technology, brand-new developments that can gradually spread around the world. Here, the speed at which diffusion occurs will influence resource use and population. However, it is also important not to neglect the role of appropriate or intermediate technology. This is largely viewed in the context of aid by a donor country, whereby the level of technology and the skills required to service it are properly suited to the conditions in the receiving country. NGOs have often been much better at directing aid towards sustainable development than government agencies have. The selective nature of such aid has targeted the poorest communities using appropriate technology and involved local people in decision-making. Appropriate technology can have a considerable influence on resources such as water supply and food production.

End-of-topic questions

1 a What does the term 'demographic transition' mean? [2]

   b Examine the differences between demographic transition in MEDCs in the 19th and early 20th centuries and trends in LEDCs today. [4]

2 a What general services are provided by natural capital? [2]

   b Discuss the problems related to the protection of natural capital. [4]

3 a What are the main components of solid domestic waste (SDW)? [2]

   b Discuss the advantages and disadvantages of incineration as a strategy for dealing with solid domestic waste. [4]

Answers to self-assessment questions

Topic 1

1.01.01

It is important to note that each person will have his or her own view on these questions and there is no 'correct' answer.

1   Green politics has raised awareness of people worldwide to the plight of the rainforest, Politicians, religious leaders, local communities and environmental groups all discuss the issue, raise funds or lead campaigns to protect the rainforest, so the answer may be that green politics is very important.

2   The rate of loss of rainforest has been slowed and people now understand its importance, but it is difficult to say whether the threat has been overemphasised. It might be worth thinking about whether the rainforest has been publicised to the detriment of other less attractive but equally vital ecosystems.

3   **Discussion point:** These are local issues so, of course, your discussions will be influenced by local conditions. Consider issues such as recycling, cosmetics produced without animal testing, or use of plastic bags.

1.01.02

1   An EVS is a particular worldview or set of paradigms that shapes the way individuals or societies perceive and evaluate environmental issues. An EVS has inputs, outputs, storages and transfers.

2   Technocentrics believe that technological developments can provide solutions to environmental problems. Ecocentrics put ecology and nature as central to humanity and propose a less materialist approach to life.

3   **Discussion point:** Think about how a technocentrist would suggest methods to solve the issue of climate change.

1.01.03

1   There is no one answer to this; each of us has our own viewpoint.

2   People will respond according to their understanding of an issue. This will come from education, experience, religious and sociopolitical concepts.

3   D.

4   **Discussion point:** What do you understand by a 'good' life? Does it include the natural world?

1.02.01

1   **a**   The systems approach enables us to study an ecosystem as a whole, examining the feeding relationships and cycling of nutrients and so on together, rather than separately.

   b   In engineering a whole system such as an engine, its inputs (fuel) and outputs (work done) and its performance can be examined rather than just the functioning of, for example, the gears.

2

System	Open	Closed	Isolated
Inputs	energy and mass	energy	none
Outputs	energy and mass	energy	none

3   **Discussion point:**

1.02.02

1   **Discussion point:** A transfer usually involves a flow through a system; transformations occur when there is a change of state (e.g. from water to vapour).

2   Input – sunlight; output – heat; storage – biomass (or other suitable examples).

3   Advantages: complex systems can be simplified; predictions can be made; different scenarios can be considered. Disadvantages: oversimplification; models depend on the skill of the modellers; different models may predict different outcomes for the same scenario.

4   **Discussion point:** Think about computer-generated images and models you have seen, for example in weather forecasting or predictions for changes in sea levels.

1.03.01

1   Because energy cannot be created or destroyed, energy that moves through an ecosystem is transformed into different forms. Not all of these (e.g. heat) can be used by animals in the food chain. Eventually there is not enough transferable, useful energy to sustain another link in the chain.

Answers to self-assessment questions

2 Entropy is a measure of the evenness of distribution of energy in a system. In a natural system there is more energy at the lower trophic levels and less at the highest.

3 Steady-state equilibrium is a stable situation to which a system will return after a disturbance. A static equilibrium is one in which there are no changes because there are no inputs or outputs in the system.

4 **Research idea:** There is more information on human population growth in Topic 8. Does the information you have researched agree with what is said there?

1.04.01

1 Sustainability means that resources are used and managed so that full natural replacement of exploited resources can take place.

2 Natural income is difficult to monitor because it may cross international boundaries and people using the resource may have different environmental value systems (EVSs). Laws and the cooperation of individuals are needed to protect natural income.

3 **Discussion point:** Think about whether you had heard of this report before starting this course. How accessible are such reports to ordinary people?

1.04.02

1 A report to assess the environmental, social and economic impact of a large-scale project. Its purpose is to provide decision-makers with evidence to decide whether a project should go ahead.

2 EIAs cannot predict all future outcomes, because systems may change as the project develops and new inputs occur. For example, the EIA for the Three Gorges Dam could not predict whether the risk of landslides would increase, or the effect of the project on social structure of the community after the dam had been built.

3 **Discussion point:** Consider the value of a project such as an airport or new road in relation to the land that must be used to build it. How can human needs be balanced against the needs of the environment?

1.04.03

1 Built-up land consumes resources by removing natural capital for housing, transportation and industrial sites.

2 You might have begun a recycling scheme or changed to walking to school rather than travelling by car.

3 **Research idea:** Suitable websites are http://footprint.wwf.org.uk or http://footprintnetwork.org/en/index.php/GFN/page/calculators

1.05.01

1 Pollution is the addition of a substance (by humans) to the environment at a rate that is greater than the rate at which it can be rendered harmless.

2 Pollution may affect health, metabolism, the nervous system or cause cancer or birth defects.

3 Point-source – from one easily identified source; non-point-source – from a variety of dispersed sources.

4 Natural sources – volcanoes, wildfires; human activity – burning fossil fuels, emissions, or by-products from factories.

5 **Discussion point:** This is a local issue so, of course, your discussions will be determined by your local situation.

1.05.02

1 A pollutant that remains in a food chain for a long time because it is not biodegradable. It will accumulate at higher trophic levels, as animals at higher trophic levels ingest a large number of organisms from lower levels to provide their energy needs.

2 Pollutants such as DDT do not remain in one place. They are transferred from country to country by currents in water, as animals move or when produce is exported. One country cannot deal with this problem on its own.

3 **Research idea:** The precautionary principle can be summarised in the phrase, 'If it's going to cause harm, don't do it.'

Topic 2

2.01.01

1 A group of organisms that can interbreed to produce fertile offspring.

2 The two species have a similar fundamental niche, but when both are present together they restrict their niche to a smaller size, to avoid competition. In this way both can survive.

3 Water, temperature and light (*or other suitable examples*).

4 **Research idea:** Use books and the internet to find organisms that live together to their mutual benefit.

2.01.02

1 A Leopard frogs in a stream.

2 C One species may be eliminated from that ecosystem.

3 Each species occupies a different niche.

4 **Research idea:** Consider the fact that neither species is native to Australia. What natural predators might they have?

2.02.01

1. They convert light energy into chemical energy in biomass, which is used by consumers.

2. Trophic level 2.

3. Both producers and decomposers *must* be present. (The other groups may be present but are not essential.)

4. **Research idea:** This is a local issue so, of course, your discussions will be influenced by the local situation. Use your knowledge of energy use by aquatic and terrestrial organisms to help here.

2.02.02

1. a Suitable pyramid drawn, base 15 cm.

   b There is only one tertiary consumer, because there is insufficient food to support any more in the area.

2. Carnivores are relatively rare because at each trophic level less than 10 per cent of the energy captured by the previous level is passed to the next. Energy from the Sun enters producers, but there are then three further transfers to carnivores. There is not enough energy left for large numbers of carnivores. Large carnivores must hunt for food, and they use much energy as they do.

3. Pyramids of biomass are constructed at one time of the year. At certain times in certain ecosystems there is little biomass, so a pyramid may be inverted. For example, small plants (phytoplankton) along the seashore grow rapidly in spring, when nutrients and light are abundant, but die later in the year, when a pyramid of biomass would not be pyramid shaped.

4. **Discussion point:** Pyramids of productivity show inputs and outputs; these are important factors in agriculture.

2.03.01

1. GSP is the biomass gained by consumers as they feed. NSP is the gain in biomass (per unit area per unit time) by consumers minus losses due to respiration.

2. B.

3. a 3056/21436 = 14.2%

4. b 125/3056 × 100% = 4.09%

5. **Research idea:** Consider abiotic factors and their effects on NPP.

2.03.02

1. Nitrogen in plant and animal tissues; atmospheric nitrogen; nitrate in the soil.

2. Nitrogen cycle – humans add nitrate to soils in the form of fertiliser in order to produce higher yields of food crops. Agriculture upsets the natural nitrogen cycle because biomass is harvested and removed from the cycle, so nitrate in the soil is not naturally replenished.

   Carbon cycle – burning fossil fuels and biomass removes carbon from storages and increases the amount of carbon dioxide in the atmosphere.

3. **Discussion point:** Points to consider: what is the role of decomposers in nutrient cycling? What would happen to organic waste without decomposers? What would happen in an ecosystem if carbon was not recycled?

2.03.03

1. Converting nitrogen gas from the atmosphere into a form (nitrates) that can be taken up and used by plants.

2. A forest, fishery or herd of cows, *or another suitable example*.

3. Use of resources at a rate that allows natural regeneration, for example harvesting timber at a rate that allows natural regrowth of remaining trees.

4. **Research idea:** Nutrients and biomass are held in the forest plants. If they are removed, fertility reduces rapidly.

2.04.01

1. Rainfall, insolation (sunlight) and temperature range.

2. The average temperature is high throughout the year, there is abundant sunlight for photosynthesis, and rainfall is high, providing ideal conditions for plant growth.

3. a Climate change may increase the temperature of the tundra so that it becomes confined to areas further north, closer to the Arctic.

   b Increasing global temperature may extend the range of hot, dry desert areas, as rainfall may be distributed differently across the planet.

4. **Discussion point:** Consider how organisms can feed at different trophic levels, and that trophic levels are defined by humans. An ecosystem cannot be precisely defined, but this does not mean that it is not helpful to try to define it.

2.04.02

1. Succession is the process of change that occurs over time as a pioneer group of species is replaced by intermediate and finally climax communities in an ecosystem. Zonation, on the other hand, is the spatial arrangement of populations along an environmental gradient in a given area due to their different tolerances of the environmental conditions.

Answers to self-assessment questions

2. Pioneer species are the first organisms to colonise an area of bare ground or rock. Lichens and mosses are examples of pioneer species, but pioneer species will be different in different areas, depending on the terrain.

3. Shrubs and grasses can survive in thinner soil which contains less humus. Broadleaved trees require thicker soils with more nutrients to support their larger size. This soil develops later in a succession.

4. **Discussion point:** Visitors disturb a research site and carry seeds and other debris with them on their feet. They may also damage fragile pioneer species. Without human interference, a newly established island can develop completely naturally.

2.04.03

1. K-strategists have few offspring but spend a lot of time caring for them. To survive well they require a good supply of nutrients and stable conditions. These conditions are most likely to be found in a climax community. r-strategists can reproduce quickly and make use of short-lived resources in unstable ecosystems.

2. There are fewer plants present at the start of a succession, thus GPP is low.

3. **Research idea:** More biomass is being used or consumed than is being produced; consider why this might be over the course of a year.

2.04.04

1. A stable community where productivity and respiration losses are balanced.

2. As more species colonise an area there are more interactions between them. More complex food chains and webs can form and more species can be supported, leading to an increase in biodiversity.

3. Grazing animals reduce the height of grass plants and prevent the succession in an area of grassland from developing further. Thus the area remains grassland – a plagioclimax.

4. Managed forests, meadows used for agriculture, farmland used for crops.

5. **Discussion point:** Discussion could include agriculture, clearing forests, building homes or other local issues.

2.05.01

1. There are several possible answers here.

2. Leaf shape and pattern, formation of bark, pollen grains are useful. These will be unique to a particular species. Colour, height and other features that vary continuously with a plant's age and development are not useful.

3. **Discussion point:** Consider that new knowledge is acquired all the time, and equipment to observe species becomes more sophisticated, for example microscopes are available now. Biochemistry has enabled relationships between species to be examined and tested.

2.05.02

1. 400.

2. The estimate would be inaccurate. It would probably lead to an overestimate of the true population.

3. Marking the top of the shell would make the snail more vulnerable to predators, as it would be easily spotted. The population size would probably be underestimated, as individuals were removed from it.

4. The quadrats should be placed at random locations (chosen using a grid and random number tables). The number of daisies present in each quadrat could be counted and compared. The number of quadrats used depends on the size of the field and the size of the quadrat used.

5. Lichens are very small, so the $0.01\,m^2$ ($10\,cm \times 10\,cm$) quadrat is best. Larger quadrats would be too large to cover a suitable area of tree trunk.

6. **Research idea:** Use the internet or reference books and think about how modern technology might be used.

2.05.03

1. a Use a scale of $2\,cm = 100\,g\,m^{-2}$.

   The mass of plants is represented by a bar 9.5 cm long, herbivores by a bar 0.1 cm and the carnivores by a single line.

   b Trophic level 3.

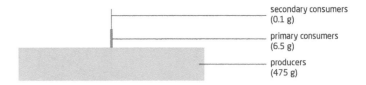

2. The water content of different species varies considerably, and water has no energy value. Dry mass gives a fair comparison of the amount of living material found in different species.

3. A = algae, C = birds.

4. **Discussion point:** Consider factors such as seasons and breeding cycles and how these might affect comparative studies.

2.05.04

1. 4.79.

2. This is a high value for the index and indicates that a good level of diversity is present in the community.

3. **Discussion point:** Consider how important it is to be logical in collecting data and how different types of ecosystems can be studied in a comparative manner.

Topic 3

3.01.01

1. A measure of the quantity of living diversity per unit area. Three types are: species diversity, habitat diversity and genetic diversity.

2. To allow two similar areas to be compared; to allow one area to be monitored over time.

3. That the area has a large number of species present and that there are even numbers of each species (the area is not dominated by one species).

4. **Research idea:** The Shannon–Wiener and Shannon–Weaver indexes and the different versions of the Simpson index are worth investigating.

3.02.01

1. There is genetic variation within species; some individuals are better adapted to the environment than others; 'fitter' individuals tend to survive and reproduce; their offspring inherit the genes that are advantageous.

2. Artificial selection occurs when humans choose which individuals will reproduce to produce offspring with certain selected characteristics; natural selection occurs in response to environmental change.

3. **Research idea:** Most religions have a view on how organisms originated. Many accept evolution.

3.02.02

1. Any species must compete with other members of its own species in order to survive and reproduce. Only those individuals with favourable characteristics are able to do so.

2. Geographic isolation involves a physical barrier (such as water or a mountain range) between individuals of a species, so they cannot breed with each other. Behavioural isolation often happens after geographic isolation; separated groups develop their own behaviour patterns, such as courtship rituals or songs, so that when they meet members of the original population, the two groups do not recognise one another.

3. **Discussion point:** Galton proposed that some humans were 'better' or 'fitter' than others. Nature and nurture can influence the way people develop.

3.02.03

1. The tectonic plate on which Australia is found has been isolated for a long time, so the species on Australia (an island) have developed in a different way from species elsewhere.

2. Formation of islands, mountain ranges or separation of the continents can cause isolation. This can lead to speciation, because interbreeding and gene flow are reduced; isolated populations adapt to their environment, so new species may form.

3. Land bridges link up areas that have been separated. Species may move to new areas and compete with organisms living there; formerly separated groups may meet again and may or may not be able to interbreed.

4. **Discussion point:** We can only gather evidence that remains about past events; we cannot go back in time to study them.

3.02.04

1. Probably five.

2. a This one is caused by human activity.

   b More than a quarter of species may become extinct, so it is just as catastrophic as previous extinctions – species do not have the opportunity to adapt quickly enough.

3. **Research idea:** Consider how data is collected. Can all species on Earth be reached and found?

3.03.01

1. Isolation of species, how specialised they are, reproductive potential, trophic level – top predators are more vulnerable.

2. Population size, population trend, geographic range, numbers of mature organisms.

3. Examples from a named species, e.g. loss of habitat, overexploitation, competition from invasive species.

4. **Research idea:** Think about the position of your organisms in their food chains. Are they keynote species? If their numbers decline, how will others be affected?

Answers to self-assessment questions

3.03.02

1. Madagascar is a poor country which must use its natural capital to provide for its people. Deforestation to provide land for poor farmers leads to a reduction in habitat for endangered species. As the human population grows, the pressure on the land increases. Invasive species, overhunting and pollution are the main ecological pressures on the country, and these are caused in part by the economic pressures.

2. Farming, habitat destruction, hunting, poaching, pollution and introduction of invasive species.

3. **Research idea:** Biodiversity hotspots include regions that have 1500 endemic plants found nowhere else and 30 per cent or less of their natural vegetation remaining. About 30 hotspots have been identified.

3.04.01

1. Ecological reasons, economic reasons and aesthetic reasons.

2. Because species do not live in places that are defined by national boundaries; and it is essential to preserve ecosystems that support one another, even if they are in different countries.

3. Similarities: both work for the conservation of habitats and preservation of living organisms.
    Differences: funding is from different sources; NGOs work using public opinion while governmental organisations can pass laws. (Or other suitable examples.)

4. **Discussion point:** Think about an organisation such as Greenpeace and how its work is publicised and how many people know about it. Is it likely that a government could be unaffected by this?

3.04.02

1. They are independent and can take decisions more quickly. They do not have to consider diplomatic concerns.

2. Size – there must be enough space for animals to have territories; shape – a protected area should have as small a perimeter as possible to minimise disturbance and intrusion from outside; edge effects – centre of the area will have different properties to the edges, organisms at the edge will have more disturbance and competition to contend with.

3. Corridors link reserves so that two smaller areas can be combined to form one larger one. They ensure that members of separated populations can interbreed and maintain a healthy gene pool.

4. Funding, community support, education and good location.

5. **Research idea:** This is a local issue so, of course, your findings and discussions will be influenced by the local situation.

Topic 4

4.01.01

1. The hydrological cycle is the natural sequence through which water passes into the atmosphere as water vapour, precipitates to earth in liquid or solid form, and ultimately returns to the atmosphere through evaporation.

2. Evapotranspiration is the sum of evaporation, sublimation and plant transpiration from the Earth's land and ocean surface to the atmosphere.

3. Advection is the horizontal movement of water in the atmosphere, in vapour, liquid or solid states in air masses. Without advection, water could not be transported from the oceans to land masses. Evaporation from water surfaces on land would not be enough to keep rivers and lakes full and provide the human population with drinking water.

4. **Research idea:** Statistics for monthly and annual precipitation should be available by city or region for most countries. If you can obtain a climate graph to illustrate the distribution of precipitation over the year, so much the better. If your school/college has its own weather-recording equipment, this information may be directly available to you.

4.01.02

1. Precipitation that falls on land and is never available for capture or storage because it evaporates from the ground or transpires from plants is called green water. The remaining water that channels into lakes, rivers, wetlands and aquifers that people can tap directly is known as blue water.

2. Because the soil is unable to take in any more water, the rain flows on the surface under the influence of gravity. This is called surface runoff or overland flow.

3. Stores are places where water is held. Examples are surface storage such as lakes, and groundwater storage.

4. **Research idea:** Consult an Ordnance Survey map or a good atlas. The most relevant scale of Ordnance Survey map is probably 1 : 50 000, where 2 cm on the map equals 1 km on the ground.

4.01.03

1. Areas that have been deforested often experience reduced infiltration, increased surface runoff and lower groundwater recharge.

2. As the area becomes more built up with impervious surfaces, infiltration declines significantly and surface runoff increases. Evapotranspiration declines due to: **a** less surface water because of high runoff, **b** the decrease in vegetation.

3. **Research idea:** If you are unaware of an example of a recent flash flood event, then an internet search should provide a number of examples from around the world. You could produce a brief bullet-point summary of your chosen flash flood event.

4.01.04

1. Differences in temperature and salinity.

2. The warm surface waters of the North Atlantic Drift give north-west Europe a more moderate climate than most regions of similar latitude. Without the warming effect of the North Atlantic Drift, north-west Europe would be on a temperature similar to that of south-east Canada, which suffers harsh winters.

3. **Research idea:** A good atlas may contain the two climate graphs you require for this piece of research. If not, an internet search should provide what you need.

4.02.01

1. For example: more than 840 000 people die each year from a water-related disease; 750 million people around the world lack access to safe water; women and children spend 140 million hours a day collecting water.

2. Two-thirds.

3. **Discussion point:** It would be interesting to fill a container with water to a weight of 20 kg and allow each member of the class to attempt to pick it up. For health and safety reasons, you should not try to walk with the full container. Just picking it up will give some indication of the efforts millions of people have to go through to provide a daily water supply for their families.

4.02.02

1. Surface irrigation systems vary widely in efficiency, at between 20 and 75 per cent. Aerial systems are considerably more efficient, at 60–80 per cent. The most efficient systems are subsurface, ranging between 75 and 95 per cent.

2. For example, for physical factors: the amount of precipitation; the seasonal distribution of precipitation; the physical ability of the surface area to store water.

For example, for human factors: the wealth of a country in terms of its ability to afford water infrastructure; the distribution of population between urban and rural areas, with the concentration of investment in water infrastructure in urban areas; the degree of contamination of rural water.

3. **Research idea:** The government department responsible for water supply and related issues should have the data you require, but it may also be generally available from other sources.

4.02.03

1. Water stress occurs when water supply is below 1700 m^3 per person per year. Water scarcity occurs when water supply falls below 1000 m^3 per person per year.

2. The world's population continues to increase significantly; increasing affluence is inflating per-capita demand for water; the increasing demands of biofuel production – biofuel crops are heavy users of water.

3. **Research idea:** The required information will be available from a range of websites including:
www.thewaterproject.org/water-in-crisis-india
www.un.org/waterforlifedecade/scarcity

4.02.04

1. Water supply is the provision of water by public utilities, commercial organisations or by community endeavours.

2. Aquifers provide approximately half of the world's drinking water, 40 per cent of the water used by industry and up to 30 per cent of irrigation water.

3. **Discussion point:** Discussion is likely to focus on the high cost of desalination plants but should also include reference to the environmental disadvantages.

4.03.01

1. Fish, prawns, mussels, seaweed (*or other suitable examples*).

2. Currents cause upwelling of nutrients from the ocean; currents carry nutrients to the coast zone; temperature of the water and light are beneficial here during certain seasons.

3. Education, advertising, labelling, news items.

4. **Research idea:** Research the types of nets and fishing methods that were used. Are tuna fish available in your area today?

4.03.02

1. Food is expensive and waste would cost the farmer money; waste food settles on the loch floor and causes a build-up

Answers to self-assessment questions

of waste; waste encourages the growth of bacteria which remove oxygen from the water (*or other suitable examples*).

2   a   Plant life is reduced beneath the sea cage due to the build-up of waste food and fish faeces.

    b   Wrasse may increase in number because the high density of salmon means there are more lice in the area.

3   Sea lice could be controlled by the use of pesticides; or by introducing wrasse (a natural predator) into the cages.

4   **Research idea:** Consumers like their salmon to be pink in colour and the farm is supplying its customers with the product that they want.

4.04.01

1   Biochemical oxygen demand is a measure of the amount of dissolved oxygen required to break down the organic material in a given volume of water aerobically.

2   Leaves and woody debris, dead plants and animals, animal manure, effluents from wastewater treatment plants.

3   Microorganisms feed on organic waste in effluent. As they remove this material from the water, their numbers fall.

4   **Research idea:** This is a local issue, so your research will be influenced by local conditions and the local situation.

4.04.02

1   An indirect measure of pollution made by assaying the impact on species within the community according to their tolerance, diversity and abundance.

2   Macro-invertebrates are easy to see, they respond to different levels of pollution and they spend all their life cycle in water, so they are exposed to pollutants for extended periods.

3   a   Poor.
    b   Excellent.
    c   Fair.

4   **Discussion point:** Think about how easy you find it to understand. Could a non-scientist use it?

4.04.03

1   If more nutrients lead to more algal growth, the increase in biomass as nutrients are taken up by plants causes a reduction in nutrients in water so equilibrium is restored.

2   Fertilisers, domestic detergents.

3   An area of ocean or fresh water where the oxygen content is so low that no living things are present.

4   **Research idea:** Visit local water systems such as a pond or stream and look for evidence of excessive plant growth.

Topic 5

5.01.01

1   A soil profile is the vertical succession down through a soil, which reveals distinct layers or horizons in the soil. A soil horizon is a specific layer that is parallel to the surface and possesses physical characteristics which differ from the layers above and beneath.

2   Leaching is a natural process by which water-soluble substances such as calcium are washed out from soil. This reduces the fertility of a soil.

3   **Research idea:** Try to find an area of bare, loose soil. In terms of organic content the most obvious evidence will be earthworms, snails and other live soil fauna or their decaying remains. For soil flora, look for decomposing leaves and other vegetation. However, your soil sample may not contain such obvious evidence as this. In this case you will be trying to identify humus, which is a dark, crumbly material rather like compost. The inorganic material in your soil will be visible grains of sand, silt or clay. Clay particles are very small and thus are very difficult to discern without magnification. However, the sticky nature of clay is easy to observe.

5.01.02

1   Because they vary as the factors and processes that influence them change.

2   For example: organic material from decaying flora and fauna; precipitation, gases and solid particles from the atmosphere; gases from the respiration of soil fauna.

3   For example: nutrients taken up by plants growing in the soil; nutrient losses through leaching; losses of soil through soil erosion and mass movement.

4   **Discussion point:** Focus on the importance of decomposition, weathering and nutrient cycling. For example, in terms of the latter, some of the plant nutrients lost as an output will return as leaf litter in humus.

5.01.03

1   The look and feel of a soil is referred to as 'soil texture' and is determined by the size and type of particles that make up the soil. While this includes organic material, it mostly refers to the inorganic material in the soil.

2   Soil texture is very important because it affects: moisture content and aeration; retention of nutrients; ease of cultivation; root penetration of crops and other vegetation.

3   Soil type can have a major impact on the viability of agricultural communities. All other factors being equal, the

better the soil, the higher the rate of primary productivity. Thus in most circumstances primary productivity would be greatest in loam soils. Clay soils would be in second place, with sandy soils trailing a very clear third.

4 **Research idea:** To find out the soil type in an area of farmland near to you, a soil map may well be available, or the local authority/council may have this information, or you could contact a local farmer. The agricultural activities may be known to you if the farmland is close by and the farming activities are obvious. If not, visiting the farm or contacting the farmer should yield this information.

5.02.01

1 Food security is when all people at all times have access to sufficient, safe, nutritious food to maintain a healthy and active life.

2 Malnutrition is insufficiency in one or more nutritional elements necessary for health and well-being.

3 With malnutrition, people are prone to a range of deficiency diseases and more likely to fall ill. People who are continually starved of nutrients never fulfil their physical or intellectual potential. Malnutrition reduces people's capacity to work, so land may not be properly tended and other forms of income may not be successfully pursued. This is threatening to lock at least some LEDCs into an endless cycle of ill-health, low productivity and underdevelopment.

4 **Research idea:** Information can be obtained from a range of relevant websites including:
www.wfp.org/hunger/malnutrition - UN World Food Programme www.msf.org.uk/malnutrition - Médecins Sans Frontières/Doctors Without Borders

5.02.02

1 Sustainable agriculture involves agricultural systems emphasising biological relationships and natural processes that maintain soil fertility, thus allowing current levels of farm production to continue indefinitely.

2 Food waste occurs towards the end of the food chain, at the retail and consumer level. In contrast, food loss occurs mainly at the front of the food chain, during production, post-harvest, and processing.

3 Twenty-eight per cent of the world's agricultural area is used to produce food that is lost or wasted; most food waste ends up in landfill; methane emissions from landfill are a significant source of greenhouse gas emissions; the carbon footprint from food waste is estimated at 3.3 billion tonnes of CO_2 equivalent each year; a very large volume of water is used each year to produce food that is lost or wasted.

4 **Research idea:** The government department likely to hold statistics on this issue may vary from country to country.

Relevant websites include:
www.worldfooddayusa.org/food_waste_the_facts
www.unep.org
www.unric.org/en/food-waste

5.02.03

1 Physical, economic, political, social/cultural.

2 From 0.45 hectares per capita to 0.23.

3 The energy in meat passes through two steps before human consumption, compared with the single step for crop production. The yield of food per unit area from lower trophic levels (e.g. crops) is greater in quantity than that from higher trophic levels (e.g. livestock), lower in cost and generally requires fewer resources.

4 **Discussion point:** Focus on the efficiency of terrestrial food-production systems. Further information can be obtained from a number of relevant websites, which include:
www.worldwatch.org/node/549
www.sustainweb.org/sustainablefood/meat_and_dairy_products_less_is_more/
www.worldwatch.org/global-meat-production-and-consumption-continue-rise

5.02.04

1 Agro-ecosystems are forms of modern farming that involve the industrialised production of livestock, poultry, fish and crops.

2 For example: very large farms; concentration on one product (monoculture) or a small number of farm products; a high level of mechanisation; low labour input per unit of production; heavy usage of fertilisers, pesticides and herbicides.

3 **Discussion point:** Local food production systems can vary from small-scale organic farming (e.g. smallholdings with hens producing free-range eggs) to very large-scale industrial farms (e.g. major battery hen units). While your and other students' perceptions may vary, people tend to be more sympathetic to smaller-scale low chemical/energy input farming than the larger-scale alternatives. However, you may well feel that large-scale commercial farming is necessary to meet the food demands of growing and more affluent populations.

5.03.01

1 a Soil degradation is a change in the soil health status resulting in a diminished capacity of the ecosystem to provide goods and services for its beneficiaries.

  b The short, tough grass is extremely patchy, with the majority of the area being bare soil. The mineral content of the soil is clear to see, but there is a general absence of humus.

Answers to self-assessment questions

2   In temperate areas, much soil degradation is a result of market forces and the attitudes adopted by commercial farmers and governments. In contrast, in the tropics, much degradation results from high population pressure, land shortages and lack of awareness. The greater climatic extremes and poorer soil structures in tropical areas give greater potential for degradation in such areas compared with temperate latitudes. This difference has been a significant factor in development or the lack of it.

3   **Research idea:** Information may be available from the department of agriculture, from local farming organisations, from farming journals, and from local farmers themselves. If the problem is severe, the national and local press are likely to have written about the problem.

5.03.02

1   a   Deforestation is the process of destroying a forest and replacing it with something else, especially with an agricultural system.

    b   Overgrazing is the grazing of natural pastures at stocking intensities above the livestock carrying capacity.

2   Salinisation is the condition in which the salt content of soil accumulates over time to above normal levels. It occurs in some parts of the world where water containing a high salt concentration evaporates from fields irrigated with standing water.

3   **Discussion point:** The discussion might recognise that some farming communities may think about short-term profit at the expense of longer-term sustainability. In LEDCs where population pressure is intense, it may be difficult for many communities to think beyond short-term survival.

5.03.03

1   Desertification is the gradual transformation of habitable land into desert.

2   Desertification is usually caused by climate change and/or by destructive use of the land.

3   **Research idea:** Desertification affects a considerable number of countries, so there is a wide choice. Individual students could select different examples and report back their findings to the class. Useful websites include:
www.greenfacts.org/en/desertification
www.un.org/fr/events/desertificationday/2008/definition.shtml

5.03.04

1   a   Soil conditioning is the process of adding materials to soil to improve soil fertility.

    b   Lime is commonly used to reduce soil acidity by increasing the pH level.

2   The planting of trees in shelter belts can do much to dissipate the impact of strong winds, reducing the wind's ability to disturb topsoil and erode particles. Shelter belts shelter the soil by reducing wind and evaporation and thus increasing soil temperature. They provide roots at the boundaries of the field, supplying valuable organic matter.

3   **Discussion point:** Show awareness of the fragility of marginal lands and the need to cultivate carefully and in a sustainable manner. Failure to appreciate the productive limits of marginal lands can quickly lead to serious soil degradation.

Topic 6

6.01.01

1   The division of the atmosphere into a number of layers in terms of temperature variation.

2   a   Temperature lapse is a decline in temperature with altitude.
    b   Temperature inversion is an increase in temperature with altitude.

3   Certain gases in the stratosphere and thermosphere absorb solar radiation.

4   **Discussion point:** Houses, factories, motor vehicles and other sources of pollution are much more concentrated in urban areas than rural areas. You might debate the contrast between the urban and rural areas closest to you, giving examples of pollution.

6.01.02

1   Water vapour, liquid water, ice.

2   a   Energy balance: the balance between incoming solar radiation and outgoing terrestrial radiation.
    b   Solar constant: the amount of solar energy received per unit area, per unit time, on a surface at right angles to the Sun's beam at the edge of the Earth's atmosphere.

3   **Discussion point:** You might refer to Figure 6.06 to explain how greenhouse gases trap a proportion of outgoing radiation from the Earth, which gives the Earth a much higher average temperature than it would otherwise have. The ability of the atmosphere to capture the Sun's warmth is essential for life on Earth.

6.02.01

1   Ultraviolet radiation has a wavelength between 100 and 400 nm, compared with between 400 and 780 nm for visible radiation.

2   Most of the ozone is in the stratosphere, with only about 10 per cent in the troposphere. The maximum concentration

of ozone is at an altitude of about 20–25 km, where it exceeds 25 millipascals. The fall in ozone concentration either side of this peak is steep on both sides.

3 **Discussion point:** Focus on the 'good' effect of stratospheric ozone, which shields the Earth from harmful ultraviolet radiation. Contrast this with the pollution problems caused by 'bad' ozone in the troposphere.

6.02.02

1 a Most ozone-depleting substances contain the halogens chlorine, fluorine and bromine.

   b They are contained in industrial products such as CFCs, hydrochlorofluorocarbons, halons and methyl bromide.

2 Volcanic eruptions.

3 **Research idea:** Various websites will provide this information including:
www.livescience.com
www.nasa.gov

6.02.03

1 Large increase in the occurrence of cataracts and sunburn; significant increase in the incidence of skin cancer; suppression of immune systems in organisms.

2 Small amounts of ultraviolet radiation are beneficial and can help prevent diseases such as rickets. However, beyond a certain point ultraviolet radiation poses a danger to human health.

3 An increase in ultraviolet radiation can decrease the productivity of phytoplankton in marine ecosystems.

4 **Research idea:** Various websites will provide this information. Such sources might include:
www.who.int/uv/health/en
www.earthobservatory.nasa.gov
http://www2.epa.gov/sunsafety/health-effects-uv-radiation-1

6.02.04

1 a A refrigerant is a substance used in a heat cycle usually including a reversible phase change from a liquid to a gas.

   b The main use of refrigerants is in refrigerators/freezers and air-conditioning systems.

2 Ammonia is a natural refrigerant that is environmentally benign in the atmosphere. It is efficient and cost-effective, with a very good safety record.

3 A gas-blown plastic is created when a plastic is 'blown' with a gas to create a foam that has a large number of voids incorporating the gas. Foam insulation is a major type of gas-blown plastic.

4 **Discussion point:** Note that there are both chemical and non-chemical alternatives to methyl bromide. Non-chemical alternatives include solarisation and crop rotation. Chemical alternatives include chloropicrin and metam sodium.

6.02.05

1 1987.

2 98 per cent.

3 It is often the case that when major agreements are signed, follow-up meetings and amendments are required to maintain international agreement. Also, more detailed scientific information may have become available since the original agreement was signed.

4 Under the Montreal Protocol, individual countries have had to show how they plan to phase out ozone-depleting substances.

5 **Discussion point:** You are likely to focus on the very significant decline in the emission of ozone-depleting substances since the Montreal Protocol was signed, but you might also discuss the illegal market in ozone-depleting substances. Internet research may yield other criticisms.

6.03.01

1 Road vehicles and other transportation account for 38 per cent of total VOC emissions. Printing/surface coating and general solvent use account for 19 per cent and 18 per cent respectively. Fourteen per cent of emissions come from other industrial processes. Residential accounts for 8 per cent, and 3 per cent is classed as miscellaneous.

2 a In the northern hemisphere, ozone levels are highest between May and September, because these are the months of most sunlight.

   b On a daily basis, ozone levels are highest between noon and early evening after the Sun's rays have had time to react with exhaust fumes from high traffic volumes in the morning rush hour.

3 It has more than doubled.

4 **Research idea:** Most large cities monitor and record levels of ozone and other major pollutants. These figures may appear with weather forecasts or in the daily press. If not, they should be available from the relevant department in the city council.

6.03.02

1 Smog is a form of air pollution that generally reduces visibility, produced by the photochemical reaction of sunlight with hydrocarbons and nitrogen oxides that have been released into the atmosphere, especially by vehicle emissions.

Answers to self-assessment questions

2 When the usual decline in temperature with altitude in the troposphere is reversed, a temperature inversion occurs, with warmer air found above colder surface air. The clear skies associated with anticyclones allow the ground to lose heat rapidly at night. The cold surface air is more dense and is prevented from rising upwards by the warmer air overlying it. Thus pollutants contained in the cold air are trapped near the ground, with their levels building up as emissions from vehicles and other sources continue to pollute the environment. Pollutants are not easily dispersed horizontally when there is very little air movement. Katabatic winds can add to the temperature inversion problem (see Figure 6.19); these are currents of cold air that blow downslope from surrounding uplands into a valley or other lowland area, thus creating or increasing the mass of cold air in a valley.

3 **Research idea:** An internet search should produce the information you require. Local newspapers may prove to be the best source. Summarise the information you gain in a fact file.

6.03.03

1 Catalytic reduction and using less air in combustion.

2 Switching from petrol to liquefied petroleum gas or compressed natural gas; implementing engine and emission controls being developed by manufacturers; cutting the distances vehicles travel by using alternative transport such as bicycles and public transport.

3 Nine of the ten most polluted cities are located in the Indian subcontinent. The remaining city is in Iran, which borders Pakistan.

4 **Discussion point:** Refer to the section in this chapter entitled 'Personal strategies to reduce air pollution' for guidance. However, you may think of strategies that are not included here.

6.04.01

1 Dry deposition is the direct uptake by the ground of pollutants in the form of particles, aerosols and gases in the absence of precipitation. In contrast, wet deposition occurs in the forms of acid rain, snow, fog and mist.

2 The burning of fossil fuels.

3 Pure water is neutral and has a pH of 7. The pH of unpolluted rainwater ranges from 5 to 6. Acid rain has a pH of less than 5.

4 **Research idea:** Wet and dry acid deposition contribute to the corrosion of metals (such as bronze) and the deterioration of paint and stone (such as marble and limestone). Dry deposition of acidic compounds can also dirty buildings, leading to increased maintenance costs.

6.04.02

1 Acid deposition can damage leaves, limiting the nutrients available to them.

2 With acid deposition, the hydrogen ions in sulfuric acid trade places with the metal ions. The hydrogen ions are retained and neutralised by the soil, but the calcium, potassium and magnesium ions are leached or washed out of the topsoil into lower inaccessible subsoil. These ions are then not available as nutrients needed for vegetation growth. Such leaching occurs naturally, but acid deposition speeds up the process.

3 Acid deposition can interfere with the ability of fish to take in oxygen, salt and nutrients. For freshwater fish, maintaining osmoregulation is vital to stay alive. Osmoregulation is the ability to maintain a state of balance between salt and minerals in the organism's tissues. Acid molecules cause mucus to develop in the gills of fish, hindering the absorption of oxygen. Some fish are unable to maintain their calcium levels when the water they swim in becomes more acidic. This can result in reproduction problems. Spring is a vulnerable time for many species, as this is the time of the year for reproduction.

4 **Research idea:** Acid rain affects crops directly and decreases soil quality to reduce yields from agriculture. Its effects are most severe near sources of sulfur dioxide and nitrogen oxides. Acid rain can damage the leaves of vegetation and cause blemishes on tomatoes and other crops. There has been some debate that crops produced under acidic conditions have lower nutritional value with fewer minerals. In richer economies, farmers can counteract the effect of acid deposition with the use of appropriate chemical additives, but for poorer farmers this is not a viable option.

6.04.03

1 Neighbouring European countries, primarily the UK, Germany and Poland.

2 The capacity of soil to neutralise some or all of the acidity of acid rainwater.

3 Granite weathers slowly and does not produce much in terms of neutralising chemicals, making the rock vulnerable to acidification.

4 **Research idea:** There are a number of useful websites including:
http://environment.nationalgeographic.com/environment
www.sciencephoto.com
Try to select as many different types of environment as you can, both urban and rural.

6.04.04

1 The Convention on Long-Range Transboundary of Air Pollution (LRTAP) was signed in 1979.

2   The Large Combustion Plant Directive (LCPD) aims at reducing sulfur emissions by giving coal-fired plants two options. They can either agree to a very limited running programme and close down by 2015, or install the equipment needed to remove sulfur from plant emissions.

3   **Discussion point:** Try to weigh up the advantages of a cleaner environment in terms of much reduced air, water and ground pollution and the redevelopment of derelict industrial sites against economic costs such as unemployment, lost production, reduced export earnings, and so on. In a higher-level discussion you might argue that the rise of the tertiary and quaternary sectors has compensated for the decline of manufacturing industry.

Topic 7

7.01.01

1   Non-renewable sources of energy are the fossil fuels (coal, oil, natural gas) and nuclear fuel.

2   Renewable energy can be used over and over again. These resources are mainly forces of nature that are sustainable and which usually cause little or no environmental pollution. Renewable energy includes hydroelectricity, biomass, wind, solar, geothermal, tidal and wave power.

3   Oil – 33 per cent, coal – 30 per cent, natural gas – 24 per cent, hydroelectricity – 6.6 per cent, nuclear energy – 4.5 per cent, and renewable energy – 1.3 per cent.

4   a   Fuelwood and charcoal are collectively called fuelwood, which accounts for just over half of global wood production.

    b   In developing countries about 2.5 billion people rely on fuelwood, charcoal and animal dung for cooking. Fuelwood provides much of the energy needs for sub-Saharan Africa. It is also the most important use of wood in Asia. In 2010, 1.2 billion people were still living without electricity.

5   **Research idea:** The BP Statistical Review of World Energy is one of the most detailed sources available for this topic. The statistics are updated every year. The energy-charting tool is an interesting way to illustrate energy production, consumption and trends.

7.01.02

1   Energy pathways are supply routes between energy producers and consumers, which may be pipelines, shipping routes or electricity cables. The political stability of these pathways is important for energy security. Some important energy pathways are vulnerable to disruption at times of political tension.

2   Some countries have built up strategic petroleum reserves so that they have a stockpile of oil to last a number of months if their oil supplies are disrupted.

3   **Discussion point:** Discussion will probably focus on the negative aspects of wind power in Table 7.03. You might think of additional disadvantages. The task will be to weigh up the balance between advantages and disadvantages and to assess why more people have become critical of this source of power.

7.01.03

1   Three main concerns about nuclear energy are:

   - the possibility of power plant accidents, which could release radiation into air, land and sea
   - the problem of radioactive waste storage/disposal – most concern is over the small proportion of 'high-level waste'; no country has yet implemented a long-term solution to the nuclear waste problem
   - high construction and decommissioning costs – recent estimates put an average price of about US$6.3 billion on a new nuclear power plant.

   Among the advantages of nuclear energy quoted by its supporters are:

   - zero emissions of greenhouse gases
   - reduced reliance on imported fossil fuels
   - nuclear power is not as vulnerable to fuel price fluctuations as oil and gas – uranium, the fuel for nuclear plants, is relatively plentiful. Most of the main uranium mines are in politically stable countries.

2   Physical factors influencing global variations in energy supply include:

   - Deposits of fossil fuels are only found in a limited number of locations.
   - Large-scale hydroelectric development requires high precipitation, major steep-sided valleys and impermeable rock.
   - Efficient solar power needs a large number of days a year with strong sunlight.

3   **Discussion point:** You might base your discussion on two or three of the bullet points immediately above these self-assessment questions. For example, you might comment on the fact that nuclear electricity has only been available (to a very small number of countries) since the mid-1950s, or refer to the fact that, in richer countries, coal has been replaced as a source of power for railways, for most homes,

Answers to self-assessment questions

and for other purposes. You might also think of other equally good examples.

7.02.01

1   a   Climate change is the long-term sustained change in the average global climate.

    b   Global warming is the increase in the average temperature of the Earth's near-surface air in the 20th and early 21st centuries and its projected continuation.

2   The 'tipping point' is the level at which the effects of climate change will become irreversible to varying degrees.

3   **Discussion point:** Your discussion is likely to focus on the fact that climate change will undoubtedly increase demands on government spending across a range of sectors and that, in pure financial terms, spending money to reduce the extent of climate change will be less expensive than having to adapt to greater degree of climate change.

7.02.02

1   Water vapour, carbon dioxide, methane, chlorofluorocarbons, nitrous oxides, ozone.

2   a   Total emissions – China, USA, India, Russia, Japan.

    b   Per-capita emissions – Australia, the USA, Saudi Arabia, Canada, South Korea.

3   **Discussion point:** The bullet-pointed information provided in the text above Self-assessment questions 7.02.02 shows that the electricity and transportation sectors together account for 60 per cent of greenhouse gas emissions in the USA. You will debate why this is so. You will also want to comment on the contributions of the industry, commercial and residential, and agricultural sectors.

7.02.03

1   Tundra ecosystems in Arctic areas are being significantly affected by temperature increase. A large area of permafrost has started to melt for the first time since it formed 11 000 years ago at the end of the last ice age. The area, which covers the entire sub-Arctic region of western Siberia, is the world's largest frozen peat bog, and scientists fear that, as it thaws, it will release billions of tonnes of methane, a greenhouse gas 20 times more potent than carbon dioxide, into the atmosphere. Scientists are putting together monitoring networks to measure the release of gases from Arctic soils.

2   Coral reefs are biologically rich ecosystems, but they are very sensitive to climate change and other forms of stress. An increase in sea temperature along with other factors such as pollution and sedimentation can effectively halt photosynthesis of the zooxanthellae (algae), resulting in the death of the living part of the coral. The death of the zooxanthellae leaves the coral in an energy deficit and without colour – a process known as coral bleaching. Large areas of coral around the world have been affected by this process.

3   This phenomenon is a change in the pattern of wind and ocean currents in the Pacific Ocean. This causes short-term changes in weather for countries bordering the Pacific, such as flooding in Peru and drought in Australia. El Niño events tend to occur every two to seven years. There is concern that rising temperatures could increase the frequency and/or intensity of El Niño events.

4   Two separate studies published in 2014 (NASA, and the University of Washington) stated that the loss of the Western Antarctic Ice Sheet is inevitable, although the collapse of the ice sheet is at least several centuries off. This will cause up to 4 m of additional sea level rise, which will change the coastline in many parts of the world. Scientists have concluded that the causes of ice loss are highly complex. It is not just due to warmer temperatures causing surface melting of the ice. Contact between the ice and the relatively warmer water at the ocean depths is a major contributory factor.

5   **Discussion point:** The website wwf.panda.org provides a wealth of information on the topic. You could produce a bullet-point summary of the impacts of climate change in the Arctic region.

7.02.04

1   Feedback is the return of part of the output from a system as input, so as to affect succeeding outputs. Positive feedback is feedback that amplifies or increases change and leads to exponential deviation away from an equilibrium. Negative feedback is feedback that tends to damp down, neutralise or counteract any deviation from an equilibrium, and promotes stability.

2   Increasing average temperatures are melting Arctic ice. Until recently, 80 per cent of solar radiation was reflected from the polar ice caps. As the area covered by ice has reduced, the area of open ocean has increased. Because oceans are darker than ice and snow, they absorb more of the Sun's energy and convert it to heat. This increases the warming effect, which melts even more ice. It is, in fact, a vicious circle known as the positive ice albedo effect.

3   The thermal inertia of the oceans is sometimes referred to as climate lag. The mass of the oceans is about 500 times that of the atmosphere. Thus the time it takes the oceans to warm is measured in decades. The greatest difficulty is in quantifying the rate at which the warm upper layers of the ocean mix with the cooler deeper waters. Because of this (and other factors), there is significant variation in estimates of climate lag.

4 **Research idea:** The Meteorological Office in the UK is an excellent source of information on most aspects of weather and climate. The section on climate feedback is presented in a clear and interesting way. You could note down any additional information that is not contained in the textbook concerning global warming.
   - concerning global warming.

7.02.05

1 Variations in the tilt of the Earth's axis; variations in the Earth's orbit around the Sun; variations in solar output.

2 **a** Global dimming is a worldwide decline in the intensity of the sunlight reaching the Earth's surface, caused by particulate air pollution and natural events.

   b Global brightening is an increasing amount of sunlight reaching the Earth's surface.

3 **Discussion point:** Prior to discussion, it might be useful to read Figure 7.21 again and make a brief note of the main points of disagreement between George Monbiot and Christopher Booker.

7.03.01

1 Climate mitigation attempts to reduce the causes of climate change, while climate adaptation strategies attempt to manage the impacts of climate change. For example: reducing global consumption of meat and dairy products; applying fertiliser more efficiently; planting fallow fields with nitrogen-fixing legume crops.

2 Solar radiation management techniques aim to reflect some of the Sun's energy back into space to try to counteract global warming. The objective of CDR techniques is to remove carbon dioxide from the atmosphere to directly reduce the enhanced greenhouse effect and ocean acidification.

3 **Discussion point:** Focus on the fact that solar radiation management techniques aim to reflect some of the Sun's energy back into space to try to counteract global warming, while the objective of CDR techniques is to remove carbon dioxide from the atmosphere to directly reduce the enhanced greenhouse effect and ocean acidification.

7.03.02

1 For example: building coastal defences to protect against rising sea levels; improving the quality of road surfaces to withstand with higher temperatures; developing drought-resistant crops.

2 The term 'adaptive capacity' describes the potential to adjust in order to minimise negative impacts and maximise any benefits from climate change. Adaptive capacity varies from place to place and can be dependent on financial and technological resources.

3 **Research idea:** Some such publications may already be known to you and your family. Your school/college library may be a good source of information. Discussion within the group could yield the titles of a number of interesting publications.

Topic 8

8.01.01

1 Birth rates: Brazil 15.0, UK 13.0.
  Death rates: Brazil 6.0, UK 8.9.

2 World: **a** 12/1000, **b** 1.2 per cent.

   MEDCs: **a** 1/1000, **b** 0.1 per cent.

   LEDCs: **a** 15/1000, **b** 1.5 per cent.

   Africa: **a** 26/1000, **b** 2.6 per cent.

   Asia: **a** 11/1000, **b** 1.1 per cent.

   Latin America/Caribbean: **a** 12/1000, **b** 1.2 per cent.

   North America: **a** 4/1000, **b** 0.4 per cent.

   Oceania: **a** 11/1000, **b** 1.1 per cent.

   Europe: **a** 0/1000, **b** 0.0 per cent.

3 China 140 years; India 47 years; Nigeria 28 years; USA 88 years.

4 **Research idea:** The government department responsible for population data is the most obvious source of information, although you will probably find it available from other sources as well. For example, in the UK, the Office for National Statistics (ONS) (www.ons.gov.uk) is the original source of such data. The relevant sections of the text above Self-assessment questions 8.01.01 provide the 'reasons' to be discussed.

8.01.02

1 Better nutrition; improved public health, particularly in terms of clean water supply and efficient sewage systems; and medical advances.

2 It takes time for social norms to adjust to the lower level of mortality before the birth rate begins to decline.

3 This happens in countries where women in particular make decisions to have only one child or no children at all, and where there is very high use of reliable contraception.

4 Critics of the model see it as too Europe-centric. They argue that many LEDCs may not follow the sequence set out in the model. It has also been criticised for its failure to take into account changes due to migration.

Answers to self-assessment questions

5 **Research idea:** For some countries, the two selected population pyramids might show different stages of population transition. However, significant changes can still be seen in population pyramids for countries within the same stage of demographic transition, for example countries in stage 4 where the population has aged considerably over the last 30 years or so.

8.01.03

1 Such policies have lowered the death rate through better public health and sanitation, agricultural development, and improved service infrastructure.

2 Economic growth allows greater spending on health, housing, nutrition and education, which is important in lowering mortality and in turn reducing fertility. Education, especially improvements in female literacy, is the key to lower fertility. With education comes a knowledge of birth control, greater social awareness, more opportunity for employment and a wider choice of action generally.

3 Such countries are concerned about the socioeconomic implications of population ageing, the decrease in the supply of labour, and the long-term prospect of population decline.

4 **Discussion point:** The current figures for the total fertility rate for the country in which you live will be available from national sources, but also from The World Population Data Sheet published by the Population Reference Bureau (www.prb.org). In terms of government encouragement to have more or fewer children, the relevant sections above Self-assessment questions 8.01.03 should provide the basis for discussion.

8.01.04

1 Population projections are the prediction of future populations based on the present age–sex structure, and with present rates of fertility, mortality and migration.

2 Population projections form the basis for a range of government and intergovernmental policies. It is important, for example, for governments to know: how many children will need places in primary schools, secondary schools and universities; how many older people will require public pensions; how much food will need to be grown/imported to feed the population; how many houses need to be built in the future.

3 The development of computing and the massive increase in computing capacity has allowed increasingly complex permutations to be included in projections. Such technological developments have also allowed frequent updating of projections to reflect changes in trends in fertility, mortality and other relevant factors.

4 **Research idea:** The information should be available from the government department responsible for population statistics. The following websites may also be useful:
www.unfpa.org/world-population-trends
www.un.org/en/development/desa/population

8.02.01

1 A resource can be defined as any aspect of the environment that can be used to meet human needs.

2 Natural capital refers to the source of supply of resources and services that are derived from nature. Natural income is the annual yield from sources of natural capital.

3 **Discussion point:** Think how well endowed or otherwise your country is in terms of forests, soils, cropland, water and sources of energy and minerals. You could produce a bullet-point summary.

8.02.02

1 a Renewable natural capital comprises living species and ecosystems. It is self-producing and self-maintaining and uses solar energy and photosynthesis to produce food and chemical energy.

  b Replenishable natural capital consists of stocks of non-living resources. Examples are the atmosphere, fertile soils and groundwater. Such resources are dependent on energy form the Sun for renewal.

  c Non-renewable natural capital consists of subsoil assets such as coal, oil copper and diamonds. Such resources are depleted as they are consumed.

2 Recyclable resources can be used over and over, but must first go through a process to prepare them for reuse.

3 **Discussion point:** For example, in some countries mines (coal, tin, etc.) have been closed because there is not enough of the resource left to make mining worthwhile. Forests may have been cleared to leave little remaining woodland. Soils may have been degraded to a considerable degree. Groundwater might be depleted.

8.02.03

1 The development of the nuclear power industry in the UK and other countries found a new use for uranium, which significantly increased its value as a resource. The electrification of the railway system, which was once totally steam-driven, was a major change that significantly reduced the demand for coal.

2 Attitudes to plastic bags and the packaging of goods in general are changing in terms of the resources like oil which are used up and the amount of waste created.

3 **Discussion point:** Discussion with parents and grandparents would be very valuable here. However, you might also think of examples within your own lifetime. For example, 20 years ago your parents might have used coal as their main source of energy; now their source of energy might be electricity, oil or natural gas.

8.02.04

1 The intrinsic value of the environment or any other entity is the value that the entity has in itself. Intrinsic value is usually contrasted with extrinsic or use value.

2 Yellowstone in the western USA, 1872.

3 **Discussion point:** Your selected natural environment might be a lake, a stretch of coastline, a forest, a hill or mountain, a glacier or an area of open countryside. It may be somewhere where you live or used to live, or somewhere you visited on holiday. It could also be somewhere you have seen in the cinema, on television or in a magazine.

8.02.05

1 Environmental sustainability means meeting the needs of the present without compromising the ability of future generations to meet their needs.

2 The ESI benchmarks the ability of nations to protect the environment over the next several decades.

3 **Discussion point:** Your discussion is likely to focus on the concept of destination footprint and to break this down into its individual components such as air travel, water use and nature of natural environments.

8.03.01

1 SDW is waste produced by households as opposed to other sectors of the economy.

2 As living standards develop, the nature of SDW changes. The proportion of waste that is organic/biomass declines as waste from higher value products becomes more important. For example, e-waste is a significant contributor to household waste in MEDCs, but only makes a minor contribution in very poor nations.

3 **Research idea:** As domestic waste is a current issue, relevant data should be available from your local authority (council). You would expect to find significant changes over time, for example an increase in e-waste. You would also expect to find that recycling had increased considerably.

8.03.02

1 Electronic waste, or e-waste, is discarded electrical or electronic devices and their parts.

2 The e-waste problem is becoming particularly serious in developing countries due to: lack of legislation and enforcement; lack of controlled take-back systems; informal sector dominance in recycling; lack of awareness by government, institutions and the general public; illegal importation of e-waste from developing countries, often using false documentation.

3 Two huge accumulations of waste in large areas of the Pacific Ocean brought together by the North Pacific Subtropical Gyre. Gyres are regions of the oceans where water rotates in a large circular pattern.

4 **Discussion point:** It should be an interesting exercise to total the e-waste for your class. Try to subdivide e-waste into different categories.

8.03.03

1 Landfill is a disposal site where solid waste is buried between layers of dirt in such a way as to reduce contamination of the surrounding land. It involves using a natural depression in the landscape or, more usually, digging a large and deep pit.

2 Incinerators are expensive to build and operate. Incineration requires a high input of energy, and the tall chimney stacks are viewed by most people as a blot on the landscape. The movement of heavy goods vehicles to and from incinerators is also considerable. The ash produced has to be disposed of in landfill. The environmental and health concerns over incineration are long-standing, with sulfur dioxide, nitrogen dioxide, nitrous oxide, carbon dioxide, chlorine, dioxin and particulates being emitted by the process of incineration. However, considerable technological advances have significantly lowered the emission of pollutants into the atmosphere.

3 Composting is where organic material that has been decomposed is recycled as a fertiliser because of its high nutrient value.

4 **Discussion point:** The relevant text prior to Self-assessment questions 8.03.03 clearly outlines the advantages and disadvantages of incineration. This is very much a value judgement. If you have ever lived near an incinerator, this might be a considerable influence on your perceptions.

8.04.01

1 Carrying capacity is the maximum number of a species, or 'load', that can be sustainably supported by a given area.

2 For example: human ingenuity has enabled resource substitution to overcome problems when a particular resource has become depleted, by coming up with a replacement; the resource requirements of local human populations vary significantly because of different lifestyles and levels of development; technological developments can impact considerably on resource requirements and

Answers to self-assessment questions

availability; affluent local populations can import resources from other geographical areas.

3 **Discussion point:** This discussion will focus very much on personal perceptions, as some people will think differently from others. Those living in densely crowded large cities may think that their environment is very close to carrying capacity (or even beyond it), while in rural areas perceptions might be very different.

8.04.02

1 Product stewardship is an approach to environmental protection in which manufacturers, retailers and consumers are encouraged or required to assume responsibility for reducing a product's impact on the environment.

2 Substitution is the use of common and thus less valuable resources in place of rare, more expensive resources.

3 **Discussion point:** The inclination to reuse is partly governed by level of income, with people on modest incomes more likely to not want to throw something away that they can find an alternative use for. An example might be an old sink being reused as a plant container. However, people on higher incomes might reuse because they are environmentally conscious. The technical ability to make minor modifications to a product can also be a factor.

8.04.03

1 a The ecological footprint is a sustainability indicator that expresses the relationship between population and the natural environment. It sums the use of natural resources by a country's population.

  b A global hectare (gha) is the equivalent of 1 hectare of biologically productive space with world average productivity.

2 Built-up land; fishing ground; forest; grazing land; cropland; carbon footprint.

3 **Discussion point:** A useful discussion could look at what might happen in terms of the individual components of the ecological footprint, such as carbon footprint and forested land. Then you might consider the situation of significantly greater pressure on all components. With intense pressure on resources, the results could be starvation, migration and war.

Answers to end-of-topic questions

Topic 1

1. a A model is a simplified description, that shows the structure or working of a system [2]

   b i Models can track the changes in climate temperature; correlate them with the proportion of carbon dioxide in the atmosphere: these data can be extrapolated to predict future trends. [3]

   ii (*any 3*) Effectiveness of carbon-emission reduction on different processes can be monitored and predictions made; but data available will be relatively short term; therefore may not be reliable; new technology may improve our ability to reduce carbon emissions. [3]

2. a Negative feedback is a self-regulating method which leads to the maintenance of a steady state, whereas positive feedback leads to increasing change in a system. [2]

   b

   [Diagram: Positive feedback loop showing Temperature → Permafrost Thaw → Carbon dioxide and methane released into atmosphere → Temperature, with + signs at each stage and a central +] [4]

   c Natural systems are open and stable systems; negative feedback maintains stability of these systems; without internal control an ecosystem could not be self-sustaining [2]

3. a i Energy cannot be created or destroyed but it can be converted from one form to another. [1]

   ii Output = 7650 kJ (10 000 − 1000 = 9000; 15% of 9000 = 1350; to output = 9000 − 1350 = 7650) [2]
   [1 mark for answer + 1 mark for working out]

   b i 25 units; To next trophic level/trophic level 3/to carnivores

   ii Lost as heat

4. Inputs: (*any 3*) light, rain, seeds, labour; soil; water; technology

   Processes: (*any 3*) planting, ploughing, harvesting; respiration

   Outputs: (*any 3*) wheat, heat, carbon dioxide, oxygen, food, income, waste products
   [1 mark for each correct answer]

5. a Despite apparently caring for the environment, the individual is still driving a large car which is producing more greenhouse gas; the message is that individuals should consider all aspects of their life styles. [2]

   b EVS is a particular worldview or set of paradigms that shapes the way a person (or society) evaluates environmental issues. [1]

   c (*any 2*) A technocentrist would seek technological or innovative solutions; such as carbon dioxide capture; and/or storage; seek new technology to reduce carbon emissions. [2]

   (*any 2*) An ecocentrist would highlight the overuse of fossil fuels; seek restrictions on emissions; aim for sustainability; monitor the production of carbon dioxide; encourage individuals to impose restraints on emissions. [2]

6. a The first quotation seems to indicate that humans can use the resources of the Earth however they please; they should not be shared out by any one powerful person. This view might be supported by a technocentrist who would seek to use all available resources.

   An ecocentrist would probably agree with the second quotation since they believe that people should seek to minimise their impact on the Earth and use only what is absolutely necessary. *Or other suitable arguments.*
   [1 mark for suggestions; 2 marks for justifying the viewpoint]

   b i (*any 5*) Wealthier countries are more likely to invest in environmental protection because: it is expensive to adapt to environmentally sustainable technology such as wind power; carbon trading will benefit MEDCs because they can afford to buy from LEDCs; LEDCs may have less effective environmental laws than MEDCs; because they want to encourage companies to set up; and to provide employment;

433

Answers to end-of-topic questions

Many LEDCs have smaller ecological footprints than MEDCs; people in LEDCs are often more environmentally aware (e.g. native peoples of the rainforest); people in LEDCs are more dependent on the environment so it is important to help protect it (e.g. Madagascar).

[1 mark for each point clearly made and justified]

ii The value of education to a society varies from society to society. One society may value environmental education whereas another may regard the environment only as a source of resources.

Some societies regard a pleasant view as a luxury which all should be able to enjoy, whereas others consider that survival is more important (*OWTTE*). **[2]**

Topic 2

1 a Mouse, cricket, rabbit, deer (*any three*). *[1 mark each]*

b Mountain lions, because rabbit and deer are their only sources of food. **[1]** Deer breed more slowly than rabbits and their population is likely to be smaller. The hawk might also be affected, but it has two alternative sources of food (snakes and mice). **[1]**

c *[2 marks both lines must be shown correctly]*

d One, the hawk. **[1]**

2 a There is the greatest amount of living plant biomass in the tropical rainforest, and the least in the coniferous forest. The tropical rainforest contains twice as much as the same area of coniferous forest. **[1]**

b The tropical rainforest produces more than three times as much new plant material per unit area as the deciduous forest, and more than four times as much as the coniferous forest. **[2]**

c Not all the new plant material (biomass) produced in a year becomes leaf litter. Some of it will be eaten by herbivores and not fall to the ground. Some of it, such as wood, will remain as part of the plant and add to its overall biomass. **[2]**

d *Any three of the following points*:

The increase in living biomass in the rainforest is greater than in either of the two other forests. **[1]**

Rainforest produces a 6 per cent increase in new plant material, whereas the deciduous forest produces only 2 per cent and the coniferous forest 2.6 per cent. **[1]**

If plant biomass was harvested from the rainforest, it would recover more quickly than the other two systems. **[1]**

The quantity of material taken would have to be sustainable – that is, not exceed the amount that the forest could replace (a sustainable yield). **[1]**

The rainforest has the highest productivity, but other factors such as accessibility and conservation issues might make it less useful. **[1]**

3 a An ecosystem is a community of interdependent organisms and the physical (abiotic) environment in which they live. **[2]**

b i A pyramid of biomass is a diagram used to show the amount of living material present at each trophic level in an ecosystem. It is constructed by measuring the amount of dry mass in samples of the organisms present. **[2]**

ii Abiotic factors are non-living components of an ecosystem; examples include temperature, rainfall, soil pH and soil type (*or other suitable examples*). **[2]**

iii Succession is the process in which communities in a particular area change over a period of time, so that the appearance of the whole area evolves and changes. **[2]**

4 a If weeds are not removed from the sample areas, the pea plants are unable to grow and their yield is very small because of competition from the weeds. If weeds are removed, pea plants grow well and the yield is high; weeds compete with the peas for nutrients and light; weeds are likely to be faster growing than the peas and to outcompete them if they are not removed.

[any 3, 1 mark each]

b Farmers must remove weeds from their fields to increase their yields, but weeding takes time and uses energy (if it is mechanised) or chemicals, which may be expensive; weeding is important, but as the graph shows it is not necessary to remove every weed, as the yield with six weeks of weeding is high; farmers must balance cost and benefit. *[1 mark for each of the 3 statements]*

c Biomass can be estimated by sampling areas of the field; a random number grid and quadrats should be used to select areas so that human bias is not a factor; **[1]** all the weeds [including roots] in the sample quadrat should be removed, rinsed and dried to constant mass in an oven; **[1]** the mass of weeds in the large field can be obtained by multiplication. **[1]**

5 a Increase, because there will be more organisms present as a succession proceeds, and more will die and be recycled adding nutrients. **[2]**

b Increase – as the succession develops, more niches will be available as the complexity of the environment increases. **[2]**

c Increase – more organisms and diversity will result in more growth of those organisms and more biomass. **[2]**

d Productivity – the energy produced by produced by producers – will increase and the biomass gained by consumers will also increase. [2]

[in each case 1 mark for the correct answer and 1 for the explanation]

6 a i *r*-strategists produce large numbers of offspring and have a short lifespan, whereas *K*-strategists are long-lived and produce few offspring. [2]

ii (*any 2*) suitable example (frogs, fish, weed plant); advantages: can take advantage of unstable or changing environments; parents do not care for their young; large numbers produced so that a few survive to maturity. [2]

iii *K*-strategists tend to reproduce slowly and have few offspring, so take a long time/may not recover following a disturbance. [1]

b i

Time / min	Population numbers
0	1
20	2
40	4
60	8
80	16
100	32
120	64
140	128
160	256

[1]

ii Graph with axes correctly labelled: time/min; population number; graph shows an exponential increase in the population.

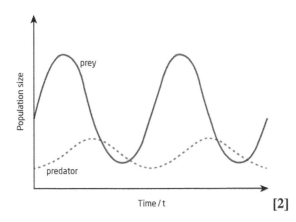

[2]

iii The curve would level off; enter the lag phase and then the plateau phase; as nutrients ran out/waste accumulated; the population would stop increasing; eventually bacteria would die. [3]

Topic 3

1 a i The tectonic plate on which Australia is found separated from other continents over 50 million years ago. The Australian land mass has indigenous species that have developed differently from species elsewhere because of this separation. The result is a range of species with unusual characteristics that are not found anywhere else in the world. [2]

ii (*any 3*) Isolation can lead to speciation; because interbreeding and exchange of genes with other populations is not possible; isolated populations adapt to their environment; so new species may form (e.g. the Galapagos Islands/Australia). [3]

iii A species tends to produce more offspring than are able to survive; the best adapted (fittest) outcompete other individuals; the fittest survive and breed; passing on their genes to the next generation; over time more individuals in the population have these favourable traits and the population gradually changes; eventually, if the population is isolated, it may become different from the original species. [4]

b i Species diversity is the variety of species per unit area, including both the number and relative abundance of species. [1]

ii Habitat diversity is the range of different habitats in an ecosystem. [1]

2 a Species-based conservation can favour high-profile species; it is less successful for unattractive organisms; but preserving the habitat of a high-profile species means conserving habitat, which will benefit all species; some species can be maintained in zoos or seed banks. [3]

b Answers must be related to a *specific* ecosystem. (*any 5*) Hunting for food; collecting specimens for medicine or pets; harvesting plants or fish; habitat loss or degradation; introduction of non-native (invasive) species; establishment of monoculture; natural hazards (e.g. volcanoes, tsunamis). [5]

c A large reserve contains more habitats; and species than a small one; a large reserve has less edge effect; large reserves have greater populations of individual species than small ones so genetic diversity is greater; a rounded shape is better than a long, thin shape of the same area because it has lower edge effect. [4]

d Advantages: (*any 2*) corridors allow gene flow between two protected areas; they allow seasonal movements to take place; they reduce accidents on roads which separate protected areas. [2]

Disadvantages: (*any 2*) can allow hunters to move easily between protected areas; may allow the spread of disease or pests; can cause an increase in edge effect. [2]

Answers to end-of-topic questions

3 a i (*any 3*) Narrow geographical range; small or declining population; low population density; low reproductive potential; few populations of the species; very specialised feeding habits. [3]

      ii Past extinctions caused by abiotic events such as volcanoes, earthquake or meteor strike, current extinctions due to biotic (human) factors; past extinctions occurred over a long period of time, current extinction is happening in hundreds not millions of years. [2]

   b i (*any 3*) Population size, numbers of mature individuals; geographic range; quality of habitat; probability of extinction. [3]

      ii (*any 3*) Whale population reduced by hunting; hunting has reduced the numbers of breeding adults so population not able to replenish its numbers; disturbances to feeding grounds caused by oil wells/drilling; whales have been caught in fishing gear; collisions with ships; underwater noise; or other suitable examples. [3]

4 a The group in the area which received extra food. [1]

   b If predators were excluded as well as the extra food, the mass was lower, so with food only there was some predation and the remaining squirrels each received more food. [1]

   c Heavier females are likely to be more successful in reproduction and will influence the population size. [1]

   d 22 ± 1. [1]

   e The numbers of squirrels per hectare in the two areas where food had been provided fell to the same values as in the control area and the area where predators had been excluded. [2]

   f Capture–mark–recapture. [1]

Topic 4

1 a Stores are places where water is held. Transfers are where water is flowing through the drainage basin system. [2]

   b Urbanisation can considerably increase the proportion of precipitation that is converted to surface runoff. Urbanisation replaces permeable vegetated surfaces with impermeable surfaces of concrete, tarmac, brick and tiles. Where open spaces exist in urban areas in parks, commons and other open land, the soil is often heavily compacted due to high recreational use.

   Urban systems are designed to move water off buildings, roads and other surfaces as quickly as possible. Think of the way buildings are designed to move water away rapidly – pitched roofs, gutters, water downpipes and drains. Most roads have a camber, or curvature to the surface, that helps water drain off them into drains at the sides of roads, rather than pooling in the centee of the road and making driving difficult.

   All these actions drastically reduce infiltration and increase surface runoff. As a result, water moves into river systems much more quickly than before, increasing the risk of flooding. The increasing size of many urban areas has increased the potential for flash floods. Flash flooding happens when precipitation falls so fast that the underlying ground cannot cope, or drain it away, fast enough. With flash flooding there is often little time between the precipitation falling and flash flooding occurring. Flash floods are capable of inundating roads, undermining buildings and bridges, tearing out trees, and scouring new channels. [4]

2 a Water security is the capacity of a population to safeguard sustainable access to adequate quantities of acceptable quality water for sustaining livelihoods, human well-being and socioeconomic development, for ensuring protection against water-borne pollution and water-related disasters, and for preserving ecosystems in a climate of peace and political stability. [2]

   b The main physical factors influencing access to safe drinking water are: the amount of precipitation in a region; the seasonal distribution of precipitation; the physical ability of the surface area to store water; the rate of evapotranspiration; the density of surface access points to water; the ease of access to groundwater supplies if they exist.

   The main human factors are: the wealth of a country in terms of its ability to afford water infrastructure; the distribution of population between urban and rural areas, with the concentration of investment in water infrastructure in urban areas; socioeconomic differences in urban areas – affluent urban districts invariably have better access to safe water than poor districts have; the degree of contamination of urban and rural water supplies; civil war and international conflict. [4]

3 a A coastal area of the Gulf of Mexico where the water contains so little oxygen that living organisms cannot survive. It is a dead zone. [1]

   b Effluent from the land poured into the Gulf of Mexico and caused eutrophication, which in turn led to increased biochemical oxygen demand and eventually the death of organisms that used to live there. [3]

4 a There is a correlation between the toxic releases and income of the residents in the areas where the toxic releases occur. There are more toxic releases in areas where incomes are low, and fewer in wealthier areas. [2]

b Toxic releases tend to come from factories and industrial areas, and it is likely that lower income households will be located nearby because workers in the factories live there; factories tend to be built close to other factories, and this increases the releases from those areas. **[2]**

c Education of factory management; legislation to regulate release of toxic materials; cleaning of waste before it is released (*or other suitable examples*). **[3]**

5 a Excess nutrients cause excess growth of algae (algal blooms); algal blooms deprive lower layers of water of light, so plants die; numbers of bacteria increase to decompose the algae, and use oxygen from the water so that fish and other species also die.

b Humans release nitrates and phosphates into waterways as they use fertilisers or produce sewage.

c It reduces biodiversity; it changes the types of species that occur in an aquatic system; it reduces oxygen content of the water so that only pollution-tolerant organisms can survive.

Topic 5

1 a The Food Security Risk Index is attempting to show the degree of food security risk according to four classes from low risk to extreme risk. **[2]**

b The countries with high and extreme high risk are mainly located in Africa, the Middle East, Asia and Latin America. **[3]**

c The countries at low risk are in North America, most of Europe, Australia, New Zealand, Japan and Chile. **[3]**

2 a Soil is the naturally occurring unconsolidated material on the surface of the Earth that has been influenced by parent material, climate, macroorganisms and microorganisms, and relief, all acting over a period of time. It is thus a mixture of inorganic mineral particles and organic material from decomposed flora and fauna that covers the underlying bedrock, and in which a wide variety of terrestrial plants grow. **[4]**

b Loam is composed of sand, silt and clay in about 40–40–20 per cent concentrations respectively, but this can vary. Different proportions of sand, silt and clay give rise to types of loam soils: sandy loam, silty loam, clay loam, sandy clay loam, silty clay loam, and loam.

Loams have greater cohesion than sandy soils and hold together better when a handful is picked up. They are soft and rich to the touch. Generally containing more nutrients and humus than sandy soils, loams have better infiltration and drainage than silty soils. While drainage is good, this type of soil retains sufficient amounts of water. Loams are easier to cultivate than clay soils and thus a very popular option with farmers and gardeners. Loamy soils may be wet in winter as water tables rise but are usually well drained in summer. **[3]**

3 a Deforestation and overgrazing are major processes that have had a huge impact on soil fertility.

Deforestation occurs for a number of reasons, including the clearing of land for agricultural use, for timber, and for other activities such as mining. Such activities tend to happen quickly, whereas the loss of vegetation for fuel wood, a massive problem in many developing countries, is generally a more gradual process. Deforestation means that rain is no longer intercepted by vegetation, with rain splash loosening the topsoil and leaving it vulnerable to removal by overland flow.

Overgrazing is the grazing of natural pastures at stocking intensities above the livestock carrying capacity. Population pressure in many areas and poor agricultural practices have resulted in serious overgrazing. This is a major problem in many parts of the world, particularly in marginal ecosystems. **[4]**

b Various cropping techniques can be employed to reduce soil degradation. These include:

- Contour ploughing – a tried and trusted technique which prevents or diminishes the downslope movement of water and soil.
- Terracing. Where slopes are too steep for contour ploughing, terracing may be practised. Here the steep slope is converted into a series of flat steps with raised outer edges (bunds). Some studies have concluded that terracing can reduce erosion 20-fold.
- Converting land from arable to pastoral uses – the planting of grass helps to bind soil particles together, reducing the action of wind and rain compared with the effect on bare soil surfaces.
- Including grasses in crop rotations.
- Leaving unploughed grass strips between ploughed fields.
- Keeping a crop cover on the soil for as long as possible, thus minimising the 'bare soil' period.
- Selecting and using farm machinery carefully – in particular, avoiding where possible the use of heavy machinery on wet soils, to prevent damage to the soil structure, and using low ground pressure set-ups on machinery when available.
- Leaving the stubble and root structure in place after harvesting.

All of these techniques have a proven track record in reducing soil degradation, with very few disadvantages.

Wind-reduction techniques have been employed over a long period to conserve soil. The planting of trees in shelter belts and the use of hedgerows can do much to dissipate the impact of strong winds, reducing the wind's

Answers to end-of-topic questions

ability to disturb topsoil and erode particles. Shelter belts shelter the soil by reducing wind and evaporation and thus increasing soil temperature. They provide roots at the boundaries of the field, supplying valuable organic matter. Hedgerows provide a habitat for a whole range of animal life, adding to the general fertility of the fields or parcels of land they surround. Research in the Philippines and elsewhere has shown that hedgerows have proved very effective in reducing soil erosion.

Strip cultivation can also reduce wind erosion significantly. Strip cultivation is the growing of crops in a systematic arrangement of strips across a field. The strips are arranged at right angles to the direction of the prevailing wind for maximum effectiveness. The strips are arranged so that a close-growing crop or strip of grass is alternated with a strip with less protective cover. [4]

4 a The FAO estimates that one-third of food produced annually for human consumption worldwide is lost or wasted along the chain that stretches from farms to food-processing factories, marketplaces, retailers, restaurants and household kitchens. This is enough sustenance to feed 3 billion people. Twenty-eight per cent of the world's agricultural area is used to produce food that is lost or wasted. However, the problem does not end here. Most food waste ends up in landfill, representing a large part of municipal solid waste. Methane emissions from landfill are a significant source of greenhouse gas emissions. The carbon footprint from food waste is estimated at 3.3 billion tonnes of CO_2 equivalent released into the atmosphere each year. The total volume of water used each year to produce food that is lost or wasted is equivalent to three times the volume of Lake Geneva. [3]

b Food waste occurs towards the end of the food chain, at the retail and consumer level. In general, the richer the nation, the higher its per-capita rate of food waste. In contrast, food loss occurs mainly at the front of the food chain, during production, post-harvest and processing. Food loss is less prevalent in MEDCs than in LEDCs. The latter tend to lack the infrastructure to deliver all of their food in good condition to consumers. In Africa, with limited storage facilities, refrigeration and transportation, 10–20 per cent of the continent's sub-Saharan grain is lost to hazards such as mould, insects and rodents. Facing similar challenges, India loses an estimated 35–40 per cent of its fruit and vegetables. [4]

Topic 6

1 Most clouds form in the troposphere, where they reflect and absorb a significant amount of incoming solar radiation. Cloud albedo is a measure of the reflectivity of a cloud. High values mean that a cloud can reflect more solar radiation. Cloud albedo can vary from less than 10 per cent to more than 90 per cent. The overall albedo of the Earth is about 30 per cent. This means that 30 per cent of incoming solar radiation is reflected back into space. It has been estimated that, if all clouds were removed from the atmosphere, global albedo would fall to around 15 per cent. Thus global cloud cover has a clear overall cooling effect on the planet.

However, although cloud cover reduces the level of insolation reaching the Earth's surface, it also reduces the amount leaving it. Rainforest areas astride the equator characterised by thick cloud cover experience days of about 30 °C and nights of about 20 °C. In contrast, temperatures in desert regions, where cloud cover is little, might reach 40 °C during the day and drop to around freezing point at night. This huge difference is because humid air absorbs heat by day and retains it at night. [4]

2 a Most of the ozone in the atmosphere is in the stratosphere, with only about 10 per cent in the troposphere. The ozone layer is an area of high concentration in the stratosphere, with a maximum concentration at an altitude of about 20–25 km. The fall-off in ozone concentration either side of this altitudinal range is steep on both sides. The ozone in the troposphere is concentrated near the Earth's surface. [2]

b It absorbs most of the biologically damaging ultraviolet sunlight, allowing only a limited amount to reach the Earth's surface. [1]

c Ozone depletion can be the result of natural causes such as volcanic eruptions. However, the impact of human activity has caused by far the greatest destruction of ozone. Ozone-depleting substances, which are generally the result of human economic activity, have had a considerable impact on stratospheric ozone. Most of these substances contain the halogens chlorine, fluorine and bromine. The halogens are contained in industrial products such as CFCs, hydrochlorofluorocarbons, halons and methyl bromide.

Halogenated organic gases are very stable under normal conditions, but they can liberate halogen atoms when exposed to ultraviolet radiation in the stratosphere. These atoms react with monatomic oxygen and slow the rate of ozone re-formation. Such pollutants enhance the destruction of ozone, thus disturbing the equilibrium of the ozone production system. [3]

d The Montreal Protocol on Substances that Deplete the Ozone Layer was negotiated and signed in September 1987. The Montreal Protocol has reduced the worldwide stock of CFCs and other ozone-destroying compounds by 98 per cent. CFCs have largely disappeared from the computer manufacturing process, polystyrene packing materials and disposable cups, refrigerators, car and home air conditioners, aerosol sprays, fire extinguishers, degreasing compounds and foam ingredients in furniture. The use of other ozone-depleting substances has also declined significantly.

The Montreal Protocol was one of the first international agreements that included the potential use of trade sanctions to achieve the objectives of the treaty. It also offered significant incentives for non-signatory countries to sign the agreement. [4]

3 Most cases of serious photochemical smog are associated with temperature inversions close to the Earth's surface. A temperature inversion is where temperature increases with altitude, trapping the cooler and more dense polluted air below.

Topography may also be a significant factor. Urban areas located in valleys with surrounding upland areas tend to experience low air circulation and a high level of accumulation of air pollutants compared with cities situated in more open landscapes.

The pattern of wind can affect the frequency of the replacement of local air with fresh air from outside the urban area. Calm atmospheric conditions associated with anticyclones mean that photochemical smog tends to stagnate over urban areas for some time. [4]

Large-scale burning associated with deforestation can contribute significantly to smog in urban areas.

4 a Acid deposition refers to the mix of air pollutants that together lead to the acidification of freshwater bodies and soils. It can be subdivided into dry deposition and wet deposition. [1]

   b Acid deposition does not usually kill trees quickly. Rather, it is more likely to weaken trees by damaging their leaves, limiting the nutrients available to them, or poisoning them with toxic substances slowly released from the soil. Sulfur dioxide interferes with the process of photosynthesis. When acid deposition is frequent, leaves tend to lose their protective waxy coating. Leaves and pine needles turn brown and fall off. Once trees are weak, they can be more easily attacked by diseases or insects that ultimately kill them. Weakened trees may also become injured more easily by cold weather. Coniferous trees are often most at risk from acid rain.

   High aluminium concentrations in soil due to acid deposition can prevent the use of nutrients by plants. Acid rain which has seeped into the ground can poison vegetation with toxic substances that are slowly absorbed through the roots. [3]

Topic 7

1 Energy security can be defined as the uninterrupted availability of energy sources at an affordable price. It depends on an adequate and reliable supply of energy providing a degree of independence. An inequitable availability and uneven distribution of energy may lead to conflict.

As energy demand has increased, reliance on energy imports has risen. For many countries energy imports account for more than half of their energy requirement, and for some countries they account for over half of their energy requirement. Apart from concerns over the price of energy, there are increasing concerns that supplies might be disrupted for political reasons. Countries such as the USA and China have built up strategic petroleum reserves to tide themselves over for a few months if normal oil supplies are disrupted. There are worries about the vulnerability of energy pathways – the supply routes between producers and consumers – which may be pipelines, shipping routes or electricity cables. For example, most Middle East oil exports go by tanker through the Strait of Hormuz, a relatively narrow body of water between the Persian Gulf and the Gulf of Oman. Roughly 30 per cent of the world's oil supply passes through the strait, making it one of the world's strategically important 'chokepoints' (a route that it would not be difficult to block in times of conflict). [4]

2 a Between 1990 and 2009 total emissions increased by 451.4 MMT carbon dioxide equivalents, a relative increase of 7.3 per cent. [1]

   b Deindustrialisation, with many heavy industries shifting to newly industrialised countries; increasing investment in emissions-control technology. [2]

   c Despite a reduction in the average emissions per vehicle, the considerable increase in the number of vehicles on US roads has resulted in this significant increase in emissions. [2]

   d A high per-capita demand for electricity in the USA; heavy reliance on burning fossil fuels, particularly coal, for electricity production. This is a polluting industry that needs to invest much more in controlling emissions. [2]

3 Geoengineering is deliberate large-scale intervention in the Earth's natural systems to counteract climate change. Geoengineering techniques can be grouped into two categories: solar radiation management and CDR. Solar radiation management techniques include stratospheric aerosols, albedo enhancement and space reflectors. CDR techniques include afforestation, ambient air capture, CCS and biochar. The effectiveness and cost of such techniques are likely to vary significantly. Geoengineering is a science that is very much in its infancy. While some techniques are viable with current technology, most are still at the theoretical stage with research still in progress. [4]

Topic 8

1 a Demographic transition is the historical shift of birth and death rates from high to low levels in a population. [2]

Answers to end-of-topic questions

b There are a number of important differences in the way that LEDCs have undergone population change compared with the experiences of most MEDCs before them. In LEDCs, birth rates in stages 1 and 2 were generally higher. About 12 African countries currently have birth rates of 45/1000 or over. Twenty years ago, many more African countries were in this situation. The death rate in LEDCs fell much more steeply, and for different reasons. For example, the rapid introduction of Western medicine, particularly in the form of inoculation against major diseases, has had a huge impact on lowering mortality. However, AIDS has caused the death rate to rise significantly in some countries, particularly in sub-Saharan Africa.

Some LEDCs had much larger base populations, and thus the impact of high growth in stage 2 and the early part of stage 3 has been far greater. No countries that are now classed as MEDCs had populations anywhere near the size of India and China when they entered stage 2 of demographic transition. For those LEDCs in stage 3, the fall in fertility has also been steeper. This has been due mainly to the relatively widespread availability of modern contraception with high levels of reliability.

Also, the relationship between population change and economic development has been much more tenuous in LEDCs. [4]

2 a The four general services provided by natural capital are: provisioning services (products from ecosystems); regulating services (benefits from ecosystem processes); cultural services (non-material benefits from ecosystems); and supporting services (services that allow other services to operate). [2]

b Natural capital provides the resources and wealth required for human populations to develop. Thus there is intense competition for many of the resources that make up the Earth's natural capital. Many aspects of natural capital are at, or have gone beyond, the limit of their sustainable use. This is because short-term benefit has often taken precedence over long-term sustainability.

Research has shown that economic growth in an increasing number of countries has been adversely affected by rapid decline of one or more aspects of natural capital. It is becoming increasingly important to understand the relationship between economic growth and natural capital so that intelligent decisions can be made in terms of sustainable development policies.

As we come to appreciate the value of natural capital more than at any time in the past, the need to conserve it and if possible restore it has moved up the political agenda. Improving our understanding of the relationships and trade-offs among forests, soil, biodiversity, water and food production, among other key ecosystem components, is driving new developments in applied scientific research. [4]

3 a The solid waste from households includes the following classes of material: paper, cardboard packaging, glass, metal (mainly cans and foil), plastics, organic waste and e-waste. [2]

b Incineration requires a very limited land area compared with landfill, and can be operated in virtually all weather conditions. The process reduces the volume of refuse very significantly. Incineration takes up much less land than landfill. It also has a good resource recovery rate. Over 400 kilowatt hours of energy can be produced by incinerating 1 tonne of waste.

On the debit side, incinerators are expensive to build and operate. Incineration requires a high input of energy, and the tall chimney stacks are viewed by most people as a blot on the landscape. The movement of heavy goods vehicles to and from incinerators is considerable. The ash produced has to be disposed of in landfill. The environmental and health concerns over incineration are long-standing, with sulfur dioxide, nitrogen dioxide, nitrous oxide, carbon dioxide, chlorine, dioxin and particulates being emitted by the process of incineration. However, considerable technological advances have significantly lowered the emission of pollutants into the atmosphere. Thus the environmental impact of incineration varies considerably around the world depending on the age of incinerators.

Incineration is often not feasible in LEDCs because of the high initial construction costs and also because wastes do not have a high enough calorie content to sustain the incineration process – more costly fuel must be added. [4]

Answers to case study questions

Topic 1

1.01.01

1 Consider whether the use of social media is as important as the physical presence of protesters.

2 Some people are willing to risk their lives for the environment; think about their EVSs. If they take these risks, should we be prepared to accept some damage to property?

3 Use the internet, social media or reference books to help you. It is important to put forward your point of view and argue your case logically.

1.01.02

1 Australia has an EVS that views whales as important and worthy of conservation (ecocentric). Whaling nations argue that they have always hunted whales for food (anthropocentric).

2 History, culture, education and media influence an EVS.

3 Whales do not belong to any nation or territory but range freely in the oceans. Without international cooperation, protecting them would be impossible.

4 In your discussions, consider factors such as how people are exposed to information about whaling. Is it in school, in the media or from their daily experiences?

1.02.01

1 Inputs are light and heat (energy only).

2 The people inside had insufficient oxygen to sustain them and were unable to produce sufficient food.

3 You should discover that Biosphere 2 is used for education and smaller research projects on the water cycle, soil studies and climate change.

1.02.02

1 Suggestions might include: computer technology that can control the application of fertilisers or the planting of seeds in selected locations to maximise their benefit; medical treatment for animals, for example vaccinations, use of growth hormones (controversial), and any other suitable examples.

2 In monocultures, pests can devastate a crop if they become established, because there are no barriers to prevent them; there is damage to the natural environmental system, which may lead to the exclusion of useful natural predators; when petroleum products become expensive, the price of food will also rise (or other suitable examples).

3 Inputs 100 years ago were lower, usually involving manual labour, use of animals such as horses and oxen, and natural fertilisers such as manure. Outputs were also lower due to lack of pesticides, antibiotics and machinery. Waste was also higher for the same reasons.

1.03.01

1 As the numbers of a predator increase, so does the pressure on the population of a prey species. This means the amount of food available to the predator decreases and its numbers fall. The consequence is that prey numbers increase and feed back to the predator population, which can also increase.

2 If the numbers of prey species fell too low, the predators might starve and there would be insufficient numbers left to breed and replenish the population. In an extreme case, the predator would become extinct in the area or move away to find new sources of food. Human interference could also upset the feedback in the system and prevent it from returning to equilibrium (*or other suitable suggestions*).

3 Possible adaptations include speed, acute hearing or sight, and toxin production for predators, and camouflage, speed, senses, and immunity to a predator's toxins for prey.

1.04.01

1 Fishing downstream has been affected; erosion has been greater than anticipated; a major earthquake has been linked to the construction project; people downstream have been affected by pollution. The EIA did not accurately predict these consequences.

2 Costs: social and environmental displacement of ecosystems and human societies, erosion, earthquakes, seismic activities (possibly); loss of biodiversity, arable land, historic sites; local people and those downstream affected by pollution; drought in Shanghai.

Benefits: large-scale production of power for China's industrial development, hydroelectric power produces cleaner energy, increasing wealth of citizens and the country. EIA predicted some of these factors but could not predict the enormous need for ecological protection and pollution control or the droughts and earthquakes that have occurred.

Answers to case study questions

3 Strengths: an EIA can ensure that environmental impacts are considered at some point in the decision-making process, they can make sure that projects conform to appropriate standards, and can protect human health and safety.

Weaknesses: decision-making is done at a level above where the EIA is carried out, and so it is difficult to assess how the EIA has affected the decision; valued resources may not be protected, and irreversible changes can take place even with an EIA.

1.05.01

1 Oil is not biodegradable, so has chronic long-term effects. When people change the oil in their cars, they are also likely to change the filters, so targeting both together is a useful strategy.

2 Small businesses, garages and repair shops, individuals who are environmentally aware.

3 Start by looking for initiatives by manufacturers, and laws requiring reduction in packaging and recycling of waste.

1.05.02

1 Because DDT accumulates in a food chain. Plants absorb a small amount of DDT; as they are eaten by consumers, the consumers take in and accumulate more DDT and so, when eagles (top consumers) feed on numbers of fish, they accumulate even more DDT and their bodies contain toxic amounts.

2 Because there was no firm scientific evidence at that time.

3 These birds are also carnivores and also consume fish, which are likely to contain accumulations of the poison.

Topic 2

2.01.01

1 Topography, salinity and temperature are the most important abiotic factors, but others will be influential.

2 The habitat of the barnacle is the physical environment in which it lives – an exposed rocky shore where is it covered and uncovered by the tide. A niche, on the other hand, is not just the *Chthamalus* habitat but also the ways in which it interacts with *Balanus* and the physical environment.

3 Human interference; habitat destruction, for example as the result of a severe storm; introduction of another invasive species; or any other suitable example.

2.01.02

1 A J-curve showing exponential growth.

2 There were no natural predators and food was plentiful, so the population continued to expand.

3 The invasive species have characteristics that make them more successful in their new habitats than the indigenous species. As a result, native species are outcompeted and die.

2.02.01

1 Caterpillar populations increased because their natural predator, the wasp, was killed by DDT. More caterpillars ate more of the roofs and caused them to collapse.

2 DDT accumulated in the food chain: insect → gecko → cats. Hence the amount consumed by the cats was fatal.

3 Rats would have been killed and eaten by the gecko-eating cats, so as cats died, rat numbers increased. Rats do not eat geckos, so would not have high levels of DDT in their bodies. The new cats helped the villagers by eating the rats.

2.02.02

1 Possible answers: In this case humans were badly affected by the poison, so Minamata Bay is well known and quoted by environmental campaigners; people now understand how dangerous mercury (and other heavy metals) are to the natural world; the location of factories and their outfalls is probably considered more carefully in most countries now (*or other suitable examples*).

2 It takes time for the poison to accumulate to dangerous levels and cause neurological problems; the economy of the region depended on the factory, so people were reluctant to blame the factory; there was limited understanding at the time about the effects of cumulative poisons.

3 One interesting area is the use of microorganisms in bioremediation, but there are other aspects too.

2.03.01

1 Herbivores feed on plants, and these contain a lot of indigestible material such as cellulose, which is lost in faeces. The food of carnivores is more nutritious and less of the food taken in is wasted.

2 Herbivores tend to stand still for long periods as they feed; they may also sit and digest their food. Carnivores on the other hand must use energy to capture and kill their prey, and this means there are greater respiratory losses.

3 Less energy is used to maintain body temperature, so less respiration occurs and heat loss from respiration is lower. Aquatic animals do not have large skeletons, so energy is not needed to build them; water supports the body of a jellyfish.

2.03.02

1 Natural cycling of carbon and nitrogen is interrupted because leaf litter and dead plants are not broken down by decomposers; ash contains nutrients, but these are rapidly

used up by growing crops, and since the crops are harvested the nutrients are not replaced.

2   Indigenous communities are likely to have lived in (for example) a forest region for many generations. They use a cycle of cultivation so that the first plot used to grow crops is not reused for a long time. Thus it can recover its fertility between uses.

3   This is positive feedback.

2.04.01

1   Graph showing exponential rise followed by a levelling off.

2   a   Spiders – carried on the wind or on the bodies of birds.

    b   Mosses – spores carried on the tides from Iceland.

    c   Plants – seeds carried on the sea or in the waste of birds or on their bodies.

3   Seabirds carry organisms, materials (such as nesting material), and also deposit waste which acts as fertiliser for the developing soil. If they nest, they also leave decaying material after the breeding season and disturb the terrain with their activities, mixing up soil and nutrients.

4   Answers should suggest an increase in the number of species and a general increase in their size and diversity.

2.05.01

1   The calculations show that, even though there are fewer organisms in pond 2, the numbers of individuals of each species are more even than in pond 1 and the pond is not dominated by any one species.

2   More individual organisms were collected in pond 1, but two species were more dominant than the other. This pond is less diverse than pond 2.

3   Netting organisms on the surface and at different depths in the ponds, using the same type of net and method of sampling in each case.

Topic 3

3.01.01

1   Species richness (or number of species), and evenness.

2   An increase in either richness or evenness leads to an increase in diversity. More species mean higher diversity, but if species are not evenly distributed, diversity falls.

3   Species richness of plants is positively correlated with net primary productivity (NPP), and both are greater in tropical forests than temperate forests. You may also like to research the relationship between species richness and latitude.

3.02.01

1   The land was drier and open grassland developed.

2   In open grassland there was more food, but the presence of grazing animals attracted predators. The horses evolved the ability to eat grass and also to run faster.

3   This research project (at Copenhagen University) will not be complete until 2016, but already it seems that Przewalski's horse is more closely related to modern horses than had been thought, and the two species diverged around 50 000 years ago.

3.03.01

1   Legislation on Mauritius might have helped protect the birds, but many people would argue that sailors required fresh supplies of food in the days before preservation and refrigeration of foods. Protection of certain regions of the island for dodo breeding, where hunting was banned, would probably have saved them. The introduction of predators and poor weather could not have been anticipated or prevented.

2   At the time of the dodo, photography was not available to record the birds. Drawings were made and specimens collected which may have been damaged by the time they reached scientists. As the bird was so unusual, it was difficult for predictions to be made about it from other specimens.

3   Possible suggestions: other flightless birds did not live on small islands, they might not have been good to eat, they were not passive like the dodo, they had not suffered habitat loss.

3.03.02

1   A longer life means that adult fish are more likely to accumulate toxic levels of persistent pesticides; long-lived individuals may survive in isolated areas where there are no potential mates and where access by new fish is impossible; long-lived species like this do not spawn every year, so reproductive success is reduced.

2   Civil disruption such as war reduces availability of food and leads to more poaching and overfishing; inability of governments to enforce laws protecting the fish; disagreements about conservation matters between the different countries through which rivers flow.

3   Possible answers might include the following. Similarities – both fish are used for food, so have been put under pressure by human population increase; both have suffered habitat loss in recent decades partly due to human interference with waterways. Differences – eels are not long lived whereas sturgeon are; eels migrate and live in both fresh and sea water; eels return to the same spawning grounds each year.

Answers to case study questions

3.03.03

1. The development of fast vehicles that could travel across the sands made the animals easier to chase down. High-powered, long-range weapons made them easy to kill.

2. Zoos and nature parks that keep small herds of animals provide a wider gene pool than remains in a very small wild population. Gene registers enable zoos to collaborate with other institutions and make sure that a wide variety of genes is present in the animals that are chosen to breed; females can be nurtured as they give birth so that larger numbers of young survive.

3. Students might consider jerboas (not threatened), desert tortoise (vulnerable), addax (critically endangered), slender-horned gazelle (endangered), Saharan cheetah (critically endangered). Many other options are possible.

3.04.01

1. No, Paine's experiment showed that the food web contained few predators but that these are crucial to the stability of the system.

2. The keystone species is not the dominant (most abundant) species in this food web. This is because it is a top predator; there are always fewer top predators than other trophic levels in a food web (see Topic 2).

3. To be sure that the results were valid and not simply due to an effect in one part of the world.

3.04.02

1. Laws must be enacted and enforced; local people must engage with the conservation programme (*or other suitable suggestions*).

2. They are charismatic species that encourage funding and visitors, which then support the conservation of other species present.

3. Ecotourism is important for funding, education and outreach. Local people, governments and tourists all benefit economically.

Topic 4

4.01.01

1. The two catchments lie in the upland massif of mid-Wales and are characterised by rolling hills. The geology, soils, topography and precipitation are similar in both river basins. The geology comprises slates, mudstones and sandstone rocks, which are generally classified as impervious. The climate is wet, with up to 2500 mm of precipitation falling on the highest ground. Although the sources of both rivers are close together, the land use in the two river basins varies significantly. The River Wye flows over moors and grassland, while the River Severn flows through an area of coniferous forest.

2. The peak flow in the River Severn (about $70 \, m^3 s^{-1}$) is significantly less than that of the River Wye (about $95 \, m^3 s^{-1}$). This is almost entirely due to the much higher rate of interception in the upper part of the drainage basin of the River Severn, because geology and precipitation are similar in both basins. After peak flow occurs, discharge in the River Severn falls off more slowly due to less rapid runoff.

3. A new urban area would reduce infiltration and increase surface runoff. Thus, the rising and falling limbs of the hydrograph for the River Severn would both become steeper.

4.02.01

1. Seventy per cent of California's runoff originates in the northern one-third of the state, but 80 per cent of the demand for water is in the southern two-thirds.

2. Northern California; the Sierra Nevada mountains; the Colorado River.

3. Although the river was originally committed to delivering 20.35 trillion litres every year, its annual flow has averaged only 17.25 trillion litres since 1930. Also, demand has escalated with population growth and rising living standards. The river now sustains around 25 million people and 820 000 ha of irrigated farmland in the USA and Mexico.

4.03.01

1. The fishery had been overfished to such an extent that there were no adult fish left to reproduce. No monitoring was taking place to assess the problem. Fishermen thought that, because the fishing grounds had always been so rich, they would always be so.

2. Factory ships are large and take huge quantities of fish; they are fitted with sonar to locate shoals of fish; they may take species indiscriminately, so young fish that are not required are killed and do not replenish the population. Large drag nets may damage the ocean floor and lead to more habitat loss.

3. Maximum sustainable yield is the largest yield or catch of fish that can be taken from the stock in the oceans without endangering the population.

4.04.01

1. Sewage is the major source of nitrate and phosphate pollutants which were causing eutrophication in the lake.

2. Biomanipulation removed the species (tilapia) that were feeding on zooplankton which would otherwise have reduced the quantity of phytoplankton (algae) in the

lake. As a result, the zooplankton numbers increased and zooplankton were able to restore equilibrium to the lake.

3 Healthier environment for nearby residents; potential use of the lake for recreation; potential use of the lake for fishing tilapia under controlled conditions.

Topic 5

5.01.01

1 For example: soil constraints; mountainous landscapes; land degradation; urban encroachment.

2 USA, India, Russia, China.

3 China has about 20 per cent of the world's population but only 7 per cent of its arable farmland, and it is becoming an increasing net importer of food. The current per-capita cultivated farmland in China is about 0.092 hectares, which is only about 40 per cent of the global average.

5.02.01

1 In the western provinces of Alberta, Saskatchewan and Alberta.

2 Soil degradation poses a considerable challenge. The loss of habitat for wild native species and the loss of such biodiversity in the long term has greatly concerned scientists. Another issue receiving much attention is the high level of fertiliser use and its environmental impact, particularly in terms of greenhouse gas emissions. Canada has one of the highest per-capita usages in the world of nitrogenous fertiliser, an issue that environmental groups have protested about. The first GM crops were planted in the prairies in the mid-1990s. Environmental groups such as Greenpeace have been constant critics pointing to a mounting range of evidence about the adverse impact of GM production.

3 Large-scale farming of this nature has had a significant impact on the environment. Soil degradation is a major issue, as is the loss of biodiversity. The almost total dominance of commercial farming in the prairies has left little in terms of habitat for wild native species. There is concern about the loss of such biodiversity in the long term. Another issue receiving much attention is the high level of fertiliser use and its environmental impact, particularly in terms of greenhouse gas emissions. The use of GM crops on the prairies since the mid-1990s has been a constant issue of criticism from environmental groups.

5.02.02

1 The rice is grown on very small plots of land using a very high input of labour. Rice cultivation by small farmers is sometimes referred to as 'pre-modern intensive farming' because of the traditional techniques used, in contrast to intensive farming systems in MEDCs, such as market gardening, which are very capital intensive.

2 The use of higher-yielding varieties of rice in recent decades has increased production significantly. The average padi rice yield in India increased from 1.5 tonnes per hectare in 1960 to over 3.5 tonnes per hectare in 2013.

3 Rice cultivation is a significant source of atmospheric methane. Methane is 20 times more potent as a greenhouse gas than carbon dioxide. The high water requirement of rice cultivation is another major issue.

5.03.01

1 Traditional soil-management strategies have included crop-fallow rotation, ripping and strip farming. However, a number of changes have been in evidence over the last decade or so. Traditional monoculture cereal cropping systems that rely on frequent summer-fallowing and use of mechanical tillage for weed control on fallow areas and for seedbed preparation are being replaced by extended and diversified crop rotations together with the use of conservation tillage (minimum- and zero-tillage) practices. Including oilseed and pulse crops in rotations that have traditionally been cereal monoculture has reduced the frequency of summer fallow.

2 The traditional practice of turning the soil before planting a new crop is a leading cause of soil degradation. An alternative is no-till farming, which minimises soil disruption. Here, famers leave crop residue on the fields after harvest, where it acts as a mulch to protect the soil and provide nutrients. To sow seeds, farmers use seeders that penetrate through the residue to the undisturbed soil below.

3 The expertise in such techniques may not be available and the cost of implementing such measures might also be a major obstacle.

5.03.02

1 Among the problems faced by the farmers of Arumeru are steep slopes, erosion hazards, variable soils and low and unreliable rainfall. Soil fertility and nutrient status are generally precarious after a significant period of soil degradation, much of it associated with population pressure.

2 The PLEC project has attempted to diversify the soil-management methods used by farmers in Arumeru to raise soil productivity and enhance soil conservation. The three basic soil-conservation principles of the project are: minimum soil disturbance or, if possible, no tillage at all; soil cover, permanent if possible; crop rotation. By following

Answers to case study questions

these principles the objective is to enhance soil fertility by improving water retention, increasing soil organic matter and reducing soil degradation.

3   Farmers have responded to their soil problems with a mixture of management practices that reflect the resources they have available, traditional knowledge and the awareness of new techniques provided by PLEC. While conservation agricultural practices in Arumeru are still at a relatively early stage, the PLEC project has concluded that there has been enough evidence of successful implementation to undertake similar work in other areas.

Topic 6

6.01.01

1   **a** Albedo: the proportion of radiation that is reflected by a surface.

   b Relative humidity: the amount of water vapour present in air, expressed as a percentage of the amount needed for saturation at the same temperature.

2   Clouds reflect and absorb a significant amount of incoming solar radiation. The overall albedo of the Earth is about 30 per cent. This means that 30 per cent of incoming solar radiation is reflected back into space. It has been estimated that, if all clouds were removed from the atmosphere, global albedo would fall to around 15 per cent. Thus, global cloud cover has a clear overall cooling effect on the planet.

3   Clouds form when air rises in the atmosphere. The molecules in rising air gradually expand, because air pressure decreases with altitude. As the air molecules move further apart in the process of expansion, energy is used up and the air cools. As air cools, its relative humidity increases. If cooling is sufficient for the relative humidity to reach 100 per cent, the air is said to be saturated and the process of condensation begins. Condensation produces water droplets and/or ice crystals, which form clouds. The temperature at which saturation occurs in a parcel of rising air is known as the dew-point temperature. Thus the base of a cloud is formed at the dew-point temperature. The top of a cloud marks the level in the atmosphere where the rising air is no longer warmer than the air around it, having cooled to the temperature of its surroundings.

6.02.01

1   The preconditions for the hole develop in late winter when polar stratospheric clouds absorb nitric acid/nitrogen oxides that would help to slow ozone loss. The formation of the ozone hole occurs in early spring as chlorine that has accumulated in the winter is released. This reacts with sunlight and destroys ozone. With nitrous oxides absent, ozone depletion is rapid. Warming in the spring and early summer brings about the breakdown of the ozone hole.

2   The ozone hole in 2010 was considerably smaller than it was in 1998. In 1998 it reached a maximum of almost 27 million km^2. In contrast, in 2010 its maximum extent was approximately 20 million km^2.

3   Chile and New Zealand.

6.03.01

1   There may be debate about which pollutant or combination of pollutants should be used to assess the degree of pollution in each city, as rank order can vary according to the measure(s) used. The number and location of monitoring stations and the time(s) of year that samples are taken may also affect results. In some cities the variations in levels of pollution are much greater than in others. City governments are aware that a reputation for high levels of pollution might deter tourism and business investment.

2   Transport is the main source of $PM_{2.5}$, NO_x and VOC emissions. It is by far the largest source of NO_x, at 67 per cent, and VOC, at 63 per cent. In contrast, transport is a minor source of SO_2 emissions. Road dust is the main source of PM_{10} emissions, and this is of course linked to the volume of road traffic. Power plants are responsible for over half of SO_2 emissions and almost 30 per cent of CO_2 emissions. Diesel generator sets are the second largest source of NO_x emissions (17 per cent).

3   The relocation of some factories to the outskirts of the city temporarily lowered pollution, but industry surrounding the city remains a significant source of pollution. Likewise, the benefits of the switch to compressed natural gas-based vehicles about a decade ago has diminished with the huge recent increase in traffic levels.

Investment in public transport has been slow to develop in Delhi compared with other cities of a comparable size. The Delhi Metro Rail only opened in December 2002, with an 8.3 km rail line, and has since been extended to 190 km with the completion of phase 2 in 2011. Construction of phase 3 is now underway, to add a further 103 km. Another significant development has been the Delhi bus rapid transit, which opened along a 5.6 km initial corridor in 2008.

Shutting coal power plants, promoting motor-less transport, and strict penalties for those violating pollution control norms are among the suggestions that the government is looking at to improve the air quality in the city over the next five years.

In 2014, the Indian environment ministry launched a national air quality index that will rank 66 Indian cities. It will give real-time information on air quality to put pressure on local authorities to take concrete steps to reduce pollution. The index will describe associated health risks in a colour-coded manner that can be understood by everyone. It is likely that this measure will significantly increase public awareness of the problem.

6.04.01

1 The Pearl River delta region, an area the size of Belgium, is located in south-east China. The Pearl River drains into the South China Sea. Hong Kong is located at the eastern extent of the delta, with Macau situated at the western entrance.

2 The high concentration of factories and power stations and the growing number of vehicles in the province is the source of the problem. The rapid increase in the number of motor vehicles has been of particular concern in recent years. Apart from increasing emissions from vehicles: a few cities in Guangdong have been burning more coal to produce power, compensating for a decline in power transmission from western China; the persistence of acid rain in Shaoguan and Qingyuan results mainly from the various polluting industries, such as cement and ceramics, which have moved into these two cities in recent years.

3 Guangdong has formulated and issued a series of measures to combat regional acid deposition and other forms of air pollution. These measures include: reducing reliance on fossil-fuel power plants and reducing the sulfur content in fuels; forbidding new cement plants, ceramics factories and glassworks; installing particulate matter control devices for cement plants and industrial boilers; upgrading air pollutant emission standards for boilers; upgrading motor vehicle emission standards; encouraging 'green' public transportation.

Topic 7

7.01.01

1 China's energy mix is dominated by coal, which accounted for 69 per cent of total energy consumption in 2011. In second place was oil, accounting for 18 per cent. The remaining contributors to Chinese energy consumption were: hydroelectric power (6 per cent), natural gas (4 per cent), nuclear (1 per cent), other renewables (1 per cent).

2 China became a net importer of oil in the mid-1990s. Since then the gap between consumption and production has steadily widened. China now consumes more than twice the amount of oil it produces.

3 China is developing a strategic petroleum reserve due to: its concerns about energy security; its increasing reliance on oil imports; and to protect itself to a certain extent from fluctuations in the global oil price, which can arise for a variety of reasons. The plan is for China to build, in three phases, facilities that can hold 500 million barrels of crude oil by 2020. This will be equivalent to about 90 days' supply.

7.01.02

1 The three fossil fuels together contributed 87 per cent of UK primary energy consumption in 2011. Natural gas was the single most important source of energy (38 per cent), followed by oil (33 per cent) and coal (16 per cent). The remaining contributors to energy consumption were nuclear (8 per cent), bioenergy and waste (4 per cent), and wind and hydroelectric power (1 per cent).

2 The energy mix of the UK has changed significantly in the past due to: changes in resource availability; technological progress; the relative cost of different sources of energy; consumer behaviour.

3 Dependency on imported energy has risen as: UK production of oil and gas in the North Sea has fallen rapidly; domestic coal production has continued its long-term decline; some nuclear power stations have closed, having reached the end of their productive lives, with others due for closure in the next decade.

7.02.01

1 For example: the mean monthly temperature is below 1 °C over the whole country between November and March; late spring and early autumn frosts leave a short growing season of 80–100 days in the north and 120–140 days in the south; the average annual precipitation is only 251 mm, ranging from 400 mm in the north to less than 100 mm in the Gobi Desert. In comparison, London's average annual precipitation is 580 mm.

2 The increase in average temperature of 2.1 °C assessed by Mongolian meteorologists and 0.7 °C by international estimates are both above the average global increase during this period. During the last ten years of the timeline, above-average precipitation was recorded for only one year. This is of great concern in a country that has large areas of desert and semi-desert. In general, winter precipitation is increasing and summer precipitation is decreasing. As most of Mongolia's precipitation falls during the summer, this trend is disturbing. The occurrence of droughts is increasing.

3 Overgrazing, deforestation, soil erosion, desertification.

7.03.01

1 Seven per cent.

2 The loss of many industries, particularly heavy industries to NICs; improvements in energy efficiency in industrial processes.

3 Energy use in the transport sector increased by 50 per cent between 1970 and 2012 with the steady rise in the number of vehicles on the UK's roads. In 1970, the number of cars on the country's roads was 10 million compared with more than 27 million today.

Answers to case study questions

Topic 8

8.01.01

1 1979.

2 China's total fertility rate has fallen from 4.77 births per woman in the 1970s, just before the one-child policy was introduced, to 1.64 in 2011.

3 The policy has had a considerable impact on the sex ratio, which at birth in China is currently 119 boys to 100 girls. This compares with the natural rate of 106:100. This is already causing social problems, which are likely to multiply in the future. Selective abortion after prenatal screening is a major cause of the wide gap between the actual rate and the natural rate. But, even if a female child is born, her lifespan may be sharply curtailed by infanticide or deliberate neglect. A paper published in 2008 estimated that China had 32 million more men aged under 20 than women.

8.01.02

1 1939.

2 For example: longer maternity and paternity leave; higher child benefits; improved tax allowances for larger families.

3 In 2014, France's total fertility rate was 2.0 compared with 1.9 in the UK, 1.4 in Germany and Italy, and 1.3 in Spain.

8.02.01

1 'The tragedy of the commons' is a term used to explain what has happened in many fishing grounds. Because the seas and oceans have historically been viewed as common areas, open to everyone, the capacity of fishing vessels operating in many areas has exceeded the amount of fish available. The result has been resource depletion. The global fishing fleet is two to three times larger than the oceans can sustainably support.

2 Fifty-three per cent of the world's fisheries are fully exploited; 32 per cent are overexploited, depleted or recovering from depletion; several important fish populations have declined to such an extent that their survival is threatened; every year billions of unwanted fish and other animals (dolphins, marine turtles, seabirds, sharks, etc.) die because of inefficient, illegal and destructive fishing practices.

3 The CFP has evolved over a number of stages. Current measures to conserve fish stocks include: taking a long-term approach by fixing total allowable catches on the basis of fish stocks; introducing accompanying conservation measures; setting recovery plans for stocks below the safe biological limits; managing the introduction of new vessels and the scrapping of old vessels in such a way as to reduce the overall capacity of the EU fleet; measures to neutralise the socioeconomic consequences of fleet reduction; measures to encourage the development of sustainable aquaculture.

The number of small fish caught is limited by: minimum mesh sizes; closure of certain areas to protect fish stocks; the banning of certain types of fishing; recording catches and landings in special log books.

Supporters of the CFC argue that it makes a strong contribution in the quest for sustainable fishing. However, environmentalists believe that short-term economic and political concerns override the objective of sustainability.

8.03.01

1 18 410 tonnes of domestic refuse is generated by the city every day.

2 By 2014, Beijing had 37 waste disposal facilities with a daily processing capacity of 22 000 tonnes. This included: nine waste transfer stations, four incineration plants, 16 sanitary landfill sites, six composting plants and two food waste treatment plants.

3 The Lujiashan incinerator, which is the world's largest, began operating at full capacity in Beijing in late 2014. It can process 3000 tonnes of household waste a day. This is about one-sixth of the daily domestic waste generated in the city.

8.04.01

1 China is about 25 per cent above the per-capita globally available biocapacity, whereas the USA's ecological footprint is about four times this level. In comparison, the footprint for Bangladesh is extremely low, at less than half of global available biocapacity.

2 For example, the USA's carbon footprint accounts for over two-thirds of its overall ecological footprint. The fraction of the overall ecological footprint made up by the carbon footprint is considerably less for China and lower still for Bangladesh. While the carbon footprint is the largest component for both the USA and China, in Bangladesh it is second to the footprint from cropland. The extremely high per-capita usage of energy in all sectors of the economy largely explains the USA's carbon footprint.

3 The ecological footprints for all three countries have changed between 1961 and 2010. In the USA, after a period of relative stability the country's ecological footprint declined modesty to narrow the gap with biocapacity. For China it has been a very different picture, with a rapidly rising ecological footprint over the last decade. The ecological footprint in Bangladesh has been rising for the past two decades at a significant rate, although not as fast as that for China.

4 You could arrange the ecological footprints in rank order, calculate the average, and discuss reasons for variation from the average.

Glossary

Abiotic factors Non-living components of an ecosystem which influence the system and organisms within it. Examples include light, pH, rainfall and temperature.

Acid deposition The mix of air pollutants that together lead to the acidification of freshwater bodies and soils.

Acidification Change in the chemical composition of soil which may trigger circulation of toxic metals.

Adaptive capacity The capacity of a system to adapt if the environment where the system exists is changing.

Advection The horizontal movement of water in the atmosphere, in vapour, liquid or solid states in air masses (i.e. wind-blown movement).

Agricultural technology The application of techniques to control the growth and harvesting of animal and vegetable products.

Agro-ecosystem The form of modern farming which involves industrialised production of livestock, poultry, crops and fish. It is typically large scale and capital intensive.

Agro-industrialisation Industrialised farming that is typically large scale and capital intensive.

Air quality index An indicator of air quality, based on air pollutants that have adverse effects on human health and the environment.

Albedo The proportion of solar radiation that is reflected by a particular body or surface.

Algal bloom A rapid increase or accumulation in the population of algae in a river or a lake. It often has the appearance of a green sludge which can be very dense and cover a considerable area.

Anoxic water An area of seawater or fresh water which is depleted of dissolved oxygen.

Anthropocentric A human-centred value system that places humans as the central species and assesses the environment from an exclusively human perspective.

Anticyclone An area of high atmospheric pressure due to subsidence in the atmosphere and characterised by clear skies and calm weather conditions.

Anti-Malthusians These are resource optimists who believe that human ingenuity will continue to conquer resource problems, pointing to so many examples in human history where, through innovation or intensification, humans have responded to increased numbers.

Aquaculture The farming of aquatic organisms such as fish, crustaceans, molluscs and aquatic plants.

Arid conditions Those where precipitation is less than 250 mm per year.

Atmosphere An envelope of gas that surrounds the Earth becoming increasingly thinner with distance from the Earth's surface and held in place by the Earth's gravitational pull.

Background extinction rate The natural extinction rate of all species. Scientists estimate that it should be about one species per million per year or up to 100 species per year.

Bioaccumulation Build-up of toxic compounds in plants and in the fatty tissue of fish, birds, and animals that eat plants.

Biocapacity The capacity of an area or ecosystem to generate an ongoing supply of resources and to absorb its wastes.

Biochemical oxygen demand (BOD) The amount of oxygen required by aerobic microorganisms to decompose the organic matter (e.g. sewage) in a sample of water.

Biodegradable pollution Pollution from a substance that breaks down naturally in the environment; for example, certain types of waste paper will naturally decay.

Biodiversity The quantity of living (biological) diversity per unit area. It includes species diversity, habitat diversity and genetic diversity.

- **Genetic diversity** is the range of genetic material present in the population of a species or its gene pool.
- **Habitat diversity** is the number of ecological niches or range of different habitats that are present per unit area of a biome, ecosystem or community. If habitat diversity is conserved, this usually leads to the conservation of both species and genetic diversity.
- **Species diversity** is the variety of species per unit area. It includes both the number of species which are present and their relative abundance.

Biomagnification The increase in concentration of persistent pollutants along a food chain.

Biomass The mass of organic material in an organism or ecosystem. Biomass is usually expressed as dry weight per unit area. Water is excluded because it is not organic and has no energy value.

Biome A group of ecosystems that share similar climatic conditions and therefore similar patterns of vegetation.

Biosphere The part of the Earth inhabited by organisms that extends from the upper atmosphere to the depths of the Earth's crust.

Biotic factors The living components of the ecosystem which influence the system and organisms within it. Examples include competition, disease and predation.

Biotic index A scale that measures the quality of an environment by indicating the types of organism present in that environment.

Blue water The proportion of precipitation that collects in water courses (in rivers, lakes, wetlands and as groundwater) and is available for human consumption.

Bore hole A hole typically drilled by machine and relatively small in diameter.

Business-as-usual The scenario for future patterns of production and consumption which assumes that there will be no major changes in attitudes and priorities.

Carbon credit A permit that allows an organisation to emit a specified amount of greenhouse gases. Also called an emission permit.

Glossary

Carbon footprint The total set of greenhouse gas emissions caused directly and indirectly by an individual, organisation, event or product (UK Carbon Trust 2008).

Carbon trading Dealing in permits between companies that have not used up the level of emissions they are entitled to and companies that would otherwise pollute above their entitlement.

Carrying capacity The largest number of individuals in a population that the resources in the environment can support for an extended period of time.

Child mortality The number of deaths of children aged under 5 per 1000 live births in a given year.

Climate A region's long-term weather patterns.

Climate change Long-term sustained change in the average global climate.

Climate adaptation refers to strategies that attempt to manage the impacts of climate change.

Climate mitigation Strategies that attempt to reduce the causes of climate change.

Climax community The final stable community of a succession. It remains unless there is further disturbance.

Clouds Visible masses of very fine water droplets or ice particles suspended in the atmosphere.

Commercial farming Farming for profit, where food is produced for sale in the market.

Community A group of populations which live in the same habitat and interact with one another. For example, the community of an orang-utan includes populations of crocodiles, tigers and leopards as well as the trees, shrubs and all the other biotic components of the forests in which they live.

Community energy Energy produced close to the point of consumption.

Compost Organic material that has been decomposed and recycled as a fertiliser because of its high nutrient value.

Condensation The process by which water vapour in the air is changed into liquid water.

Consumer An animal that feeds on plants or other animals and obtains its energy and nutrients from them.

Contour ploughing A pattern of ploughing that ensures that the ridges and furrows are at right angles to the slope, preventing moisture from running downhill and thus reducing erosion considerably.

Coral bleaching A process whereby coral communities lose their colour, due either to the loss of pigments by microscopic algae (zooxanthellae) living in symbiosis with their host organisms (polyps) or to loss of the zooxanthellae themselves.

Crude birth rate (CBR) (generally referred to as 'the birth rate') The number of births per thousand population in a given year. A very broad indicator that does not take into account the age and sex distribution of the population.

Crude death rate (CDR) (generally referred to as 'the death rate') The number of deaths per thousand population in a given year. A broad indicator that is heavily influenced by the age structure of the population.

Crust The thin, outermost layer of the Earth. The land we live on, known as the continental crust, is formed of this solid rock layer. Below the sea, the crust is known as the oceanic crust and it forms the bed of the ocean.

Dead zone An area in an ocean or freshwater where there is not enough oxygen to support life.

Decomposers The bacteria and fungi which feed on dead and decaying material and recycle nutrients in the ecosystem.

Deflected succession A succession in which human activity results in the naturally occurring climax community being replaced by a plagioclimax.

Deforestation The process of destroying a forest and replacing it with something else, especially by an agricultural system.

Deindustrialisation The shift of manufacturing industry from MEDCs to lower-cost NICs and LEDCs.

Demographers Professionals who study the characteristics (e.g. composition, distribution, trends) of human populations.

Demographic transition The historical shift of birth and death rates from high to low levels in a population.

Demography The scientific study of human populations.

Desalination The conversion of salt water into fresh water by the extraction of dissolved solids.

Desertification The gradual transformation of habitable land into desert.

Destination footprint The environmental impact caused by an individual tourist on holiday in a particular destination.

Developing steady-state equilibrium *See* Equilibrium.

Diet The kinds of food that a person, animal or community habitually eats.

Diversity A function of the number of different species and the relative numbers of individuals of each species.

Doubling time The number of years it would take a population to double its size at its current growth rate.

Dust storm Severe windstorm that sweeps clouds of dust across an extensive area, especially in an arid region.

Ecocentric A nature-centred value system that views people as being under nature's control rather than in control of it.

Ecological efficiency The percentage of energy assimilated in one tropic level that is available to the next.

Ecological footprint A sustainability indicator, which expresses the relationship between population and the natural environment. It accounts for the use of natural resources by a country's population.

Ecological pyramids Diagrams used to provide a picture of the quantities of organisms present at each trophic level in an ecosystem.

Ecological services Attributes of the natural environment that often involve essential processes such as the water cycle and photosynthesis.

Economic water scarcity When a population does not have the necessary monetary means to access an adequate source of water.

Ecosystem A community of interdependent organisms and the physical (abiotic) environment in which they live.

Ecosystem services A form of natural income derived from natural capital.

Ecotourism A specialised form of tourism where people experience relatively untouched natural environments such as coral reefs, tropical forests and remote mountain areas, and ensure that their presence does no further damage to these environments.

Edaphic factors The physical, chemical and biological properties of soil (e.g. water content, organic content, texture and pH).

Effective precipitation The amount of precipitation that is actually added and stored in the soil.

Emergent properties Features of a system that cannot be present in the individual component parts.

Energy balance The balance between incoming solar radiation and outgoing terrestrial radiation.

Energy intensity A measure of the energy efficiency of a nation's economy. It is calculated as units of energy per unit of GDP.

Energy ladder The improvement of energy use due to rising household incomes.

Energy mix The relative contribution of different energy sources to a country's energy production/consumption.

Energy pathways Supply routes between energy producers and consumers which may be pipelines, shipping routes or electricity cables.

Energy poverty A lack of access to modern energy services due to insufficient income.

Energy security Uninterrupted availability of energy sources at an affordable price.

Enhanced greenhouse effect The result of human activities which increase the concentration of naturally occurring greenhouse gases; it leads to global warming and climate change.

Entropy A measure of the evenness of energy distribution in a system.

Environmental Impact Assessment (EIA) A report prepared before a proposed large-scale project to assess the possible positive or negative impact that the project may have on the environment. An EIA should include social and economic aspects as well as effects on the environment.

Environmental sustainability Meeting the needs of the present without compromising the ability of future generations to meet their needs.

Environmental value system (EVS) A particular worldview or set of paradigms that shapes the way individuals or societies perceive and evaluate environmental issues.

Epidemiological transition The change from mainly infectious diseases, still common in LEDCs, to the degenerative diseases, which have become the main cause of death in MEDCs.

Equilibrium A state of balance which exists between the different parts of any system.

- **Stable equilibrium** is an equilibrium that tends to return to the same equilibrium after a disturbance.
- **Static equilibrium** is an equilibrium in which there are no changes over time because there are no inputs and outputs to the system.
- **Steady-state equilibrium** (aka dynamic equilibrium) is an equilibrium that allows a system to return to its steady state after a disturbance.
  - **Developing steady-state equilibrium** is a steady-state equilibrium which is developing over time (e.g. in a succession).
- **Unstable equilibrium** is an equilibrium that forms a new and different equilibrium after a small disturbance.

Eutrophication Natural or artificial addition of nutrients (nitrates and phosphates) to a body of water resulting in depletion of the oxygen content. Human activity can accelerate the process by the addition of sewage, detergents and agricultural fertilisers.

E-waste (electronic waste) Discarded electrical or electronic devices and their parts.

Evaporation The process of water in a liquid state changing to a gaseous state (water vapour) due to an increase in temperature.

Evapotranspiration The sum of evaporation, sublimation and transpiration from land and ocean surfaces to the atmosphere.

Evolution A gradual change in the genetic characteristics of a population.

Exponential growth This occurs when the growth rate of a mathematical function is proportional to the function's current value. This means an increase in number or size at a constantly growing rate. It is also called geometric growth.

External factors Factors regulating populations from outside the population (e.g. predation or disease).

External forcings Processes both outside and within the atmosphere that can force changes in climate.

Externality The side-effects, positive and negative, of an economic activity that are experienced beyond its site.

Extinction The point when a species ceases to exist or the last known individual of the species dies.

Feedback The return of part of the output from a system as input to the same system, so as to affect succeeding outputs.

Fertile soil Soil that is rich in the nutrients necessary for basic plant nutrition, including nitrogen, phosphorus and potassium.

Fertility rate The number of live births per 1000 women aged 15–49 years in a given year.

First law of thermodynamics Energy cannot be created or destroyed but can be converted from one form to another.

Flash flood Flood caused by heavy or excessive rainfall in a short period, generally less than six hours. Raging torrents of water can rip through river beds, urban streets or mountain canyons, sweeping everything before them. They can occur after minutes or a few hours of excessive rainfall.

Food chain The relationship between organisms in a system in which one organism is food for another. The first organism in the chain (the producer) makes its own food, usually by photosynthesis.

Food loss Food that does not reach consumers due to losses occurring mainly at the front of the food chain, during production, post-harvest and processing.

Glossary

Food miles The distance food travels from the farm where it is produced to the plate of the final consumer.

Food security When people at all times have access to sufficient, safe, nutritious food to maintain a healthy and active life.

Food waste Food that does not reach consumers due to losses toward the end of the food chain, at the retail and consumer level.

Fossil fuels Fuels consisting of hydrocarbons (coal, oil and natural gas) formed by the decomposition of prehistoric organisms in past geological periods.

Fuel poverty When a low-income household is living in a home that cannot be kept warm at a reasonable cost.

Fuelwood Wood and charcoal used to supply energy.

Geoengineering The deliberate large-scale intervention in the Earth's natural systems to counteract climate change.

Geopolitics Political relations among nations, particularly relating to claims and disputes pertaining to borders, territories and resources.

Geothermal gradient The rate at which temperature rises as depth below the surface increases.

Global brightening An increase in the amount of sunlight reaching the Earth's surface.

Global dimming Worldwide decline of the intensity of the sunlight reaching the Earth's surface, caused by particulate air pollution and natural events (e.g. volcanic ash).

Global hectare (gha) The equivalent of one hectare of biologically productive space with world average productivity.

Global warming The increase in the average temperature of the Earth's near-surface air in the 20th and early 21st centuries and its projected continuation.

Goods Marketable commodities such as oil, copper, grain and timber.

Governmental organisation A group which follows the policies of one or more governments and is funded by them. Compare to NGO.

Green revolution The introduction of high-yielding seeds and modern agricultural techniques in less economically developed countries.

Green water The proportion of total precipitation absorbed by soil and plants, then released back into the air.

Greenhouse gas sink A process, activity or mechanism that removes a greenhouse gas from the atmosphere.

Grey water Water that has already been used for one purpose, but can possibly be reused for another purpose.

Gross productivity (GP) The total gain in energy or biomass per unit area per unit time (including that which is lost to respiration).

- **Gross primary productivity (GPP)** is gained through photosynthesis in producers.
- **Gross secondary productivity (GSP)** is gained through absorption by consumers.

Habitat The environment in which a species usually lives. The habitat of an orang-utan is the rainforests of Borneo and Sumatra.

Heatwave A prolonged period of excessively hot weather.

Humus A dark crumbly substance that is formed when organic material breaks down; it is very fertile for plant growth.

Hydrograph A graph showing the rate of flow (discharge) over a certain period past a specific point in a river. The rate of flow is expressed in cubic metres per second.

Hydrological cycle The natural sequence through which water passes into the atmosphere as water vapour, precipitates to Earth in liquid or solid form, and ultimately returns to the atmosphere through evaporation.

Ice sheet A thick layer of ice covering extensive regions of the world, notably Antarctica and Greenland.

Incineration The process of burning waste material in a furnace at high temperatures so that only ashes, gas and heat remain.

Indicator species Species that, by their presence or absence, can be indicative of polluted water (or other systems).

Inner core The Earth's layer which is the solid, hot centre made of iron and nickel.

Inorganic compounds Compounds of mineral origin.

Insolation Heat energy from the Sun consisting of the visible spectrum together with ultraviolet and infrared rays.

Internal factors Factors regulating populations from within a population or species (e.g. density-dependent fertility and the size of breeding territory).

Intraspecific competition Competition for resources between members of the same species.

K-strategists Organisms which have few offspring but invest a large amount of time in caring for them so that most of them survive.

Key An identification tool using a series of steps each involving just one decision. At each step, another choice is given.

Keystone species A species that has a disproportionate effect on the structure of a community.

Landfill A disposal site where solid waste is buried between layers of dirt in such a way as to reduce contamination of the surrounding land.

Land tenure The ways in which land is or can be owned.

Leaching A natural process by which water-soluble substances such as calcium are washed out from soil. This reduces the fertility of a soil.

LEDC Less economically developed country.

Limiting factor A resource for which the demand is greater than the supply.

Low-carbon economy Country where significant measures have been taken to reduce carbon emissions in all sectors of the economy.

Macro-invertebrates Animals without backbones that can be seen with the naked eye. Examples include water fleas, dragonfly nymphs, mayflies and worms.

Malnutrition Insufficiency in one or more nutritional elements necessary for health and wellbeing.

Mantle The layer of the Earth that surrounds the outer core and is semi-molten rock called magma. In the upper areas, this rock is hard, but deeper down it is softer.

Mass extinctions Times when the Earth loses more than three-quarters of its species in a geologically short interval.

Maternal mortality rate The annual number of deaths of women from pregnancy-related causes per 100 000 live births.

Maximum sustainable yield (MSY) This is equivalent to either the net primary or net secondary productivity of the system. In the context of fisheries, it is the largest proportion of fish that can be caught without endangering the population.

MEDC More economically developed country.

Microgeneration Generators producing electricity with an output of less than 50 KW.

Natural capital The source or supply of resources and services that are derived from nature.

Natural decrease What happens to a population when the number of births is lower than the number of deaths.

Natural greenhouse effect The property of the Earth's atmosphere by which long wavelength heat rays from the Earth's surface are trapped or reflected back by the atmosphere.

Natural income The annual yield from sources of natural capital.

Natural selection The proposed key mechanism which leads to the formation of new species.

Negative feedback Change in a system that stabilises the system and allows it to eliminate any deviation from the preferred conditions.

Neo-Malthusians These are resource pessimists who follow the general arguments of the Reverend Thomas Malthus with regard to population growth.

Net productivity (NP) The gain in energy or biomass per unit area per unit time remaining after the deduction of losses through respiration (i.e. the biomass available to consumers at subsequent trophic levels).

- **Net primary productivity (NPP)** is the gain by producers in energy or biomass per unit area per unit time minus respiratory losses.
- **Net secondary productivity (NSP)** is the gain in energy or biomass per unit area per unit by consumers minus respiratory losses.

Newly industrialised country (NIC) A country that has undergone rapid and successful industrialisation since the 1960s.

Niche The particular environment and 'lifestyle' that a species has. It includes factors such as the place where the organism lives and breeds, as well as its food and feeding method, activity patterns and interactions with other species. A niche is unique to each species and offers the exact conditions that a species needs or has become adapted to. Two different species cannot occupy exactly the same niche.

Non-biodegradable pollution Pollution from a substance that cannot be broken down within an organism or trophic level.

Non-governmental organisation (NGO) A group that is funded by individuals or independent groups.

Non-point-source pollution Pollution that is dispersed in nature and cannot be attributed to a single source. Also called sustained pollution.

Non-renewable energy sources Energy sources, mainly fossil fuels, that take millions of years to form. As consumption increases, these resources are depleted and they will eventually run out.

Non-renewable natural capital Consists of subsoil assets such as coal, oil, copper and diamonds. Such resources are depleted as they are consumed.

Optimum population The population that achieves a given aim in the most satisfactory manner. This is generally viewed in economic terms as the population that would produce the highest average standard of living.

Optimum rhythm of growth The level of population growth that best utilises the resources and technology available. Improvements in the resource situation or/and technology are paralleled by more rapid population growth.

Organic compounds Carbon-containing compounds, excluding carbon dioxide and carbonates, that are found in the bodies of living organisms.

Outer core The Earth's layer which surrounds the inner core and is made up of liquid iron and nickel.

Overfishing A level of fishing resulting in the depletion of the fish stock.

Overgrazing Grazing of natural pastures at stocking intensities above the livestock carrying capacity.

Ozone A faintly blue-tinged odourless gas with the capacity to absorb ultraviolet radiation and convert it into heat energy.

Ozone-depleting substances Substances that cause the deterioration of the Earth's protective ozone layer.

Parent rock The upper layer of rock on which soil forms under the influence of biological and biochemical processes and human activity. The properties of the parent rock are changed in the process of soil formation through the effect of other soil formation factors, but to a large extent the properties of the parent rock still determine the properties of the soils.

Peak oil production The year in which the world or an individual oil-producing country reaches its highest level of production, with production declining thereafter.

Pedogenesis The process of soil development.

Percentage frequency The percentage of the total number of quadrats sampled that a particular species is found in.

Percentage cover The percentage of the total area within a quadrat which is covered by the species of interest.

Photochemical smog A form of air pollution which generally reduces visibility, produced by the photochemical reaction of sunlight with hydrocarbons and nitrogen oxides that have been released into the atmosphere, especially by vehicle emissions.

Photosynthesis The process by which green plants make their own food using water and carbon dioxide.

Physical water scarcity When physical access to water is limited.

Phytoplankton Microscopic organisms that live in aquatic environments (either freshwater or marine) and start aquatic food chains.

Pioneer community The first group of organisms to colonise a bare area of land.

Glossary

Plagioclimax A climax community that has replaced the naturally occurring climax community as result of human activity.

Plate tectonics The theory that the Earth's outer covering is divided into several plates that move over the rocky inner layer above the core.

Point-source pollution Pollution that can be traced to a single source such as an oil refinery, a power station or a chemical plant. Sometimes called incidental pollution.

Pollutant Substance that causes pollution.

- **Primary pollutant** A substance that is active as soon as it is emitted (e.g. smoke, CO).
- **Secondary pollutant** A substance that is formed from a primary pollutant that has undergone a change (e.g. sulfuric acid).

Pollution The addition of any substance or form of energy (e.g. heat, sound, radioactivity) to the environment at a rate faster than the environment can accommodate it by dispersion, breakdown, recycling or storage in some harmless form.

- **Acute pollution** A single isolated incident such as an oil spill.
- **Chronic pollution** Long-term pollution such as emissions from a factory.

Population A group of organisms of the same species which live in the same area at the same time and which are able to interbreed.

Population density The number of members of a species per unit area.

Population momentum The tendency for population growth to continue beyond the time that replacement level fertility has been achieved because of a relatively high concentration of people in the childbearing years. This situation is due to past high fertility rates which results in a large number of young people.

Population policy The measures taken by a government aimed at influencing population size, growth, distribution, or composition.

Population pressure The situation in which population per unit area exceeds the carrying capacity.

Population projection The prediction of future populations based on then present age–sex structure, and with present rates of fertility, mortality and migration.

Population pyramid A bar chart, arranged vertically, that shows the distribution of a population by age and sex.

Population structure The composition of a population, the most important elements of which are age and sex.

Positive feedback Change in a system that leads to more and greater change.

Potable water Water that is free from impurities, pollution and bacteria, and is thus safe to drink.

Precipitation Water that falls to Earth from the atmosphere (e.g. rain, snow, sleet, hail, dew and frost).

Pressure gradient The change in atmospheric pressure per unit of horizontal distance.

Primary productivity The production of energy by autotrophs (producers), usually measured as biomass per unit area per unit time.

Producer An autotrophic organism that produces its own food from inorganic materials.

Product stewardship A system of environmental responsibility whereby producers take back a product, recycling it as far as possible, after the customer has finished with it.

Productivity The conversion of energy into biomass in a given period of time.

Pro-poor strategies Development strategies that result in increased net benefits for poor people.

Proved reserves of oil Quantities of oil that geological and engineering information indicates with reasonable certainty can be recovered in the future from known reservoirs under existing economic and operating conditions.

Quadrat A sampling device, usually a square of fixed size which may be $0.25\,m^2$ ($0.5\,m \times 0.5\,m$) or $1\,m^2$ depending on the area being sampled.

***r*-strategists** Organisms which have a relatively short lifespan during which they reproduce once and produce large numbers of offspring.

Rainwater harvesting The accumulation and storage of rainwater for reuse on site, rather than allowing it to run off.

Rate of natural change The difference between the birth rate and the death rate.

Recycling The concentration of used or waste materials, their reprocessing, and their subsequent use in place of new materials.

Reductionist approach to a system reduces the complex interactions within it to their constituent parts, in order to study them; whereas a systems approach considers the whole system and the interactions between the various components.

Recyclable resource One that can be used over and over, but must first go through a process to prepare it for reuse.

Recycling deserts Areas where rates of recycling are well below the national or regional average.

Refrigerant A substance used in a heat cycle and usually including a reversible phase change from a liquid to a gas.

Regolith The irregular cover of loose rock debris that covers the Earth.

Relative humidity The amount of water vapour present in air expressed as a percentage of the amount needed for saturation at the same temperature.

Renewable natural capital Comprises living species and ecosystems. It is self-producing and self-maintaining and uses solar energy and photosynthesis to produce food and chemical energy.

Renewable (sustainable) energy sources Sources of energy such as solar and wind power that are not depleted as they are used.

Replacement level fertility The level at which each generation has just enough children to replace the adults in the population. Although the level varies for different populations, a total fertility rate of 2.12 children is usually considered as replacement level.

Replenishable natural capital Stocks of non-living resources. Examples are the atmosphere, fertile soils and groundwater. Such resources are dependent on energy from the Sun for renewal.

Repowering Replacing first-generation wind turbines with modern multi-megawatt turbines which give a much better performance.

Reserves-to-production ratio (R:P) The reserves remaining at the end of any year are divided by the production in that year. The result is the length of time that those remaining reserves would last if production were to continue at that level.

Resilience The tendency of a system to maintain stability and resist tipping points.

Resource depletion The consumption of non-renewable, finite resources which will eventually lead to their exhaustion.

Respiration The process in which food, often in the form of glucose, is broken down to release the energy it contains.

Salinisation The condition in which the salt content of soil accumulates over time to above normal levels; occurs in some parts of the world where water containing high salt concentration evaporates from fields irrigated with standing water.

Second law of thermodynamics In isolated systems entropy tends to increase.

Secondary productivity The biomass gained by heterotrophs (consumers) as they feed. It is usually measured as biomass per unit area per unit time.

Semi-arid conditions Those where precipitation is less than 500 mm per year.

Seral stages The different stages of succession.

Sex ratio The number of males per 100 females in a population.

Shale oil Oil from reserves (sometimes called tight oil reserves) in shales and other rock formations from which it will not naturally flow freely.

Shelter belt A barrier of trees and shrubs that protects against the wind and reduces erosion.

Simpson diversity index An index that takes into account both the number of different species present (the species richness) and the abundance of each species. If the habitat has similar population sizes for each species present the habitat is said to have 'evenness'.

Society An arbitrary group of individuals who share some common characteristics such as location cultural background, religion or value system.

Soil A mixture of inorganic mineral particles and organic material from decomposed flora and fauna that covers the underlying bedrock, and in which a wide variety of terrestrial plants grow.

Soil conditioners Materials added to soil to improve soil fertility.

Soil degradation The physical loss (erosion) and the reduction in quality of topsoil associated with nutrient decline and contamination.

Soil horizon A specific layer of soil that is parallel to the surface and possesses physical characteristics which differ from the layers above and beneath.

Soil profile The vertical succession down through a soil which reveals distinct layers or horizons in the soil.

Soil texture The relative content of particles of various sizes (e.g. sand, silt and clay) in the soil.

Soil toxicity The extent of the presence of toxic chemicals in a soil.

Solar constant The amount of solar energy received per unit area, per unit time on a surface at right-angles to the Sun's beam at the edge of the Earth's atmosphere.

Solid domestic waste (SDW) Waste produced by people living in their homes (as opposed to other sources of waste such as agriculture, industry and the service sector). It is effectively what households put in their bins.

Speciation The process that establishes a new species which is fertile but which can no longer breed with the original species.

Species A group of organisms that can interbreed to produce fertile offspring.

Strategic petroleum reserves Large reserves of oil held by countries including the United States and China to tide them over for a few months or so if normal oil supplies are disrupted.

Strip cultivation Growing crops in a systematic arrangement of strips across a field, usually at right angles to the direction of the prevailing wind. The strips are arranged so that a close-growing crop or strip of grass is alternated with a strip with less protective cover in order to reduce wind erosion.

Sublimation The process of changing from a solid to a gas without a liquid phase.

Subsistence farming The most basic from of agriculture where the product is consumed entirely or mainly by the family who work the land or tend the livestock.

Substitution The use of common and thus less valuable resources in place of rare, more expensive ones (e.g. replace-ment of copper by aluminium in a variety of products).

Succession The process in which communities in a particular area change over a period of time so that the appearance of the whole area evolves and changes.

- **Primary succession** begins when an area of bare ground or rock is colonised for the first time.
- **Secondary succession** takes place after an area of land has been cleared (e.g. by a fire or a landslide) and soil is already present.

Sustainable yield (SY) The rate at which capital can be exploited without depleting the original stock or affecting its potential replenishment.

Sustainability The use and management of resources so that full natural replacement of exploited resources can take place.

Sustainable agriculture Agricultural systems emphasising biological relationships and natural processes, which maintain soil fertility thus allowing current levels of fram production to continue indefinitely.

System An assemblage of parts and the relationships between them which enable them to work together to form a functioning whole.

- **Closed system** is a system that exchanges energy but not matter across the boundaries of the system.
- **Isolated system** is a system that exchanges neither energy nor matter with its environment.
- **Open system** is a system that exchanges both matter and energy within its surroundings across the boundaries of the system.

Glossary

Systems approach A way of visualising a complex set of interactions in ecology, society or another system.

Technocentric A technologically based value system that believes that the brain power of humans will enable us to control the environment.

Temperature inversion An increase in temperature with altitude.

Terracing Conversion of a steep slope into a series of flat steps with raised outer edges so that cultivation can be practised.

Thermal expansion Rise in sea level caused by expansion of water molecules due to increase in sea and ocean temperatures.

Thermal stratification The division of the atmosphere into a number of layers in terms of temperature variation.

Tipping point (climate change) The point at which damage caused to global systems by climate change becomes irreversible.

Tipping point (general) The minimum amount of change within a system that will destabilise it and cause it to reach a new equilibrium or stable state.

Total allowable catch The maximum quantity of a particular type of fish that may be caught each year in total or by the fishing vessels of individual nations.

Total fertility rate The average number of children that would be born alive to a woman during her lifetime, if she were to pass through her child-bearing years conforming to the age-specific fertility rates of a given year.

Tragedy of the commons The idea that common ownership of a resource leads to over-exploitation as some nations will always want to take more than other nations.

Transect A tape or rope stretched across a sample area. At suitable intervals, populations can be sampled and measurements made of abiotic factors along the transect line to assess the changing distribution of a plant or animal species.

Transfers Transfers involve flow through a system and a change in location.

Transformations Transformations lead to an interaction within a system and the formation of a new end product or they may involve a change of state.

Transpiration The process by which plants absorb water through the roots and transport it to the leaves from where it is lost as water vapour.

Trophic level The position an organism occupies in a food chain as a result of its feeding habits. The term also defines a group of organisms in a community which feed at the same position in a food chains.

Turnover time The time taken for a water molecule to enter and leave part of the hydrological system.

Ultraviolet (UV) radiation Invisible electromagnetic rays with a wavelength shorter than that of visible light but longer than that of X-rays.

Upwelling The process by which water rises from beneath the surface to replace surface water pushed away by winds blowing across the ocean.

Volatile organic compounds (VOCs) Mainly synthetic chemicals that play a major role in the formation of photochemical smog.

Waste hierarchy A classification of waste management strategies according to their desirability from disposal (least desirable) through energy recovery, recycling, reuse, to waste prevention.

Water pollution Contamination of a source of water making it unsuitable for use.

Water security The capacity of a population to safeguard sustainable access to adequate quantities of acceptable quality water for sustaining livelihoods, human well-being and socio-economic development, for ensuring protection against water-borne pollution and water-related disasters, and for preserving ecosystems in a climate of peace and political stability.

Water supply The provision of water by public utilities, commercial organisations or community endeavours.

Water-scarce area An area where water supply falls below 1000 cubic metres per person a year.

Water-stressed area An area where water supply is below 1700 cubic metres per person per year.

Weather The atmospheric conditions at a specific place at a particular point in time.

Well A shaft, relatively large in diameter, sunk into the ground by hand or machinery to obtain water.

Zonal classification of soils The subdivision of the world into broad soil regions based primarily on differences in climate.

Index

abiotic components 51, 53–54
    measuring 108–11
abiotic factors 105
accuracy 115
ACFOR scale 107
acid deposition 290–2
    on forests 293
    effect of geology 294–5
    international agreements 297–8
    nutrient effect on soil 293
    pollution management strategies 295–6
    regional effect 294–5
    replace, regulate and restore model 295–6
    on soil, water and living organisms 292–3
    toxic effect on fish 293
acid rain 11, 42, 290–1
acidification 242, 243, 294
acute pollution 41
adaptation 343–7
adaptive capacity 344
advection 163
aerobic respiration 63
aerosols, alternatives to CFCs in 274
afforestation 339
age and sex pyramids 360–2
Agenda 21 387
agricultural technology 229
agriculture, greenhouse gas emissions and 338–9
agro-ecosystem 233
agro-industrialisation 233, 234
agro-system 233
air freight 234
air pollution 39, 43, 264, 279
    urban, management 284–9
air quality index 281
albedo 12, 259, 320, 322
algal bloom 207, 208
allopatric speciation 126
Alpine tundra 91
ambient air capture 340
anaerobic digestion (AD) 400
anaerobic respiration 63
anemometer 110
anoxic water 200
Antarctic Circumpolar Current 169
Antarctica 268-9, 270
    temperature changes in 328
anthropocentric system
    (anthropocentrism) 5, 6, 7
anticyclone 282
anti-Malthusians 354
aquaculture 195–6
aquaphytoplankton 189
aquatic biomes 88
aquatic food 188-9
aquatic primary production 189
aquatic production, harvesting 189
aquatic productivity 189–90
aquatic systems, exploitation of 191–3
    international and national policy 192
    local and individual action 193
    new technology and illegal fishing 193
aquifers 178–80
    replenishing 184
Arabian oryx (*Oryx leucoryx*) 142–3
arable land 221
Arctic
    environmental changes in 327
    tundra 91–92
arid conditions 177
arid regions 177
Arumeru, Arusha, Tanzania 252–4
atmosphere 214, 256–7
    composition 258-9
    impact of human activities on 264–5
    structure 259–60
atmospheric and oceanic circulatory systems 317–18
atmospheric moisture 260
atmospheric pressure 257–8
autotrophs 65, 77
Azotobacter 83

background extinction rate 136
Bacon, Francis 217
bald eagle 47
barnacles 57
Basle Convention 393
Beaufort scale 110, 111
Beluga sturgeon (*Huso huso*) 141–2
Bhopal disaster 3, 4
bioaccumulation 70–72, 73
biocapacity 406
biochar 340
biochemical oxygen demand (BOD) 200–1
biodegradable pollution 41
biodiversity 119
    genetic diversity 119, 120
    habitat diversity 119, 120
    species diversity 119, 120
biodiversity hotspot 121
biodiversity values, measurement 147–8
bioenergy 340
biofuels 304
biogeochemical cycle 22
biological mechanical treatment (BMT) 400
biomagnification 70–71
biomass 65, 84
    measuring 111–12
biome 16, 17, 86
    comparison 92
    distribution, structure and productivity 86–88
    shifting 325
    tropical 143–4
biosphere 18, 214
Biosphere 2, 19
biotic components 51, 55–58
biotic factors 105
biotic index 121, 203

blue water 165
bore holes 181
botanic gardens 154
BP oil spill, Gulf of Mexico (2010) 4–5, 46, 301
Brundtland report 32
buffer zone 158
built-up land 37
business-as-usual scenario 384, 408

calorimeter 112
Canada
    oil sands 308
    prairies 249–52
cancer villages, China 43
Cancun (Mexico) agreements of 2010 346
capital, defining 372–3
captive breeding programmes 154–5
capture-mark-release-recapture technique 105
carbon capture and storage or sequestration (CCS) 340
carbon credit 12, 338
carbon cycle 82, 335
carbon dioxide 43, 320, 323, 324
carbon dioxide capture 12, 13
carbon dioxide removal (CDR) 339–41
carbon footprint 13, 406
carbon geoengineering 339
carbon sequestration 12, 13
carbon sinks 335
carbon trading 338
carbon uptake 37
carnivores 65, 66, 76-7
carrying capacity 58, 60, 350, 388, 402–4
    energy and material use and 404–5
Carson, Rachel 34
    Silent Spring 3, 46, 47
'charismatic' species 155
chemical energy 75
Chernobyl disaster 3–4, 311
China
    anti-natalist policy 365–6
    cancer villages 43
    energy sources 313–14
    Pearl River Delta region, acid deposition 296–7
    Three Gorges Dam 36
    waste disposal in Beijing 401
Chitwan National Park 157, 158–9
chlorofluorocarbons (CFCs) 263, 268, 272, 273, 323
    alternatives to, in aerosols 274
chlorophyll 63, 65, 75
chloropictin 273
chronic pollution 41
CITES 153-4, 155
Citizen Monitoring Biotic Index 203–4
clay soils 219
Clean Air Act 1970 (USA) 43
    amendments 43
climate 317, 318

Index

climate adaptation 336, 337
climate change 11, 12, 134, 170, 180, 319–21
   differences of opinion 347
   local and personal initiatives 347
   human health concerns 328–9
   international agreements 345
climate graphs 164
climate lag 331, 332
climate mitigation 336–42
   reducing energy consumption 336
climax community 93, 94–95, 96
   factors affecting 100–2
   human interference in 100–2
climograph 87
closed system 18, 19
clouds 260–1
   formation 261
   seeding 184–5
   types 261–2
coal 301, 315
coastal inundation 327
cod, Atlantic (*Gadus morhua*) 194–5
commercial farming 235-6, 249–54
community 62, 120–1
community energy 338
competition 55, 57
complexity theory 96
compost 397
composting 397
condensation 164
conservation 146
   approaches to 151–6
   arguments for 147–9
   species-based conservation strategies 153–6
   water 185
conservation areas 151–3
   buffer zones 152
   edge effects 151–2
   management 152
   size 151
   wildlife corridors 152
consumers 64, 65, 82
continental drift 131–2
continental formation and movement 131–2
contour ploughing 248
Convention on Long-Range Transboundary Air Pollution (LRTAP) 297
Copenhagen Climate Change Conference (2009) 346
coral bleaching 325
cornucopians 6
crop growing areas 325–6
cropland 37
crude birth rate (CBR) 354
crude death rate (CDR) 354
crust, Earth 129, 130
cultivation techniques 248
culture 368

dams 180–1
Darwin, Charles 123–4, 217
   On the Origin of Species 124
DDT (dichlorodiphenyltrichloroethane) 3, 42, 46–7, 67, 71
dead zones 208

decomposers 64, 65, 75, 82
deep ecologists 6, 11
Deepwater Horizon oil spill *see* BP
deflected succession 100
deforestation 84, 101, 145, 241, 382–3
deindustrialisation 297
Delhi, air pollution 287–9
demographers 350
demographic transition 358
   in LEDCs 360
   model of 358–60
desalination 12, 183-4
desert 91, 92
desertification 187, 245–6
destination footprint 385
desulfurization 43
detritivores 65
developing steady-state equilibrium 26
development policies
   epidemiological transition 364
   population and 364
Diamond, Jared 34
dichotomous key 103–4
diet 231
digestion 65
disability-adjusted life years (DALYs) 271
dissolved oxygen 53, 109
diversity 113
diversity indices 121, 122
DNA 129
dodo (*Raphus cucullatus*) 134, 137, 140–1
doubling time 357
drainage basin system 165, 166
Durban Platform for Enhanced Action 346
dust storm 246
dynamic equilibrium 266

E-waste (electronic waste) 392–3
Earth
   crust 129, 130
   inner core 129
   mantle 129
   outer core 129
   axis, climate change and 320
   orbit, climate change and 320
ecocentrism(ecocentric) 5–7, 11, 12, 13
ecological deficit 13
ecological efficiency 75
ecological footprints 11, 12–13, 37–8, 405–9
   calculation 409–10
   comparisons 410–12
   population-technology-resources and 412–14
economic water scarcity 177, 178
ecosystem 5, 51-2, 62
   inputs and outputs 23
   services 32, 376
ecotourism 158–9, 385
edaphic factors 100
eduction 368
effective precipitation 215
El Ninõ 326
electricity access deficit 303
electromagnetic spectrum 74
emergent properties 18

emissions trading 43
energy, variable patterns over time 312–13
Energy and Resources Institute 277
energy balance 262, 263
   UK 315–16
energy consumption 302–3
   UK 343
energy content, measuring 112
energy efficiency 337
energy flow, human impact on 84, 85
energy flow diagrams 75–76
energy intensity 343
energy ladder 303
energy mix 300
energy pathways 308
energy poverty 300
energy resources 300–3
energy security 300
energy sources, choice of 311–12
energy transfers 69–70, 74–77
enhanced greenhouse effect 321
entropy 25, 64
Environmental Impact Assessments (EIAs) 34–35, 36
environmental managers 6
Environmental Protection Agency 3
environmental sustainability 384
   attitudes to 387
   development and 386–7
Environmental Sustainability index 385–6
environmental value systems (EVSs) 2, 7–8
epidemiological transition 364
equilibrium 331
eutrophication 206–7
   impacts of 207–8
evaporation 163
evapotranspiration 164
evolution 123–5
   of horse 125
exponential growth 60, 353–4
extended producer responsibility 404
external factors 195
external forcings 333
externality 320
extinction 135, 144
extinction rates 135–-6

feedback 26, 331
feedback systems 26–28
feeding activities 52
Ferrel cells 318
fertile soil 214
fertilisers 82, 84, 247
fertility rate 355–7, 367
field techniques, choosing and evaluating 116
financial capital 373
fisheries 190–1
fishery-management strategies 194
fishing 190–1
fishing grounds 37
flash floods 168
flightless birds 131
flow velocity 53, 109
flowmeter 109, 110
flows 21

flue gas 43
Food and Agriculture Organization (FAO) 221, 227, 228
food chains 65-6, 69–71, 72, 75
food loss 227
food miles 234–5
food production and distribution, adverse influences 223–5
food production systems 228–9
   economic factors 228-9
   greater consumer power 229
   physical factors 228
   political factors 229
   social/cultural factors 229
   terrestrial 231–2
food security 224
food shortages, short- and long-term effects 225
food waste 227, 228
food webs 62, 66, 89, 96, 100
Foresight: *Global Food and Farming Futures* (2011) 34
forest 37
   biomes 89
   cultural value 381
   water management 185
fossil fuels 11, 12, 82, 83, 301, 339
France: pro-natalist policy 367
freshwater biomes 88
freshwater ecosystems 53–54
Friends of the Earth 335
fuel poverty 300
fuelwood 302, 303
Fukushima nuclear accident 5
fundamental niche 52, 57

Gaia hypothesis 20
Galapagos finches 128
garbage patches 393–4
gas-blown plastic 273
genetic diversity 119, 120
geoengineering 339
geographical isolation 126–7
geopolitics 307
geothermal energy 304
geothermal gradient 304
glaciers 162
Global Assessment of Human-induced Soil Degradation (GLASOD) 240
global brightening 333–5
global dimming 333–5
global fishing 375–6
global food crisis 224–5
global hectare (gha) 406
global warming 223, 321, 322
   arguments of prominent individuals 333
   feedback in 331–2
   human perceptions 335
   natural vs anthropogeniuc causes 332–3
   negative feedback in 331
   positive feedback in 331
glucose 64
Gondwanaland 131
Gothenburg Protocol 297
governmental organisations 149
granite 214

grassland 90
Gravestone Project 298
grazing land 37
Great Pacific garbage patch 393–4
green politics 8
green revolution 233, 234
green water 165
greenhouse effect 262–3
greenhouse gas sink 336
greenhouse gases 262, 263
   effects of 323
   geographical sources 323–4
   global temperature and 320-1
   human activities and 322
Greenpeace 3, 9, 335, 393
   Save the Whales campaign 3, 10, 197
Greenpeace 393
grey water 185
grey wolves (*Canis lupus*) 155
gross primary productivity (GPP) 77
gross productivity (GP) 79
gross secondary productivity (GSP) 79
groundwater 162
growth curves, phases of 59

Haber process 83
habitat 52
habitat diversity 119, 120
Hadley cells 318
halogenated organic gases 267-8
hasification 400
HCFCs 273
heat energy 24
heatwave 326
herbivores 65, 76–77
herbivory 56
heterotrophs 65
holism 20
horse, evolution of 125
human capital 373
humus 215
hydroelectricity 302, 303
hydrograph 167
hydrological cycle 163–4
   shared 165–6
   solar energy and 163–4
hydrology 167–8
hydrothermal vents 88

ice caps 162
ice sheet 327
identification of organisms 103–4
incineration 397, 400
Inconvenient Truth, An 4
India: Country Programme 277–8
indicator species 203
infant mortality 359, 360, 361, 362, 363, 365, 368
infiltration 165
infrared waves 74
inner core 129
inorganic compounds 82
insolation 260, 262
Intergovernmental Panel on Climate Change (IPCC) 328, 333, 336, 341, 343, 345

internal factors 195
international conventions on biodiversity 151
International Council for Exploration of the Sea (ICES) 192
International Federation of Arts Councils and Culture Agencies (IFACCA) 34
International Union for Conservation of Nature (IUCN) 134, 137, 193
International Whaling Commission (IWC) 3, 197
interspecific competition 55
Inter-Tropical Convergence Zone 318
intraspecific competition 55, 58
intrinsic value of environment 380
irrigation water 175
isolated system 18, 19

J-curve 58–59
Julie's Bicycle 34

K-strategists 97, 98
katabatic winds 284
key 103–4, 205
keystone species 155, 156
kinetic energy 24
Kyoto Protocol 345, 346

Lago Paranoá, Brazil 210
land per capita 231
land tenure 229
landfill 395–7
Large Combustion Plant Directive (LCPD) 297
laws of thermodynamics
   first 24, 70, 112
   second 25
leaching 215
LEDC 2, 39, 43
Levi-Strauss, Claude 24
light 74
light energy 63, 75
light intensity 54, 110
light pollution 39
lime 247
limestone 82
limiting factor 58
Lincoln index 105–6
lithosphere 94, 214
loam soils 219
local soil 220
logging 382–3
longwave radiation 320
Longworth trap 105, 106
loss-on-ignition (LOI) method 110
Lovelock, James 20
low-carbon economy 338

Maasai Mara National Reserve, Kenya 382
macro-invertebrates 200, 201
Madagascar 145–6
magma 130
malnutrition 225
Malthusian growth model 353
mantle 129
manufactured capital 373
margin of error 115

Index

marginal lands, ploughing of 248–9
marine biomes 88
marine ecosystems 53
marine food chain 188
marine mammals, harvesting 197–8
Marine Stewardship Council (MSC) 193
mass extinctions 132
    causes 133, 134, 136
        human activities and 134, 136
        risk 136–7
maternal mortality rate 364, 365
maximum sustainable yield (MSY) 80, 190–1, 387–8
meat consumption 232
mechanical biological treatment (MBT) 400
MEDC 2 7, 13, 43, 46
mercury poisoning 72
mesosphere 260
metam sodium 274
methane 323, 338–9
methyl bromide 273–4
microgeneration 338
Millennium Development Goals 34, 364
Millennium Ecosystem Assessment (MA) 33
Minamata Bay, Japan, disaster 2, 72
models of systems 21–23
Monbiot, George 333, 334
Mongolia, climate change in 329–30
Montreal Protocol 272, 273, 275–8
moorland 101, 102
moraine 94
mutualism 56–57
myomatosis 60

National Adaptation Programme of Action (NAPA) 346
National Oceanic and Atmospheric Administration (NOAA) 267
national parks 380–2
natural capital 31, 373–4
    changes over time 378–9
    critical 375
natural decrease 355
natural gas 301, 302
natural greenhouse effect 321
natural income 32, 373–4
natural selection 123, 124
Nature of Things, The 34
negative feedback 27, 28, 30, 331
neo-Malthusians 353, 354
Neolithic Revolution 350
net primary productivity (NPP) 77, 78
net productivity (NP) 79
net secondary productivity (NSP) 79
newly industrialised country (NIC) 232
niche 52
nicotine 73
Nile Water Agreement (1929) 187
nitrates 83, 207
nitrifying bacteria 83
nitrites 83
Nitrobacter 83
nitrogen cycle 83
nitrogen oxides 285, 323, 338–9

Nitrosomonas 83
noise pollution 39
non-biodegradable pollution 71
non-governmental organisations (NGOs) 149, 186–7
non-point-source pollution 41, 42
non-renewable energy sources 300
non-renewable natural capital 377
North Atlantic deep water (NADW) 169
North Atlantic Drift 170
Northeast Greenland National Park 381
nuclear energy 302
nuclear power 310–11, 315
nutrient cycling, human impact on 84, 85

objectivity 116
ocean alkalinity enhancement 340
ocean circulation systems 169–70
ocean conveyor belt 169–70
ocean currents 319–20
ocean fertilisation 340
oil 301, 305–8
    consumption 306
    demand 305–7
    geopolitics and 307–8
    peak production 307
    price 306
    shale 307
oil sands 308
open system 18, 19
optimum population 366, 404
optimum rhythm of growth 404
organic compounds 82
organic farming 233–4
Our Common Future (1987) (Brundtland report) 32
outer core 129
overfishing 375
overgrazing 241
oxygen, dissolved 53, 54
ozone 265–7, 323
    halogenated organic gases and 267–8
    UV radiation and 265–7
ozone-depleting substances 272–3
    illegal market in 278
ozone layer, hole in 268–9, 270

Pangaea 131
paradigm shift 119, 132
parasitism 56
parent rock 213
peak oil production 307
pedogenesis 239
percentage cover 107, 108
percentage frequency 107, 108
percolation 165
persistent pollution 42
personal value systems 14
pesticides 45, 71, 82
pH 53, 108, 290, 291
phosphates 207
photochemical smog 282–4
photosynthesis 62–63, 75
physical water scarcity 177, 178

phytoplankton 68
Pilot Analysis of Global Ecosystems (PAGE) 177
pioneer communities 93
pitfall trap 105, 106
plagioclimax 100, 102
plastic waste 14
plate tectonics 129–30, 132
point-source pollution 40-1
polar cells 318
pollutant
    distribution 43
    primary 42, 279, 280
    secondary 42, 279, 280
pollution 39–40, 198, 279
    acute 41
    chronic 41
    cleaning up and restoring ecosystems 45–46
    human activity, changing 43–45
    major sources 42–43
    management 43–46, 394–5
    regulating and reducing 45
population 55
population, global
    growth rate 357
    rapid increase in 350–1
    recent demographic change 352
population density 107
population momentum 352
population policy 365
population pressure 353, 354
population projections 369–70
population pyramids 361-2
population structure 360, 363
population-technology-resource relationships 412–14
positive feedback 27–28, 331
potable water 173
prairies 90, 101, 102
precipitation 164, 165
precipitation effectiveness (effective precipitation) 215
predation 55
predator-prey system 30
pressure gradient 257
primary pollutant 42, 279, 280
primary productivity 77
primary succession 93–94
pro-poor strategies 387
producers 64–65
product stewardship 404
production to respiration ratio (P:R ratio) 96
productivity 75, 77–81
protected areas 151–3, 157–9
proved reserves of oil 305
pyramids of biomass 68–9, 77, 111
pyramids of numbers 67–68, 77
pyramids of productivity 69–70, 77
pyrolysis 400

quadrats 105, 106–7

r-strategists 97, 98
rabbits 60
radio waves 74

Rainbow Warrior 3
rainforest 101, 120, 230
    as ecosystem 143
    loss of 8, 143–4
    see also deforestation
rainwater-harvesting systems (RHS) 186
random error 115
rate of natural change 355
realised niche 52
recyclable resources 377
recycling 398
Red List (Red Data Book) (IUCN) 137–9
red tide 208
reductionist approach 18
reflection 75
refrigerants, recycling 272–3
regolith 215
reintroduction programmes 154–5
relative humidity 261
reliability 115
religion 11
renewable (sustainable) energy sources 300, 301
renewable energy 302, 303–4, 339
renewable natural capital 376, 382–3
replacement-level fertility 355
replenishable natural capital 377
repowering 309
reproductive isolation 126–8
    behaviour 127
    other causes 127
    timing 126–7
reserves-to-production ratio (R-P) 305, 306
reservoirs 180–1
resilience of system 29
resources 379
    defining 372–3
    depletion 377
respiration 62, 63–64
restoration of natural capital (RNC) 374
reverse-osmosis systems 184
Rhizobium 83
rice 236, 237–8
Rio Declaration 387
Rio Earth Summit (1992) 140, 345, 385, 387
Rio+20 (Earth Summit 2012) 387
RNC Alliance 374
Rome Summit (2009) 223–4

S-curve 58–59
salinisation 242, 243
salinity 53, 108
sandy soils 218–19
Sankey diagrams 76–77
savannah 90
SAVE FOOD initiative 228
Secchi disc 109
secondary pollutant 42, 279, 280
secondary productivity 79
secondary succession 94–95
seed banks 154
self-reliance soft ecologies 6
semi-arid conditions 177
semi-arid regions 177

seral stages 93
sere 94
Severn, River 167–8
sex ratio 363
shale oil 307
shelter belts 247
shifting cultivation 230
Simpson diversity index 113, 114, 120, 121
Simpson's index 122
Simpson's reciprocal index 122
slash and burn agriculture 85, 230
slope 54, 110
    gradient and length 243
social capital 373
society 7
soil 213
soil buffering capacity 295
soil conditioners 220, 247
soil conservation 247–9
soil degradation 240–6
    biological 244
    causes 242
    chemical 243–4
    climate and land use change 244–5
    consequences of 245–7
    desertification 245–6
    extent 240–1
    local 246
    physical 242
    processes 241–2
    as threat to food security 246–7
soil drainage 54
soil ecosystems, succession and 239
soil erodibility 243
soil fertility 65
soil formation 214, 217
soil horizon 215
soil management systems 249–54
soil mineral content 54
soil moisture 54
soil moisture content 110
soil particle size 54
soil pollution 243
soil profile 213, 215–16
soil separates 218
soil surface roughness 243
soil systems 213
soil texture 218
soil toxicity 244
soil type 220–1
solar constant 263
solar geoengineering 339
solar maximum 320
solar minimum 320
solar panel 312
solar power 304
solar radiation 87, 262–3, 320
solarisation 273
solid domestic waste (SDW)
    energy from 398–400
    household production 391
    management practices 399–400
    pollution and 394–5
    transport 398

    types of 389–91
    waste diary 392
spatial habitat 52
speciation 123–5, 131
species 51, 52
    abundance, measurement of 107–8
    diversity 119, 120
    estimates 135–6
stable equilibrium 26
static equilibrium 26
steady-state equilibrium 25
    developing steady-state equilibrium 26
    unstable equilibrium 26
steppes 90
Stern Review 320–1
storages 21
strategic petroleum reserves 308
stratopause 260
stratosphere 260
stratospheric ozone depletion, history of 277
strip cultivation 248
struggle for survival 124
subduction 130
sublimation 164
subsistence farming 237, 249–54
substitution 405
succession 93
    energy flow and productivity 96–97
    primary 93–94
    secondary 94–95
sulfur dioxide 43
sunspots 320
Surtsey 99–100
survival of the fittest 124
survivorship curves 98
survivorship strategies 97–100
sustainability 32, 33, 383–4
sustainable agriculture 227
sustainable development 32–34, 384–8
Sustainable Fisheries Partnership 193
Sustainable Food Systems Programme (SFSP) 227–8
sustainable scale 374
Sustainable Scale Project 278
sustainable systems 31–32
sustainable yield (SY) 79, 80, 387–8
Suzuki, David 34
system 16
    scale of 17–18
    flows in 21
    storages in 21
    models 21–23
    closed 18, 19
    isolated 18, 19
    open 18, 19
systematic error (bias) 115
systems approach 16, 17, 18

tavy 145
technocentrism 5, 6, 7, 11–12, 13
temperate forest 89, 90, 92
temperature 54, 109
temperature inversion 259, 283–4
terracing 220, 248

Index

terrestrial ecosystems 54
terrestrial food production systems, sustainability of 226–7
  antibiotics 227
  commerce vs subsistence food production 227
  fertilisers 226
  industrialisation, mechanisation and fossil fuel use 226
  legislation 227
  pest control 227
  pollinators 227
  scale 226
  seed, crop and livestock choices 226
  water use 226
theory 125
thermal expansion 327
thermal pollution 39
thermal stratification 259–60
thermohaline currents 169
thermosphere 260
Thoreau, Henry David 34
Three Gorges Dam, China 36
Three Mile Island disaster 311
throughflow 165
tidal power 304
tipping point (climate change) 321, 322
tipping point (general) 29
Toronto Conference (1988) 345
total allowable catch 194
total fertility rate 355–7
tourism 384–5
tragedy of the commons 375
transect 107
transfers 20
transformations 62, 20–21
transpiration 164
transport, waste 398
trophic levels 65–67
tropical rainforests 89–90, 92
troposphere 259, 318
tropospheric ozone 280, 281
tundra 91, 92, 325
turbidity 53, 109
turnover time 163

UK energy consumption 343
ultraviolet radiation (UV) 74, 260
  living tissues and 270–1
  impact on humans 271
  impact on plant and animal life 271
  ozone and 265–7
  UV-A 267
  UV-B 267, 270
  UV-C 267
UNDP 277
United Nations Collaborative Programme on Reducing Emissions from Deforestation and Forest Degradation in Developing Countries (UN-REDD) 339
United Nations Educational, Scientific and Cultural Organization (UNESCO) 99
United Nations Environment Programme (UNEP) 149–50, 227, 228, 271, 275
United Nations Framework Convention on Climate Change (UNFCCC) 345
United Nations revision of future population 370–1
United Nations Watercourses Convention 187
units of energy 112
unstable equilibrium 26
upwelling 170
upwelling zones 189
urban air pollution, management 284–9
urban areas, greenhouse gases in 345
urbanisation 368

value system 335
Venezuela, oil sands in 308
virtual water 187
vitamin A 73
volatile organic compounds (VOCs) 280, 286

Waldo, Ralph 34
Waste and Resources Action Programme (WRAP) 228, 397
waste hierarchy 396
waste transport 398
water, fresh
  distribution and storage 162–3
  external development assistance 186–7
  global water crisis 171–3
  increasing demand and unequal access 173–6
water conservation 11, 185–6
water erosion 243
water flow velocity 53
water footprint 174
water infiltration and surface runoff 166–8
  impact of agriculture 167
  impact of deforestation 167
  impact of urbanisation 168
water management 11
  forest 185
  pollution 209
water pollution 43, 179–80
  management strategies 209
  monitoring 198–9
  organic 201–2
  types 198
water redistribution 181–3
water resources 11–12
  international conflict 187
water scarcity 177–8
water security 171–2
water supplies 180–2
water temperature 53, 54, 109
water vapour as greenhouse gas 323
water-scarce areas 177
water-stressed areas 177
WaterAid 187
Watt, James 28
wave action 53, 109
wave energy 304
wavelength 75
weather 317, 318
weather patterns 326
weathering, enhanced 340
wells 181
whaling 3, 8, 10, 197
wind energy 304
wind erosion 243
wind power 308–10
wind speed 54, 110
wind turbine 309, 312
wind-reduction techniques 247–8
World Commission on Environment and Development 32
World Health Organization (WHO) 271
World Travel & Tourism Council (WTTC) 385
  Green Globe 385
World3 402
WWF 150, 382
Wye, River 167–8

X-rays 74

Yellowstone National Park 381, 382

zonal classification of soils 214, 215
zonation 95
zoos 154–5